Doppler Radar
and
Weather Observations

Doppler Radar and Weather Observations

RICHARD J. DOVIAK

DUŠAN S. ZRNIĆ

National Severe Storms Laboratory
National Oceanic and Atmospheric Administration
Norman, Oklahoma
and
Departments of Electrical Engineering and Meteorology
University of Oklahoma
Norman, Oklahoma

1984

ACADEMIC PRESS, INC.
(Harcourt Brace Jovanovich, Publishers)

Orlando   San Diego   San Francisco   New York   London
Toronto   Montreal   Sydney   Tokyo   São Paulo

COPYRIGHT © 1984, BY ACADEMIC PRESS, INC.
ALL RIGHTS RESERVED.
NO PART OF THIS PUBLICATION MAY BE REPRODUCED OR
TRANSMITTED IN ANY FORM OR BY ANY MEANS, ELECTRONIC
OR MECHANICAL, INCLUDING PHOTOCOPY, RECORDING, OR ANY
INFORMATION STORAGE AND RETRIEVAL SYSTEM, WITHOUT
PERMISSION IN WRITING FROM THE PUBLISHER.

ACADEMIC PRESS, INC.
Orlando, Florida 32887

United Kingdom Edition published by
ACADEMIC PRESS, INC. (LONDON) LTD.
24/28 Oval Road, London NW1 7DX

Library of Congress Cataloging in Publication Data

Doviak, Richard J.
  Doppler radar and weather observations.

  Includes index.
  1. Radar meteorology. I. Zrnič, Dusan. II. Title.
QC973.5.D68   1984   551.6'353   83-15840
ISBN 0-12-221420-X

PRINTED IN THE UNITED STATES OF AMERICA

84 85 86 87   9 8 7 6 5 4 3 2 1

Contents

PREFACE	xi
ACKNOWLEDGMENTS	xiii
LIST OF SYMBOLS	xv

1 Introduction   1
         References   3

2 Electromagnetic Waves and Propagation

2.1	Waves	4
2.2	Propagation Paths	7
	2.2.1 Refractive Index of Air	7
	2.2.2 Refractivity N	9
	2.2.3 Spherically Stratified Atmospheres	11
	References	20

3 Principles of Radar

3.1	The Doppler Radar (Transmitting Aspects)	21
	3.1.1 The Electromagnetic Beam	24
	3.1.2 Antenna Gain	26
3.2	The Target	26
	3.2.1 Scattering Cross Section	26
	3.2.2 Doppler Shift	28
3.3	Attenuation	29
	3.3.1 Attenuation by Rain	30
	3.3.2 Attenuation by Cloud Droplets	30
	3.3.3 Snow Attenuation Rate	32
	3.3.4 Gaseous Attenuation Rate	32
3.4	The Doppler Radar (Receiving Aspects)	33
	3.4.1 The Radar Equation	34
	3.4.2 The Received Waveform (In-Phase and Quadrature Components)	35

	3.5	Practical Considerations	38
		3.5.1 System Noise Temperature	38
		3.5.2 Bandwidth	42
		3.5.3 Filtered Waveform	42
		3.5.4 Signal-to-Noise Ratio; Matched Filters	44
	3.6	Ambiguities	44
		References	46

4 Weather Echo Signals

4.1	The Echo Sample	48
4.2	Signal Statistics	49
4.3	The Weather Radar Equation	51
	4.3.1 The Range Weighting Function	52
	4.3.2 Resolution Volume	57
	4.3.3 Reflectivity Factors	58
4.4	Signal-to-Noise Ratio for Distributed Targets	59
4.5	Correlation of Echo Samples along the Range Time Axis	60
	References	61

5 Doppler Spectra of Weather Echoes

5.1	Spectral Analysis of Weather Signals	62
	5.1.1 Discrete Fourier Transform	62
	5.1.2 Convolution and Correlation	67
	5.1.3 Power Spectrum of Random Sequences	71
	5.1.4 Bias, Variance, and the Window Effect	73
	5.1.5 Expressing Spectral Estimates in Terms of the True Spectrum	75
	5.1.6 Variance of the Periodogram	80
5.2	Doppler Weather Spectrum and Its Relation to Reflectivity and Radial Velocity Fields	81
5.3	Velocity Spectrum Width, Shear, and Turbulence	87
	References	90

6 Meteorological Radar Signal Processing

6.1	Spectral Moments	91
6.2	Radar Signal Processing	92
6.3	Echo Sample Intensity Estimation	94
	6.3.1 Sample Time Averaging	94
	6.3.2 Range Time Averaging	102
6.4	Mean Frequency Estimators	103
	6.4.1 Autocovariance Processing: The Pulse Pair Processor	103
	6.4.2 Spectral Processing	107
6.5	Estimators of the Spectrum Width	108
	6.5.1 Autocovariance Processing	108
	6.5.2 Spectral Processing	111

	6.6	Minimum Variance Bounds	113
	6.7	Performance on Data	115
		References	119

7 Considerations in the Observation of Weather

7.1	Range Ambiguities	121
	7.1.1 Probability of Obscuration by a Single Cell	124
	7.1.2 Obscuration by a Squall Line	126
7.2	Velocity Ambiguities	129
7.3	Echo Coherency	130
7.4	Techniques to Extend the Unambiguous Range and Velocity and to Reduce the Loss of Information from Overlaid Echoes	132
	7.4.1 Phase Diversity	132
	7.4.2 Spaced Pairs with Polarization Coding	133
	7.4.3 Staggering the PRT to Increase the Unambiguous Velocity	135
	7.4.4 Interlaced Sampling	139
	7.4.5 Correcting Aliased Velocities	140
7.5	Methods to Decrease the Acquisition Time	142
	7.5.1 Frequency Diversity	143
	7.5.2 Random Signal Transmission	143
7.6	Effective Antenna Pattern of a Scanning Radar	146
7.7	Antenna Side Lobes	150
7.8	Ground Clutter and Its Suppression	153
7.9	Spectral Artifacts	156
	7.9.1 Quantization and Saturation Noise	158
	7.9.2 Amplitude and Phase Imbalance	161
	7.9.3 Phase Jitter	163
7.10	Detection of Weakly Scattering Weather Targets	163
	7.10.1 Pulse Compression	164
	7.10.2 Complementary Codes	167
	7.10.3 FM cw Doppler Radar	168
	References	177

8 Rain Measurements

8.1	Drop Size Distributions	179
	8.1.1 Cloud Drop Distributions	181
	8.1.2 Raindrop Size Distributions	184
8.2	Terminal Velocity of Drops	187
8.3	Rainfall Rate, Reflectivity, Attenuation, and Liquid Water Content	188
	8.3.1 Liquid Water Content	189
	8.3.2 Reflectivity Factor Z	190
	8.3.3 Rainfall Rate	193
	8.3.4 Attenuation Rate K	195
8.4	Single-Parameter Measurement to Estimate the Rainfall Rate	199
	8.4.1 R,Z Relations	199
	8.4.2 Attenuation Method	202

8.5	Dual Parameter Measurement to Estimate the Rainfall Rate		205
	8.5.1 Rain Parameter Diagram		206
	8.5.2 Dual Wavelength Method		208
	8.5.3 Dual Polarization		212
	8.5.4 Rain Gauge and Radar		224
8.6	Distribution of Hydrometeors from Doppler Spectra		231
8.7	Summary		233
	References		234

9 Observations of Winds, Storms, and Related Phenomena

9.1	Thunderstorm Structure	241
9.2	Observations with Two Doppler Radars	249
	9.2.1 Reconstruction of Wind Fields	250
	9.2.2 Observation of Tornadic Storms	253
	9.2.3 Errors in Synthesized Wind Fields	260
9.3	Linear Wind Measurements with a Single Doppler Radar	260
	9.3.1 Least Squares Fitting of the Wind Field	264
	9.3.2 Analysis on a Circular Arc	265
	9.3.3 Analysis on a Complete Circle	266
	9.3.4 Analysis on Sections of a Conical Surface	273
	9.3.5 Analysis within a Volume: The VVP Method	273
	9.3.6 Prestorm Observations	274
9.4	Nonlinear Wind	278
	9.4.1 Vertical Wind	278
	9.4.2 Waves	280
9.5	Weather Phenomena Observed with a Single Doppler Radar	281
	9.5.1 Vortices	281
	9.5.2 Severe Storms	285
	9.5.3 Doppler Spectra of Tornados	294
	9.5.4 Downdrafts	301
	9.5.5 Gust Fronts	304
	9.5.6 Lightning	312
	9.5.7 Hurricanes	317
	References	320

10 Measurement of Turbulence

10.1	Statistical Theory of Turbulence	324
	10.1.1 Turbulence Spectra and the Correlation Function	325
	10.1.2 Structure Functions (Locally Homogeneous Fields)	329
	10.1.3 Structure and Correlation Functions from Similarity Assumptions	331
	10.1.4 Chandrasekhar's Theory	333
10.2	Spatial Spectra of Point and Average Velocities	336
	10.2.1 Filtering by the Resolution Volume	336

Contents ix

		10.2.2 Variance of Point and Average Velocities	342
		10.2.3 Turbulence Parameters from a Single Radar	343
		10.2.4 Turbulence Parameters from Two Doppler Radars	344
	10.3	Doppler Spectrum Width and Eddy Dissipation Rate	345
	10.4	Doppler Spectrum Width in Severe Thunderstorms	347
		References	357

11 Echoes from the Precipitation-Free Turbulent Troposphere

	11.1	Reflection, Refraction, and Scatter: Coherence	359
	11.2	Formulation of the Wave Equation for Inhomogeneous and Turbulent Media	361
	11.3	Solution for Fields Scattered by Irregularities	365
	11.4	Fraunhofer Scatter	371
		11.4.1 Discussion and Examples	374
		11.4.2 Expected Scattered Power Density	378
	11.5	Fresnel Scattering	388
		11.5.1 Correlation Length Shorter Than the Fresnel Length	388
		11.5.2 Correlation Length Comparable to or Larger Than the Fresnel Length	391
		11.5.3 Backscattering from Anisotropic Irregularities	397
	11.6	Structure Constant of the Refractive Index	400
		11.6.1 Dependence of the Structure Constant on the Height	406
		11.6.2 Inertial Subrange	411
		11.6.3 Criteria for Measurement of the Velocity of Refractive Index Irregularities	414
	11.7	Observations of Clear-Air Reflectivity	416
	11.8	Observations of Wind, Waves, and Turbulence in Clear Air	423
		11.8.1 Wind Profiling	423
		11.8.2 Kinematic Structure of the Convective Boundary Layer	426
		References	433

Appendix A Geometric Relations for Rays in the Troposphere

	A.1	Integral Solutions for Ray Path in a Spherically Stratified Medium	437
	A.2	Relating a Target's Apparent Range and Elevation Angle to Its True Height and Great Circle Distance	438
		References	439

Appendix B Correlation between Echo Samples as a Function of Sample Time 440

Appendix C   Correlation of Echoes from Spaced Resolution Volumes

    C.1   Echo Sample Correlation versus Range Difference $c\delta\tau_s/2$ 443
    C.2   Correlation of Echoes from Azimuthally Spaced Resolution Volumes   445

Appendix D   Geometric Optics Approximation to the Wave Equation   448

Appendix E   Derivation of Green's Function   450

        Reference   451

INDEX   453

Preface

To be able to observe remotely the internal motions of a tornadic thunderstorm that presents a hazardous threat to human communities is an impressive experience. We were fortunate to have entered the field of radar meteorology at a time when the use of Doppler radar was rapidly growing. Such advances were made possible by the general availability of inexpensive digital hardware, facilitating the implementation of theory developed in the early years of radar. The Doppler weather radar techniques developed by radar engineers and meteorologists may soon find applications in the National Oceanic and Atmospheric Administration's (NOAA) NEXRAD (next generation radar) program. A network of Doppler weather radars is planned to replace the present aging radar system used by the National Weather Service (NWS). Improvements in the techniques to provide warnings of tornadoes and other hazardous phenomena continue to be made. Doppler weather radar has already found a home in several television stations that broadcast early warnings of storm hazards.

To a large extent this book is based on lectures given in a course on radar meteorology taught by both authors at the University of Oklahoma. A considerable portion of Chapter 11 is derived from a graduate course in wave propagation through random media given earlier by R. J. Doviak at the University of Pennsylvania. Material from this book has also been used in a one-week course on radar meteorology offered nationally by the Technology Service Corporation. The opportunities available to us at the National Severe Storms Laboratory were vital for the pursuit of research in the diverse disciplines required to develop and apply Doppler radar in remote sensing of severe thunderstorms. Such research fostered comprehensive and detailed treatment of the theory and practice of radar design, digital signal processing techniques, and interpretation of weather observations.

In this book we have aimed to enhance radar theory with observations and measurements not available in other texts so that students can develop an understanding of Doppler radar principles, and to provide practicing engineers and meteorologists with a discussion of timely topics. Thus we present

Doppler radar observations of tornado vortices, hurricanes, and lightning channels. In order to better relate radar observations to weather events commonly observed by eye, radar data fields are correlated with photographs of the physical phenomena such as gust fronts, downbursts, and tornadoes.

While our focus is on meteorology, the theory and techniques developed and discussed here have applications to other geophysical disciplines. Propagation in and scatter from media having a random distribution of discrete targets or from media described by continuous temporal and spatial random variations in their refractive properties also occur in the nonstormy atmosphere and in the ocean, where waves or turbulence, or both, can be generated by intrinsic physical phenomena, by vehicles (such as aircraft, ships, and rockets) traversing the media, or by perturbations purposely created to examine media characteristics. Radar specialists, who often are more interested in the detection and tracking of vehicles and who treat storms as annoying clutter, should find the observations of weather phenomena and the characterization of their properties described herein useful in design studies aimed to maximize target echoes and to minimize interference by weather.

Acknowledgments

Invaluable contributions have been made to this book by many individuals and organizations over the decade during which we have been involved in weather radar research. Edwin Kessler, Director of the National Severe Storms Laboratory (NSSL), is principally responsible for supporting our endeavor; and our colleagues (engineers, technicians, and meteorologists) at NSSL have contributed in countless ways to this documentation of an aspect of the science of weather observation. Furthermore, our associates in NOAA's Environmental Research Laboratories in Boulder, Colorado; in the National Center for Atmospheric Research (NCAR); in the Air Force Geophysical Research Laboratories; and at NASA's Atmospheric Research Facility on Wallops Island, Virginia, have unselfishly provided much of their time, experience, and counsel. The Doppler weather radar project at NSSL has also received support from our associates in other U.S. Government agencies such as the Federal Aviation Administration and the Department of Energy. The Departments of Electrical Engineering and Meteorology at the University of Oklahoma have provided an exceptionally pleasant environment for lecturing and the opportunity to interact with students, who were the principal motivators for this book. The section on linear wind measurements in Chapter 9 was contributed by Albert Koscielny of NSSL. We thank Earl Gossard of NOAA's Wave Propagation Laboratory for his detailed review of Chapter 11 and Richard Strauch, of the same laboratory, for the many constructive comments he made throughout the book. Personnel from the Field Observing Facility of NCAR have given helpful criticism in their review of Chapters 1–9. We are particularly indebted to Joy Walton and Mickey Tyo for their tireless effort and exceptional efficiency in preparing the manuscript through many revisions. The printing of the color plates was made possible by a contribution from the Cooperative Institute for Mesoscale Meteorological Studies, Norman, Oklahoma.

List of Symbols

The following symbols are those used most frequently.

a_e	Effective earth radius
A_e	Effective aperture area of the antenna
B_n	Noise bandwidth
B_6	Receiver–filter bandwidth, 6-dB width
c	Speed of light, 3×10^8 m s^{-1}
C_n^2	Structure parameter of refractive index
D	Diameter of the antenna system
D_e	Diameter of an equivalent volume spherical raindrop
D_0	Median volume diameter
D_{0t}	Parameter in the truncated drop size distribution
E	Electric field intensity
f	Frequency
f_d	Doppler shift
f_N	Nyquist frequency
$f^2(\theta, \phi)$	Normalized one-way power gain of radiation pattern
g	Gravitational constant (9.81 m s^{-2})
g_t, g_r	Power gain of transmitting and receiving antenna
$I(\mathbf{r}, \mathbf{r}_1)$	Illumination weighting function for resolution volume V_6
$I(t)$	In-phase component of the complex signal
k	Attenuation rate (m^{-1})
k	Boltzman constant [1.38×10^{-23} (W s K^{-1})]
k_c	Attenuation rate due to clouds (m^{-1})
k_g	Gaseous attenuation rate
k_r	Wind shear in the r direction
k_θ	Wind shear in the θ direction
k_ϕ	Wind shear in the ϕ direction
K	Attenuation rate (dB km^{-1})
K_r	Attenuation rate due to rain (dB km^{-1})
K_s	Attenuation rate due to snow (dB km^{-1})
l	One-way propagation loss due to scatter and absorption

ln	Natural logarithm
log	Logarithm to base 10
l_r	Finite bandwidth receiver loss factor
L_r	$10 \log l_r$ (dB)
m	Complex refractive index of water
M	Number of echo samples along sample time axis
M	Liquid water content
M_I	Number of independent samples
n	Atmosphere's refractive index
N	Refractivity $= (n-1) \times 10^6$
N	White noise power
N_s	Surface refractivity ($N_s = 313$)
$N(D)$	Drop size distribution
P_r	Echo power
P_t	Peak transmitted power
P_w	Partial pressure of water vapor
$P(\tau_s)$	Instantaneous weather echo power
Q_w	Total water content
$Q(t)$	Quadrature phase component of the complex signal
r	Range to target
r_a	Unambiguous range
r_t	Vortex radius
r_6	6-dB range width of sample volume
$\mathbf{r}_0$	Vector range to the resolution volume V_6 center
R	Rainfall rate
R	Gas constant for dry air (287.04 m² s⁻² K⁻¹)
$R(T_s)$	Autocorrelation of $V(nT_s)$
S	Echo (signal) power
$S_n(f)$	Normalized power spectral density
SNR	Signal-to-noise ratio
T	Absolute temperature (K)
T_s	Pulse repetition time (PRT) or sample time interval
v	Velocity
v_a	Unambiguous velocity limits
v_r	Radial component of velocity (Doppler velocity)
V_6	Resolution volume
$V(kT_s)$	kth complex signal sample
w	Vertical velocity
w_t	Terminal velocity
$W(r)$	Range weighting function
Z	Reflectivity factor
Z_e	Effective reflectivity factor

List of Symbols

Z_{DR}	Differential reflectivity
α	Antenna rotation rate
δ	Wind direction
ε	Eddy dissipation rate
η	Target reflectivity cross section per unit volume
η_0	377-Ω space impedance
θ	Angular distance from the beam axis
θ_e	Elevation angle
θ_1	One-way beamwidth between half-power points
λ	Electromagnetic wavelength
ρ	Mass density of air
$\boldsymbol{\rho}$	Distance in lag space
ρ_w	Density of water
σ_a	Absorption cross section
σ_b	Backscatter cross section
σ_s	Spectrum width due to shear
σ_t	Spectrum width due to turbulence
σ_t	Extinction or attenuation cross section
σ_v	Doppler velocity spectrum width
σ_α	Spectrum width due to antenna rotation
σ_θ	Second central moment of the two-way power pattern
τ	Pulse width
τ_s	Range time delay
ϕ	Azimuth
ϕ_a	Apparent pattern width
ω	Angular frequency
ω_d	Doppler shift (rad s^{-1})

1
Introduction

Radars were developed to detect and determine the range of aircraft by radio techniques ("radar" is an acronym for radio detection and ranging), but as they became more powerful, their beams more directive, their receivers more sensitive, and their transmitters coherent, radars have been successfully applied in mapping the earth's surface and weather, and their signals have reached out into space to explore surface features of our planetary neighbors. The radar beam penetrates thunderstorms and clouds to reveal, like an x-ray photograph, the inside structure of these otherwise unobservable events. Recently pulsed-Doppler radar techniques have been applied to map severe storm structures with remarkable success, showing in real time the development of incipient tornado cyclones. Such observations should enable weather forecasters to provide better forecasts and warnings and researchers to understand the life cycle and dynamics of storms.

By far the most comprehensive treatment of radar techniques is found in the collected works compiled by M. I. Skolnik in his *Radar Handbook* (1970). Battan's text (1973) on weather radar applications is probably the most widely used by meteorologists, and Atlas (1964) also gives a concise and informative review of many weather radar topics. Both of these works emphasize the electromagnetic scatter and absorption by hydrometeors. A book by Nathanson (1969) emphasizes the "total radar environment" as well as radar design principles. The radar environment as defined by Nathanson is also the source of unwanted reflection (clutter) from the sea and land areas. (Precipitation is said to produce clutter when aircraft are the targets of interest.) Thus, precipitation echoes are comprehensively treated. The anomalous propagation of radar signals enhances ground clutter. A good general reference on the propagation of electromagnetic waves through the stratified atmosphere is the book by Bean and Dutton (1966).

We attempt here to provide a solid foundation in weather radar principles and to relate the radar parameters and signal characteristics to the target's meteorological properties. Chapter 2 describes the effect that the atmosphere has on the path of the signal. We lightly touch on subjects comprehensively treated elsewhere (e.g., the scattering properties of hydrometeors), present techniques used in extracting a target's properties from its echoes, and relate radar parameters and echo power to the weights given to the hydrometeor target backscatter cross section.

To build the foundation, in weather radar principles in Chapter 3 we trace the radar signal path from the transmitter, through the antenna, along the beam to the target, and on its return to the receiver, highlighting along the way the important aspects of the signal properties. We use pulsed-Doppler radar and consider a discrete target to develop the radar principles and the radar equation. These will be seen to apply naturally to the case of non-Doppler radars that are extensively used in the present national network of weather radars.

We then extend these principles, in Chapter 4, to the more complex weather target that is a conglomerate of discrete targets producing a continuous stream of echoes with randomly fluctuating amplitude and phase. We show the origin of these fluctuations and develop the weather radar equation.

We treat the discrete Fourier transform in Chapter 5 and apply it to weather echo signals so as to make a connection between the Doppler spectrum and shear and turbulence of flow in a fluid. In Chapter 6 we treat the weather Doppler spectrum and outline the signal processing methods used to derive the principal moments of the Doppler spectrum, emphasizing results rather than processor details.

The very important topic of range and Doppler velocity ambiguities as they pertain to distributed targets, as well as other considerations in observing weather, are presented in Chapter 7. The limitations imposed by antenna sidelobes, ground clutter, signal decorrelation, and power are discussed, together with techniques to mitigate these limitations. Thus a comprehensive treatment of pulse compression and the Doppler processing of frequency modulated signals is given. We develop the theory needed to explain commonly encountered artifacts due to signal processing and show that antenna rotation coupled with signal averaging produces a broadened apparent pattern.

The physics behind a variety of methods of rainfall estimation is discussed in Chapter 8. These methods are divided into single- and multiple-parameter techniques, depending on the number of independent measurements.

A brief introduction to storm structure, in Chapter 9, is followed by examples of wind fields, obtained from the analysis of Doppler radar data on storms. Photographs are provided of several of the significant observable events associated with storms. The important research subject of data analysis from more than one Doppler radar is briefly discussed. Multiple Doppler data synthesis to map the wind field with high resolution confirms the interpretation of single Doppler signatures of severe weather events.

Although much of the discussion of thunderstorms is focused on their hazards, one should not be led to believe that thunderstorms bring only misery. Each storm can release on the order of 10^{10} kg of beneficial rain water and, at the cost of about 31 cents per ton, the water is worth nearly a million dollars if properly stored and distributed. We need to learn methods by which losses due to storms can be lessened while their benefits continue. Proper warning and the protection of life and property are the first defense. The modification of storms to reduce their hazards without a loss of rain appears to be a long-term effort. Each storm releases energy at the rate of about 10^7 MW of latent heat (Sikdar et al., 1974). This prodigious amount of energy spread over large volumes would indeed be difficult to control.

In Chapter 10 the theory of turbulence is reviewed, with emphasis on topics applicable to the radar measurements. Spatial spectral representation, filtering by the resolution volume, and examples of eddy dissipation rates are presented. The contributions of turbulence and shear to the Doppler spectrum width in a severe storm are examined.

A theory based on Fourier spectral representations is developed in Chapter 11 to explain radar echoes from clear air refractive index irregularities. Existing theories are extended to develop a formulation for the Fresnel scatter that is observed by radars probing optically clear air. These theories are amply illustrated with specific examples and are used to explain actual observations. Waves and turbulence in the earth's convective boundary layer are revealed by the use of the Doppler weather radar to observe echoes from refractive index irregularities. Radar reflectivity is related to the dynamics of atmospheric flow, and the potential of weather radars for mapping the kinematic structure of the atmosphere is discussed.

REFERENCES

Atlas, D. (1964). Advances in radar meteorology. *Adv. Geophys.* **10**, 317–478.

Battan, L. J. (1973). "Radar Observation of the Atmosphere." Univ. of Chicago Press, Chicago, Illinois.

Bean, B. R., and Dutton, E. J. (1966). "Radio Meteorology," Natl. Bur. Stand., Monogr. 92, Supt. Doc. U.S. Govt. Printing Office, Washington, D.C.

Nathanson, F. E. (1969). "Radar Design Principles." McGraw-Hill, New York.

Sikdar, D. N., Schlesinger, R. E., and Anderson, C. E. (1974). Severe storm latent heat release: Comparison of radar estimate versus a numerical experiment. *Mon. Weather Rev.* **102**, 455–465.

Skolnik, M. I., ed. (1970). "Radar Handbook." McGraw-Hill, New York.

2

Electromagnetic Waves and Propagation

To understand the remote sensing of weather by radar requires knowledge of a few basic properties of electromagnetic waves and the effects that the atmosphere has on these waves as they propagate between the radar and the target volume. This chapter reviews fundamental wave concepts and presents elementary theories that describe wave propagation. Useful formulas are derived that quantify some of the important effects that the environment has on the radar's capability to locate and detect weather targets.

2.1 WAVES

Electromagnetic or radio waves from radar are electric **E** and magnetic **H** force fields that propagate through space at the speed of light and interact with matter along their paths. These interactions cause the scattering, diffraction, and refraction also common to visible electromagnetic radiation (light). These waves, focused into beams by the antenna system, have sinusoidal spatial and temporal variations; the distance or time between successive wave peaks of the electric (magnetic) force defines the wavelength λ or wave period (i.e., the reciprocal of the frequency f, in hertz). These two important electromagnetic field parameters are related to the speed of light c:

$$c = \lambda f = 3 \times 10^8 \quad \text{m s}^{-1}. \tag{2.1}$$

Microwaves are electromagnetic forces having spatial wavelengths between 10^{-3} and 10^{-1} m, whereas visible radiation has a wavelength of about 6×10^{-7} m. The upper end (0.01–0.1 m) of the microwave band is used by weather and aircraft surveillance radars (see Fig. 2.1).

The electric field wave far from the transmitting antenna has a time t and range r dependence generally given by

$$E(r, \theta, \phi, t) = \frac{A(\theta, \phi)}{r} \cos\left[2\pi f \left(t - \frac{r}{c}\right) + \psi\right] \quad \text{V m}^{-1}, \tag{2.2a}$$

2.1 Waves

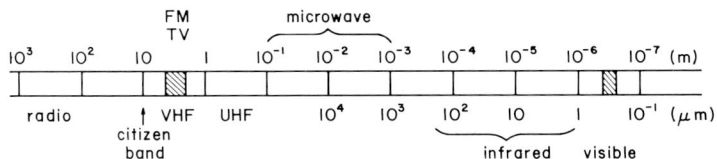

Fig. 2.1. The electromagnetic spectrum, showing the location of the microwave band relative to the radio frequency and optical bands.

where A depends upon θ, ϕ (the direction of **r** from the radiation source), and ψ is usually an unknown but constant phase. The dependence (2.2a) of **E** and **H** on r, t, θ, and ϕ is characteristic of all electromagnetic waves propagating in space devoid of matter, be they radio waves or light. Equation (2.2a) approximates well at weather radar frequencies the properties of waves propagating through our atmosphere. Because a force has direction, **E** is a vector quantity and (2.2a) represents one component of the electric field. The waves propagate in the direction of **r**; that is, an observer's range r must increase at a rate c in order to stay on a wave crest ($t - r/c = $ const). The vectors **E**, **H** are always perpendicular to **r** and to each other when r is large compared to the antenna dimensions.

Because the principal factors characterizing a periodic electric field are the amplitude $A(\theta, \phi)/r$ and phase $2\pi f(t - r/c) + \psi$, it is convenient and instructive to use complex number or phasor notation to describe these parameters. The electric field, (2.2a) is then expressed as

$$E = \frac{A(\theta, \phi)}{r} \exp\left[j2\pi f\left(t - \frac{r}{c}\right) + j\psi\right], \tag{2.2b}$$

which, according to Euler's formula, can be represented by a two-dimensional phasor diagram (Fig. 2.2). This diagram clearly describes the time and space dependence of amplitude and phase to within an integral multiple of 2π. **E** and **H** are real, measurable quantities, and it is understood that we need to take the real part of (2.2b), or of any other phasor we may introduce, in order to obtain the real time and space dependence. The time dependence of the phase is of paramount importance in understanding the principles of Doppler radar.

Another important electromagnetic field quantity is the time-averaged power density $S(r, \theta, \phi)$:

$$S(r, \theta, \phi) = \frac{1}{2} \frac{EE^*}{\eta_0} = \frac{A^2(\theta, \phi)}{2\eta_0 r^2} \quad \text{W m}^{-2}, \tag{2.3}$$

where * denotes complex conjugation. It can be shown that $\frac{1}{2}EE^*$ is the time average of the square of (2.2a). The factor η_0 is the wave impedance (in space, or in earth's atmosphere at radar wavelengths, η_0 is a constant equal to 377 Ω)

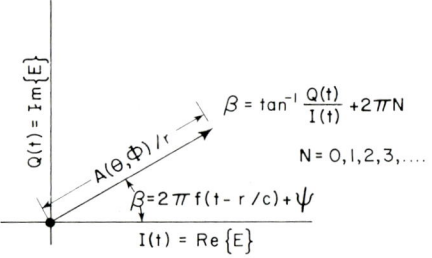

Fig. 2.2. Phasor diagram. Re$\{E\}$ is a real part of the complex electric field E and Im$\{E\}$ the imaginary part.

and is the ratio of the electric to the magnetic field amplitude. Time averages represented by (2.3) are averages of power over a cycle or period f^{-1} of the wave, and (2.3) is usually called the *peak power density* of an electromagnetic packet of waves if power is *pulsed* (i.e., transmitted in a burst of energy). The significance of $S(r, \theta, \phi)$ is that it represents the power density flowing outward from a source, either continuously or in bursts, and products of S with areas, later to be specified, represent the power that is received, absorbed, scattered, and so on.

Prior to the 1930s most remote probing of our physical environment was performed at the short wavelength electromagnetic radiation visible to the human eye (visual observations and recording of script), whereas long wavelengths (i.e., 10–300 m) of the radio band were used to measure properties of the heavily ionized air (ionosphere) surrounding the earth between 10^5 and 3×10^5 m above ground.

The spatial resolution of observations, i.e., the discrimination between two adjacent similar objects, is dependent on the wavelength and antenna size. The angular resolution or diameter of the first null in a circular diffraction pattern is well approximated by

$$\Delta\theta \approx 140\lambda/D \quad \text{deg} \tag{2.4}$$

(Born and Wolf, 1964, p. 415), where D is the diameter of the antenna system. Thus, radio waves (i.e., $\lambda = 10^2$–10^3 m) require huge antenna installations to achieve an angular resolution of a few degrees—rather poor for optical antenna systems. For example, the human eye has an angular resolution of about $0.02°$. It is evident that remote radio sensing, even at microwave frequencies, is characterized by poor resolution compared to optical standards.

The essential distinguishing feature favoring microwaves for weather radars is their property to penetrate rain and cloud and thus to provide a view inside showers and thunderstorms, day or night. Rain and cloud do attenuate microwave signals, but only slightly (for $\lambda \geq 0.05$ m) compared to the almost complete extinction of optical signals. Raindrops scatter electromagnetic energy, and the portion scattered constitutes the signal whose characteristics are diagnosed to determine storm properties. Scattered signal strength measures rain intensity, and the time rate of phase change (Doppler shift) measures the raindrop speed in the **r** direction.

2.2 PROPAGATION PATHS

Before we proceed to the principles of radar, it will be useful to discuss the subject of propagation. In free space, waves propagate in straight lines because everywhere the dielectric permittivity ε_0 and magnetic permeability μ_0 are constants related to speed of propagation $c = (\mu_0 \varepsilon_0)^{-1/2}$. However, the atmosphere's permittivity is vertically stratified, and therefore microwaves do not propagate along straight lines; often the radar beam is refracted (bent) back to the surface (anomalous propagation), causing distant ground clutter to be sometimes falsely interpreted as weather echoes. It is common practice today to make simple corrections for refractive effects, but although these work well most of the time, there exist atmospheric conditions that require sophisticated methods to determine the path of the radar signal.

The path of radar signals depends principally on the change in height of the atmosphere's refractive index n, or relative permittivity $\varepsilon_r = \varepsilon/\varepsilon_0 = n^2$. We shall show how the refractive index is related to temperature, pressure, and water vapor content. Then, given a vertical profile of these meteorological variables, one can determine the path to and strength of the radar signal at the target. Because microwave scattering occurs from refractive index variations of small scale (i.e., centimeter sizes), we shall find the expressions developed herein useful when we discuss, in Chapter 11, radar echoes from clear air.

2.2.1 Refractive Index of Air

The refractive index is proportional to the density of molecules and their polarization. Molecules that produce their own electric field without external forces are called *polar*. Polar molecules possess a permanent displacement of opposing charges within their internal structure, thus causing a dipole electric field that reaches far beyond the molecule's interatomic spacing. The water vapor molecule is polar. Although dry air molecules do not possess a permanent dipole moment, they become polarized when an external electric force (e.g., radar signal) is impressed on them. Without external forces, polar molecules have their dipole moments (with direction along the axis connecting the centers of opposing charge) randomly oriented due to thermal agitation. External forces can align these molecules so that their dipole fields add constructively to enhance the net electric force acting on each molecule. Thus the electric force acting on any molecule is the sum of the external electric field plus that produced by the polarized molecules.

The permittivity of a gas depends only on the density N_v of molecules and a factor α_T proportional to the molecule's level of polarization. It can be shown that

$$(\varepsilon_r - 1)/(\varepsilon_r + 2) = N_v \alpha_T / 3. \tag{2.5}$$

By a remarkable coincidence, this relation was found independently at almost the same time by two scientists of almost identical names, H. A. Lorentz and L. Lorenz; accordingly, Eq. (2.5) is called the Lorentz–Lorenz formula (Born and Wolf, 1964, p. 87). It applies to gases at all but extreme pressures. For air, ε_r is very near unity (i.e., $\varepsilon_r \simeq 1.000300$) and relative permeability μ_r is unity, so

$$\mu_r \varepsilon_r = \varepsilon_r = n^2 = 1 + N_v \alpha_T. \tag{2.6}$$

Laboratory measurements of ε_r can be used to deduce α_T. For a gas that contains a mixture of molecules of types (1), (2), ...,

$$n^2 = 1 + N_v^{(1)} \alpha_T^{(1)} + N_v^{(2)} \alpha_T^{(2)} + \cdots. \tag{2.7}$$

Now, the number density of molecules of any type is proportional to the mass density ρ of the gas:

$$N_v = \rho/M, \tag{2.8}$$

where M is the mass of the molecule under consideration. The mass density, pressure, and temperature obey the equation of state:

$$\rho = \rho_0 (273/T)(P/1013), \tag{2.9}$$

where ρ_0 is the mass density at standard temperature and pressure (0°C and 760 mm Hg), P is the pressure in millibars, and T is the temperature in degrees Kelvin. The number of molecules per unit volume of any gas at fixed temperature and pressure is independent of the gas type (Avogadro's law). At standard temperature and pressure we have a number equal to

$$N_{v0} = 2.6873 \times 10^{25} \quad \text{m}^{-3} \tag{2.10}$$

and

$$N_v = N_{v0} \frac{273}{1013} P/T. \tag{2.11}$$

Thus (2.7) can be written

$$n^2 = 1 + N_{v0} \frac{273}{1013 T}(P_1 \alpha_T^{(1)} + P_2 \alpha_T^{(2)}). \tag{2.12}$$

2.2 Propagation Paths

T is outside the parentheses because we assume thermal equilibrium, and P_1 is the partial pressure of gas 1. For the troposphere we need to consider only the contribution from dry air (a nonpolar gas) and water vapor (a polar gas). For dry air

$$n^2 = 1 + C_d P/T, \tag{2.13}$$

where C_d is a constant. For a combination of dry air and water vapor, (2.7) has the form

$$n^2 = 1 + C_d P_d/T + C_{w1} P_w/T + C_{w2} P_w/T^2, \tag{2.14}$$

where the last term is the contribution to the refractive index from the permanent dipole moment of the water vapor molecule. P_d and P_w are the partial pressures of dry air and water vapor. The constant parameters (e.g., N_{v0}) are contained in the constants C.

2.2.2 Refractivity N

Because the relative permittivity ε_r and refractive index n of the atmosphere are so near unity at microwave frequencies, it becomes convenient to introduce a different measure of the refractive properties of air. The *refractivity N* is defined as

$$N \equiv (n - 1) \times 10^6. \tag{2.15}$$

From (2.6) one can express n as

$$n = [1 + (\varepsilon_r - 1)]^{1/2}. \tag{2.16a}$$

Now, $\varepsilon_r - 1$ is small compared to 1, and expansion of (2.16a) gives the linear relation

$$N = \tfrac{1}{2}(\varepsilon_r - 1) \times 10^6 \tag{2.16b}$$

between N and ε_r. Using Eq. (2.14) and the preceding equations, N can be written in the form

$$N = C'_d P_d/T + C'_{w1} P_w/T + C'_{w2} P_w/T^2. \tag{2.17}$$

Bean and Dutton (1966) give a survey of the various measurements and estimates of the constants C'. From their work we have

$$C'_d \simeq 77.6 \quad \text{K mbar}^{-1}, \tag{2.18a}$$

$$C'_{w1} = 71.6 \quad \text{K mbar}^{-1}, \tag{2.18b}$$

$$C'_{w2} = 3.7 \times 10^5 \quad \text{K}^2 \text{ mbar}^{-1}. \tag{2.18c}$$

Equation (2.17) can be approximated to an accuracy of about $0.1N$ by the simplified form

$$N = (77.6/T)(P + 4810P_w/T), \qquad (2.19)$$

where P is the total pressure in millibars and T is in degrees kelvin.

For example, given a relative humidity of 60% and $T = 17°C$ at sea level,

$$P_w \simeq 10 \text{ mbar}, \qquad P \simeq 1000 \text{ mbar}, \qquad T \simeq 300 \text{ K}.$$

Thus

$$N \simeq 0.26 \times (10^3 + 1.6 \times 10^2) \simeq 300$$

and

$$n = 1 + N \times 10^{-6} = 1.000300.$$

It is apparent that the refractive index of the atmosphere differs very little from that of free space. Nevertheless, a change in n in the fifth and sixth significant digits is sufficient to have a measurable effect on electromagnetic wave propagation and scattering.

Both pressure and temperature usually decrease with height from sea level up to about 10 km, at which altitude the temperature begins to remain relatively constant for several kilometers (Fig. 2.3). In the troposphere the fractional decrease in pressure is larger than that for temperature, so N normally decreases with altitude. When the rate of decrease in N exceeds a certain value (i.e., $dN/dh \leq -157 \text{ km}^{-1}$) electromagnetic beams are bent toward the surface of the earth (i.e., trapped), as we shall demonstrate next. This condition is usually brought about by inversion layers, that is, layers of atmosphere in which the temperature *increases* with height, thus causing the

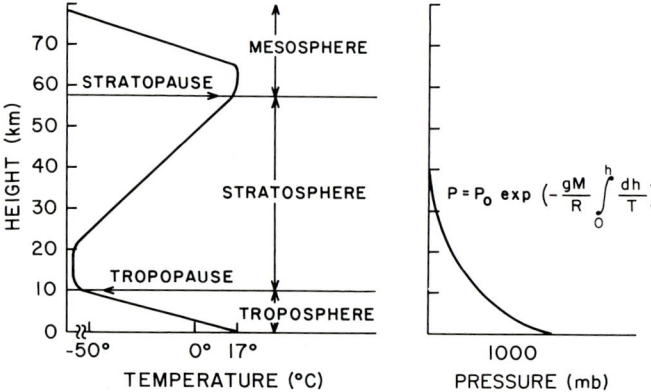

Fig. 2.3. Dependence of the temperature and pressure on height. (R is the universal gas constant and M is the molecular weight in atomic mass units.)

2.2 Propagation Paths

slope dN/dh to be more negative. In addition to systematic smooth variations in atmospheric properties, there are small-scale fluctuations in temperature, pressure, water vapor content, etc. that cause N to have small-scale variations. Electromagnetic wave scattering occurs from these refractive index fluctuations. (The properties of the scattered waves will be given in Chapter 11.) We shall ignore for now these fluctuations and consider the effects of a spherically stratified refractive index (i.e., N that is dependent only on height) on electromagnetic propagation through the lower altitudes of the troposphere.

2.2.3 Spherically Stratified Atmospheres

For most applications we can assume temperature and humidity to be horizontally homogeneous so that the refractive index is a function of R only (see Fig. 2.4). At microwave frequencies it is permissible to assume that wave fronts propagate along rays, analogously to optical propagation. In Appendix A we show that the ray path in a spherically stratified atmosphere is given by the intergral

$$s(h) = \int_0^h \frac{aC \, dh}{R[R^2 n^2(h) - C^2]^{1/2}}, \qquad (2.20a)$$

$$C = an(0) \cos \theta_e, \qquad (2.20b)$$

where $s(h)$ is the great circle distance (along the earth's surface) to a point directly below the ray at height h above the surface, and a is the earth's radius (see Fig. 2.4). The refractive index $n(h)$ is assumed to be smoothly changing with h so that ray theory is applicable. The elevation angle θ_e is that of the ray at the transmitter, and $n(0)$ is the refractive index at that location. It can also be shown that (2.20) is a solution to the exact differential equation

$$\frac{d^2 h}{ds^2} - \left(\frac{2}{R} + \frac{1}{n}\frac{dn}{dh}\right)\left(\frac{dh}{ds}\right)^2 - \left(\frac{R}{a}\right)^2 \left(\frac{1}{R} + \frac{1}{n}\frac{dn}{dh}\right) = 0, \qquad (2.21)$$

which describes the ray path (Hartree et al., 1946).

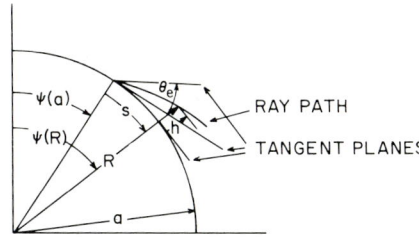

Fig. 2.4. Ray path in a spherically stratified atmosphere.

2.2.3.1 Effective Earth's Radius Model

Under the condition that the ray is nearly parallel to the surface, $(dh/ds)^2 \ll 1$. Furthermore, if we restrict our attention to ray paths in the first 10 or 20 km of the atmosphere, $h \ll a$, $R \approx a$, $n \approx 1$, so that (2.21) reduces to

$$\frac{d^2h}{ds^2} = \frac{1}{a} + \frac{dn}{dh}. \tag{2.22}$$

The curvature of the ray path is

$$C_0 \equiv \left[R^2 + 2\left(\frac{dR}{d\psi}\right)^2 - R\frac{d^2R}{d\psi^2} \right] \Big/ \left[R^2 + \left(\frac{dR}{d\psi}\right)^2 \right]^{3/2}, \tag{2.23}$$

but $s = a\psi$, and using these simplifying assumptions we find that

$$C_0 \simeq 1/a - d^2h/ds^2. \tag{2.24}$$

Substitution of (2.24) into (2.22) shows that the curvature of the ray is

$$C_0 = -dn/dh. \tag{2.25}$$

It is obvious from (2.22) that if $dn/dh = -1/a$, then dh/ds is constant and equal to zero if θ_e were zero. That is, the ray travels parallel to the earth's surface. This is a condition for trapping electromagnetic beams or rays whose initial elevation angle is zero. The slope or gradient of refractivity dN/dh needed to trap a ray with $\theta_e = 0$ is -157 km^{-1}. Electromagnetic beams launched at higher elevation angles require larger N gradients in order to be trapped.

The ray trapped for $\theta_e = 0$ has a curvature of $1/a$, but its curvature relative to the earth is zero. We are then led to conclude that the ray's curvature C_e *relative to the earth* is

$$C_e = -1/a - dn/dh. \tag{2.26}$$

One can also consider the ray path to be straight; the negative of (2.26) is then the earth's curvature relative to the straight ray. The radius of curvature a_e for the earth's surface is then

$$a_e = -\frac{1}{C_e} = \frac{1}{1/a + dn/dh} = k_e a, \tag{2.27}$$

where

$$k_e \equiv \frac{1}{1 + a(dn/dh)}.$$

Equation (2.27) shows that if dn/dh is constant, then the earth can be considered to have an effective radius a_e and the ray paths to be straight lines.

2.2 Propagation Paths

Thus, we can easily compute the ray height versus arc distance s for this model. The height above ground is

$$h = k_e a \left[\frac{\cos \theta_e}{\cos(\theta_e + s/k_e a)} - 1 \right]. \tag{2.28a}$$

The following two equations relate h and s to radar-measurable parameters, the range r and θ_e:

$$h = [r^2 + (k_e a)^2 + 2rk_e a \sin \theta_e]^{1/2} - k_e a, \tag{2.28b}$$

$$s = k_e a \sin^{-1} \left(\frac{r \cos \theta_e}{k_e a + h} \right). \tag{2.28c}$$

Researchers have found that the gradient of n is typically $-1/4a$, so the effective radius of the earth is

$$a_e = \tfrac{4}{3} a. \tag{2.28d}$$

Although the effective earth's radius model conveniently determines beam height as a function of range or arc length, two limitations need to be discussed:

(1) n is linearly dependent on h.
(2) The development of Eq. (2.27) assumed $dh/ds \ll 1$, which imposes a limit on the use of an effective earth's radius.

The gradient of the refractive index is not always a constant, and we have particularly severe departures from linearity when there are strong temperature inversions or large moisture gradients. Furthermore, the refractive index cannot decrease linearly without bound, because at large heights it must asymtotically approach unity. The unrealistic profile of n assumed by the model with an earth's radius of $\tfrac{4}{3}$ is contrasted in Fig. 2.5 with a realistic dependence of n on h, as given by a reference atmosphere model that agrees quite closely with measured N data. Both models assume surface refractivity $N_s = 313$.

It is obvious that for $h \geq 2$ km there is a considerable difference between the N values. We may well wonder whether the effective earth radius model would be useful for ray paths above 2 km.

For weather radar applications it can be shown that the earth radius of $\tfrac{4}{3} a$ model can be used for all θ_e if h is restricted to the first 10–20 km and if n has a gradient of $-1/4a$ in the first kilometer of the atmosphere. Figure 2.6 shows a comparison of ray paths for $a_e = \tfrac{4}{3} a$ and an exponential reference atmosphere where

$$n(h) = n_s e^{-0.1439h}. \tag{2.29}$$

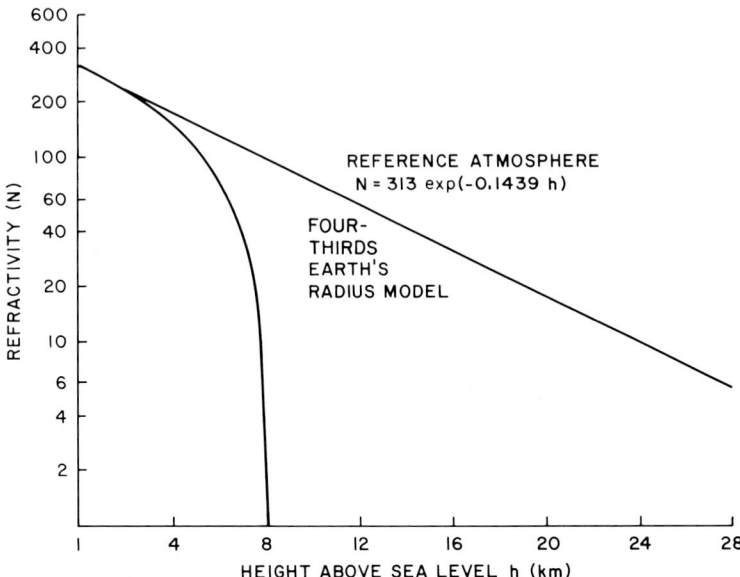

Fig. 2.5. Dependence of refractivity on height for a reference atmosphere contrasted with that implied by the effective earth's radius $a_e = \frac{4}{3}a$.

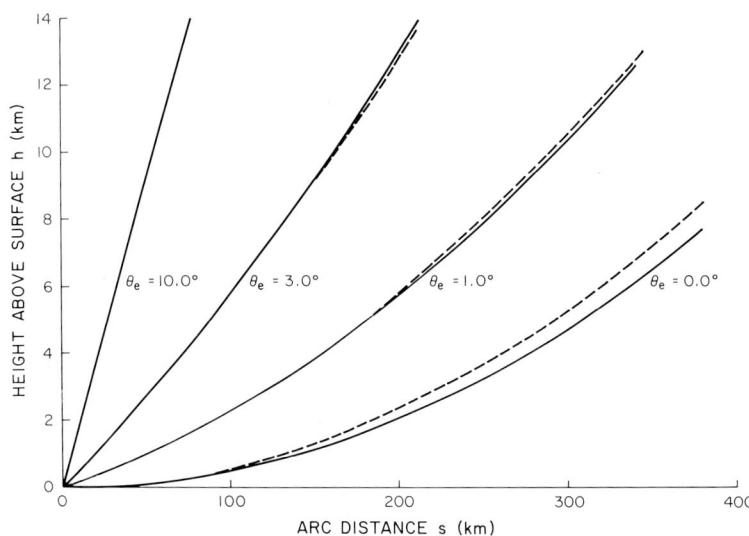

Fig. 2.6. Comparison of the ray paths for an $a_e = \frac{4}{3}a$ model and an atmosphere with an exponentially dependent refractive index. N_s: ———, exponential atmosphere; ---, $\frac{4}{3}$ earth's radius.

2.2 Propagation Paths

We can see from Fig. 2.6 that, although the difference in n is large for $h \geq 2$ km, the difference in the ray paths is no greater than 1 km at a range of 250 km for $\theta_e = 0$, and the difference in the ray path height decreases rapidly as θ_e is increased from 0. Furthermore, most weather radar beamwidths are of the order of 1°, and errors in height are small compared to the beamwidth at these ranges. We therefore conclude that if the refractive index is well represented by (2.29), use of effective radius $a_e = \frac{4}{3}a$ predicts beam height with sufficient accuracy for conventional weather radar applications.

What are the effects of large temperature inversions? The following section discusses the ray paths in the lower atmosphere when there is an anomously large gradient of refractivity.

2.2.3.2 Ground-Based Ducts; Reflection Height

We shall show how ray paths can be determined for refractive index profiles that depart considerably from those associated with the exponential reference atmosphere. We assume that the refractive index profile can be approximated by a piecewise linear model of N versus h. Consider the refractive index profile depicted in Fig. 2.7. This model shows a large gradient $(dN/dh = -300$ km$^{-1})$ of refractivity for the first 100 m and thereafter a gradient associated with an effective earth's radius of $\frac{4}{3}a$.

When $h \ll a$, the integrand of (2.20) can be linearized with respect to h and thus integrated to produce the following relation for 0 km $\leq h \leq$ 0.1 km:

$$s(h) = [(\cos \theta_e)/(1 + \beta_0 a)]\{[a^2 \sin^2 \theta_e + 2a(1 + \beta_0 a)h]^{1/2} - a \sin \theta_e\}, \quad (2.30)$$

where $\beta_0 = -3 \times 10^{-4}$ km^{-1} and we have substituted 1 for $n(0)$. There is a like expression for 0.1 km $\leq h$, where the gradient of n is β_1;

$$s'(h') = [(\cos \theta'_e)/(1 + \beta_1 a')]$$
$$\times \{[a'^2 \sin^2 \theta'_e + 2a'(1 + \beta_1 a')h']^{1/2} - a' \sin \theta'_e\}. \quad (2.31)$$

$s'(h')$ is the arc distance from the point of emergence of the ray from the layer, h' is the height of the ray above h_1, $a' = a + h_1$, and θ'_e is the angle

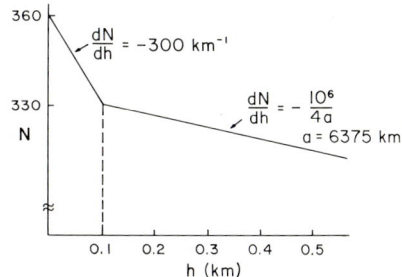

Fig. 2.7. Refractivity N profile for a model atmosphere in which there is a strong ground-based temperature inversion.

made by the ray at $h_1 = 0.1$ km. This angle is

$$\theta'_e = \tan^{-1}(dh/ds) \quad \text{at} \quad h_1 = 0.1 \text{ km}. \tag{2.32}$$

Differentiating (2.30) and substituting into (2.32), we obtain

$$\theta'_e \simeq \tan^{-1}\{[a^2 \sin^2 \theta_e + 2ah_1(1 + \beta_0 a)]^{1/2}/a \cos \theta_e\}. \tag{2.33}$$

Because $\beta_0 a < -1$, the angles $\theta_e \leq \theta_p$ cause the radical to be imaginary, where

$$\theta_p = \sin^{-1}[-2h_1(1 + \beta_0 a)/a]^{1/2} \tag{2.34}$$

is the penetration angle. All rays having $\theta_e \leq \theta_p$ are reflected within the layer $0 \text{ km} \leq h \leq 0.1$ km. The height h_r of the ray at the point of reflection is obtained by solving for h_r in the differential equation

$$\frac{dh}{ds} = \frac{[a^2 \sin^2 \theta_e + 2ah_r(1 + \beta_0 a)]^{1/2}}{a \cos \theta_e} = 0, \tag{2.35}$$

obtained by differentiating (2.30). The solution of (2.35) is

$$h_r = (-a \sin^2 \theta_e)/2(\beta_0 a + 1). \tag{2.36}$$

In this example $\theta_p = 0.31°$, and rays having an elevation angle less than $0.31°$ are trapped in the layer. A ray having $\theta_e = 0.2°$ has a height of reflection $h_r = 43$ m, and reflection occurs at an arc distance obtained by solving (2.30):

$$s(h_r) = (-a \cos \theta_e \sin \theta_e)/(\beta_0 a + 1), \tag{2.37}$$

which in this example is 24.4 km. The ray returns to earth at an arc distance $2s(h_r)$. Thus a target at a distance of 49 km, which would not be visible to radar under normal propagation conditions, becomes visible if the beam elevation angle is $0.2°$ for the given refractivity profile. A few sample ray paths for this case are shown in Fig. 2.8. The ray path for $h \geq 0.1$ km is obtained using an effective earth's radius of $\frac{4}{3}a$ and θ' from (2.33) as the initial elevation angle of the ray emerging above the layer h_1.

Also apparent from Fig. 2.8 is the effect of inversion on spatial resolution. For example, suppose we observe targets at a distance of 50 km using a radar with $0.2°$ angular resolution and pointed at an elevation angle of $0.3°$. The vertical beamwidth in the absence of inversion would have been 175 m, but in the presence of inversion the beamwidth broadens to 270 m. Such broadening not only leads to deterioration of the resolution but can also result in erroneous measurements of target cross section because the power density at the target is reduced.

Experimental data (Fig. 2.9) clearly show the significance of surface based ducts. These data were taken over desert terrain and reported by Day and Trolise (1950). The nighttime cooling of the earth results in a surface-based

2.2 Propagation Paths

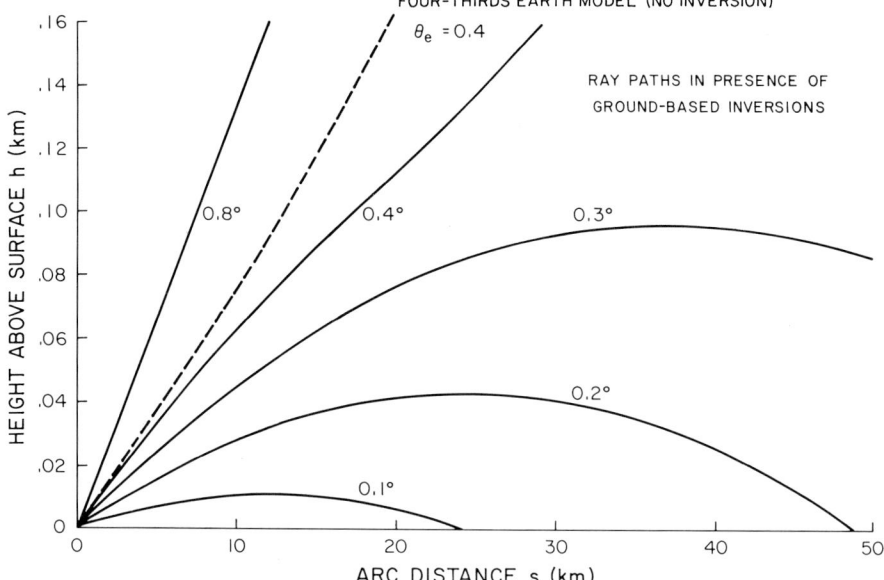

Fig. 2.8. Ray paths in an atmosphere modeled as shown in Fig. 2.7. A surface-based inversion exists in the first 100 m of height.

inversion, thus causing enhanced reception by a receiver located 43 km from a transmitter. Both the receiving and transmitting antennas were on the surface. The 0-dB attenuation relative to free space indicates that the receiver and transmitter antenna are within radio line of sight, although geometrically the receiver is below the transmitter's horizon. Figure 2.9a shows the sudden enhancement of signals when solar heating is extinguished. Figure 2.9b shows a series of refractive index profiles expressed in terms of B, where

$$(B - 1) \times 10^6 = N + 38.7h(\text{km}). \tag{2.38}$$

In this form, the N values associated with the standard gradient, 38.7 km$^{-1} \simeq 10^6/4a$, are removed in order to highlight those changes in refractive index profile that depart markedly from the standard. Although the depth over which the gradient is strongly negative is rather shallow ($\simeq 50$ m) the effect on the transmission path is pronounced. In the daytime the lowest several hundred meters of the atmosphere are well mixed and the refractive index is nearly linear with height. If the atmosphere were standard, plots of $(B - 1) \times 10^6$ would be vertical lines. Thus in the case shown the lower atmosphere has a refractive index lapse rate more positive than standard during the daytime, and for the data shown on Fig. 2.9 $k_e \simeq 1.2$ rather than 1.33.

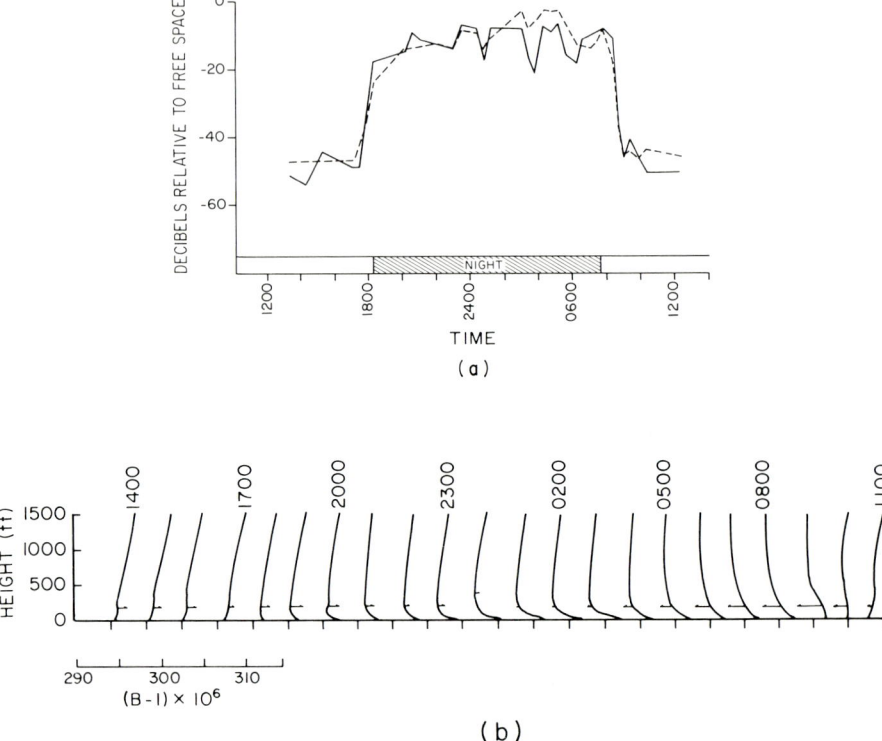

Fig. 2.9. (a) Intensity of a microwave signal at a receiver located 43 km from a transmitter. Both transmitter and receiver are on the surface. Intensity (in decibels) is measured relative to the signal level that would be received if transmitter and receiver were 43 km apart along a straight line-of-sight path. ---, $\lambda = 9.1$ cm; ———, $\lambda = 3.2$ cm. (Modified from Day and Trolise, 1950.) (b) Diurnal variations of height profiles of refractivity expressed in B units [see Eq. (2.38)]. The arrows indicate the value $(B - 1) \times 10^6 = 295$. Average B curves; 5–6, February 1946. (Modified from Day and Trolise, 1950.)

A striking example of echoes in the presence of strong ground-based inversion is shown on Fig. 2.10a. These anomalous propagation data were obtained with a 10-cm radar at Wallops Island, Virginia. The radar beamwidth is about 0.4°, and the elevation angle was 0.5°. Judging from the continuous extent of echoes, we conclude that parts of the beam must have been grazing along the ground. Clearly visible on the figure are islands and waterways in the Chesapeake Bay as well as the bridge–tunnel complex. For comparison, the chart of the area and the radar location are drawn on Fig. 2.10b.

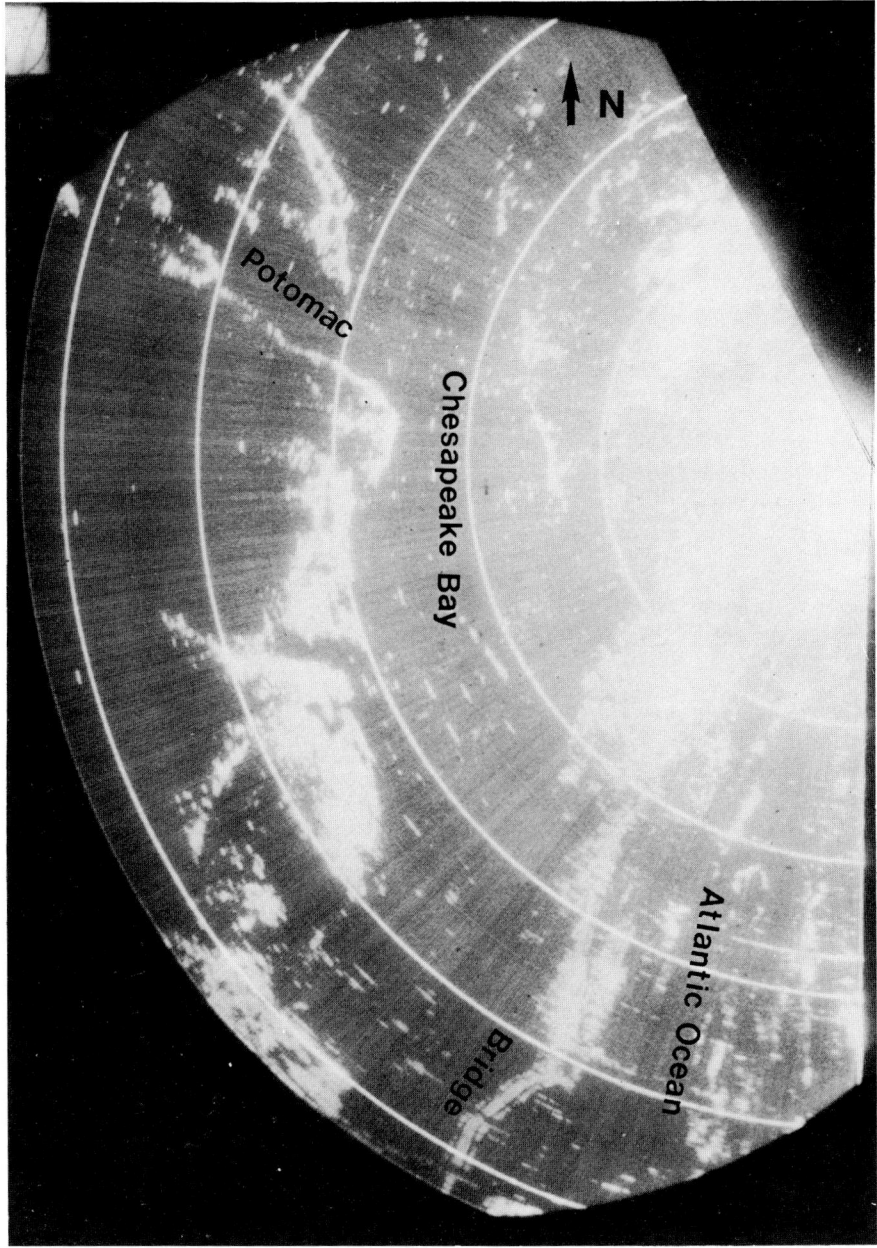

Fig. 2.10a. A plan position indicator (PPI) display of ground targets made visible because of a strong ground-based temperature inversion. Circular arcs are range marks spaced 10 nautical miles apart.

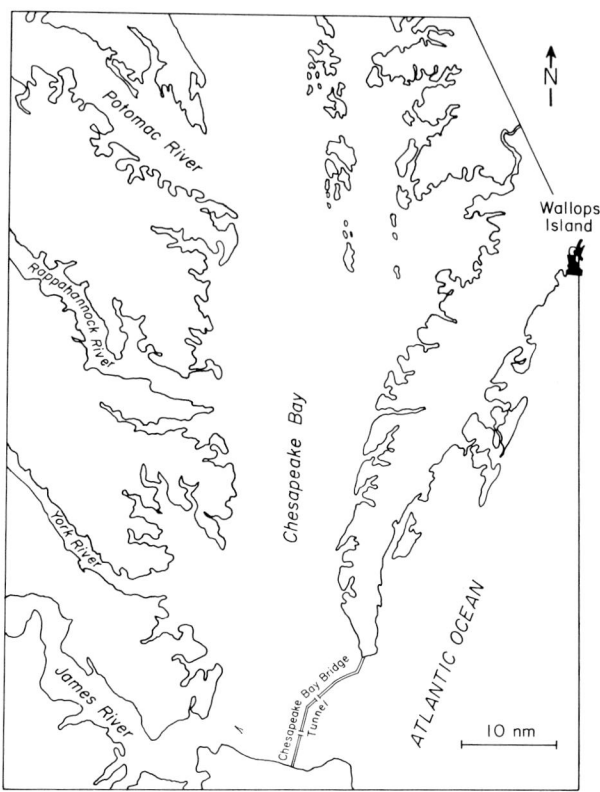

Fig. 2.10b. Map of the area scanned by the radar beam.

REFERENCES

Bean, B. R., and Dutton, E. J. (1966). "Radio Meteorology," Natl. Bur. Stand., Monogr. 92, Supt. Doc. U.S. Govt. Printing Office, Washington, D.C.

Born, M., and Wolf, E. (1964). "Principles of Optics," 2nd ed. Macmillan, New York.

Day, J. P., and Trolise, L. G. (1950). Propagation of short radio waves over desert terrain. *Proc. IRE* **38,** 165–175.

Hartree, D. R., Michel, J. G. L., and Nicolson, P. (1946). Practical methods for the solution of the equations of tropospheric refraction. "Meteorological Factors in Radio Wave Propagation," pp. 127–168. Physical Society, London.

3
Principles of Radar

We now describe the important elements of a pulsed-Doppler radar, with particular emphasis on its application to measurement of weather.

3.1 THE DOPPLER RADAR (TRANSMITTING ASPECTS)

Figure 3.1 shows in block diagram the principal components of a pulsed-Doppler radar. A high-power amplifier (a vacuum tube amplifier called a Klystron is usually used because it produces signals of fine spectral purity) is turned on and off by the pulse modulator to transmit a train of high peak power microwave pulses having duration τ of about 1 μs (10^{-6} s) with spacing at the pulse repetition time (PRT) T_s, the sampling time interval. The pulse of peak power density can be represented as $S(r, \theta, \phi)U(t - r/c)$, where

$$U(t - r/c) = \begin{cases} 1, & r/c \leq t \leq (r/c + \tau), \\ 0 & \text{otherwise,} \end{cases} \quad (3.1)$$

This power density illuminates objects (targets) as it propagates within a narrow beam, and a tiny fraction of this radiation is scattered by these targets toward a receiver located in most cases, at the transmitter site. Furthermore, for economic reasons the same antenna is shared by the transmitter and receiver.

The transmit–receive (TR) switch connects the transmitter to the antenna during the time τ, whereas the receiver (i.e., mixer–detector plus amplifiers) is connected during the time interval $T_s - \tau$, the *listening period*. The TR switching is not performed instantaneously, and there is a period of time (usually a few tens of microseconds) wherein the receiver does not have full sensitivity for detection.

When targets are small (compared to $c\tau$), few, and well separated (no two targets within $r \pm c\tau/2$), the returned signals or echoes are replicas of the transmitted pulse. These echoes are modified by the receiver to enhance the

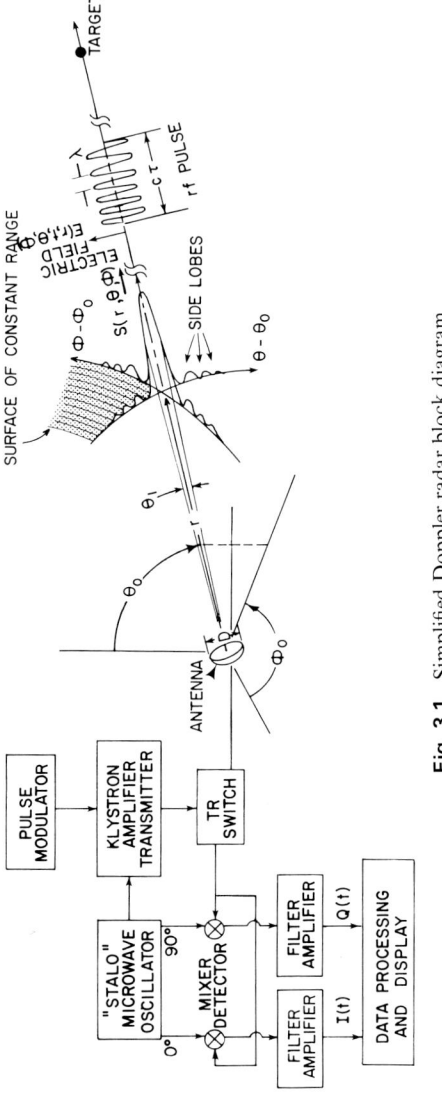

Fig. 3.1. Simplified Doppler radar block diagram.

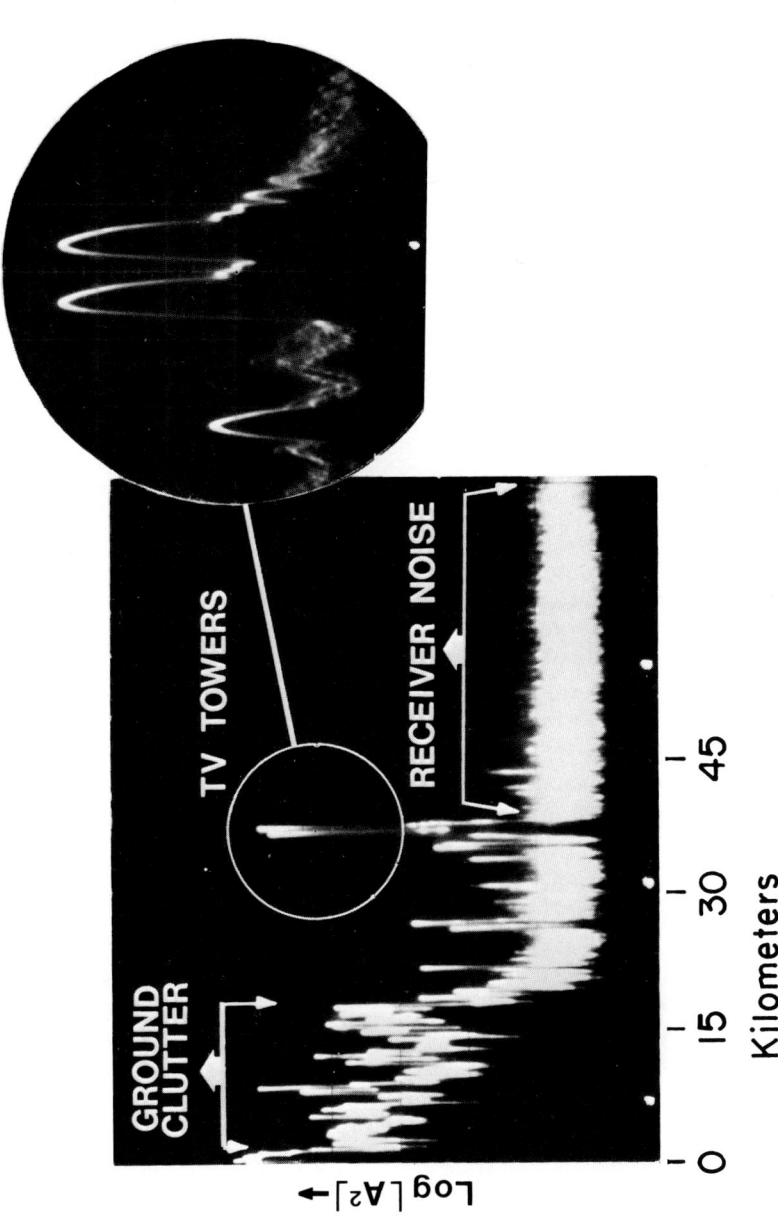

Fig. 3.2. Time delay and strength of echoes from discrete targets (e.g., TV towers at about 40 km) from the radar site. Ground clutter is a conglomerate of indistinguishable echoes from targets on the ground. The vertical scale is proportional to the logarithm of the echo power; the time scale is 50 µs/division. The transmitter pulse width τ is 1.2 µs. Inset shows an expanded view of the tower echoes, with a scale of 5 µs/division.

signal-to-noise ratio (see Section 3.5.4) and then displayed on video equipment to show their relative strengths and delay referenced to the time of the transmitted pulse (Fig. 3.2). This echoing principle was first applied in the late 1920s to measure remotely the properties of the ionosphere, and in 1934 a 10-m wavelength radar was first used by the Naval Research Laboratory to locate an airplane. The use of microwaves in radars did not become practical until early in 1940, when a powerful and efficient transmitting tube, the magnetron, was developed. The first detection of storms by microwave radar was made in England in 1941. There is an excellent historical review by Atlas (1964) of early developments in radar meteorology.

The development of high-power and high-gain Klystron amplifiers in the 1950s made practical the generation of microwaves that are phase coherent from pulse to pulse, a requirement for pulsed-Doppler radars if the velocity of targets is to be measured. Radar signals are phase coherent from pulse to pulse if the time between wave crests of successive transmitted pulses is fixed or known, the latter condition being applicable if stable magnetron oscillators are used. Because magnetrons are not normally phase-locked to injected signals, the phase of the sinusoids is random from pulse to pulse and needs to be measured and recorded.

3.1.1 The Electromagnetic Beam

The microwave pulse leaves the antenna in an essentially collimated beam of diameter D equal to that of the antenna–reflector (Fig. 3.3). However, because of diffraction, the electromagnetic beam begins to spread at a range $r = D^2/\lambda$ into a conical one having an angular spread given by (2.4). The beamwidth θ_1 (Fig. 3.1) is commonly specified as the angle within which the microwave radiation is at least one-half its peak intensity (3-dB width).

The radiation pattern $S(\theta, \phi)$ describes the angular distribution of power density that emanates from the antenna. It is impossible to confine all the energy into a narrow conical beam, and some of it inevitably falls outside the main beam lobe into side lobes (Fig. 3.1), the main lobe being that region where S monotonically decreases. Usually the power density in any side lobe is less than $\frac{1}{100}$ th of the peak density in the main lobe. Furthermore, the sum total of power in side lobes can often be held to just a few percent of that transmitted within the main lobe (Sherman, 1970).

The antenna–reflector is usually a paraboloid of revolution and is illuminated by a source located at the focal point (Fig. 3.3). The illumination is made to be nonuniform across the reflector in order to reduce sidelobe levels, and often its intensity versus distance ρ from the axis has the dependence $[1 - 4(\rho/D)^2]^2$. In this case the normalized power density pattern $f^2(\theta)$ has the following angular dependence symmetrical about the reflector axis

3.1 The Doppler Radar (Transmitting Aspects)

Fig. 3.3. The National Severe Storms Laboratory's weather radar antenna–reflector, shown inside its protective geodesic radome (radar cover dome). The reflector, a paraboloid of revolution, has a diameter of 9.14 m. The radiation source is the horn at the end of the curved (black) waveguide. The tubes extending to the right support the source and waveguide.

(Sherman, 1970):

$$f^2(\theta) = \frac{S(\theta)}{S(0)} = \left\{ \frac{8J_2[(\pi D \sin \theta)/\lambda]}{[(\pi D \sin \theta)/\lambda]^2} \right\}^2, \qquad (3.2a)$$

where θ is the angular distance from the beam axis and J_2 is the Bessel function of second order. This formula describes quite accurately the radiation pattern in the angular region containing the first few side lobes, but beyond that actual radiation patterns often have larger contributions from power scattered by imperfections in the reflector and structures supporting the source. When the beamwidth is small compared to 1 rad, Eq. (3.2a) shows that the 3-dB beamwidth θ_1 is

$$\theta_1 = 1.27\lambda/D \quad \text{rad}, \tag{3.2b}$$

and the width between the first nulls becomes $3.27\lambda/D$ (rad).

The microwave energy is within a $c\tau$-thick spherical shell that expands (i.e., propagates) at a speed c. Thus, the power density $S_i(\theta, \phi)$ incident on targets decreases inversely with r^2, although the power P'_t transmitted through any enclosing sphere is constant. This is the reason for the $1/r$ dependence of E in (2.2). Because of losses in the antenna, its transmission lines, and its protective radome, the power P_t delivered to the antenna's input port is larger than P'_t.

3.1.2 Antenna Gain

If P'_t were radiated equally in all directions, S would be equal to $P'_t/4\pi r^2$. However, the antenna focuses radiation into a narrow angular region wherein the peak radiation intensity S_p is many times stronger than $P'_t/4\pi r^2$. The ratio

$$S_p/(P'_t/4\pi r^2) \equiv g'_t \tag{3.3}$$

defines the maximum directional gain of the antenna. Measurement of P'_t is difficult, and the antenna engineer therefore measures the power P_t delivered to the antenna's input port and S_p at distances far (i.e., $r > 2D^2/\lambda$) from the antenna. In this case the computed gain g_t accounts for losses of energy associated with the antenna system (e.g., radome and waveguide). Then the incident radiation power density at range r is given by

$$S_i(\theta, \phi) = P_t g_t f^2(\theta, \phi)/4\pi r^2, \tag{3.4}$$

where $f^2(\theta, \phi)$ is the normalized [i.e., $f(\theta, \phi) = 1$ at θ_0, ϕ_0; Fig. 3.1] power gain pattern, and g_t is simply the antenna power gain along the beam axis.

3.2 THE TARGET

3.2.1 Scattering Cross Section

The *cross section* σ of a target is an apparent area that intercepts a power σS_i, which is assumed to be radiated isotropically, to produce at the receiver a power density

$$S_r = S_i \sigma(\theta', \phi')/4\pi r^2 \tag{3.5}$$

3.2 The Target

equal to that scattered by the actual target. The area σ is sometimes called the *differential cross section* to distinguish it from the total cross section, which is proportional to the total power scattered by the target. The polar angles θ', ϕ' to a receiver are referenced to a polar axis connecting the target and the transmitter with the target at the origin. The definition (3.5) shows that the scatterers do not radiate power isotropically, and hence the target cross section $\sigma(\theta', \phi')$ can depend on the relative location of the transmitter and receiver. It is easy to deduce that the apparent cross section may have no resemblance to the target's physical cross section. In fact, thin metallized fibers (chaff) of real cross section small compared to λ^2 can have scattering cross sections many times larger than their physical area. For example, consider a chaff element (needle) of 0.1-mm diam that has length l parallel to the electric field vector and that is resonant (e.g., $l = \lambda/2$) at the 10 cm wavelength of an incident wave. It has a maximum backscatter cross section $\sigma_{bm} = 0.857\lambda^2$ equal to 8.6×10^{-3} m^2 (Nathanson, 1969, p. 223), whereas the geometric cross section is only 5×10^{-6} m^2. For needles not perpendicular to $\mathbf{S}_i$ and oriented so that $\mathbf{E}$ is not parallel to the axes, $\sigma_b = \sigma_{bm}(1 - \sin^2\psi/\sin^2\theta)\cos^2[(\pi/2)\cos\theta]$. The angle ψ is the angle between the chaff axis and the plane of polarization (i.e., the plane containing $\mathbf{S}_i$ and $\mathbf{E}$), and θ is the angle of $\mathbf{S}_i$ relative to the axis.

On the other hand, a large plain sheet of conductor can have an extremely small backscatter cross section when it is oriented so its normal is not along the line to the radar. The cross section can be many orders of magnitude smaller than the conductor's area projected on a plane perpendicular to the line of sight.

The backscatter cross section σ_b of a water drop of diameter D small compared to λ (i.e., $D \leq \lambda/16$) is

$$\sigma_b = (\pi^5/\lambda^4)|K_w|^2 D^6, \tag{3.6}$$

where $K_w = (m^2 - 1)/(m^2 + 2)$ and $m = n - jn\kappa$ is the complex refractive index of water. The refractive index is n, and κ is the attenuation index (Born and Wolf, 1964, p. 613). Some authors (Battan, 1973) define an absorption coefficient $k = n\kappa$ and list its value as a function of wavelength and temperature. The magnitude $|K_w|$ is proportional to the induced electric dipole moment of the drop. As the drop diameter gets larger relative to the wavelength, higher-order moments of the electrical vibrations in the drop must be considered in order to describe adequately the scattered fields. $|K_w|^2$ varies between 0.91 and 0.93 for wavelengths between 0.01 and 0.10 m and is practically independent of temperature (Battan, 1973, p. 39). Ice spheres have $|K_w|^2$ of about 0.18 (for a density 0.917 g cm^{-3}), a value independent of temperature as well as wavelength in the microwave region. The backscatter cross section (3.6) is called the Rayleigh approximation because it has a wavelength dependence similar to the scattering cross section of atmospheric molecules whose diameters are small compared to the optical wavelengths, and (3.6) shows that waves at shorter λ are more strongly scattered—a fact that Rayleigh used to explain why the sky is blue.

If the target is small compared to wavelength, the scattered energy is radiated nearly isotropically. Large targets have a much more directive scatter radiation pattern, and often more radiation flows in directions other than back to the transmitter. For example, scattered power density in the direction of **r** (forward scatter) can be 100–1000 times larger than that returned in the direction of the source (known as the Mie effect; Born and Wolf, 1964).

There is an abundance of experimental and theoretical work that relates a particle's cross section to its shape, temperature, size, and mixture of phases (e.g., water-coated ice spheres). These works are reviewed by Battan (1973) and Atlas (1964).

3.2.2 Doppler Shift

When the pulse of radiation impinges on a target (e.g., a raindrop), it forces molecular vibrations in synchronism with the time-changing electric and magnetic fields. If the drop is stationary or moving along a surface of constant range (i.e., r is fixed; see Fig. 3.1), its molecules will have vibrations at the frequency of the radiation field; if it is moving toward the transmitter at velocity v, its vibrational frequency is higher by v/λ because the target molecules experience more rapid fluctuations of electric and magnetic force. The vibrating molecules themselves generate electromagnetic fields, which in turn radiate outward from the target. For monostatic radar, for which the transmitter and receiver are colocated, the frequency of scattered radiation is Doppler-shifted by an amount

$$f_d = -2v_r/\lambda, \tag{3.7}$$

where v_r is the radial component of the velocity—positive being away from the radar. We see that the factor of 2 in (3.7) is the result of a two-step increase in the frequency. First, the target's electric vibrational frequency is increased by v_r/λ; second, the frequency of its radiation field in the direction of the receiver is increased by v_r/λ. Similar reasoning shows that there is no Doppler shift in the scattered electromagnetic field along a straight line between the transmitter and target at points beyond the target (i.e., the forward-scatter path). A mathematical development leading to (3.7) is given in Section 3.4.2.

3.3 ATTENUATION

Were it not for electromagnetic energy absorption by water or ice drops, radars with shorter wavelength radiation would be much more in use because of their superior spatial resolution [Eq. (2.4)]. Short-wavelength (e.g., $\lambda = 3$ cm) radars suffer echo power loss that can be 100 times larger than that of radars operated with $\lambda \geq 10$ cm.

Each drop absorbs an amount of power P_L that can be expressed simply

$$P_L = \sigma_a S_i, \tag{3.8}$$

where σ_a is the apparent area that intercepts from the incident radiation a power equal to the power dissipated, as heat, in the drop. There is no simple formula for σ_a applicable to drop diameters found in moderate-to-heavy rain, even if diameters satisfy the condition $D \leq \lambda/16$ (see Section 8.3.4). Thus one needs to resort to higher-order terms in the series solution formulated by Mie for the sphere scatter problem.

A wave suffers power loss both from energy absorption and scatter. Analogous to the backscatter cross section, there is a *total* scatter cross section σ_s that accounts for the total power scattered by a drop. This is proportional to the integral over a sphere of the scattered power density $S_r(\theta', \phi')$; hence for small spherical drops $\sigma_s = 2\sigma_b/3$ (Battan, 1973). Thus the total power extracted from a wave is proportional to the sum $\sigma_a + \sigma_s$.

If we assume that the presence of drops within an elemental volume $\Delta V(r)$ does not significantly alter the incident power density S_i within this volume (i.e., we can neglect the scattering of the scattered field—the Born approximation) then the power density change ΔS_i in a wave propagating a short distance Δr through the volume is

$$\Delta S_i = -\frac{\Delta r}{\Delta V} \sum_{n=1}^{N} (\sigma_{an} + \sigma_{sn}) S_i, \tag{3.9}$$

where the summation extends over all N drops within ΔV and where σ_{an}, σ_{sn}, are the absorption and scatter cross section, respectively, of the nth particle. The rate of change in power density is then

$$\lim_{\Delta r \to 0} \left(\frac{\Delta S_i}{\Delta r}\right) = \frac{-1}{\Delta V} \sum_{n=1}^{N} (\sigma_{an} + \sigma_{sn}) S_i. \tag{3.10}$$

In the limit $\Delta r \to 0$, S_i can be considered a constant within ΔV and hence can be placed outside the summation. Then the power density at any range r is the integral solution of (3.10),

$$S_i(r_2) = S_i(r_1) \exp\left(-\int_{r_1}^{r_2} k \, dr\right), \tag{3.11}$$

where $k \equiv (\Delta V)^{-1} \sum_n (\sigma_{an} + \sigma_{sn})$ is the attenuation rate. This rate, expressed in decibels per kilometer is

$$K \equiv \frac{d}{dr_2}\left[10 \log \frac{S(r_1)}{S(r_2)}\right] = 4.34 \times 10^3 k \quad \text{dB km}^{-1} \tag{3.12}$$

when k has units of inverse meters.

Combining (3.4), (3.5), and (3.11) we deduce that the echo power density S_r at the radar antenna is

$$S_r(r, \theta, \phi) = P_t g_t f^2(\theta, \phi) l^2 \sigma_b / (4\pi r^2)^2 \quad \text{W m}^{-2}, \tag{3.13a}$$

$$l = \exp\left(-\int_0^r (k_g + k)\, dr\right), \tag{3.13b}$$

where l is the one-way loss factor due to gaseous k_g and droplet k (either cloud or precipitation) attenuation, which we discuss in the next few pages.

3.3.1 Attenuation by Rain

Values of K for various rainfall rates are tabulated by Bean and Dutton (1966, pp. 292–296). Attenuation losses have been calculated by Burrows and Attwood (1949) for the raindrop size distribution observed by Laws and Parsons (1943). A least squares fit applied to the logarithms of the Burrows and Attwood data yields the following one-way attenuation rates at a temperature of 18°C (Fig. 3.4):

$$K_r = \begin{cases} 0.000343 R^{0.97} & (\text{dB km}^{-1}), \quad \lambda = 10 \text{ cm}, \quad (3.14a) \\ 0.0018 R^{1.05}, & \lambda = 5 \text{ cm}, \quad (3.14b) \\ 0.01 R^{1.21}, & \lambda = 3.2 \text{ cm}, \quad (3.14c) \end{cases}$$

where R is the rainfall rate in millimeters per hour. At other temperatures a multiplicative correction to (3.14) must be applied. The correction factor is independent of the rainfall rate for a 10-cm wavelength, and an empirical formula valid for temperatures between 0° and 40°C is presented in Fig. 3.4. At shorter wavelengths the dependence on rainfall rate and temperature is tabulated by Burrows and Attwood (1949).

3.3.2 Attenuation by Cloud Droplets

Cloud and fog particles (liquid or ice) generally are not detected with weather radars but nevertheless can cause significant attenuation of radar signals. When drops are very small (i.e., $D \ll \lambda/16$; see Fig. 8.10) the absorption

3.3 Attenuation

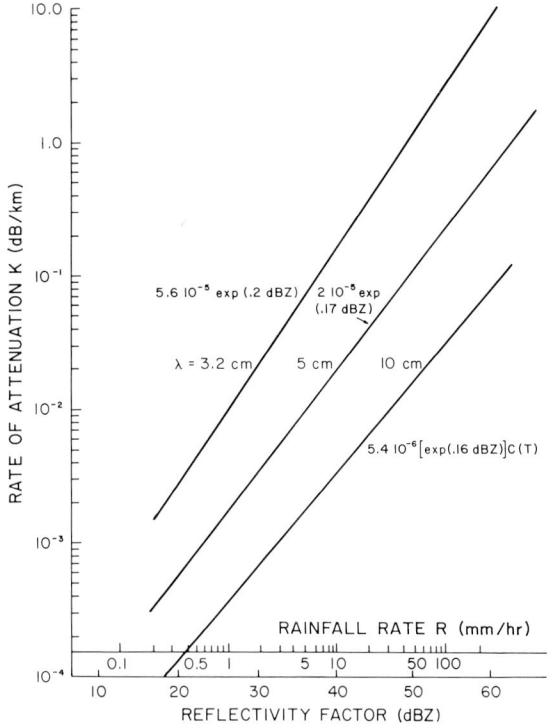

Fig. 3.4. Rate of attenuation (one-way) for propagation through rain showers versus rainfall rate ($T = 18°C$). The Laws and Parsons drop size distribution is assumed. The relation between the rainfall rate and drop size distribution is developed in Chapter 8, and the reflectivity factor is defined in Section 4.3.3. Temperature adjustment factor $C(T) = 2 \exp(0.035T)$. $dBZ = 10 \log(400 R^{1.4})$. ($dBZ$ = decibel units of Z.)

cross section is

$$\sigma_a = \frac{\pi^2 \operatorname{Im}(-K_w)}{\lambda} \sum_{n=1}^{N} D_n^3. \qquad (3.15)$$

Losses resulting from absorption are therefore proportional to $\sum D_n^3$, i.e., to the liquid water density M (g m^{-3}) along the propagation path. Thus the attenuation rate k_c in a cloud can be expressed

$$k_c = k_1 M, \qquad (3.16)$$

where k_1 is defined as the attenuation coefficient per unit density. Because the total scatter cross section σ_s is proportional to the sixth power of the drop diameter it is usually small compared to σ_a for cloud droplets if $D_n \ll \lambda/16$. However, in rain at low rates and for wavelengths less than 1 cm, σ_s

can be larger than σ_a (see Fig. 8.12). As the total water content Q_w in ΔV is $\pi\rho \sum_n D_n^3/6$ (ρ is the water density in grams per cubic meter), and assuming that $k_c = \sigma_a/\Delta V$, one finds $M = Q_w/\Delta V$, which, substituted into (3.16), produces

$$k_c = (6\pi/\lambda\rho)M \, \text{Im}(-K_w) \simeq M 10^{-8}/\lambda^2 \quad (\text{m}^{-1}), \quad (3.17)$$

where λ is in meters.

The approximation is good to an accuracy of 5% at wavelengths of 5–10 cm and temperatures of 18°C where $\text{Im}(-K_w)$ depends inversely on λ. In order to obtain k_c in decibels per kilometer, one simply multiplies (3.17) by 4.34×10^3, as in (3.12). Tables of attenuation rates for other wavelengths and temperatures are given by Battan (1973, p. 70) using the data of Gunn and East (1954).

Although attenuation through ice clouds is usually negligibly small, it cannot always be ignored for water clouds because the attenuation rate is then generally an order of magnitude larger. When drop diameters are small compared to the wavelength, attenuation due to absorption becomes independent of drop size distribution and depends, as we have shown only on the cloud water density. Note, however, that the backscatter cross section (3.6) depends on the sixth power of the drop diameter. Thus clouds between the radar and a distant precipitating storm can produce measurable attenuation even at $\lambda = 10$ cm (e.g., 0.9 dB for two-way propagation through 50 km of 1.0 g m^{-3} liquid-water clouds), which may cause underestimates of rain in a distant storm. These estimates could go uncorrected because clouds containing only droplets may not always be detected by the weather radar.

3.3.3 Snow Attenuation Rate

Battan (1973) gives the following formula for attenuation by dry snow at 0°C:

$$K_s = 3.5 \times 10^{-2}(R^2/\lambda^4) + 2.2 \times 10^{-3}(R/\lambda) \quad \text{dB km}^{-1}, \quad (3.18)$$

where the first term is due to scattering and the second to absorption, R is in millimeters per hour of melted snow, and λ is measured in centimeters. At a wavelength of 10 cm, K_s can be safely neglected, but at shorter wavelengths it might be appreciable, especially if the rate R is large. Wet snow and water-coated ice attenuate more than dry ice, but because of the irregular shapes present there are no simple formulas. Graphs of attenuation by dry and water-coated ice spheres are given by Battan (1973).

3.3.4 Gaseous Attenuation Rate

Besides attenuation due to rain and cloud droplets, there is attenuation due to energy absorbed by the atmosphere's molecular constituents, mainly

3.4 The Doppler Radar (Receiving Aspects)

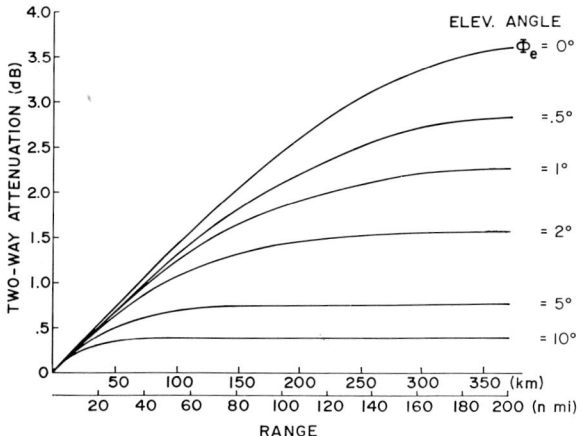

Fig. 3.5. Rate of (two-way) attenuation for propagation through a standard atmosphere ($\lambda = 10$ cm). The Central Radio Propagation Laboratory's (CRPL) exponential reference atmosphere (Fig. 2.5) is used to compute ray paths, and the pressure–temperature profile is based on the International Civil Aviation Organization (ICAO) standard atmosphere.

water vapor and oxygen. This gaseous attenuation rate k_g is not negligible when targets are far away ($r \geq 60$ km) and beam elevation is low, even at $\lambda = 10$ cm if accurate cross section measurements are required.

The gaseous absorption loss depends not only on the propagation path length but also on the region of the troposphere penetrated (Blake, 1970). This dependence for the standard atmosphere is illustrated in Fig. 3.5, which has been plotted for the case $\lambda = 10$ cm from curves published by Blake (1970). An empirical formula that approximates the two-way attenuation K_g for elevation angles $\theta_e < 10°$ and slant ranges $r < 200$ km is

$$K_g = [0.4 + 3.45 \exp(-\theta_e/1.8)]$$
$$\times (1 - \exp\{-r/[27.8 + 154 \exp(-\theta_e/2.2)]\}) \quad \text{dB}. \quad (3.19)$$

This formula approximates the theoretical loss curves (Fig. 3.5) to within 0.2 dB.

3.4 THE DOPPLER RADAR (RECEIVING ASPECTS)

The *echo power* P_r (in watts) collected by the antenna system from a wave scattered by a target at r, θ, ϕ, is

$$P_r = S_r(r, \theta, \phi) A_e(\theta, \phi), \quad (3.20)$$

where A_e is the effective aperture area of the antenna for radiation from direction θ, ϕ. We shall prove that

$$A_e = (g_r\lambda^2/4\pi)f^2(\theta, \phi), \tag{3.21}$$

where g_r is the gain of the receiving antenna, equal to g_t if the transmitting antenna is used for echo reception, and P_r is measured at the same location in the antenna system as P_t. Consider two antennas 1 and 2 at a distance r from each other, and assume that antenna 1 is radiating while 2 is receiving the power:

$$P_{r2} = [P_{t1}g_1 f_1^2(\theta, \phi)/4\pi r^2]A_{e2}. \tag{3.22a}$$

Interchanging the roles of the two antennas produces

$$P_{r1} = [P_{t2}g_2 f_2^2(\theta, \phi)/4\pi r^2]A_{e1}. \tag{3.22b}$$

If the transmitted powers are made equal, $P_{t2} = P_{t1}$, then (by reciprocity) the receiving powers must be equal; i.e., $P_{r1} = P_{r2}$. From this we find that

$$A_{e1}/g_1 f_1^2(\theta, \phi) = A_{e2}/g_2 f_2^2(\theta, \phi) = \text{const.} \tag{3.23}$$

Because the constant in (3.23) is independent of antenna type, one can calculate it for the simplest possible antenna, say, the dipole, for which one obtains $\lambda^2/4\pi$ (Jordan and Balmain, 1968). Thus, Eq. (3.21) follows. For large-aperture antennas and targets along the beam, A_e is almost equal to the physical area of the antenna reflector. For example, the weather radar antenna shown in Fig. 3.3 has $A_e = 42$ m^2, whereas its physical area is 65.7 m^2. The significantly smaller area A_e is caused principally by a taper in the illumination of the reflector, which in the reception mode is equivalent to weighting the power density reflected from the parabolic surface. Furthermore, radome and waveguide losses cause A_e to be smaller than the physical area.

3.4.1 The Radar Equation

Combining (3.13a), (3.20), and (3.21), we arrive at the radar equation for a discrete target having backscatter cross section σ_b:

$$P_r = P_t g^2 \lambda^2 l^2 \sigma_b f^4(\theta, \phi)/(4\pi)^3 r^4, \tag{3.24}$$

where we have substituted $g_t = g_r = g$ because the same antenna is used for transmitting and receiving. This equation relates echo power P_r to the radar parameters and target location.

Here is an example of the tremendous sensitivity of weather radar for the detection of some extremely small targets. The National Severe Storms Laboratory's 10-cm (i.e., $\lambda = 0.1$ m) radar receiver can detect echo power as weak as 10^{-14} W; the transmitted peak power is nearly 10^6 W, and antenna gain is 4×10^4. Thus, using (3.24), we see that at a range of 20 km, a cross section as small as

$$\sigma_b(\text{minimum}) = 2 \times 10^{-7} \text{ m}^2$$

can be detected. That is, a single water drop with a diameter of 6.3 mm could be detected at 20 km. It would not take, therefore, a very large number of drops to provide a detectable signal at larger range (e.g., 200 km); moreover the transmitted power and receiver performance of this radar is only moderate in relation to advanced systems in use today (such as deep-space probes). Nonmeteorological targets such as birds and insects have sufficiently large backscatter cross section to make them easily visible to the radar. For example, the cross sections of birds can be as large as 10^{-2} m^2 (sea gull), and a housefly has a cross section of 10^{-5} m^2.

3.4.2 The Received Waveform (In-Phase and Quadrature Components)

If there is a single target and the receiver bandwidth is sufficiently large (see Section 3.5.2), the echo signal voltage $V(t)$ replicating the transmitted waveform of the electric field **E** is proportional to it.

$$V(t, r) = A\{\exp[j2\pi f(t - 2r/c) + j\psi]\}U(t - 2r/c), \qquad (3.25)$$

where $2r$ is the total path traversed by the incident and scattered waves, A is now the complex voltage (containing the phase shift produced by the scatterer) at the input to the mixer detector (Fig. 3.1), and, as before, $U = 1$ when its argument is between 0 and τ, zero otherwise. Range time $2r/c$ specifies, in units of time, the target location. The total phase

$$\beta = 2\pi f(t - 2r/c) + \psi \qquad (3.26)$$

of the echo is dependent on t as well as r.

An important receiver component is the STALO (stabilized local oscillator; Fig. 3.1), which oscillates at the transmitted frequency f plus (or minus) a small but fixed offset frequency f_Δ. A portion of its signal is summed with the echo signal, whose strength is weak relative to the STALO's. For this condition the sum of the two signals can be well approximated by an expression in which the STALO signal is modulated by another that exactly replicates the echo signal in both phase and amplitude shape except that its frequency is f_Δ (Rideout, 1954, p. 311). The summed signals are applied to a nonlinear (usually square law) device whose output contains sum and difference frequencies and harmonics of these. Filtering is used to retain the modulating signal and to remove the STALO plus harmonic signals. This summing, nonlinear operation and filtering is called *heterodyning* if $f_\Delta \neq 0$.

Usually f_Δ is selected to be an intermediate frequency (e.g., 30 MHz) much higher than the frequencies contained in the spectrum of $V(t)$. Thus the voltage $V_0(t)$ at the output of the filter (Fig. 3.1) is

$$V_0(t) = A \exp\{-j[(4\pi r/\lambda) - \psi]\} U(t - 2r/c) e^{j 2\pi f_\Delta t} \tag{3.27}$$

if losses are ignored. Heterodyning serves only to convert (shift) the carrier frequency without affecting the modulation envelope. Receivers that have intermediate frequency amplifiers (i.e., $f_\Delta \neq 0$) are called *superheterodyne* receivers. However, in order to have signals that can be handled by digital data processing and video display equipment, it is necessary to convert the intermediate frequency to the baseband at which $f_\Delta = 0$. This double conversion eliminates the increased noise levels found in *homodyne* (direct conversion) receivers. For simplicity Fig. 3.1 shows homodyning, so that the STALO frequency is the same as the transmitted frequency; i.e., $f_\Delta = 0$. In this case the signal is detected directly without its passing through an intermediate frequency. The STALO signal is continuous wave so that whenever echoes arrive a STALO signal is mixed with them.

With further discussion restricted to the homodyne receiver, Doppler radar usually has two mixers (without which the direction of the target motion, toward or away, cannot be determined); in one the STALO signal is phase shifted by 90° prior to mixing so that its rectified and filtered output is

$$V_0(t) = A \exp\{-j[(4\pi r/\lambda) - \psi - (\pi/2)]\} U(t - 2r/c). \tag{3.28}$$

The actual signal from one mixer is the imaginary part of (3.28) and from the other the real part of (3.28), which are

$$I(t) = (|A|/\sqrt{2}) U(t - 2r/c) \cos(4\pi r/\lambda - \psi + \psi_i), \tag{3.29a}$$

$$Q(t) = (-|A|/\sqrt{2}) U(t - 2r/c) \sin(4\pi r/\lambda - \psi + \psi_i), \tag{3.29b}$$

the in-phase $I(t, r)$ and quadrature $Q(t, r)$ components, respectively, of the signal $V(t, r)$. The phase shift ψ_i is produced by the target upon scattering. For convenience we again ignore losses, and we use a factor of $1/\sqrt{2}$ in (3.29) so that the sum of I^2 and Q^2 equals the input power $|A|^2/2$ averaged over a cycle of the microwave signal.

Echoes from stationary targets have signals in which the phase $\gamma = -(4\pi r/\lambda) + \psi$ is time independent. If r increases with time, the phase decreases and the time rate of phase change,

$$\frac{d\gamma}{dt} = -\frac{4\pi}{\lambda} \frac{dr}{dt} = -\frac{4\pi}{\lambda} v_r = \omega_d, \tag{3.30}$$

is the Doppler shift (in radians per second), a result we had previously deduced [Eq. (3.7)].

3.4 The Doppler Radar (Receiving Aspects)

It is relatively easy to see from (3.29) that, for usual radar conditions (i.e., $\tau \approx 10^{-6}$ s) and target velocities of the order of tens of meters per second, the change in the trigonometric functions is extremely small during the time $U(t - 2r/c)$ is nonzero. Thus we measure the target phase shift over the longer time $T_s \approx 10^{-3}$ s from echo pulse to echo pulse rather than during a pulse period. Because of this the pulsed-Doppler radar behaves as a phase- as well as an amplitude-sampling device; samples are at $t = \tau_s + (n - 1)T_s$, where τ_s is the time delay between the nth transmitted pulse and its echo. τ_s is called the *range time* because it is proportional to range (i.e., $\tau_s = 2r/c$). It is convenient to introduce another time scale, the *sample time*; that is, time is incremented in discrete steps of length T_s, the sample time, after $t = \tau_s$ (see Fig. 4.1 for a display of both time axes). It is important to realize that the echo phase and amplitude changes are examined in sample-time space at the discrete instants $(n - 1)T_s$ for a target at range time τ_s. Figure 3.6 shows how samples of $I(t)$ and $Q(t)$ change for stationary and moving targets. It is evident that if $Q(t)$ is negative and increasing in a positive direction while

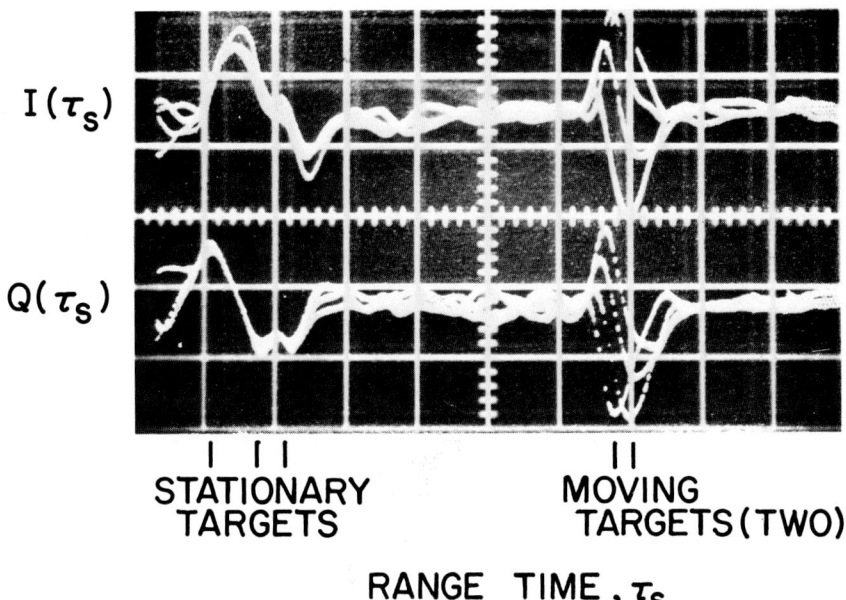

Fig. 3.6. Samples of the in-phase $I(\tau_s)$ and quadrature phase $Q(\tau_s)$ components of the echo signal for moving and stationary targets. Successive sampling intervals (four T_s's) have been overlaid onto one another to show the relative change of amplitude from echo sample to echo sample of targets at range time τ_s. The vertical scale is linear in voltage, and the range-time scale τ_s is 2 μs/division. (See Fig. 4.1 for an idealized representation.)

$I(t)$ is positive, the Doppler shift is positive (i.e., the angle γ increases in a counterclockwise direction), and the target is moving toward the radar.

3.5 PRACTICAL CONSIDERATIONS

The preceding discussion treats an ideal radar to facilitate an understanding of radar principles, but it omits some important limitations in radar measurements.

3.5.1 System Noise Temperature

In an ideal, noiseless radar receiver, there is no lower limit in detecting the weakest echo signals. However, all receivers generally can detect echo power only above some limit imposed by the random voltage from the thermal agitation of electrons in the receiver (in Fig. 3.1 it is the mixer–detector). Some high-performance radar receivers have parametric amplifiers or other low-noise amplifiers prior to the mixer–detector so that microwave emission from the earth's surface and atmosphere can exceed the noise generated by the receiver.

Noise considerations are often unimportant for weather radars that observe storms, because the echo signal power is much larger than noise. However, when Doppler weather radars are used to measure wind in clear air, as discussed in Chapter 11, the noise contributions (from components between the receiver and target such as transmission lines, radomes, or the TR switch) have paramount importance because the echo power can be smaller than the noise power. Even when a signal is weaker than the noise, the processing techniques described in Chapter 6 can extract Doppler velocity information.

The noise power at some convenient reference point in the radar system (e.g., the receiver input) has contributions from several sources. Radiation from space (cosmic noise) and from the oxygen and water vapor molecules in the atmosphere is intercepted by the antenna and, after being attenuated by the radome and transmission lines (between the antenna and receiver), is present at the receiver input. Transmission lines and other components in the path between the antenna and receiver input not only attenuate external radiation but, because they are at a temperature well above absolute zero, also generate noise that passes to the receiver input. The noise power generated by an attenuating component is given by (Dicke et al., 1946)

$$P_n = kTB_n(1 - l), \qquad (3.31)$$

where l, the loss factor ($l < 1$), is the ratio of the power out to the power in of

3.5 Practical Considerations

the device; T is the temperature of the component in degrees kelvin; $k = 1.38 \times 10^{-23}$ W s K^{-1} is the Boltzmann constant; and B_n is the noise bandwidth of the component. Usually power loss is expressed in decibels, and then $l = \log^{-1}(-L/10)$. The power loss is caused by the thermal agitation of electrons in the walls of the device that confine or guide the electromagnetic signals. Even though transmission lines (i.e., waveguides) attenuate noise from sources outside the antenna and may reduce the noise at the input to the receiver, they attenuate the signal as well. Although the noise power at the receiver might be less than the noise power without the attenuating waveguide, what really matters is the echo signal-to-noise ratio (SNR) and any device that adds noise will decrease the SNR at the receiver's input. It is therefore important to use transmission lines having small attenuation.

The thermal noise power at the receiver input is spread over bandwidths that are large compared to the radar receiver's bandwidth, so much of the noise power is filtered. It is accepted practice to express the radar noise level at the reference point in terms of a system temperature

$$T_{sy} \equiv P_N / k B_n, \qquad (3.32)$$

where B_n is now the noise bandwidth of the receiver and P_N is the effective noise power at the reference point, which is usually at the input to the receiver.

Not only do the components in front of the receiver contribute noise, but the receiver, which is comprised of a low-noise amplifier (LNA), mixer, etc. (see Figs. 3.1 and 3.7), adds noise to the signals that it amplifies. A figure of merit for an LNA is its noise temperature T_R, which is referenced to the LNA's input when the LNA is considered as a noiseless device whose output is connected to noiseless components. Thus, although the noise power added by the LNA is not really present at its input terminals, we add this noise to that contributed by components before the LNA. Thus T_{sy} includes T_R even

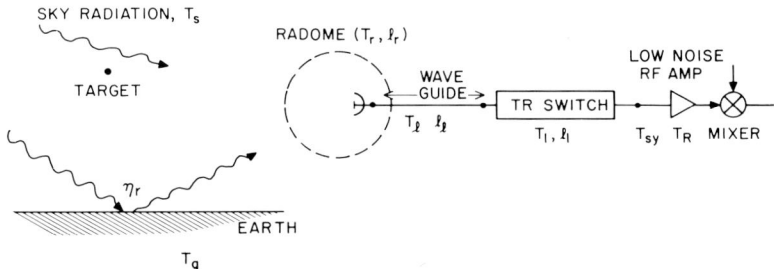

Fig. 3.7. Schematic of the absorbing elements that need to be considered when estimating the system noise temperature T_{sy} at the receiver input. The waveguide path may also contain other components, such as rotary joints, that add additional losses to the transmission path. Furthermore, rain and clouds add to the sky noise temperature.

though P_N is not entirely present at the reference point. In practice the LNA is not connected to noiseless devices, so the LNA's output supplies power into subsequent components that add noise. However, if the gain of a LNA is sufficiently high to amplify the equivalent noise power at its input terminal to levels far above the additional noise due to subsequent mixer or amplifiers, then the noise contributed by them can be ignored.

Figure 3.7 shows components of the transmission path between the target and the reference point (input to LNA) that need to be considered in computing T_{sy}. Based on (3.31) and (3.32) we arrive at the following expression for T_{sy}:

$$T_{sy} = T_s l_r l_l l_1 + T_r l_l l_1 (1 - l_r) + T_l l_1 (1 - l_l) + T_1 (1 - l_1) + T_R, \quad (3.33)$$

where T_s is the sky noise temperature due to cosmic and atmospheric radiation and T_r, T_l, and T_1 are the temperatures of the radome, transmission line, and TR switch, respectively. These latter temperatures can usually be set equal to 290 K, which approximates the temperature of the environment. The loss factor associated with each element is shown in Fig. 3.7.

The contribution to receiver noise from the sky temperature is a function of the direction in which the antenna points, because cosmic radiation is nonuniformly distributed over angular space (it is maximum along the galactic plane). Furthermore, when the antenna's beam is vertically directed, the absorption in the thin blanket of the earth's atmosphere is small and so is the atmosphere radiation intercepted by the antenna. As the beam points toward the horizon, however, more of the absorbing layer is within the field of view of the antenna. Thus both sources of sky noise have an angular dependence, which is summarized in Fig. 3.8. Figure 3.8 gives the sky noise temperature T_s for an idealized antenna having the sun in a unity-gain side lobe but without earth directed side lobes. Above 50 MHz but below 1000 MHz, cosmic or galactic noise dominates other noise sources. Above about 1000 MHz, cosmic noise becomes negligible and there remains only a contribution from oxygen and water vapor absorption and reradiation, which occur at their respective resonance peaks of 60 and 22 GHz. Although the oxygen resonant peak is farther from the weather radar frequencies, most atmospheric noise comes from the oxygen molecules.

Although the main lobe may point at a relatively cool sky, a real antenna has side lobes that are directed at a relatively warm and reflecting earth. Thus cosmic, atmospheric, and earth radiation contribute to an effective sky noise temperature

$$T'_s = T_s(1 - \chi) + \chi(1 - \eta_r)T_g + T_s \chi \eta_r, \quad (3.34)$$

where χ is the fraction of the antenna's power pattern subtending the ground

3.5 Practical Considerations

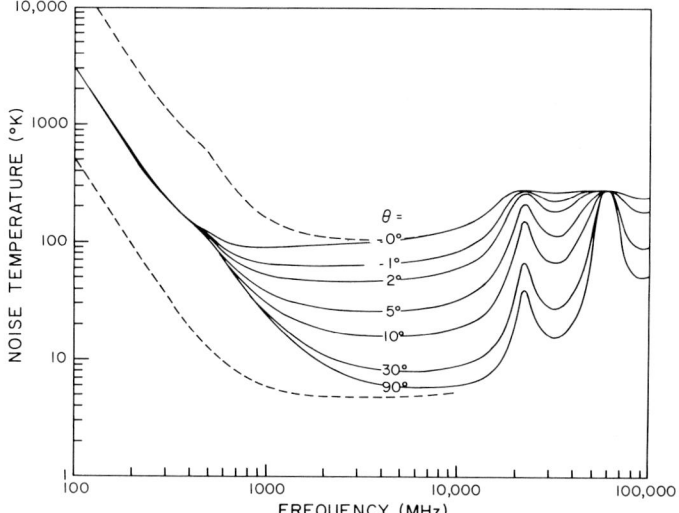

Fig. 3.8. Sky noise temperature of an idealized antenna (lossless, no earth-directed side lobes) located on the earth's surface. Solid curves are for geometric-mean galactic temperature, sun noise ten times quiet level, sun in unity-gain side lobe, cool temperate-zone troposphere, and 2.7 K cosmic blackbody radiation. The upper dashed curve is for maximum galactic noise (center of galaxy, narrow-beam antenna), sun noise 100 times quiet level, zero elevation angle, and other factors the same as for the solid curves. The lower dashed curve is for minimum galactic noise, zero sun noise, and elevation angle of 90°. The maxima at 22.2 and 60 GHz are due to water-vapor and oxygen absorption resonances. From "Radar Handbook," edited by M. I. Skolnik. Copyright © 1970 McGraw-Hill Book Company. Used with the permission of McGraw-Hill Book Company.

at temperature T_g, and η_r is the fraction of incident noise power reflected from the earth. A suggested conventional value for χ is 0.125, which results if an isothermal earth is viewed over a π-steradian solid angle by side lobes averaging -3 dB gain (Blake, 1970). For example, consider an antenna pointed at a relatively high angle, and suppose that the main beam and first side lobe, which contain about 90% of the radiated power, "see" an average temperature of 20 K due to oxygen absorption. The other 10% of side lobe reception "view" a warm earth at 300 K that is assumed to reflect 50% of the incident radiation. Then the effective sky temperature is

$$T'_s = 0.9 \times 20 + 0.5 \times 0.1 \times 300 + 0.5 \times 0.1 \times 20 = 35 \quad \text{K}.$$

Thus considerable radiation could be intercepted by ground-directed side lobes, and so T'_s should be used in place of T_s in (3.33).

Radome losses deserve further comment. These losses are due to absorption by the protective membrane and its supporting structure, as well as to scatter from the support frame. The scatter loss will result in increased side lobe levels. Thus radome scatter loss contributes to system noise temperature only through the antenna radiation pattern and cannot be considered part of l_r in (3.33).

3.5.2 Bandwidth

The echo pulse, as with any signal, can be decomposed into Fourier spectral coefficients. The receiver amplifies only a band of these spectral or frequency components; that is, the receiver has a finite bandwidth. The receiver's filter (Fig. 3.1) response $G(f)$ is usually a monotonically decreasing function of frequency, and the filter's bandwidth B_6 is best specified as the frequencies within which the attenuation of power, $G^2(f)$, is less than one-fourth of its highest level—its 6-dB width. Noise bandwidth B_n is not so simply specified because it depends on $G(f)$ as well as on the noise distribution $N(f)$, but in practice it is nearly equivalent to the 3-dB width if noise is *white*, i.e., independent of frequency (Kraus, 1966, p. 265).

The larger B_6 is, the better is the fidelity of the echo pulse shape, but noise power increases in proportion to B_n. Usually it is not important to detect the return echo with nearly perfect fidelity; nor is it imperative to excessively reduce noise. Rather, a compromise between the two conflicting effects is reached. We shall see later that both filter bandwidth and transmitter pulse width determine the range-dependent weight given to the scatterers' cross section.

An important measurement made with radar is the range to the target. The radar often needs to resolve targets when they are closely spaced and have largely different backscatter cross sections. For a noiseless receiver of infinite bandwidth, targets are resolved if their spacing is wider than one-half the spatial pulse width (i.e., $c\tau/2$), and targets so spaced return echoes that arrive at separate and distinct time intervals. Finite-bandwidth receivers distort the echo pulse by reverberating echo power within the receiver in amounts decreasing with time after the peak echo signal; an example is television tower echoes (Fig. 3.2). This causes the receiver output pulse to have widths larger than τ, and weak echoes could then be masked by strong ones.

3.5.3 Filtered Waveform

It is demonstrable that the filter output echo $I(t)$ or $Q(t)$ has a time waveform approximated by

$$I(t) = \tfrac{1}{2} I_0 \{ \text{erf} \left[a B_6 (t + \tau/2) \right] - \text{erf} \left[a B_6 (t - \tau/2) \right] \}, \qquad (3.35)$$

where I_0 is the prefilter amplitude $A/\sqrt{2}$ [(3.29)] of a rectangular echo having

3.5 Practical Considerations

a width τ (see Fig. 3.9); $a = \pi/(2\sqrt{\ln 2})$; erf[ ] is the error function, given by

$$\text{erf}(x) = \frac{2}{\sqrt{\pi}} \int_0^x e^{-t^2} \, dt; \qquad (3.36)$$

and $t = 0$ is the time at which the output echo is maximum. This solution is for an assumed filter frequency response $G(f)$ described by the Gaussian function

$$G(f) = e^{-(4 \ln 2) f^2 / B_6^2}. \qquad (3.37)$$

Although (3.35) approximates the pulse shape well if the filter response is well approximated by (3.37), it has the defect of producing signal for all t: It does not show a starting time. This is because the Gaussian function is practically approached only in the limit of an infinite pole filter (i.e., an unbounded number of cascaded elementary electrical circuits), which would cause an infinite propagation delay through the filter. Nevertheless, (3.35) is instructive because it shows the interrelation between B_6, τ, and the range time τ_s (we relate τ_s to t in Section 4.3.1). Furthermore, on a radar one can measure accurately the time delay to the peak response, and (3.35) describes the actual response quite well about this peak. It is readily seen that when B_6 is much larger than τ^{-1}, the filtered echo pulse is nearly a scaled replica of the transmitted pulse.

The filter bandwidth and shape specify a range-dependent function that weights the scatterers' cross sections that contribute to the sample of echoes

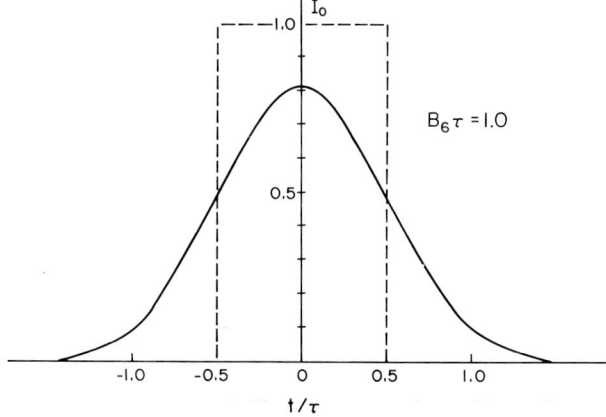

Fig. 3.9. Output (solid line) of Gaussian response filter for a rectangular pulse input (dashed lines). The delay in the output relative to input is not shown. The filter width B_6 is matched to pulse width τ (i.e., $B_6 \tau = 1.0$).

from distributed targets such as rain (Chapter 4). Often B_6 is chosen significantly larger than would be required to produce an acceptable echo response in order to allow for frequency instabilities in incoherent radar transmitters. A filter can also reject interference from extraneous sources, and smaller B_6 makes the receiver more selective. A filter with rectangularly shaped frequency response is much more selective than the Gaussian filter, but its time response to echo pulses may be objectionable because of larger reverberating signals. (These mask nearby weak echoes and, in rain showers, smear gradients of rain intensity.) A Gaussian filter is often a good compromise between frequency selectivity and a time response having echo power quickly decaying after the peak signal time.

3.5.4 Signal-to-Noise Ratio; Matched Filters

An important echo parameter is the ratio of its peak signal power, $S(0) \equiv I^2(0) + Q^2(0)$, to noise power. This signal-to-noise ratio, readily obtained from (3.32) and (3.35) is

$$\text{SNR} = [I^2(0) + Q^2(0)] \operatorname{erf}^2[aB_6\tau/2]/kT_{sy}B_n. \quad (3.38)$$

For a Gaussian filter shape $B_n = 1.06B_6/\sqrt{2}$, and SNR is maximized when

$$B_6 = 1.04/\tau. \quad (3.39)$$

It can be shown that the SNR is maximized in general when the filter response is matched to the prefilter echo spectrum (Nathanson, 1969, p. 277). To simplify the filter hardware and to achieve better filtering of extraneous signals, a matched filter is often only approximated in practice. Taylor and Mattern (1970, pp. 5–25) demonstrate that for a wide variety of filter responses (3.39) optimizes the SNR, and condition (3.39) is therefore considered to match the echo spectrum.

3.6 AMBIGUITIES

Range and velocity ambiguities are inherent in Doppler radars. Often Doppler radars are operated with uniform PRT (i.e., T_s) so that when targets have range r larger than $cT_s/2$, their echoes for the nth transmitted pulse are received after the $(n + 1)$th pulse is transmitted (Fig. 3.10). Therefore, these echoes are received during the same time interval that targets at $r < cT_s/2$ return echoes from the $(n + 1)$th pulse. Thus, the range r to the distant target may appear to have a value $r' = r - (N_t - 1)r_a$, which is ambiguous. (N_t designates the trip or $cT_s/2$ interval of the target, and $r_a = cT_s/2$ is that

3.6 Ambiguities

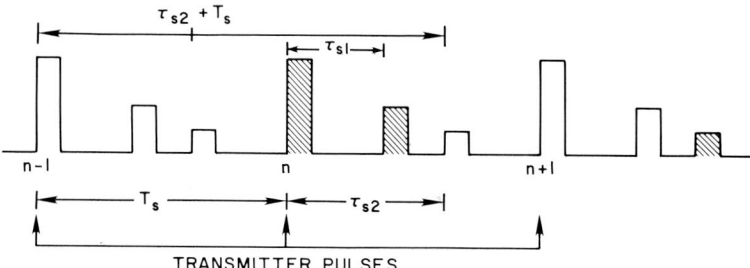

Fig. 3.10. Range-ambiguous echoes. The nth transmitted pulse and its echoes are cross-hatched. This example assumes that the larger echo at delay τ_{s1} is unambiguous in range but the smaller echo, at delay τ_{s2}, is ambiguous. This second-trip echo, which has a true range delay $T_s + \tau_{s2}$, is due to the $(n-1)$th transmitted pulse.

range within which all targets must lie in order to have their ranges unambiguously measured.) We emphasize that r_a does not necessarily limit the range to which the pulsed-Doppler radar can achieve useful measurement. If its STALO (Fig. 3.1) is phase coherent over many T_s intervals, the radar can accurately measure velocities of targets beyond r_a.

The second ambiguity relates to measurement of target velocity. As discussed in Section 3.4.2, the target's phase γ is sampled at intervals T_s and its change $\Delta\gamma$ over the interval T_s is a measure of the Doppler frequency $f_d = \omega_d/2\pi$. Unfortunately, given a set of sampled phases (computed from I and Q samples), we cannot relate them to one unique Doppler frequency. As Fig. 3.11 shows, the same set of samples could have resulted from any one

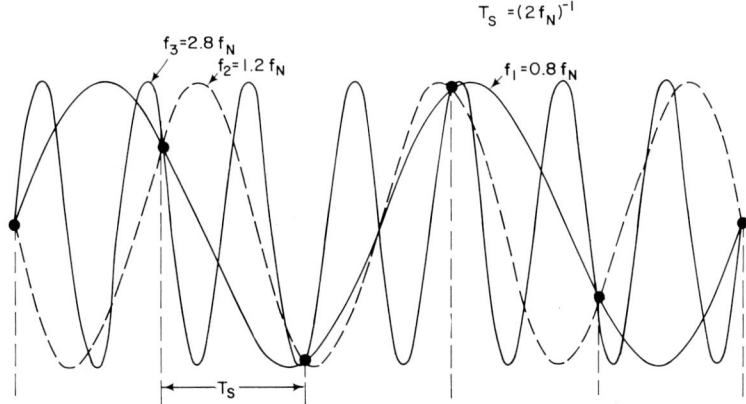

Fig. 3.11. Signals at three different Doppler frequencies that yield, when sampled, the same set of data. These Doppler frequencies are aliases of each other.

of three signals having different Doppler frequencies. All such signals that fit the sample data set are called *aliases*, and $f_N = (2T_s)^{-1}$ is the Nyquist (or folding) frequency. All Doppler frequencies between $\pm f_N$ are the principal aliases, and a frequency higher than f_N is ambiguous with those between $\pm f_N$. Thus, target radial speeds must lie within the unambiguous velocity limits $v_a = \pm \lambda/4T_s$ to avoid ambiguity. However, the transmission of pulses at two interlaced pulse repetition frequencies (PRFs) and signal processing can extend the unambiguous velocity (see Chapter 7).

REFERENCES

Atlas, D. (1964). Advances in radar meteorology. *Adv. Geophys.* **10**, 317–478.
Battan, L. J. (1973). "Radar Observation of the Atmosphere." Univ. of Chicago Press, chicago, Illinois.
Bean, B. R., and Dutton, E. J. (1966). "Radio Meterology," Natl. Bur. Stand., Monogr. 92, Supt. Doc. U.S. Govt. Printing Office, Washington, D.C.
Blake, L. V. (1970). Prediction of radar range. *In* "Radar Handbook" (M. I. Skolnik, ed.), Chapter 2. McGraw-Hill, New York.
Born, M., and Wolf, E. (1964). "Principles of Optics," 2nd ed. Macmillan, New York.
Burrows, C. R., and Attwood, S. S. (1949). "Radio Wave Propagation," p. 219. Academic Press, New York.
Dicke, R. H., Beringer, R., Kyhl, R. L., and Vane, A. B. (1946). Atmospheric absorption measurements with a microwave radiometer. *Phys. Rev.* **70**, 340–348.
Gunn, K. L. S., and East, T. W. R. (1954). The microwave properties of precipitation particles. *Q. J. R. Meteorol. Soc.* **80**, 522–545.
Jordan, E. C., and Balmain, K. G. (1968). "Electromagnetic Waves and Radiating Systems." Prentice-Hall, Englewood Cliffs, New Jersey.
Kraus, J. D. (1966). "Radio Astronomy." McGraw-Hill, New York.
Laws, J. O., and Parsons, D. A. (1943). The relation of raindrop-size to intensity. *Trans. Am. Geophys. Union* **24**, 452–460.
Nathanson, F. E. (1969). "Radar Design Principles." McGraw-Hill, New York.
Rideout, V. C. (1954). "Active Networks." Prentice-Hall, Englewood Cliffs, New Jersey.
Sherman, J. W. (1970). Aperture-antenna analysis. *In* "Radar Handbook" (M. I. Skolnik, ed.), Chapter 9. McGraw-Hill, New York.
Skolnik, M. I., ed. (1970). "Radar Handbook." McGraw-Hill, New York.
Taylor, J. W., and Mattern, J. (1970). Receivers. *In* "Radar Handbook" (M. I. Skolnik, ed.), Chapter 5. McGraw-Hill, New York.

4

Weather Echo Signals

Chapter 3 delineated the radar signal properties of waves scattered from discrete targets. It has been assumed that no more than one target lies in the range interval $r_i \pm c\tau/2$, where r_i is the range to the ith target and that the target dimensions are small compared to $c\tau$. Targets satisfying these conditions are called *point targets*. Weather echoes are composites of signals from a very large number of hydrometeors each of which can be considered a point target. Collectively they are designated *distributed targets* or *clutter*. After a delay (the round trip propagation time between the radar and the near boundary of the scatter volume), echoes are continuously received over a time interval equal to twice the time it takes the microwave pulse to propagate across the volume containing scatterers. Because one cannot resolve the individual targets, we resort to sampling, at discrete range-time delays τ_s, the composite echo that forms a complex voltage $V(t) = I(t) + jQ(t)$ (Fig. 4.1).

Each voltage sample is a weighted composite of discrete echoes from all the scatterers, with weights determined by the radiation pattern $f^2(\theta, \phi)$ and a range weighting function $W(r)$ dependent on the product of receiver bandwidth and transmitted pulse width $B_6\tau$. These weighting functions determine a resolution volume in space where targets contribute significantly to the echo sample at τ_{s1}.

Echo signals from each scatterer in the resolution volume constructively or destructively (depending on the relative phases of the echoes) interfere with each other to produce a composite echo sample $V(\tau_{s1})$, and the random size and location of scatterers cause the amplitude and phase of $V(\tau_{s1})$ to be random variables. The echo sample at another range time delay τ_{s2} is a composite signal from scatterers in a different resolution volume, and hence we expect $V(\tau_{s1})$ to differ from $V(\tau_{s2})$. Thus, $V(\tau_s)$ fluctuates as τ_s increases (Fig. 4.1) even when the scatter density is spatially uniform. The correlation between $V(\tau_s)$ samples taken at different τ_s is related to the radial dimensions of the resolution volume and the spacing $\delta\tau_s$ between samples.

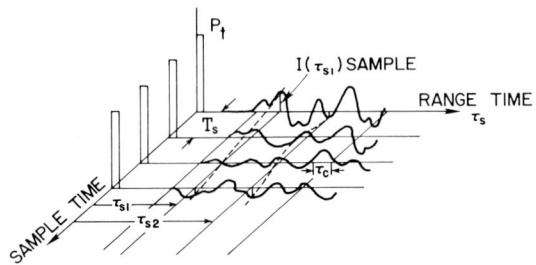

Fig. 4.1. Idealized traces for the in-phase component $I(\tau_s)$, representing echoes from distributed targets. Each trace represents echoes from a single transmitted microwave pulse P_t. Instantaneous samples are taken at τ_{s1}, τ_{s2}, etc. The dashed line indicates a probable time dependence of the sample at τ_s if the sampling rate T_s^{-1} is increased and there are no ambiguous targets. τ_c depicts the signal correlation time along τ_s and is related to $B_6\tau$. Samples at fixed τ_s taken at T_s intervals are used to construct the Doppler spectrum for scatterers located about the range $c\tau_s/2$.

In this chapter we determine the statistical properties of the echo samples $V(\tau_s)$ and instantaneous power $P(\tau_s)$, relate the average of successive $P(\tau_s)$ samples to the backscatter cross section of targets within the resolution volume, and develop a form of the weather radar equation that accounts for range weighting. Chapter 5 discusses the correlation of echo samples versus T_s, whereas here we show the autocorrelation of $V(\tau_s)$ versus lag $\delta\tau_s$.

4.1 THE ECHO SAMPLE

As mentioned previously, the echo sample $V(\tau_s)$ is a composite

$$V(\tau_s) = \frac{1}{\sqrt{2}} \sum_i A_i W_i e^{-j4\pi r_i/\lambda} \qquad (4.1)$$

of signals, where $|A_i|/\sqrt{2} = (I_i^2 + Q_i^2)^{1/2}$ is the prefilter echo amplitude [see Eq. (3.29)] of the ith scatterer located at r_i, ϕ_i, θ_i and W_i is the corresponding range weighting function. We shall not yet be specific about the functional form of W_i. For now it suffices to say that W_i has a range dependence such that it principally weights those targets residing near a range r determined by τ_s (i.e., $r = c\tau_s/2$). Furthermore, the antenna weighting factor $f^2(\theta, \phi)$ is assumed to be part of A_i. We can ignore the constant phase ψ given in (3.27). We also assume target velocity sufficiently small that W_i is independent of it. This is tantamount to assuming that the Doppler shift is small compared to B_6; otherwise, fast-moving targets would return signals at frequencies that fall outside the bandwidth of the receiver. Weather targets never move at speeds to shift echoes outside the receiver bandwidth.

4.2 Signal Statistics

The sample, at range time τ_s, of echo power averaged over a cycle of the transmitted frequency f is proportional to [see Eq. (2.3)]

$$P(\tau_s) = VV^* = \frac{1}{2}\sum_{i,k}^{N_s} A_i A_k^* W_i W_k^* \exp[j4\pi(r_i - r_k)/\lambda]$$

$$= \frac{1}{2}\sum_i |A_i|^2 |W_i|^2 + \frac{1}{2}\sum_{i \neq k} A_i A_k^* W_i W_k^* \exp[j4\pi(r_i - r_k)/\lambda]. \quad (4.2)$$

We have summed the power in the I and Q channels, and therefore the factor of $\frac{1}{2}$ in front of VV^* required for the power in either the I or Q channels of real signals is not present in (4.2). The instantaneous echo power $P(\tau_s)$ is for one transmission, and N_s is the number of scatterers. An instantaneous power is considered to be the power averaged over any one cycle of the radar frequency. When scatterers move relative to one another, the echo sample at range time τ_s differs for each transmitted pulse, and the amount of change depends on the sampling time interval T_s (the interval between transmitted pulses) and the relative velocity of the scatterers. The equation defining $P(\tau_s)$ for successive transmission will have the same form, but the r_i and r_k will have changed owing to scatterer motion. If scatterers within a sample volume move randomly a significant fraction of a wavelength (e.g., $\lambda/4$) between successive transmissions, each successive echo sample $V(\tau_s)$ (spaced T_s apart) will be uncorrelated. In order to make coherent Doppler measurements of the scatterer's mean radial speed, the time T_s between successive samples must be small enough that successive echoes, at fixed delay τ_s, are correlated.

The first sum in (4.2) is a constant independent of the scatterer's position, whereas the second represents the fluctuating portion of the instantaneous power. Fluctuations are caused by the target displacement, which changes the phase of each elemental echo. Although the second sum can be significantly larger than the first [it has $N_s(N_s - 1)$ contributions, compared to N_s for the first term] for some echo samples, its average over many successive samples (i.e., its sample time average) approaches zero as the number of samples increases without limit; this is because the average of the complex exponential term tends to zero. The first sum is then the sample time mean power $\bar{P}(\tau_s)$. An accurate estimate of this term is important because it relates to the estimates of liquid water in the resolution volume.

4.2 SIGNAL STATISTICS

The I and Q components of the echo's sample are random variables if the scatterers' positions are unpredictable. Let us consider two consecutive echoes spaced T_s seconds apart. The first is given by (4.1), and the second one can be written

$$V(\tau_s, T_s) = \frac{1}{\sqrt{2}} \sum_i |A_i W_i| \cos \gamma_i - j \frac{1}{\sqrt{2}} \sum_i |A_i W_i| \sin \gamma_i, \quad (4.3\text{a})$$

where

$$\gamma_i = 4\pi r_i/\lambda + 4\pi v_i T_s/\lambda + \psi_i. \quad (4.3\text{b})$$

$v_i = \Delta r_i/T_s$ is the average radial velocity needed to move the ith scatterer by Δr_i, and ψ_i contains the phase due to scattering and the phase of W_i. Because the range extent of the resolution volume is much larger than the wavelength ($c\tau/2 \gg \lambda$), and there are many scatterers, it is natural to expect the first term of (4.3b) to be uniformly distributed between $-\pi$ and π. Even though the distribution of $4\pi r_i/\lambda$ need not be uniform, its width usually spans many intervals of 2π so that multiple aliasing (of phases into the unambiguous 2π interval) causes the distribution across 2π to be, for all practical purposes, uniform. Therefore, regardless of the v_i or ψ_i distributions, the phases γ_i are also uniformly distributed. This follows because the distribution of the sum of two random variables is obtained after convolving (on a circle from $-\pi$ to π) the two individual distributions. Because one of them is uniform, the distribution of the sum will always be uniform.

Now we are in a position to apply the central limit theorem to the real and imaginary parts of (4.3a). The theorem states that a sum of independent random variables tends to have a Gaussian distribution if their number is large and none of the variables is dominant (i.e., much larger than the rest). Both conditions are certainly true for hydrometeors, and thus the $I(\tau_s, T_s)$ and $Q(\tau_s, T_s)$ have Gaussian distribution with zero mean. It is worth noting that

$$I(\tau_s, T_s) = \frac{1}{\sqrt{2}} \sum_i |A_i W_i| \cos \gamma_i = |V(\tau_s, T_s)| \cos[\theta(\tau_s, T_s)], \quad (4.4\text{a})$$

$$Q(\tau_s, T_s) = -\frac{1}{\sqrt{2}} \sum_i |A_i W_i| \sin \gamma_i = |V(\tau_s, T_s)| \sin[\theta(\tau_s, T_s)]. \quad (4.4\text{b})$$

The equality in (4.4) follows because a sum of sinusoids can always be expressed as a sinusoid with a phase $\theta(\tau_s, T_s)$ and an amplitude factor (*envelope*) $|V(\tau_s, T_s)|$. However, this does not mean that I and Q have pure sinusoidal variation with time.

In addition to being Gaussian, the in-phase and quadrature components are independent random variables (see Appendix B). Therefore the joint probability distribution of I and Q is the product of the individual probabilities:

$$p(I, Q) = (1/2\pi\sigma^2) \exp(-I^2/2\sigma^2 - Q^2/2\sigma^2). \quad (4.5)$$

where σ^2 is the mean square value of I (equal to that for Q).

From (4.4) and (4.5) one can obtain using well-established procedures the distributions of $|V|$, θ, and the power $P(\tau_s)$ (Papoulis, 1965, p. 418). It can be shown that the phase θ is independent of the amplitude $|V|$ and is uniformly distributed, while the amplitude $|V| = (I^2 + Q^2)^{1/2}$ has Rayleigh probability density:

$$p(|V|) = (|V|/2\sigma^2)\exp(-|V|^2/2\sigma^2). \tag{4.6}$$

Because the power $P(\tau_s)$ is proportional to $I^2 + Q^2$, it follows from (4.5) that $P(\tau_s)$ is exponentially distributed with density

$$p(P) = (1/2\sigma^2)\exp(-P/2\sigma^2) \tag{4.7}$$

and a mean value $\bar{P}(\tau_s) = 2\sigma^2$. Constants of proportionality and impedance factors that make the transition from (4.6) to (4.7) dimensionally correct have been ignored and will be ignored henceforth. Radar receivers have gains and losses, and therefore the receiver needs to be calibrated with a known input power in order accurately to relate the output $I^2 + Q^2$ to the echo power.

We emphasize that, although I and Q are independent random variables, the stochastic processes $I(\tau_s, nT_s)$ and $Q(\tau_s, nT_s)$ are not independent. This means that in general $E[I(\tau_s, mT_s)Q(\tau_s, kT_s)] \neq 0$ for $k \neq m$ [see Appendix B, Eq. (B.13), for the proof]. The correlation between two successive samples of the complex signal will be appreciably different from zero only if the distribution of $4\pi v_i T_s/\lambda$ in (4.3b) is narrow compared to 2π (see Appendix B), which is equivalent to saying that the distribution of the v_i is narrow compared to $\lambda/2T_s$ (i.e., the radar's Nyquist velocity).

4.3 THE WEATHER RADAR EQUATION

We now relate the sample time average of echo power $\bar{P}(\tau_s)$ to the radar parameters and target cross section. The contribution to the average echo power from each scatterer is, from (4.2),

$$P_i = \tfrac{1}{2}|A_i|^2|W_i|^2, \tag{4.8}$$

where $\tfrac{1}{2}|A_i|^2$ is the prefilter echo power and hence can be directly expressed in terms of radar parameters and target cross section through use of (3.24). Thus the sample time average power at delay τ_s is

$$\bar{P}(\tau_s) = \frac{P_t g^2 \lambda^2}{(4\pi)^3} \sum_i \frac{l_i^2 \sigma_{bi} f^4(\theta_i, \phi_i)|W_i|^2}{r_i^4}. \tag{4.9}$$

We now consider an elemental volume ΔV that contains many hydrometeors. The summation of σ_{bi} over this volume normalized to ΔV defines the *reflectivity*

$$\eta \equiv (\Delta V)^{-1} \sum_{\Delta V} \sigma_{bi}, \tag{4.10}$$

which is the scatter cross section per unit volume. The terms $l_i, f(\theta_i, \phi_i), r_i,$ and W_i are assumed not to vary significantly over this elemental volume. Replacing the sum by an integration, because elemental volumes have a size that is small compared to the spatial extent of the weighting functions, we have the following form of the weather radar equation:

$$\bar{P}(\mathbf{r}_0) = \frac{P_t g^2 \lambda^2}{(4\pi)^3} \int_0^{r_2} \int_0^{\pi} \int_0^{2\pi} \frac{\eta(\theta, \phi, r) l^2}{r^4} f^4(\theta, \phi) |W(r_0 - r)|^2 \, dV, \quad (4.11)$$

where

$$dV = r^2 \, dr \, \sin\theta \, d\theta \, d\phi,$$

θ, ϕ are angular positions relative to the mainbeam axis whose vector radius $\mathbf{r}_0$ is directed to the location of maximum weight (i.e. the center of the resolution volume). The upper limit for the r integration does not extend to infinity because targets beyond some range r_2 cannot return echoes soon enough to contribute to the echo sampled at delay τ_s. In the next section we shall be more specific about the range of the limits of integration for r and the functional form of $W(r)$. For now we need only recognize that $W(r)$ must decrease faster than r^2 for small r in order for the integral to be bounded at the lower limit (in general the r dependence vanishes as $r \to 0$).

In general η, l are functions of $\mathbf{r}_0$ but let us assume that $f^4(\theta, \phi)|W(r)|^2$ has a scale (resolution volume dimensions) such that the reflectivity and attenuation can be considered constant over the region that contributes most to $P(\tau_s)$. We also assume insignificant contribution from regions outside. Shortly we shall be more precise as regards resolution volume size. Furthermore, we assume that the range to the resolution volume is large compared to its range extent. Thus we can approximate (4.11) by

$$\bar{P}(\mathbf{r}_0) \simeq \frac{P_t g^2 \lambda^2 l^2 \eta}{(4\pi)^3 r_0^2} \int_r |W(r)|^2 \, dr \int_0^{2\pi} \int_0^{\pi} f^4(\theta, \phi) \sin\theta \, d\theta \, d\phi. \quad (4.12)$$

When antenna patterns are circularly symmetric and with Gaussian shape, it can be shown (Probert-Jones, 1962) that

$$\int_0^{\pi} \int_0^{2\pi} f^4(\theta, \phi) \sin\theta \, d\theta \, d\phi = \pi \theta_1^2 / 8 \ln 2, \quad (4.13)$$

where θ_1 is the 3-dB width (in radians) of the one-way pattern.

4.3.1 The Range Weighting Function

We now need to determine the weighting function $|W(r)|^2$. Assume that the scatterers are randomly distributed in space, and let us sum the echo

4.3 The Weather Radar Equation

voltages from elemental volumes over θ and ϕ at constant range so as to produce an effective voltage V_r per unit range interval. That is, a shell of thickness dr at range r returns an echo pulse of incremental amplitude $V_r \, dr$ to the receiver's input.

Targets in a vanishingly thin spherical shell at range r have all the characteristics of point targets and hence will generate an echo voltage shape that, at the receiver input, replicates the transmitted one. Thus the receiver's response to this incremental echo pulse will have a range time dependence of the form

$$W(\tau_s - \tau_d) V_r(\tau_d) \, dr, \quad (4.14a)$$

$$W = 0, \quad \tau_s < \tau_d, \quad (4.14b)$$

where $\tau_d = 2r/c$ and the exact form of W depends on the receiver filter response (see, e.g., Section 3.5.3). The inequality (4.14b) expresses the fact that no signal can be present at the receiver output until the pulse transmitted at $\tau_s = 0$ makes its trip to the target and the echo returns, with a delay τ_d.

Moreover, as expressed in Section 3.5.3 finite receiver bandwidth causes the response $W(\tau_s - \tau_d)$ to have nonzero values even after the input signal has gone to zero. Thus the filter output voltage sampled at any range time delay τ_s receives contributions from all the shells between range zero and $c\tau_s/2$, although the shells nearest to the radar contribute a vanishingly small voltage increment to samples at increasingly large τ_s.

Therefore the voltage sampled at τ_s is

$$V(\tau_s) = \sum_{i=1}^{M} W[(s-i)\,dt] V_r(i\,dt) \, dr, \quad (4.15)$$

where $\tau_s = s\,dt$, $\tau_d = i\,dt$, $dt \equiv 2\,dr/c$, and $M\,dr = c\tau_s/2$. We replace the sum with an integral to obtain

$$V(\tau_s) = \int_0^{r_s = c\tau_s/2} W(r_s - r) V_r(r) \, dr. \quad (4.16)$$

Because of propagation delays in transmission lines from the antenna to the receiver, and because a receiver of finite bandwidth requires a finite time to reach maximum signal, the receiver output peaks τ_r seconds after the leading edge of the echo pulse (from a point target) arrives at the antenna (Fig. 4.2). Thus τ_r is the radar delay. This delay adds significantly to the time it takes the echo to propagate from the target to the receiver output, and therefore it must be subtracted from the measured delay τ_s. Thus a voltage sample at τ_s receives maximum weight from a target at range

$$r_0 = c(\tau_s - \tau_r)/2. \quad (4.17)$$

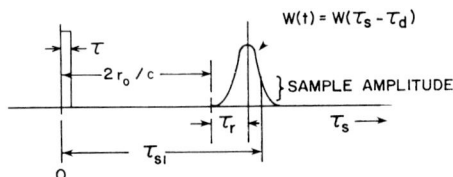

Fig. 4.2. The receiver's voltage response to an input echo that replicates the transmitted pulse shape. The point target is located at $r_0 = c\tau_d/2$, and τ_r is the radar delay.

Conversely, we can state that, for the voltage increments contributing to the sample at τ_s, the receiver's filter effectively gives less weight to those scatterer's displaced in range relative to r_0. Thus a scatterer receives a weight that has the functional dependence

$$W(r_0 - r + c\tau_r/2) \tag{4.18}$$

with maximum at $r = r_0$. In view of the radar delay and the requirement to determine the location of targets that contribute most to the voltage sample at delay τ_s, the integral (4.16) can be rewritten in terms of r_0 and τ_r using (4.17):

$$V(\tau_s) = \int_0^{r_0 + c\tau_r/2} W(r_0 - r + c\tau_r/2) V_r(r)\, dr. \tag{4.19}$$

The radar delay τ_r and range weighting function $W(r)$ are important parameters that must be known by the radar meteorologist. Included in τ_r should be all other system delays (e.g., the propagation time between the transmitter and antenna) that together make up the total radar delay τ_r from the transmitted pulse starting location to the antenna and back to the receiver output.

From Fig. 4.2 it becomes quite simple to realize how the delays can be measured and even how the range weighting function can be determined. It suffices to insert a replica of the transmitter pulse into the receiver and to observe the shape of the response that is the *mirror image* of the weighting function. In another way, given the receiver's impulse response $h(t)$, it is known that its response to any input pulse shape is the convolution of $h(t)$ with the input.

Consider the receiver to have a frequency transfer described by a Gaussian function, the transmitted pulse to be rectangular, and assume all transmission line delays are zero. Consequently, from (3.35) we conclude that each scatterer has a range-dependent weight

$$W(r) = \tfrac{1}{2}\{\operatorname{erf}[(2aB_6/c)(r_0 - r + c\tau/4)] \\ - \operatorname{erf}[(2aB_6/c)(r_0 - r - c\tau/4)]\} \tag{4.20}$$

for the sample taken at delay τ_s. In Section 3.5.2 we noted that a perfect

4.3 The Weather Radar Equation

Gaussian filter has infinite delay so, strictly speaking, we cannot reference r to a starting time $\tau = 0$ and sample time τ_s. However, (4.20) should describe accurately the weighting function about any range $r = r_0$ where it is maximum. In order to obtain weights for those targets that contribute significantly to the echo sampled at time τ_s, given the practical realization of the Gaussian filter, we need only substitute (4.17) for r_0 with τ_r a *measured* delay of the radar.

Note that, in the limit of infinite bandwidth, the radar delay is one-half the transmitted pulse width (i.e., $\tau_r \to \tau/2$) if transmission line delays are absent (see Fig. 4.2). Thus we have an equivalent radar delay $\tau/2$ even though the signal experiences no delay in the radar. This apparent contradiction arises because we have assigned a range to a scatterer corresponding to the time its echo gives maximum weight to the signal sampled at delay τ_s. Although (4.20) may not be the best means of assigning a range to an isolated point target (because we do not know if the echo amplitude is maximum at the sample time τ_s), it is a proper formula for locating the range to those distributed scatterers that give maximum weight to the signal sampled at τ_s.

Next it is desirable to express the range weighting function $W(r)$ as a product of a receiver loss factor l_0 and a weighting function $f_w(r)$ whose peak is normalized to unity. This is in order to obtain a form of the weighting function analogous to the product of antenna gain squared, g^2, and pattern function $f^4(\theta, \phi)$.

The range weighting of the target cross section is a function both of transmitted pulse width τ and receiver bandwidth B_6, analogous to the two measurable parameters g and θ_1 for angular weighting. Thus we define

$$\int_0^\infty W^2(r_0 - r)\,dr = \int_0^\infty l_0^2 f_w^2(r_0 - r)\,dr \equiv l_r c\tau/2, \qquad (4.21)$$

where l_r is the echo power loss due to finite receiver bandwidth.

For Gaussian frequency transfer it can be shown that

$$l_0 = \mathrm{erf}(b), \qquad (4.22\mathrm{a})$$

$$f_w(r_0 - r) = [1/2\,\mathrm{erf}(b)]\{\mathrm{erf}(x + b) - \mathrm{erf}(x - b)\}, \qquad (4.22\mathrm{b})$$

where

$$b = B_6 \tau \pi / 4\sqrt{\ln 2}, \qquad a = \pi/2\sqrt{\ln 2}, \qquad (4.22\mathrm{c})$$

$$x = (2aB_6/c)(r_0 - r). \qquad (4.22\mathrm{d})$$

We can obtain an analytic solution for l_r by approximating $\mathrm{erf}(y)$ by $\tanh(y)$. We find

$$l_r = \coth(aB_6 \tau) - 1/aB_6 \tau. \qquad (4.23)$$

The receiver loss $L_r = -10 \log l_r$ due to a finite bandwidth is plotted in Fig. 4.3 along with an exact numerical solution of (4.21). From Fig. 4.3 we see that if $B_6 \tau \geq 1$, then (4.23) has an error less than 0.6 dB. However for $B_6 \tau < 1$ we need the exact solution for accurate calibration. Although $f_w^2(r)$ broadens at smaller $B_6 \tau$, l_0^2 decreases rapidly, so in the limit the loss of signal power due to finite receiver bandwidth is, for $B_6 \tau \ll 1$,

$$L_r \to -10 \log(\tfrac{1}{3} a B_6 \tau). \tag{4.24}$$

Thus the echo sample power decreases linearly with B_6 for $B_6 \tau \ll 1$. This may seem disastrous, but what really counts is the signal-to-noise ratio (see Section 4.4). Furthermore, we note that, in the limit as $B_6 \tau \gg 1$, $l_r \to 1$ and $\int W^2 \, dr \to c\tau/2$, so (4.12) agrees with the Probert–Jones radar equation.

Nathanson and Smith (1972) examined the perfectly matched filter receiver and deduced L_r to be 1.8 dB. For $B_6 \tau = 1$, corresponding to a practical matched filter (i.e., Gaussian frequency response and a rectangular pulse), a numerical integration of the exact function $W^2(r)$ shows $L_r \simeq 2.3$ dB, or about 0.5 dB more than the value obtained with a perfectly matched filter receiver.

We now have a final general form of the weather radar equation:

$$\bar{P}(\mathbf{r}_0) = \frac{P_t g^2 \lambda^2 l^2(\mathbf{r}_0) \eta(\mathbf{r}_0) l_r}{(4\pi)^3 r_0^2} \frac{\pi \theta_1^2}{8(\ln 2)} \frac{c\tau}{2}. \tag{4.25}$$

Equation (4.25) shows the dependency of averaged echo power $\bar{P}(\mathbf{r}_0)$ on measurable quantities for the radar.

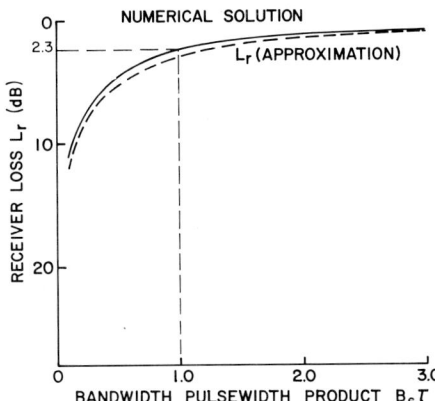

Fig. 4.3. Receiver signal power loss L_r (dB) due to a finite bandwidth. The receiver frequency transfer is Gaussian, and the echo pulse is rectangular. The solid curve is a numerical solution of the exact response function, and the dashed curve is obtained from an analytical approximation [Eq. (4.23)].

4.3.2 Resolution Volume

For many reasons it is useful to define the resolution volume V_6 to be that circumscribed by the 6-dB contour. The 6-dB width of $f^4(\theta, \phi)$ (the two-way antenna pattern function) is often taken to represent the angular resolution of V_6. In an analogous manner we define the 6-dB width of f_w^2 to represent the range width r_6 of V_6. The solution of the transcendental equation for r_6 is not easy, but we can approximate the error function by a hyperbolic tangent function. This approximation gives an analytic solution that shows the interrelation between receiver bandwidth B_6 and pulse width τ. The solution is

$$r_6 = (1/aB_6\tau)(\tfrac{1}{2}c\tau)\cosh^{-1}[2 + \cosh(aB_6\tau)]. \tag{4.26}$$

The range width r_6 is plotted versus $B_6\tau$ in Fig. 4.4. Also shown is a numerical solution of the exact transcendental equation. We see that as the bandwidth–pulse width product gets large, the 6-dB range width approaches $c\tau/2$. Thus the resolution volume is a cylinder having a 6-dB diameter $r\theta_1$ and a range extent given by (4.26). Outside this resolution volume, reflectivity is weighted by factors $< \tfrac{1}{4}$ due to either antenna pattern, receiver response, or both. The size of the resolution volume is 0.1 km³ for $\theta_1 = 1°$, $r = 50$ km, $B_6\tau = 1$, and $\tau = 1\,\mu\text{s}$. We must be cautious about ignored contributions to echo power from regions outside V_6 because weather targets have cross sections that vary by many orders of magnitude. Thus a collection of weakly scattering targets in V_6 may contribute less power than strong scatterers outside V_6.

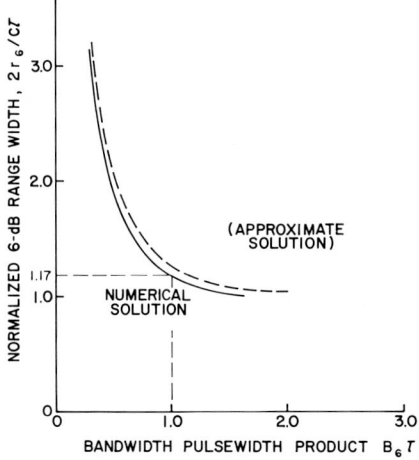

Fig. 4.4. Normalized 6-dB range width, $2r_6/c\tau$, of the resolution volume versus receiver bandwidth–pulse width product. The numerical solution to the transcendental equation is the solid line, and the dashed line shows an approximate analytic solution, given by (4.26).

4.3.3 Reflectivity Factors

Radar meteorologists need to relate the *reflectivity* η, which is general radar terminology for the scattering cross section density, to factors that have meteorological significance. If scattering particles are known to be spherical and have diameters that are small compared to the wavelength (i.e., in the Rayleigh approximation), we can substitute (3.6) into (4.10) to obtain

$$\eta = \frac{\pi^5}{\lambda^4} |K_w|^2 Z, \tag{4.27a}$$

where

$$Z \equiv \frac{1}{\Delta V} \sum_i D_i^6 \tag{4.27b}$$

is the *reflectivity factor*. Whenever the Rayleigh approximation does not apply, and this is usual for short-wavelength ($\lambda < 10$ cm) radars measuring thunderstorm precipitation, it is accepted practice to write

$$\eta = (\pi^5/\lambda^4) |K_w|^2 Z_e, \tag{4.28}$$

where Z_e is the *effective reflectivity factor*. Because values of Z commonly encountered in weather observations span many orders of magnitude, radar meteorologists use a logarithmic scale dBZ = 10 $\log_{10} Z$ (where Z is in units of mm^6/m^3). Precipitation produces dBZ values ranging from near 0 dBZ to values somewhat larger than 60 dBZ in regions of heavy rainfall and hail (see Chapters 8 and 9).

The use of reflectivity factors alone does not really relate radar echo power to meteorologically significant factors such as rainfall rate or liquid water content because there is one more essential parameter that needs to be known in addition to the phase (i.e., liquid, solid, or a mixture such as melting ice). This is the particle size distribution (see Chapter 8). Although the choice of η or Z_e to be used in the radar equation should be equally weighted, radar meteorologists prefer Z_e.

Incorporating (4.28) into (4.25), we have the weather radar equation that gives the average echo power in terms of Z_e (in cubic meters):

$$\bar{P}(\mathbf{r}_0) = \pi^3 P_t g^2 l^2 l_r \theta_1^2 c\tau |K_w|^2 Z_e / 2^{10} (\ln 2) \lambda^2 r_0^2, \tag{4.29}$$

where all units are in the MKS system; θ_1 is in radians, and g is dimensionless. However, it is common to express Z_e in units of (millimeters)^6 per cubic meter, θ_1 in degrees, r_0 in kilometers, λ in centimeters, τ in microseconds, P_t in watts, and $\bar{P}(\tau_s)$ in milliwatts. Thus in units conventional to the radar meteorologist, the weather echo power is

$$\bar{P}(\text{mW}) = \frac{\pi^5 10^{-17} P_t(\text{W}) g^2 l^2 l_r \tau(\mu s) \theta_1^2 (\text{deg}) |K_w|^2 Z_e (\text{mm}^6/\text{m}^3)}{6.75 \times 2^{14} (\ln 2) r_0^2 (\text{km}) \lambda^2 (\text{cm})}. \tag{4.30}$$

4.4 SIGNAL-TO-NOISE RATIO FOR DISTRIBUTED TARGETS

The SNR for distributed targets is just (4.25) divided by $kT_{sy}B_n$. Again consistent with the approximation (4.23), we find that the SNR depends on $B_6\tau$ as

$$\text{SNR} = \frac{C_0}{r^2}\left[\coth(aB_6\tau) - \frac{1}{aB_6\tau}\right]\frac{\tau^2}{B_6\tau}, \qquad (4.31)$$

where C_0 contains those constants pertaining to the radar. We immediately note that if $B_6\tau$ is a constant, then SNR is proportional to the square of the transmitted pulse width. It is also readily seen that the maximum SNR is obtained as $B_6 \to 0$. Therefore, unlike for point target measurements, we do *not* obtain a maximum for the weather SNR at $B_6\tau \simeq 1$ (cf. Section 3.5.4).

Even though the SNR increases monotonically as B_6 decreases, the resolution r_6^{-1} worsens. However, if one constrains the resolution to be a constant, then the SNR is an optimum when the filter response is matched to the transmitted pulse. By using a slightly different definition of resolution, one can obtain the optimum signal shape (Zrnić and Doviak, 1978). That is, the resolution can be defined as the square root of the second moment of the range weighting function W^2 (squared for echo power weights). Then, for a given resolution, SNR is a maximum if the transmitted pulse has Gaussian shape matched to the filter's impulse response.

As an example, consider a desired resolution of $r_6 = 500$ m, and assume that the filter impulse response and transmitted voltage pulse are Gaussian with second central moments σ_τ^2. Because the convolution of two Gaussian functions of equal variance is also Gaussian but with double the value of the second moment, it follows that the range weighting function $W(r_0 - r)$ has a second moment σ_w^2 proportional to $2\sigma_\tau^2$ and $W^2(r_0 - r)$ has a moment proportional to σ_τ^2. Now we can find directly that

$$r_6^2 = 2c^2\sigma_\tau^2 \ln 4. \qquad (4.32)$$

Solving for σ_τ (with $r_6 = 500$ m), we find $\sigma_\tau = 1.0$ μs, and the 6-dB pulse width is 2.35 μs. The frequency response of a receiver having a Gaussian impulse response is also Gaussian, and it can be shown by taking the Fourier transform of $h(t)$ that

$$\sigma_f = 1/2\pi\sigma_\tau, \qquad (4.33)$$

where σ_f is the second moment of the amplitude spectrum, from which we find the 6-dB width B_6 to be $\sqrt{2 \ln 2}/\pi\sigma_\tau = 375$ kHz. If the same calculation is done for a practical radar with rectangular pulse and Gaussian filter, one finds from Fig. 4.4 that $\tau = 2r_6/1.17c = 2.85$ μs and $B_6 = 350$ kHz.

4.5 CORRELATION OF ECHO SAMPLES ALONG THE RANGE TIME AXIS

Sample spacing along the range time axis is usually chosen so that there are independent estimates of reflectivity and velocity along the beam. Both pulse width τ and receiver–filter bandwidth B_6 determine the correlation of these estimates, and sometimes B_6 is deliberately chosen to be small (i.e., not matched to τ) in order to observe meteorological events in a larger range interval with fewer samples along the range time axis. This approach becomes more advantageous when real-time data processing equipment limits simultaneous observations to few range time samples (as is sometimes the case for real-time Doppler spectral processors) and pulse width cannot be increased. If B_6 is matched to τ or, as in many meteorological radars, is large compared to τ^{-1}, then dimensionally small meteorological events such as tornadic vortices can be missed by samples spaced farther than the range extent of the sample volume.

In this section we examine the correlation between samples along the range time axes and determine how receiver bandwidth and transmitter pulse width affect this correlation. In Appendix C we show that the signal at the receiver's input has a correlation R_{xx} given by

$$R_{xx}(\delta\tau_s) = \begin{cases} \bar{P}(\mathbf{r}_0)[1 - |\delta\tau_s|/\tau], & |\delta\tau_s| \le \tau, \\ 0 & \text{otherwise.} \end{cases} \quad (4.34)$$

The signal $V(\tau_s)$, after passing through the filter, has a correlation $R_{vv}(\delta\tau_s)$ given by (Papoulis, 1965, p. 346)

$$R_{vv}(\delta\tau_s) = R_{xx}(\delta\tau_s) \star h^*(-\delta\tau_s) \star h(\delta\tau_s), \quad (4.35)$$

where $h(\delta\tau_s)$ is the unit impulse response of the filter and $\star$ designates the convolution operation. For a Gaussian filter it can be shown that

$$h(t) = 0.5 B_6 (\pi/\ln 2)^{1/2} \exp[-(\pi B_6 t)^2/4\ln 2], \quad (4.36)$$

and therefore

$$h^*(-\delta\tau_s) \star h(\delta\tau_s) = \int_{-\infty}^{+\infty} h^*(-\varepsilon) h(\delta\tau_s - \varepsilon)\, d\varepsilon$$

$$= (0.5)^2 B_6 (2\pi/\ln 2)^{1/2} \exp[-(\pi B_6 \delta\tau_s)^2/8\ln 2]. \quad (4.37)$$

We then find that the echo samples have a correlation R_{vv} given by the convolution of the input signal correlation R_{xx} and (4.37):

$$R_{vv}(\delta\tau_s) = \frac{a B_6 \tau \bar{P}(\mathbf{r}_0)}{\sqrt{2\pi}} \int_{-1}^{+1} \{(1 - |x|) \exp[-(a B_6 \tau)^2 (x - \delta\tau_s/\tau)^2/2]\}\, dx, \quad (4.38)$$

where $a = \pi/2\sqrt{\ln 2}$.

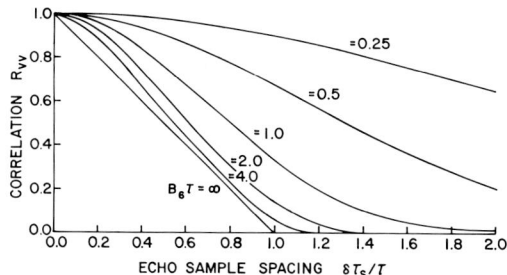

Fig. 4.5. Normalized correlation of echo samples spaced along the range time axis.

Equation (4.38) has been evaluated numerically, and the results are plotted in Fig. 4.5. The solution shows that when B_6 is more than twice τ^{-1} the echo sample correlation is principally controlled by pulse width, whereas when $B_6 < 0.5\tau^{-1}$ it is controlled by the receiver–filter 6-dB bandwidth. A useful analytic formula for correlation when $B_6 \ll \tau^{-1}$ is

$$R_{vv}(\delta\tau_s) = R_{vv}(0)e^{-(aB_6\delta\tau_s)^2/2}, \qquad (4.39)$$

where $R_{vv}(0) = aB_6\tau\bar{P}(\mathbf{r}_0)/\sqrt{2\pi}$ and $\delta\tau_s$ is the range time sample spacing.

REFERENCES

Nathanson, F. E., and Smith, P. L. (1972). A modified coefficient for the weather radar equation. *Prepr., Conf. Radar Meteorol., 15th, 1972* pp. 228–230.

Papoulis, A. (1965). "Probability, Random Variables, and Stochastic Processes." McGraw-Hill, New York.

Probert-Jones, J. R. (1962). The radar equation in meteorology. *Q. J. R. Meteorol. Soc.* **88**, 485–495.

Zrnic, D., and Doviak, R. J. (1978). Matched filter criteria and range weighting for weather radar. *IEEE Trans. Aerosp. Electron. Syst.* AES-14, 925–930.

5

Doppler Spectra of Weather Echoes

In order to measure the power-weighted distribution of velocities, a frequency analysis of $V(\tau_s, mT_s)$ [see Fig. 4.1 and Eq. (4.1)] is needed. This can be accomplished by estimating its power spectrum. It is important to bear in mind that the frequency analysis is performed along the sample time axis mT_s for samples $V(\tau_s, mT_s)$ at fixed τ_s. Thus we have discrete samples, spaced T_s apart, of a continuous random process. We shall first review the theory of discrete Fourier transforms, then discuss convolution and the correlation of random signals, and finally show the relation between the correlation function and power spectrum. Throughout this section we shall omit τ_s from the argument; and the results derived apply to any sequence of samples at fixed range $r = c\tau_s/2$.

5.1 SPECTRAL ANALYSIS OF WEATHER SIGNALS

5.1.1 Discrete Fourier Transform

Fourier methods of time series analysis are now commonly used in many branches of science and engineering. Digital computers with associated software or special-purpose machines rapidly compute the Fourier coefficients of lengthy time sequences. The *discrete Fourier transform* (DFT) of a signal sampled M times at a uniform spacing T_s (Fig. 5.1) is defined as

$$Z(kf_0) = \sum_{m=0}^{M-1} V(mT_s) e^{-j2\pi f_0 T_s mk}, \tag{5.1}$$

where $V(mT_s)$ is the complex voltage of the mth sample and $Z(kf_0)$ is the complex amplitude of the kth spectral coefficient. The inverse transform (IDFT) is

$$V(mT_s) = \frac{1}{M} \sum_{k=0}^{M-1} Z(kf_0) e^{j2\pi f_0 T_s mk}. \tag{5.2}$$

5.1 Spectral Analysis of Weather Signals

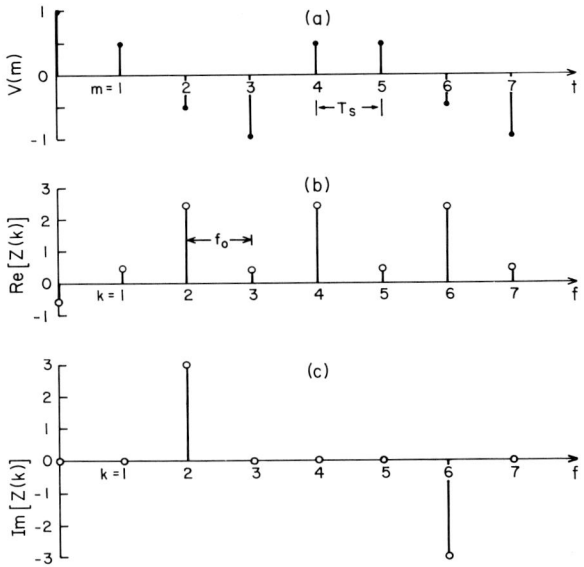

Fig. 5.1. (a) Sequence $V(mT_s)$ of real numbers. (b), (c) The complex spectral coefficients $Z(kf_0)$. The number of points $M = 8$. The magnitude $|Z(f_0)|$ is symmetric with respect to $f = 0$ and $4f_0$ (the Nyquist frequency) because the sequence $V(mT_s)$ is real. It can be seen that the sequence $V(mT_s)$ has a dominant sinusoidal component with period $4T_s$.

Note that (5.1) is an expansion in terms of multiples of the fundamental (lowest) frequency

$$f_0 = 1/MT_s, \tag{5.3}$$

and as such gives M complex Fourier coefficients (amplitudes and phases). It is easy to prove by substitution that (5.2) satisfies (5.1):

$$Z(kf_0) = \frac{1}{M} \sum_{m=0}^{M-1} \sum_{n=0}^{M-1} Z(nf_0) e^{-j2\pi f_0 T_s m(k-n)}. \tag{5.4a}$$

Interchanging the order of summation and summing the geometric progression over m, we get

$$\sum_{m=0}^{M-1} e^{-j2\pi f_0 T_s m(k-n)} = (1 - q^M)/(1 - q), \tag{5.4b}$$

where

$$q = \exp[-j2\pi f_0 T_s (k - n)]. \tag{5.4c}$$

Because (5.4b) equals M when $k = n$ and 0 otherwise, (5.4a) is indeed an identity.

We note that $Z(kf_0)$ is periodic in k:

$$Z(kf_0) = Z(lf_0) \quad \text{if} \quad l = \mathrm{mod}_M(k). \tag{5.5a}$$

Furthermore, when the sequence is real,

$$Z(-kf_0) = Z^*(kf_0). \tag{5.5b}$$

Thus, for the example in Fig. 5.1 (where $M = 8$), $Z(0) = Z(8)$, $Z(1) = Z(9)$, etc.

Likewise, the sequence $V(mT_s)$ calculated from (5.2) is periodic in m: $V(mT_s) = V[(m + M)T_s]$. We should not conclude from this property that our original sequence of echo samples is periodic. We have taken a segment, M samples long, of a much longer sequence that, for weather echoes, is definitely nonperiodic. However, it is the property of the transform we have given to repeat periodically the sample values outside the segment MT_s.

The Nyquist frequency $(2T_s)^{-1}$ is the highest frequency that can be unambiguously measured in the sampled sequence because, in order completely to specify a sinusoid, at least two samples within a period are needed (see Fig. 3.11). Because the exponential in (5.1) is cyclic, $Z(-kf_0) = Z[(M - k)f_0]$, so sometimes for display purposes the spectral coefficients are arranged so that there is an equal number ($M/2$ lines) on each side of $k = 0$. This is done in order to have positive Doppler frequencies to the right and negative ones to the left of the zero ($k = 0$) line. In such a case the power $|Z(Mf_0/2)|^2$ is split, so there are actually $M + 1$ lines.

Substituting (5.3) into (5.1) and (5.2), we can eliminate f_0 and T_s from the calculation. For simplicity we also drop them from the arguments in (5.1) and (5.2) to obtain a compact form of the DFT,

$$Z(k) = \sum_{m=0}^{M-1} V(m) e^{-j(2\pi/M)mk}, \tag{5.6a}$$

and its inverse,

$$V(m) = \frac{1}{M} \sum_{k=0}^{M-1} Z(k) e^{j(2\pi/M)mk}. \tag{5.6b}$$

In order to understand the properties of the DFT, we illustrate how it acts on a sampled complex sinusoid $A \exp[j(2\pi\alpha f_0 mT_s + \psi)]$ of frequency αf_0 and constant phase ψ.

From (5.6a) we find the Fourier coefficients

$$Z(k) = A \sum_{m=0}^{M-1} \exp\left[j\left(\frac{2\pi}{M}\alpha m + \psi\right)\right] \exp\left(-j\frac{2\pi}{M}mk\right). \tag{5.7}$$

5.1 Spectral Analysis of Weather Signals

This finite sum represents a geometric progression, as in (5.4b), and can be reduced to

$$Z(k) = A \frac{\sin[\pi(\alpha - k)]}{\sin[(\pi/M)(\alpha - k)]} \frac{\exp[j\pi(\alpha - k)]}{\exp[j(\pi/M)(\alpha - k)]} e^{j\psi}.$$

Its magnitude is

$$|Z(k)| = A \left| \frac{\sin[\pi(\alpha - k)]}{\sin[(\pi/M)(\alpha - k)]} \right|. \qquad (5.8)$$

If α is an integer, then

$$|Z(k)| = \begin{cases} 0 & \text{for } k \neq \alpha, \\ AM & \text{for } k = \alpha. \end{cases} \qquad (5.9)$$

Thus, when the input sinusoid has a period that is exactly contained in the interval length MT_s, its discrete Fourier transform consists of a single coefficient. In other words, the DFT has singled out (filtered) the sinusoid. It is to be noted that although each sample has amplitude A, the amplitude of the spectral coefficient $k = \alpha$ is AM. Samples are therefore said to *sum coherently* to give a spectral-coefficient amplitude M times the signal amplitude.

Figure 5.2a illustrates a complex sinusoid with period $4T_s$, a sample sequence $8T_s$ in duration ($M = 8$), and $\alpha = 2$. The Fourier coefficients $|Z(k)|$ are zero for $k \neq 2$, and $|Z(2)| = 8A$ (Fig. 5.2b). The region (where k is a continuous variable) about the peak of (5.8) defines the main lobe of the DFT filter, and its width is a measure of the filter's frequency selectivity. The other minor lobes are the filter's side lobes, which cause the power to be spilled into other frequency bins (k values) if α is not an integer.

When the sinusoid's frequency is not an exact multiple of the fundamental (i.e., α is a noninteger), all of the Fourier coefficients are nonzero. They are strictly a function of $\text{mod}_M (\alpha - k)$ and are largest for coefficients that are closest to α. Again the DFT filters the sinusoid; however, the magnitude of the spectral coefficient closest to α is less than MA because power has spilled over into all frequency bins (Fig. 5.3). The DFT is used to locate and isolate the many sinusoids that compose a signal. Because these sinusoids do not necessarily have period that is an integer portion of the fundamental MT_s, the power of a large-amplitude sinusoid can spill over into many frequency bins, and this may mask identification of weaker sinusoids. This occurs because one views the sequence $V(m)$ only in a finite-time or data window. The illustration of Fig. 5.3 shows an eight-point DFT of a pure sinusoid with frequency $2.5f_0$. Because the sinusoid signal reconstructed from M samples must be periodic outside the interval MT_s, there is a discontinuity in slope at $m = 0$ and $m = 8$. Because of this discontinuity, a single sinusoid cannot

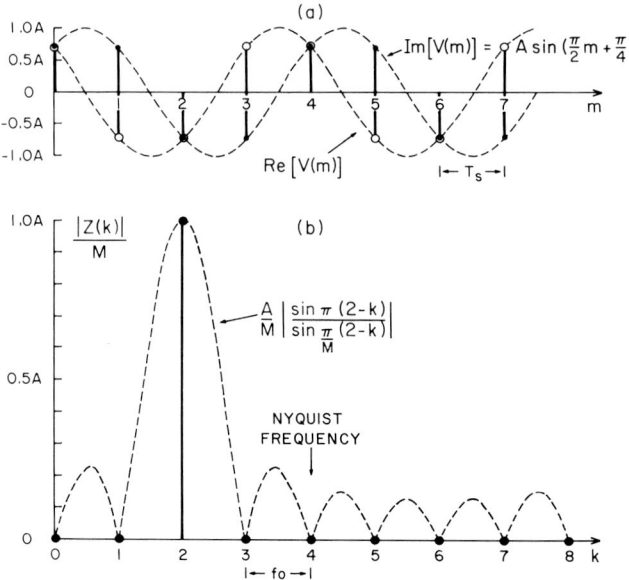

Fig. 5.2. (a) Complex sinusoid of period $4T_s$ (dashed lines). The eight samples, spaced T_s apart, embrace exactly two periods of the sinusoid. (b) The amplitudes of the Fourier coefficients for the sinusoid.

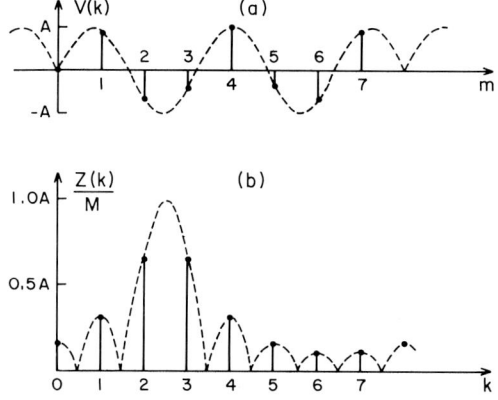

Fig. 5.3. (a) The sinusoid has a period of $3.2T_s$ and hence is not exactly contained in the interval $8T_s$. Only the real component of the complex sinusoid is sketched. (b) Because we use integers k to compute the DFT, even though the frequency of the sinusoid is 2.5 times the fundamental $f_0 = (8T_s)^{-1}$, the DFT coefficients are not synchronized with the sinusoid. Therefore power from the line at $k = 2.5$ has spilled via side lobes of the DFT filter, into all the frequency bins.

5.1 Spectral Analysis of Weather Signals

fit smoothly through all the sample points, so there must exist power in other frequency bins. Can one control the window effect? The answer is yes.

Examining Fig. 5.3 and Eq. (5.8), we notice that a spectral coefficient reaches a maximum if one allows k to be a noninteger equal to α. Thus, by slowly changing k in (5.8), we would have a coherent sum of samples wherever k reached the value equal to the unknown frequency of the sinusoid. However, this does not prevent other sinusoids from possibly spilling their power into the selected coefficient. Therefore, rather than searching for individual peaks, a more common approach is to weight the time series samples and thus avoid the abrupt transition at the beginning and end of the time sequence. This procedure always broadens the main lobe of the DFT filter, although the side lobe amplitudes are decreased so that power spilled into other bins is smaller. We shall examine the window question more thoroughly in Section 5.1.4.

Although efficient methods for calculating the DFT coefficients were discovered at the beginning of this century (Cooley *et al.*, 1967), they became widely known only after the publication of Cooley and Tukey (1965). All such algorithms are appropriately called *fast Fourier transforms* (FFT). They use advantageously the periodicity of trigonometric functions so that the total number of complex operations for calculating all M complex coefficients is proportional to $M \log_2 M$ if M is a power of 2. The number of complex multiplications is one-half that of the additions (Tretter, 1976). In contrast, the straightforward method, (5.6a), requires M^2 operations of both kinds. However, the FFT produces all M coefficients, regardless of whether one wants them. If for some reason only a few coefficients are needed, the regular DFT may be more efficient.

5.1.2 Convolution and Correlation

The Fourier transform has been found to be of great utility in evaluating the convolution and correlation of sample sequences. The output response of any linear system to any input signal can be obtained from the system's impulse response through the application of convolution (e.g., Section 4.5). Let the input sequence be $V(m)$ and the filter impulse response sequence (or weighting function) be $h(m)$. Then the output sequence $Y(l)$ is obtained from the convolution (Papoulis, 1977):

$$Y(l) = V \bigstar h = \sum_{m=-\infty}^{\infty} V(m)h(l-m) = \sum_{m=-\infty}^{+\infty} V(l-m)h(m). \quad (5.10)$$

The sequence $h(l - m)$ is a mirror image of $h(m)$ about $m = 0$ and shifted to the right by l increments. It is well known in the theory of continuous systems that a convolution has a Fourier transform that equals the product of the transforms of each signal. A similar but not identical relationship holds for discrete periodic signals:

$$Y(l) = \frac{1}{M} \sum_{k=0}^{M-1} H(k)Z(k)e^{j(2\pi/M)lk} = \sum_{m=0}^{M-1} h(m)V(l-m). \quad (5.11)$$

To prove (5.11), one should express $h(m)$ and $V(l-m)$ as inverse transforms of $H(k)$ and $Z(n)$. Then, similarly to (5.4a), the sum over m will differ from zero only if $k = n$ resulting in the identity. The difference arises because the product of $H(k)Z(k)$ is a DFT of a circular convolution (Gold and Rader, 1969). Such a convolution (5.11) operates on a periodic extension of either of the sequences $h(m)$ and $V(m)$ (Fig. 5.4).

Two deterministic sequences $U(m)$, $V(m)$ can be tested for similarity with the use of a cross correlation function:

$$R_{VU}(l) = \sum_{m} V(l+m)U^*(m). \quad (5.12)$$

Whereas the convolution operation (5.10) involves shifting the *mirror image* of a sequence and summing products, the correlation calls for shifts and

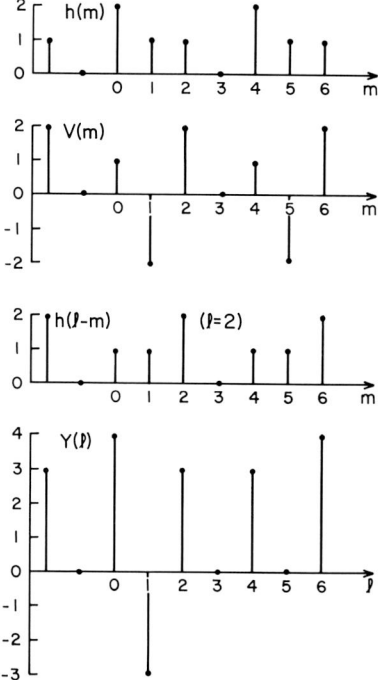

Fig. 5.4. Example of circular convolution. The number of points $M = 4$. The filter weighting function $h(m)$ and the time sequence $V(m)$ have three nonzero terms. Circular convolution involves flipping one of the sequences about $m = 0$, shifting by l increments, cross multiplying the two sequences $V(m)$, $h(l-m)$, and summing over the M products.

5.1 Spectral Analysis of Weather Signals

sums only. Therefore, the convolution will equal the correlation only if the sequence is symmetric.

A sequence with statistical properties independent of time is called *stationary*. Strictly speaking, no such sequence can represent a physical process, because all known processes continuously change. However, when the time of observation is small compared with the time it takes for significant change to occur in the statistical properties (e.g., mean value, variance, and correlation) of the process, the sequence representing the process can be considered stationary. Weather radar signals are a perfect example of such a stationary process.

In dealing with random signals, the most one can hope to accomplish is to estimate some average statistical parameters. Two useful ones are the mean and autocorrelation; if they are time invariant, the process is *wide sense stationary*.

The autocorrelation of a complex wide-sense-stationary signal $V(m)$ is defined as

$$R(l) = E[V^*(m)V(m + l)] = E[V^*(m - l)V(m)]. \qquad (5.13)$$

where $E[\ ]$ denotes the expected value of an ensemble average (Papoulis, 1965). The ensemble is a collection or set of random sequences in which $V(m)$ varies from member to member of the ensemble even though m is fixed. The correlation $R(l)$ is a measure of similarity between the sequence and its shifted conjugate. Noiselike signals have an autocorrelation with a sharp peak at the origin; in fact, the less correlated the signal samples, the sharper is the peak. Useful properties of the autocorrelation are the following:

The average signal power is

$$E[|V(m)|^2] = R(0). \qquad (5.14)$$

If $V(m)$ has a periodic component, the autocorrelation has it too. When $V(m)$ does not have a pure sinusoidal component, its autocorrelation for large lag l equals the square of the mean value:

$$\lim_{l \to \infty} R(l) = E^2[V(m)]. \qquad (5.15)$$

This is so because, as l increases, $V^*(m)$ and $V(m + l)$ become less correlated until, in the limit, the correlation of the fluctuating components vanishes and only the mean value contributes to $R(l)$. Although the autocorrelation need not vanish, the autocovariance of uncorrelated samples is zero. The autocovariance is just the autocorrelation after the mean value $E[V(m)]$ is removed from the signal. For most weather echoes, $E[V(m)]$ is zero (Section

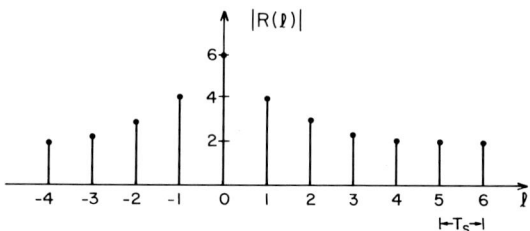

Fig. 5.5. Magnitude of an autocorrelation function. The power of a sequence is 6, and after about $3T_s$ the time samples are not correlated. The magnitude of the mean $E[V(m)]$ is $\sqrt{2}$.

4.2), but for ground targets it is not. Our discussion is illustrated by Fig. 5.5, which shows a possible autocorrelation function of an assumed stationary sequence having nonzero mean. Because the process is stationary, we can clearly see from (5.13) that $R(-l) = R^*(l)$.

At first sight it appears that the autocorrelation function is simply evaluated. Obviously, averaging over all the different realizations of the process (ensemble averaging) is not practical; fortunately, signals we deal with are often stationary, and we may use time averages to deduce the statistical parameters from *one realization* of a process. For example, scatterers are usually statistically stationary during the period of observation (typically <1 s), and they produce one member of the ensemble of all possible echo sequences that could occur under similar conditions. It is much simpler to examine one member and deduce its statistical properties. When the statistical properties of the ensemble can be deduced from sample time averages, the ensemble members $V(m)$ are said to be *ergodic*. Only then can the properties of a class be inferred from time averages of one member. Echoes from several range locations in a homogeneous atmosphere are members of an ergodic ensemble (if one ignores beam spreading and attenuation). We shall assume all our signals to be ergodic Thus, the ensemble average (5.13) is equivalent to

$$R(l) = \lim_{M \to \infty} \frac{1}{M} \sum_{m=0}^{M-|l|-1} V^*(m)V(m+l). \tag{5.16}$$

However, we usually have a finite number of samples and must form estimates $\hat{R}(l)$ (the caret is used to denote estimates), such as

$$\hat{R}(l) = \begin{cases} \dfrac{1}{M} \sum_{m=0}^{M-|l|-1} V^*(m)V(m+l) & \text{for } |l| \leq M-1, \\ 0 & \text{otherwise.} \end{cases} \tag{5.17}$$

5.1.3 Power Spectrum of Random Sequences

Besides being useful in its own right for determining the rapidity of signal change [if $R(l)$ vanishes at small l and remains small, the samples $V(m)$ are changing significantly from sample to sample], the mean power, periodicity, noisiness, etc., the autocorrelation enables us to find through its Fourier transform the frequency distribution of the random signal power. The power spectrum $S(f)$ is defined as the DFT of the autocorrelation function:

$$S(f) \equiv \lim_{M \to \infty} T_s \sum_{l=-(M-1)}^{M-1} R(l) e^{-j2\pi f T_s l}. \tag{5.18}$$

We can apply the theory of Fourier series to obtain the inverse relation

$$R(l) = \int_{-1/2T_s}^{1/2T_s} S(f) e^{j2\pi f T_s l} \, df. \tag{5.19}$$

The power spectrum has a repetitive cycle equal to T_s^{-1}, the Nyquist interval. This can be seen on substituting $f = f + T_s^{-1}$ in (5.18). Equation (5.18) and its inverse (5.19) form the Fourier transform pair and uniquely relate $R(l)$ and $S(f)$. They are completely analogous to a regular Fourier series representation of a periodic function; $S(f)$ is a continuous function of period T_s^{-1}, and the $R(l)$ are its complex Fourier series coefficients.

One may question how well the transform pair (5.18), (5.19) represents the properties of the actual echo voltage $V(mT_s)$ since $R(l)$ is evaluated only at discrete increments spaced T_s apart. To answer this question, we resort to the sampling theorem: If a function $V(t)$ contains no frequencies higher than $Mf_0/2$ cycles per second, it is *completely determined* by giving its ordinates at a series of points spaced $(Mf_0)^{-1} = T_s$ apart, the series extending *throughout* the time domain. Thus the samples $R(l)$ completely determine the signal power spectrum so long as the Nyquist frequency $(2T_s)^{-1}$ is larger than the highest frequency contained in $V(t)$. However, the relations (5.18) and (5.19) are useful even when $V(t)$ has frequencies beyond the Nyquist interval. Then $S(f)$ is an aliased spectrum that may be considerably different from the true one. Nevertheless, useful information can often be obtained from aliased power spectra when aliasing is recognized, provided the aliased shape is not too different from the actual one. We shall show soon that $S(f)$ as defined here is equivalent to the expected value of the spectral power $|Z(f)|^2$.

As defined in (5.18), $S(f)$ represents the power per hertz, from which the name *power spectral density* derives. From (5.13), $R(-l) = R^*(l)$; then, because the summation in (5.18) is symmetric about $l = 0$, it can be shown that $S(f)$ is real and positive. The power spectrum is only one of the average parameters of a signal, and it is a very useful one. However, it is not a complete measure or description of the random process.

A process that generates a particularly simple power spectrum is white noise, for which

$$S(f) = NT_s$$

and

$$R(l) = \begin{cases} N & \text{for } l = 0, \\ 0 & \text{otherwise.} \end{cases} \tag{5.20}$$

N is the total noise power in the frequency interval between $-1/2T_s$ and $1/2T_s$. In the absence of echoes, the output of a radar receiver can be described as white noise. Sources of white noise are thermally induced random motions of electrons in the rf amplifier and the mixer of the receiver, cosmic and atmospheric radiation, and quantization in the analog-to-digital converters. Quantization noise is white only if the quantization error (i.e., the difference between the quantized and actual values) is uncorrelated from sample to sample. In addition to being white, the in-phase and quadrature components of thermal noise and radiation have Gaussian amplitude distributions (very much like weather echoes). That is, the I and Q values are normally distributed about zero, and the variance of the distribution is equal to the noise power N (see Section 4.2). This is not the case for quantization noise, which has values uniformly distributed over the range equal to the quantization interval. Thus white noise, which relates to the uniformity of the power as a function of f, and the Gaussian amplitude distribution of Is and Qs are independent characteristics of the signal.

We shall examine two power spectrum estimates and prove their equivalence. First, one can compute the spectrum estimate using the definition (5.18) and the estimate $\hat{R}(l)$;

$$\hat{S}_1(f) = T_s \sum_{l=-(M-1)}^{M-1} \hat{R}(l) e^{-j2\pi f T_s l}. \tag{5.21}$$

Second, it is often computationally more efficient to use the FFT to compute $\hat{S}_2(f)$ directly from the data:

$$\hat{S}_2(f) \equiv |Z(f)|^2 T_s/M, \tag{5.22}$$

or, using (5.1),

$$\hat{S}_2(f) = \frac{T_s}{M} \left[\sum_{m=0}^{M-1} V^*(m) e^{j2\pi f T_s m} \sum_{n=0}^{M-1} V(n) e^{-j2\pi f T_s n} \right]$$

$$= \frac{T_s}{M} \left\{ \sum_{m,n}^{M-1} V^*(m) V(n) e^{-j2\pi f T_s l} \right\}, \tag{5.23}$$

where $l = n - m$ and the double sum has been abbreviated. The double sum is really an addition of the product terms in an $M \times M$ matrix (Fig. 5.6). The

5.1 Spectral Analysis of Weather Signals

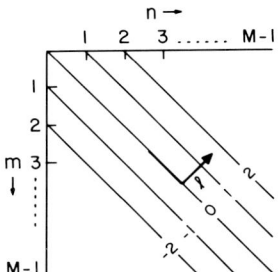

Fig. 5.6. Matrix showing how a double sum can be reduced to a single sum of the correlation estimates $\hat{R}(l)$.

diagonal is a sum of M samples of $\hat{R}(0)$, and the terms along the lines on either side of the diagonal are the $M - |l|$ correlation estimates at different lags l [multiplied by $\exp(-j2\pi f T_s l)$], as shown in Fig. 5.6. Thus we can evaluate (5.23) by summing along l, the various estimates $\hat{R}(l)$. This is equivalent to

$$\sum_{m,n}^{M-1} V^*(m)V(n)e^{-j2\pi f T_s l} = \sum_{l=-(M-1)}^{M-1} e^{-j2\pi f T_s l} \sum_{m=0}^{M-|l|-1} V^*(m)V(m+l). \quad (5.24a)$$

The inner sum on the right side of (5.24a) is a sum along lines parallel to the diagonal and is, from (5.17), equal to $M\hat{R}(l)$. Therefore,

$$\frac{T_s}{M} \sum_{m,n}^{M-1} V^*(m)V(n)e^{-j2\pi f T_s l} = T_s \sum_{l=-(M-1)}^{M-1} \hat{R}(l)e^{-j2\pi f T_s l}, \quad (5.24b)$$

and hence

$$\hat{S}_2(f) = \hat{S}_1(f), \quad (5.25)$$

which completes the proof. The estimate $\hat{S}_2(f)$ has the descriptive name *periodogram* because it was used by geophysicists to find periodicities in data.

5.1.4 Bias, Variance, and the Window Effect

Estimation theory has established several attributes of estimators. Two important ones are the *bias* and the *variance*. An estimate is said to be unbiased if its mean equals the true mean of the parameter that is estimated. To test if $\hat{S}(f)$ is biased, we examine the ensemble average of (5.23). Before doing this it is convenient to introduce an artifice: We represent $V_M(m)$, the finite segment of our complex sequence, by the product

$$V_M(m) = V(m)d(m), \quad (5.26)$$

where

$$d(m) = \begin{cases} 1 & \text{for } 0 \le m \le M-1, \\ 0 & \text{otherwise.} \end{cases} \quad (5.27)$$

The weighting sequence $d(m)$ in (5.26) is referred to as the *data window* because it reveals only a finite portion (of length M) of the otherwise infinite sequence $V(m)$, i.e., an observer sees through the window $d(m)$ the truncated series $V_M(m)$.

With $V_M(m)$ expressed by (5.26), the expectation of (5.23) is

$$E[\hat{S}(f)] = \frac{T_s}{M} \sum_{l=-(M-1)}^{M-1} \sum_{m=0}^{M-1-|l|} d^*(m)d(m+l) E[V^*(m)V(m+l)] e^{-j2\pi f T_s l}$$

$$= T_s \sum_{l=-(M-1)}^{M-1} R(l) e^{-j2\pi f T_s l} \sum_{m=0}^{M-1-|l|} \frac{d^*(m)d(m+l)}{M}$$

$$= T_s \sum_{l=-(M-1)}^{M-1} w(l) R(l) e^{-j2\pi f T_s l}. \tag{5.28}$$

To arrive at (5.28) we have assumed $V(m)$ to be stationary, which means that $E[V(m)V^*(m+l)]$ is independent of m. Therefore only the data window product is summed over m. The sum of data window products is a correlation of the data window samples and is called the *lag window* $w(l)$, because it multiplies the true autocorrelation $R(l)$. From (5.28) the lag window $w(l)$ is

$$w(l) = \frac{1}{M} \sum_{m=0}^{M-1-|l|} d^*(m)d(m+l). \tag{5.29}$$

In arriving at (5.28) and (5.29) we did not use the fact that the data window (5.27) was rectangular. Therefore (5.29) is quite general and valid for any window type. A rectangular data window produces a triangular (or Bartlett) lag window, given by

$$w(l) = \begin{cases} 1 - \dfrac{|l|}{M}, & -M \le l \le M, \\ 0 & \text{otherwise}, \end{cases} \tag{5.30}$$

which when introduced into (5.28) results in the following mean value of the spectrum estimate:

$$E[\hat{S}(f)] = T_s \sum_{-(M-1)}^{M-1} \left(1 - \frac{|l|}{M}\right) R(l) e^{-j2\pi f T_s l}. \tag{5.31}$$

Note that (5.31) equals the true spectrum (5.18) only if $M \to \infty$. Therefore the periodogram is a biased estimate of the true power spectrum.

To summarize, we have shown that a rectangular data window is equivalent to the triangular lag window. If a periodogram is obtained from a uniformly weighted time series, the result, in the mean, is the same as if the autocorrelation were weighted with a triangular window prior to the trans-

5.1 Spectral Analysis of Weather Signals

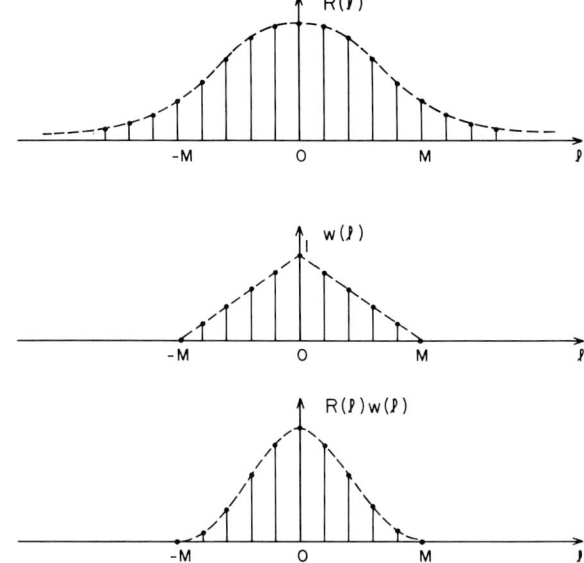

Fig. 5.7. True $R(l)$ and the windowed autocorrelation $R(l)w(l)$ functions when a rectangular data window is used to compute the periodogram.

form (5.31). This effect is illustrated in Fig. 5.7, where one can see that the largest deviation from the true autocorrelation occurs at larger values of lag l.

5.1.5 Expressing Spectral Estimates in Terms of the True Spectrum

It is not convenient to compute the bias from (5.28). One would like to have the $E[\hat{S}(f)]$ expressed in terms of the true $S(f)$. We shall prove that

$$E[\hat{S}(f)] = T_s \sum_{l=-\infty}^{\infty} w(l)R(l)e^{-j2\pi f T_s l} = \int_{-(2T_s)^{-1}}^{(2T_s)^{-1}} S(f')W(f - f')\,df', \quad (5.32)$$

where the spectral window $W(f)$ and the lag window $w(l)$ are Fourier transform pairs like (5.18) and (5.19):

$$W(f) = T_s \sum_{l=-(M-1)}^{M-1} w(l)e^{-j2\pi f T_s l}, \quad (5.33)$$

$$w(l) = \int_{-(2T_s)^{-1}}^{(2T_s)^{-1}} W(f)e^{j2\pi f T_s l}\,df. \quad (5.34)$$

We now apply the inverse Fourier transform (5.19) to both sides of (5.32). Because the left side is a Fourier transform, its inverse simply retrieves the product

$$w(l)R(l) = \int_{-(2T_s)^{-1}}^{(2T_s)^{-1}} \int_{-(2T_s)^{-1}}^{(2T_s)^{-1}} S(f')W(f - f')e^{j2\pi f T_s l}\, df'\, df. \quad (5.35)$$

After the change of variable $f - f' = \varepsilon$, and because $W(f)$ has a period T_s^{-1}, we obtain

$$w(l)R(l) = \int_{-(2T_s)^{-1}}^{(2T_s)^{-1}} W(\varepsilon)e^{j2\pi\varepsilon T_s l}\, d\varepsilon \int_{-(2T_s)^{-1}}^{(2T_s)^{-1}} S(f')e^{j2\pi f' T_s l}\, df', \quad (5.36)$$

which is an identity. Therefore, (5.32) reveals that the expected value of a periodogram is a circular convolution of a true spectrum $S(f')$ with the spectral window $W(f')$. Because it is an integration, the convolution broadens the true spectrum and produces spillage of the sharp spectral peak into the side lobes (Fig. 5.8). Because the window is periodic, the convolution operation is the same as a moving average on a circle. The lower curve in Fig. 5.8 is the expected value of the estimated power spectrum; a value of $E[\hat{S}(f_k)]$ is obtained by multiplying the upper two curves point by point, integrating the result, and repeating these steps for various values of f_k. The convolution is circular; as the window peak moves to the right, the sidelobes from its periodic extension appear in the interval $\pm 1/2T_s$. It is best to visualize the points $-1/2T_s$ and $1/2T_s$ as being tied together with all points on a circle.

Because the lag window is a correlation of a data window sample (5.29), it follows that the amplitude spectral window $D(f)$ can be computed from the Fourier transform of the data window.

The various relations and transforms between window weights and its spectra are summarized in Table 5.1 and schematically depicted in Fig. 5.9.

As an example, consider a sequence of uniform unit weights. The magnitude of its transform $|D(k)|$ is equal to Eq. (5.8) with $\alpha = 0$:

$$|D(f)| = \left|\frac{\sin(\pi T_s M f)}{\sin(\pi T_s f)}\right|, \quad (5.37)$$

where $f = kf_0$ was substituted into (5.8).

Figure 5.10 illustrates a weather signal weighted with a uniform window, i.e., unweighted samples, and one with a von Hann (raised cosine) window (Table 5.2), showing a considerable difference in the spectral domain, especially in spectral skirts. In the periodogram obtained with uniform weight applied to data samples (Fig. 5.10 RECT) we notice the slower decay of spectral coefficients in the skirts that have a shape like (5.37). Since the

5.1 Spectral Analysis of Weather Signals

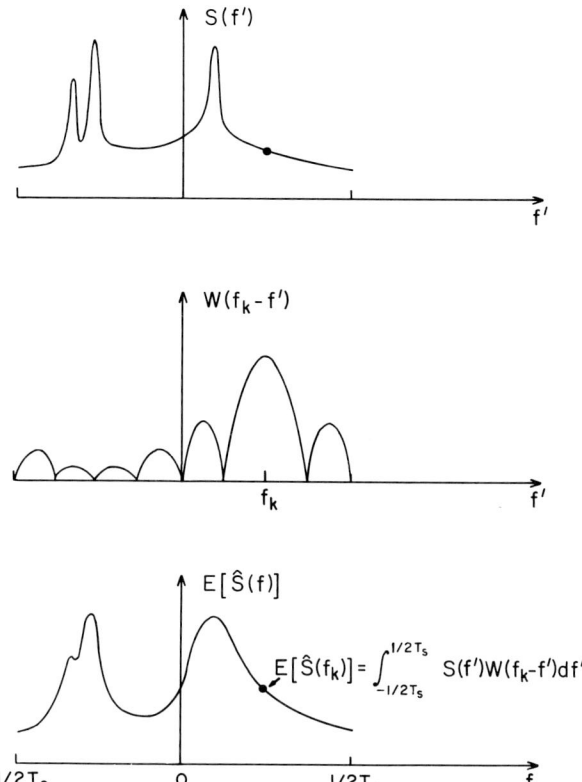

Fig. 5.8. Spectral window effect. The upper plot is a true spectrum $S(f')$. The middle is a spectral window centered at f_k, whereas the lower graph is the smeared spectrum estimate.

TABLE 5.1 Relation between Window Weights and Their Spectra[a]

Amplitude spectral window:	$D(f) = \sum_{m=0}^{M-1} d(m)e^{-j2\pi fT_s m}$
Data weights:	$d(m) = \dfrac{1}{M}\sum_{m=0}^{M-1} D(f)e^{j2\pi fT_s m}$
Lag window:	$w(l) = \dfrac{1}{M}\sum_{-(M-1)}^{M-1} d^*(m)d(m+l) = \int_{-(2T_s)^{-1}}^{(2T_s)^{-1}} W(f)e^{j2\pi fT_s l}\,df$
Power spectral window:	$W(f) = |D(f)|^2 T_s/M = T_s \sum_{l=-(M-1)}^{M+1} w(l)e^{-j2\pi fT_s l}$

[a] The data window is a sequence of weights $d(mT_s)$ spaced T_s apart.

78   **5 Doppler Spectra of Weather Echoes**

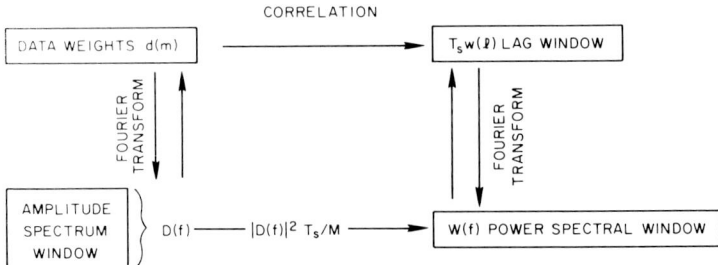

Fig. 5.9. Various relations between windows.

von Hann window has a gradual transition between no data and data points, its spectral window has a less concentrated main lobe and significantly lower side lobes. The resulting spectrum retains these properties and enables one to observe the spectra of signals as weak as 40 dB below the main peak. This is very significant when one is trying to estimate the peak winds of

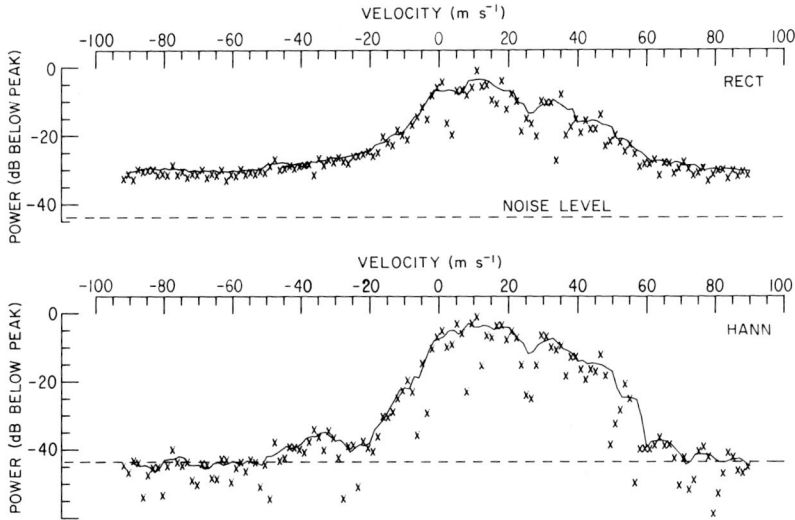

Fig. 5.10. Power spectra of weather echoes, showing statistical fluctuations in the spectral estimates (denoted by ×s). RECT signifies the spectra of unweighted echo samples, whereas HANN signifies samples weighted by a von Hann window. Total number of points is 128. Solid curves are five-point running averages of the spectral power. This spectrum is from a small tornado that touched down on 20 May 1977 at 18:53:50 in Del City, Oklahoma, at about 35 km from the radar. Azimuth: 6.1°; elevation: 3.1°; altitude 1.952 km; gate 03; SNR = 31 dB.

5.1 Spectral Analysis of Weather Signals

TABLE 5.2 Spectra of Several Data Windows

Data window type	Weighting function[a]	Amplitude spectrum window
Rectangular	$d(m) = 1, \quad 0 \leq m \leq M - 1$	$\|D(f)\| = \left\|\dfrac{\sin(\pi T_s M f)}{\sin(\pi T_s f)}\right\|$
Triangular	$d(m) = \begin{cases} \dfrac{2m+1}{M}, & 0 \leq m \leq \dfrac{M-1}{2} \\ 2 - \dfrac{2m+1}{M}, & \dfrac{M-1}{2} \leq m \leq M-1 \end{cases}$	$\|D(f)\| = \dfrac{\sin^2(\pi T_s M f)}{\sin^2(\pi T_s f)}$
von Hann	$d(m) = \dfrac{1}{2}\left\{1 + \cos\left[\left(m - \dfrac{M-1}{2}\right)\dfrac{2\pi}{M}\right]\right\}$	
Hamming	$d(m) = 0.54 + 0.46\cos\left[\left(m - \dfrac{M-1}{2}\right)\dfrac{2\pi}{M}\right]$	$\|D(f)\| = \dfrac{\sin(\pi T_s M f)}{\sin(\pi T_s f)}$ $\times \left[a - b\dfrac{\sin^2(\pi T_s M f)\cos(\pi/M)}{\sin^2(\pi T_s f) - \sin^2(\pi/M)}\right]^b$

[a] Sometimes the listed weighting functions are used as lag window coefficients $w(l)$.
[b] For the von Hann window $a = b = 0.5$, and for the Hamming window $a = 0.54$, $b = 0.46$.

tornados or other severe weather within the resolution volume; the power in spectral "skirts" due to high velocities is rather weak and would be masked by the strong spectral peaks seen through the window side lobes unless a suitable window were applied. The apparent lack of randomness in the coefficients in the spectral skirts for the rectangularly weighted data is due to the larger correlation between coefficients. This correlation is attributed to the strong spectral powers seen through the nearly constant level-window sidelobes [Zrnic, 1980].

The example shown in Fig. 5.10 is for a tornado within the resolution volume. The rectangular window power spectrum $|D(f)|^2 T_s/M$ [see (5.37)] is readily apparent at negative velocities. The dynamic range [i.e., the decibel difference between the main lobe and side lobe value of $|D(f)|^2$] for the most distant side lobe is about 30 dB. This is in contrast to the 45-dB dynamic range with the von Hann window, which better defines the true spectrum and the maximum velocity (60 m s^{-1}). The receiver noise level is indicated by the dashed lines in Fig. 5.10. It is apparent that this level is masked by the powers seen through the side lobes of the rectangular window.

Much effort was spent by investigators in devising window functions. Two that are simple to use are the von Hann data window and the Hamming window. Table 5.2 lists four window types and their transforms. Some general rules governing window use are the following:

(1) Windows with a smooth transition between zero and unit weight have spectra $W(f)$ with lower side lobes.

(2) The window effect is negligible when the window length is large compared to the lag that is required to decorrelate the data.

(3) It is not possible to reduce spectral window side lobes without increasing the width of the main lobe.

(4) Data windows usually have even symmetry; that is,

$$d(M - 1 - m) = d(m) \quad \text{for} \quad m = 0, 1, \ldots, M - 1.$$

5.1.6 Variance of the Periodogram

The probability distribution of the in-phase and quadrature components and the window type determine the variance of the periodogram. We consider in-phase and quadrature components that are zero-mean Gaussian distributed (weather echoes) and for which the fourth moment of the complex signal is (Reed, 1962)

$$E[V^*(m)V(n)V^*(k)V(l)] = E[V^*(m)V(n)]E[V^*(k)V(l)]$$
$$+ E[V^*(m)V(l)]E[V(n)V^*(k)]. \quad (5.38)$$

For this condition it can be shown (Zrnić, 1980) that the variance of a periodogram is exactly given† by

$$\text{var}[\hat{S}(f)] = E^2[\hat{S}(f)]. \quad (5.39)$$

Moreover, power spectrum coefficients at any frequency f are exponentially distributed regardless of the spectral shape, window type, or number of points in the transform. The last conclusion is bothersome because it means that the estimate does not improve with longer periodograms; only the resolution increases. It is natural to expect that the estimate could be improved by increasing the number of data points. Two roughly equivalent methods that reduce the variance are

(1) averaging short periodograms and
(2) making weighted running averages on a long periodogram.

Although no comprehensive treatment of the digital processing of complex signals is available, several books deal extensively with the processing

---

† Analogous results for a real Gaussian signal hold only in the limit of a large number of samples.

of real signals (e.g., Tretter, 1976). Many of the results valid for real signal carry over to complex signals; however, there are differences, some of which we have highlighted in the first part of this chapter.

We have emphasized the periodogram approach to power spectral estimation, which is especially suitable for signals that have an unknown spectral shape. However, the reader is cautioned that optimum spectral estimation is a complex problem and depends heavily on the type of signals and the desired results. For instance, when spectra have a shape (e.g., Gaussian; see Section 5.2) that can be described with a few parameters, it is advantageous to fit such shapes either to the periodogram (Waldteufel, 1976) or to the autocorrelation function (Sato and Woodman, 1982). Spectra that are well modeled as outputs of linear filters driven by white noise can be estimated from the filter coefficients (Cadzow, 1982). Models based on parametric fitting allow extensions of the estimated autocorrelation function beyond the limits imposed by the finite-time window. However, the amount of extractable information is not increased.

5.2 DOPPLER WEATHER SPECTRUM AND ITS RELATION TO REFLECTIVITY AND RADIAL VELOCITY FIELDS

The Doppler spectrum is a power-weighted distribution of the radial velocities of the scatterers that mostly lie within the resolution volume. The power weight depends not only on the reflectivity of scatterers but also on the weighting given to scatterers by the antenna pattern, the transmitted pulse width, and the receiver filter.

To begin we consider scatterers that produce a radial velocity field† $v(\mathbf{r}_1)$ and a reflectivity field $\eta(\mathbf{r}_1)$. Let the center of the resolution volume V_6 be at a location $\mathbf{r}_0$ (Fig. 5.11), with the following corresponding illumination function:

$$I(\mathbf{r}_0, \mathbf{r}_1) = C_1 f^4(\theta - \theta_0, \phi - \phi_0) W^2(r_0, r_1)/r_1^4 \qquad (5.40)$$

due to the antenna pattern $f(\theta, \phi)$ and the range weighting $W(r)$ functions (Chapters 3 and 4) where C_1 is a constant that can be obtained from the weather radar equation (4.11).

---

† Here and in subsequent chapters we shall wherever possible omit the subscript r and the adjective *radial*, letting v always designate the radial velocity except in those sections where it is needed to denote a Cartesian component of the wind. (The subscript r will be used for the radial Doppler velocity when necessary to avoid confusion.)

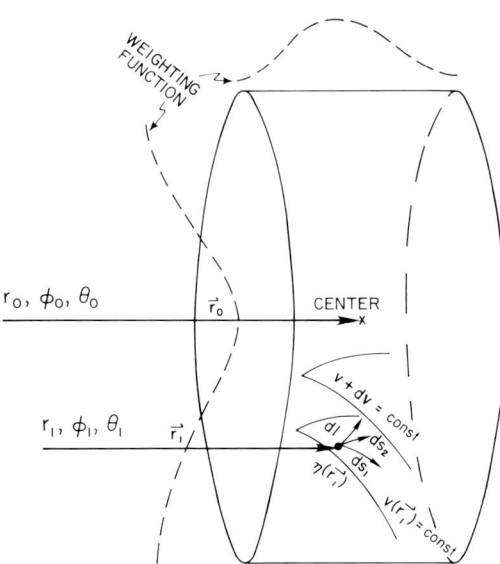

Fig. 5.11. Parameters and geometry that contribute to the weather signal power spectrum; $\eta(\mathbf{r}_1)$ is the reflectivity and $v(\mathbf{r}_1)$ the radial velocity field; $\mathbf{r}_0$ is the center of resolution volume. The weighting functions in angle and range are indicated by dashed lines.

We locate a surface of constant velocity, $v(\mathbf{r}_1) = $ const and seek the total power contribution from scatterers in the velocity range v to $v + dv$. This contribution will obviously be a sum of powers from the volume between the two surfaces v and $v + dv$. It is convenient to choose for the elemental volume the product $ds_1\, ds_2\, dl$, where ds_1 and ds_2 are two orthogonal arc-lengths, at a point $\mathbf{r}_1$ tangent to $v(\mathbf{r}_1) = $ const (Fig. 5.11). The third coordinate dl is perpendicular to the surface of constant v:

$$dl = |\mathbf{grad}\ v(\mathbf{r}_1)|^{-1}\, dv. \tag{5.41}$$

The elemental volume contributes an increment of average power in the velocity interval $v, v + dv$ equal to

$$d\bar{P}(v) = \eta(\mathbf{r}_1)I(\mathbf{r}_0, \mathbf{r}_1)|\mathbf{grad}\ v(\mathbf{r}_1)|^{-1}\, ds_1\, ds_2\, dv. \tag{5.42}$$

Finally, the integral over the surface A of constant v gives the total power in the velocity range $v, v + dv$ and is, by definition, the product of power spectrum density and dv. This is,

$$\bar{P}(\mathbf{r}_0, v) = \bar{S}(\mathbf{r}_0, v)\, dv = \left[\iint_A \eta(\mathbf{r}_1)I(\mathbf{r}_0, \mathbf{r}_1)|\mathbf{grad}\ v(\mathbf{r}_1)|^{-1}\, ds_1\, ds_2\right] dv. \tag{5.43}$$

The overbar in (5.43) denotes the mean power spectrum density.

5.2 Doppler Weather Spectrum

Equation (5.43) is fundamental and worthy of further discussion. First, the area A consists of the *union* of all *isodop surfaces* (surfaces of constant Doppler velocity) on which the radial velocity is a constant v. At each point $\mathbf{r}_1$ on the surface the reflectivity is multiplied by the corresponding weighting function. The gradient term adjusts the isodops' contribution according to their density. (That is, the closer the isodop surfaces are spaced, the smaller is the weight applied to the spectral components in the velocity interval between two isodops.)

New targets are constantly replacing old ones in the elemental volume between the isodop surfaces v, $v + dv$ within and around V_6, where the weighting function is significant, and therefore $S(\mathbf{r}_0, v)$ changes randomly. It has an exponential distribution if there are many targets in this elemental volume, because then the central limit theorem applies to the sum of the voltages contributed by the targets (i.e., the voltages I and Q have Gaussian distribution; see Chapter 4). On observing stationary weather phenomena and averaging the statistically varying periodograms, one obtains, in the limit of large averages, the spectrum given by (5.43). However, this does not imply that a Doppler spectrum uniquely specifies the velocity and reflectivity distribution in a resolution volume. On the contrary, a variety of reflectivity–velocity combinations may yield identical Dopper spectra.

To calculate the mean velocity and the spectrum width, the normalized $\bar{S}_n(\mathbf{r}_0, v)$ version of (5.43) is used:

$$\bar{S}_n(\mathbf{r}_0, v) = \bar{S}(\mathbf{r}_0, v) \bigg/ \int_{-\infty}^{\infty} \bar{S}(\mathbf{r}_0, v)\, dv. \tag{5.44}$$

The normalized power density can be considered a probability distribution of radial velocities. Note that the integral in the denominator is the total power and can be obtained from the volume integral of (5.42):

$$\bar{P}(\mathbf{r}_0) = \iiint \eta(\mathbf{r}_1) I(\mathbf{r}_0, \mathbf{r}_1)\, dV_1, \tag{5.45}$$

where $dV_1 = ds_1\, ds_2\, dl$.

The resolution-volume-averaged reflectivity $\bar{\eta}(\mathbf{r}_0)$ can be obtained by dividing (5.45) by the volume integral of the illumination function $I(\mathbf{r}_0, \mathbf{r}_1)$:

$$\bar{\eta}(\mathbf{r}_0) = \iiint \eta(\mathbf{r}_1) I_n(\mathbf{r}_0, \mathbf{r}_1)\, dV_1, \tag{5.46a}$$

where the normalized illumination function is defined as

$$I_n(\mathbf{r}_0, \mathbf{r}_1) = I(\mathbf{r}_0, \mathbf{r}_1) \bigg/ \iiint I(\mathbf{r}_0, \mathbf{r}_1)\, dV_1. \tag{5.46b}$$

The differential volume dV_1 no longer needs to be tied to the coordinates s_1, s_2, and l. Hence, dV_1 can now be $r_1^2\, dr_1 \sin\theta\, d\theta\, d\phi$, for example. Note that $\bar{\eta}(\mathbf{r}_0)$ is the illumination-weighted reflectivity, where weights are determined principally by $W^2(r_0 - r_1)$ and $f^4(\theta, \phi)$ within V_6. The integral value in the denominator of (5.46) can be found from Eqs. (4.13)–(4.21) for weather radar parameters usually met in practice.

We now turn to the mean Doppler velocity, defined as

$$\bar{v}(\mathbf{r}_0) = \int_{-\infty}^{\infty} v \bar{S}_n(\mathbf{r}_0, v)\, dv, \tag{5.47}$$

which is a combination of reflectivity (power) *and* the illumination-function-weighted velocity and could be quite different from the $I(\mathbf{r}_0, \mathbf{r}_1)$-weighted velocity. The relationship between the point velocities $v(\mathbf{r}_1)$ and the power weighted moment $v(\mathbf{r}_0)$ is obtained by substituting (5.41), (5.43), and (5.44) into (5.47):

$$\bar{v}(\mathbf{r}_0) = \iiint v(\mathbf{r}_1)\eta(\mathbf{r}_1)I(\mathbf{r}_0, \mathbf{r}_1)\, dV_1 \bigg/ \iiint \eta(\mathbf{r}_1)I(\mathbf{r}_0, \mathbf{r}_1)\, dV_1. \tag{5.48}$$

Unlike the averaged reflectivity, this is the average of point velocities weighted by both reflectivity and the illumination function. In the special case of uniform reflectivity, the $I(\mathbf{r}_0, \mathbf{r}_1)$-weighted mean velocity $\bar{v}_I(\mathbf{r}_0)$ equals the first moment of $\bar{S}_n(v)$, where

$$\bar{v}_I(\mathbf{r}_0) = \iiint v(\mathbf{r}_1) I_n(\mathbf{r}_0, \mathbf{r}_1)\, dV_1. \tag{5.49}$$

Like the reflectivity, this is the average of point velocities weighted by the normalized illumination function.

The velocity spectrum width $\sigma_v(\mathbf{r}_0)$ is obtained from

$$\sigma_v^2(\mathbf{r}_0) = \int_{-\infty}^{\infty} [v - \bar{v}(\mathbf{r}_0)]^2 \bar{S}_n(\mathbf{r}_0, v)\, dv. \tag{5.50}$$

Similarly, the width (5.50) reduces to

$$\sigma_v^2(\mathbf{r}_0) = \iiint v^2(\mathbf{r}_1)\eta(\mathbf{r}_1)I(\mathbf{r}_0, \mathbf{r}_1)\, dV_1 \bigg/ \iiint \eta(\mathbf{r}_1)I(\mathbf{r}_0, \mathbf{r}_1)\, dV_1 - \bar{v}^2(\mathbf{r}_0) \tag{5.51}$$

and corresponds to a weighted deviation of velocities from the averaged velocity.

The mean velocity (5.48) depends on the distribution of the scatterers' cross section within the resolution volume where the weighting functions are significant. Thus $\bar{v}(\mathbf{r}_0)$ cannot in general be equated to a spatial mean velocity. However, if reflectivity and illumination are symmetrical about the resolution

5.2 Doppler Weather Spectrum

volume center, and if radial wind changes linearly across V_6, then $\bar{v}(\mathbf{r}_0)$ is the true spatial average of the radial wind component.

A simplification of (5.43) occurs when the radial velocity field and the reflectivity are height invariant. Then the power spectrum reduces to

$$\bar{S}(v, \mathbf{r}_0) = \int \left[\int \eta(x_1, y_1) I(\mathbf{r}_0, x_1, y_1, z_1) |\text{grad } v(x_1, y_1)|^{-1} \, ds \right] dz_1. \quad (5.52)$$

Assuming that I is product separable in the variables x_1, y_1, z_1, the inner integral is two dimensional and sums the contributions along the line $s = s(x_1, y_1)$ on which $v(x_1, y_1)$ is constant. Now, $ds = (dx_1^2 + dy_1^2)^{1/2}$, so the integration is a surface integral with area element $ds \, dz_1$. Both (x_1, y_1) and $\text{grad } v(x_1, y_1)$ are independent of z, but the illumination function may not be. The formulation in (5.52) is useful in simulating the spectra of tornado vortices, because in this case both velocity and reflectivity are nearly independent of height, at least over the resolution volume. At each point x_1, y_1 along a strip of constant velocity, the reflectivity is multiplied by the corresponding weighting function. To account for the contributions of other infinitesimal strips within the resolution volume, integration is performed along the third (z-axis) dimension. Equation (5.52) was used to compute the spectra of model tornados and mesocyclones. These compare well with actual measurements (see Chapter 9).

It can be shown with the help of (5.52) that when wind shear and η are constant across the resolution volume, the power spectrum follows the weighting function shape. A simple example illustrates these statements.

Let us obtain the power spectrum of a radial wind with uniform shear k_ϕ in a direction perpendicular to the beam axis when scatterers produce uniform reflectivity η (Fig. 5.12). For simplicity we consider only the inner integral along isodops (a two-dimensional problem). Furthermore, we assume the beamwidth is sufficiently narrow that Cartesian distances approximate well the azimuthal arcs over the resolution volume V_6 (i.e., $x_1 = r_0 \phi$). Therefore, the pattern weight $f^4(\phi)$ in terms of x_1 [see Eq. (C.18)] is

$$f^4(x_1) = \exp\{-[4 \ln 4/(r_0 \theta_1)^2] x_1^2\}, \quad (5.53)$$

where θ_1 is the one-way 3-dB beamwidth. The rectangle (Fig. 5.12a) represents the region in space where range and antenna pattern weights are significant; this is the resolution volume. Clearly here $|\text{grad } v| = |k_\phi|$. Applying (5.52), we have

$$\bar{S}(v, \mathbf{r}_0) = (\eta/k_\phi) \, dz_1 \int I(\mathbf{r}_0, x_1, r_1) \, ds. \quad (5.54)$$

The integration along a contour of constant v constrains x_1. In general, both

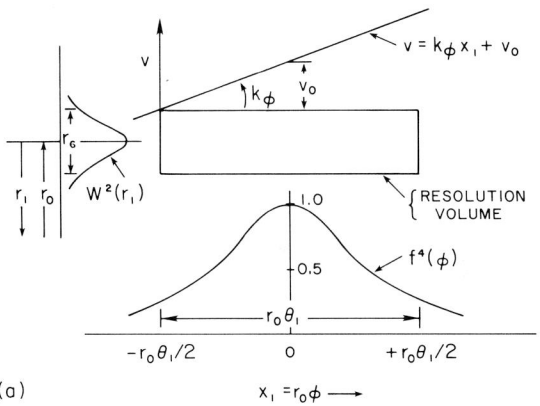

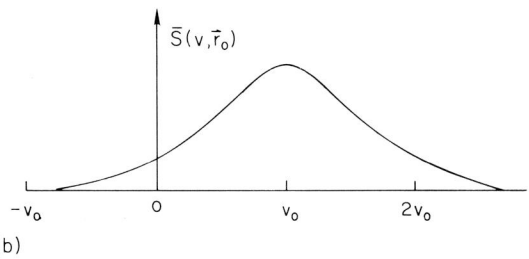

Fig. 5.12. Resolution volume within which wind shear and reflectivity are assumed to be uniform. The beamwidth (3 dB) at range r_0 is $r_0\theta_1$, and $f^4(\phi)$ is the two-way antenna pattern function. (b) The power spectrum $S(v,\mathbf{r}_0)$ in this example follows the shape of the weighting function.

x_1 and r_1 are linked by $v = \text{const}$. Thus, in terms of $v = \text{const}$, $ds = dr_1$ and

$$x_1 = (v - v_0)/k_\phi. \tag{5.55}$$

Therefore, from (5.40), the illumination function is

$$I(\mathbf{r}_0, \mathbf{r}_1, v) = C_1 f^4\left(\frac{v - v_0}{k_\phi}\right) W^2(\mathbf{r}_0, \mathbf{r}_1)\bigg/r_0^4. \tag{5.56}$$

Substituting this into (5.54), we obtain

$$\bar{S}(v, \mathbf{r}_0) = \frac{\eta \, dz_1 \, C_1}{r_0^4 k_\phi} \exp\left\{-\left[\frac{4 \ln 4}{(r_0\theta_1)^2}\left(\frac{v - v_0}{k_\phi}\right)^2\right]\right\} \int W^2(\mathbf{r}_0, \mathbf{r}_1) \, dr_1. \tag{5.57}$$

Thus, the power spectrum (Fig. 5.12b) follows the antenna pattern shape in this not very rare situation of uniform shear. Because the Gaussian shape well approximates the range and angular weighting patterns, we may infer that when Doppler spectra are Gaussian the reflectivity and radial velocity shear are somewhat uniform within V_6. However, another reason is found in other several independent mechanisms that modify the spectrum, such as the window effect, turbulence, or antenna motion. How the final spectral shape is determined by these mechanisms can be deduced by considering each separately assuming each to be independent of the others. The normalized power density $\bar{S}_n(v, \mathbf{r}_0)$ can be considered to be a probability distribution of velocities. Thus the shear in Fig. 5.12a produces a density of velocities centered about the mean v_0, (Fig. 5.12b). Let us now ignore shear and consider that advection replenishes targets within V_6. The drift of targets through V_6 causes signal decorrelation and spectral broadening about v_0, the extent of which depend on v_0 and the size of V_6. Small-scale turbulence shuffles target positions, and the turbulent velocity will produce a normalized power density centered about zero, with a shape determined by the reflectivity-weighted turbulent velocity distribution within the resolution volume. Again, this normalized power density can be assumed to be the probability density of velocities.

Consider two density functions, one, $S_{n1}(v)$, centered about v_0 and due to shear and the other, $S_{n2}(v)$, centered about zero and due to turbulence. The resultant Doppler velocity shift is caused by both these mechanisms. A theorem in probability states that the probability density for the sum of two independent random variables is the convolution of the probability densities of each variable (Papoulis, 1965, p. 189). Thus the normalized power spectrum $S_n(\bar{v})$ is

$$S_n(\bar{v}) = \int S_{n1}(v) S_{n2}(\bar{v} - v)\, dv. \tag{5.58}$$

The central limit theorem of probability states that, under certain general conditions, the convolution of a number m of probability densities approaches a normal curve as m increases, independent of the distribution of each density (Papoulis, 1965, p. 267). In fact, it is easy to show that even for small m (e.g., $m = 3$) the resultant density is well approximated by a Gaussian function (Papoulis, 1965, p. 268). Because several mechanisms contribute to real and apparent Doppler velocity shifts (scanning antenna beams cause apparent Doppler shifts through the replenishment of targets within V_6), it is no wonder that actual power spectra often appear to be Gaussian in shape.

5.3 VELOCITY SPECTRUM WIDTH, SHEAR, AND TURBULENCE

The velocity spectrum width (i.e., the square root of the second spectral moment about the mean velocity) is a function both of radar system parameters such as beamwidth, bandwidth, and pulse width and the meteorological parameters that describe the distribution of hydrometeor density and velocity within the resolution volume. Relative radial motion between targets broadens the spectrum. For example, turbulence produces random relative radial motion of drops. Wind shear can cause relative radial target motion, as can differences in the speed of fall of various size drops. There is also a contribution to spectrum width caused by the beam's sweeping through space (i.e., the radar does not receive echoes from identically weighted targets on successive samples). This change in the location of the resolution volume V_6 from pulse-to-pulse results in a decorrelation of echo samples and a consequent increase in the spectrum width σ_v (see Appendix C). The echo samples will be uncorrelated more quickly (independent of particle motion inside V_6) the faster the antenna is rotated. Thus spectrum width increases in proportion to the antenna angular velocity.

Because the cited spectral broadening mechanisms are independent of one another, the square of the velocity spectrum width σ_v^2 can be considered as a *sum* of the variances contributed by each. That is,

$$\sigma_v^2 = \sigma_s^2 + \sigma_\alpha^2 + \sigma_d^2 + \sigma_t^2, \tag{5.59}$$

where σ_s^2 is due to shear, σ_α^2 to antenna motion, σ_d^2 to different speeds of fall for different sized drops, and σ_t^2 to turbulence. Because broadening caused by the mean relative motion between the radar and targets is small except in the case of an airborne radar, σ_α will only refer to antenna rotation. We ignore the window effect because it is related to the method of spectral moment analysis and furthermore does not contribute to echo sample decorrelation as do the mechanisms considered in (5.59). These mechanisms are related to the meteorological status of the resolution volume or, for the case of σ_α^2, the rotation rate. Methods other than DFT spectral width estimation—for example, the covariance method (see Chapter 6)—would produce different biases. The significance of the width σ_v for weather radar design is discussed in Chapter 7.

The spectrum second moment σ_d^2, due to the different radial components of fall speed of the assorted-size drops, is related to the radar and meteorological parameters:

$$\sigma_d^2 = (\sigma_{d0} \sin \theta_e)^2. \tag{5.60}$$

The width σ_{d0} is caused by the spread in terminal velocity of various size drops falling relative to the air contained in V_6. Lhermitte (1963) has shown that for rain $\sigma_{d0} \simeq 1.0$ m s^{-1} and is nearly independent of the drop size distribution. The elevation angle $\theta_e = (\pi/2) - \theta_0$ is measured with respect to the beam center.

If the antenna pattern is Gaussian with a one-way half-power pattern width θ_1 and rotates at an angular velocity of α the spectrum width due

5.3 Velocity Spectrum Width, Shear, and Turbulence

solely to resolution volume displacement is (Appendix C)

$$\sigma_\alpha^2 = (\alpha\lambda \cos \theta_e/2\pi\theta_1)^2 \ln 2. \tag{5.61}$$

We shall prove that the wind shear width term σ_s is composed of three contributions:

$$\sigma_s^2 = \sigma_{s\theta}^2 + \sigma_{s\phi}^2 + \sigma_{sr}^2, \tag{5.62}$$

where the terms are due to radial velocity shear along the elevation, azimuth, and radial directions, respectively. The assumptions behind (5.62) are that shear is constant within the resolution volume and that the weighting function is product separable along the θ, ϕ, and r directions.

If the wind is linear about v_0, the speed v can be expressed

$$v - v_0 = k_x x + k_y y + k_z z, \tag{5.63}$$

where the ks are the components of shear along the various axes. Let us orient the coordinate system so that y is in the elevation direction, x along the azimuth, and z parallel to the antenna axis; these coordinates are orthogonal components of $\mathbf{r}_1 - \mathbf{r}_0$ with the origin at the tip of $\mathbf{r}_0$. From (5.63) it follows that the mean velocity $\bar{v} = v_0$ at $\mathbf{r}_1 = \mathbf{r}_0$. Now, if and only if the illumination function is product separable,

$$I(x, y, z) = W^2(z) f_\phi^4(x) f_\theta^4(y), \tag{5.64}$$

substitution of (5.63) and (5.64) in (5.51) produces

$$\sigma_s^2 = \sigma_x^2 k_x^2 + \sigma_y^2 k_y^2 + \sigma_z^2 k_z^2, \tag{5.65}$$

where σ_x^2, σ_y^2, and σ_z^2 are second moments of $f_\phi^4(x)$, $f_\theta^4(y)$, and $W^2(z)$, respectively at range r_0. For resolution volume sizes small compared to their range, distances transverse to the beam axis can be approximated by arc-lengths. If k_θ, k_ϕ, k_r are the shears in the directions θ, ϕ, r, then

$$\sigma_s^2 = \sigma_{s\theta}^2 + \sigma_{s\phi}^2 + \sigma_{sr}^2 = (r_0\sigma_\theta k_\theta)^2 + (r_0\sigma_\phi k_\phi)^2 + (\sigma_r k_r)^2, \tag{5.66}$$

where σ_θ^2 and σ_ϕ^2 are defined as the second central moments of the two-way antenna power pattern in the indicated directions and $\sigma_z^2 = \sigma_r^2$ is the second central moment of the weighting function $W^2(r)$. A circularly symmetric Gaussian pattern has

$$\sigma_\theta^2 = \sigma_\phi^2 = \theta_1^2/(16 \ln 2). \tag{5.67}$$

For a rectangular transmitted pulse and a Gaussian receiver frequency response under matched conditions (i.e., $B_6 \tau = 1$),

$$\sigma_r^2 = (0.35 c\tau/2)^2. \tag{5.68}$$

To estimate the shears needed in (5.66), one may use differences between radial velocities at adjacent elevations and azimuths in the spherical system with the radar at its center. The weather radar usually scans conically, with the vertical axis along the cone's axis. Thus it becomes difficult, because of the finite beam width and errors in velocity estimates, to measure accurately the shears k_ϕ, k_θ at the higher elevation angles. For example, as $\theta_e \to \pi/2$ one should use four measurements 90° apart in azimuth and slightly displaced from the zenith to deduce the two orthogonal shear components.

The width σ_t due to turbulence is somewhat more difficult to model. For turbulence that is homogeneous and isotropic within V_6, Frisch and Clifford (1974) have shown that σ_t^2 is related to the eddy dissipation rate ε (see Chapter 10).

A detailed discussion of the contribution of turbulence and shear to the spectrum width is presented in Chapter 10, where examples of data fields are also given.

REFERENCES

Cadzow, J. A. (1982). Spectral estimation: An overdetermined rational model equation approach. *Proc. IEEE* **70**, 907–939.

Cooley, J. W., and Tukey, J. W. (1965). An algorithm for the machine calculation of complex Fourier series. *Math. Comput.* **19**, 297–301.

Cooley, J. W., Lewis, P. A. W., and Welch, P. D. (1967). Historical notes on the fast Fourier transform. *Proc. IEEE* **55**(10).

Frisch, A. S., and Clifford, S. F. (1974). A study of convection capped by a stable layer using Doppler radar and acoustic echo sounders. *J. Atmos. Sci.* **31**, 1622–1628.

Gold, B., and Rader, C. M. (1969). "Digital Processing of Signals." McGraw-Hill, New York.

Lhermitte, R. M. (1963). "Motions of Scatterers and the Variance of the Mean Intensity of Weather Radar Signals," SRRC-RR-63-57. Sperry-Rand Res. Cent. Sudbury, Massachusetts.

Papoulis, A. (1965). "Probability, Random Variables, and Stochastic Processes." McGraw-Hill, New York.

Papoulis, A. (1977). "Signal Analysis." McGraw-Hill, New York.

Reed, I. S. (1962). On a moment theorem for complex Gaussian processes. *IRE Trans. Inf. Theory* **IT-8**, 194–195.

Sato, T., and Woodman, R. F. (1982). Spectral parameter estimation of CAT radar echoes in the presence of fading clutter. *Radio Sci.* **17**, 817–826.

Tretter, S. A. (1976). "Introduction to Discrete-Time Signal processing." Wiley, New York.

Waldteufel, P. (1976). An analysis of weather spectra variance in a tornadic storm. NOAA Technical Memorandum ERL-NSSL-76.

Zrnić, D. (1980). Spectral statistics for complex colored discrete-time sequence. *IEEE Trans. Acoust., Speech, Signal Process.* **ASSP-28**(5), 596–599.

6

Meteorological Radar Signal Processing

In this chapter proven methods of weather echo processing are presented, with emphasis on obtaining the first three spectral moments. Weather radars sample within the PRT a large number of resolution volumes (e.g., approximately 1000), and hence the optimum estimation algorithms, which require much processing, may be impractical. Simpler methods, with fewer calculations and smaller memory storage, are used. Those readers interested in applications of maximum-likelihood estimation to spectral moments are referred to Zrnić (1979a).

6.1 SPECTRAL MOMENTS

The pulsed-Doppler weather radar should supply the three most important spectrum moment estimates. These are

(1) the echo power or zero moment of the Doppler spectrum (this is an indicator of liquid water content or precipitation rate in the resolution volume),

(2) the mean Doppler velocity or the first moment of the power normalized spectra (this is equal to the mean motion of scatterers weighted by their cross sections and, for near-horizontal antenna orientations, is essentially the air motion toward or away from the radar), and

(3) the spectrum width σ_v, the square root of the second moment about the first one of the normalized spectrum (this is a measure of the velocity dispersion, i.e., shear or turbulence within the resolution volume).

Moment estimates use samples of a randomly varying signal. In the case of weather echoes, single sample estimates have too large a statistical uncertainty to yield meaningful data interpretation. Thus, a large number of echo samples (acquired during a few tens of milliseconds) must be processed to provide the required accuracy. The actual number required depends on both radar system characteristics and meteorological conditions. These include the signal-to-noise ratio, the distribution of velocities within the resolution volume, the receiver transfer function, and the number of samples processed.

Reflectivity estimation requires only power sample averaging to reduce statistical fluctuations. However, mean Doppler velocity estimation involves the Fourier transform or complex covariance calculation and requires a large amount of data processing.

The need to obtain the principal Doppler moments economically, with minimum uncertainty, and in digital format (to facilitate processing and analysis with electronic computers) has prompted the use of covariance estimate techniques. The advantages of covariance processing coupled with the new technology (e.g., medium-scale integrated circuits), have made possible the implementation of this signal processing technique on the pulsed Doppler radar.

It is the very nature of the weather echo that it imposes limitations and tradeoffs on the Doppler radar. Weather targets are distributed quasi-continuously over large spatial regions (from tens to hundreds of kilometers), the strength of weather echoes easily spans an 80-dB power range, and the signals themselves are semicoherent, i.e., are not purely sinusoidal.

6.2 RADAR SIGNAL PROCESSING

The block diagram in Fig. 6.1 is that of a typical meteorological Doppler radar receiver. The mixer has exactly the same function as the mixer in Fig. 3.1 except that here the output has a carrier frequency at the intermediate frequency (IF) whereas in Fig. 3.1 the carrier was at zero frequency. The

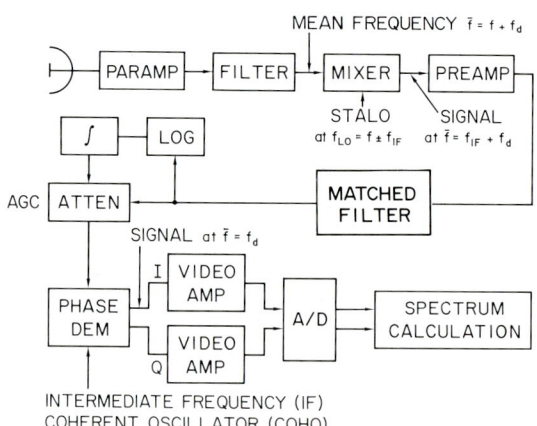

Fig. 6.1. Block diagram of the Doppler weather radar receiver.

6.2 Radar Signal Processing

automatic gain control loop (AGC) is a feed-forward type so that the average power estimate made by the integrator controls the gain during acquisition of samples for the next spectrum moment or periodogram calculations. This gain, which is constant during dwell time, matches the input signal's dynamic range to the range of the analog-to-digital (A/D) converter (otherwise many more bits would be required in the A/D, with the severe consequence of limiting the time of conversion from the analog level to the digital word). Alternative schemes in which the signal instantaneously controls the gain of the linear amplifier have also been used (Mueller and Silha, 1978).

The radio frequency (rf) filter serves the purpose of rejecting interference from other transmitters, as well as rejecting image noise, so that radar sensitivity is improved. A matched filter is either in the video (as in Fig. 3.1) or in the IF part (as in Fig. 6.1) of the receiver; direct current (dc) offset control and gain balance circuits must be available in the two video amplifiers. A second mixer, sometimes called a phase demodulator (DEM), operates in exactly the same way as the mixer in Fig. 3.1 except that it mixes an IF reference oscillator (COHO) with the IF signal that has an amplitude and phase modulation identical to the received microwave signal at frequency f. Ground clutter suppression and the calculation of spectral moments are done in the last block, which contains a special-purpose processor, an on-line computer, or both. The echo power estimate is sometimes obtained from the output of the logarithmic amplifier.

Each and every range gate produces a sequence of complex video samples $V(kT_s)$ separated by the pulse repetition time T_s and consisting of a signal part $s_k e^{j\omega_d kT_s}$ and a white noise part n_k; i.e.,

$$V(kT_s) = s_k e^{j\omega_d kT_s} + n_k, \quad k = 0, 1, \ldots, M - 1. \tag{6.1}$$

The signal sample is written as a product of two terms for later convenience. The multiplication of a time signal by $e^{j\omega_d t}$ shifts the spectrum by f_d. Thus the spectrum of s_k will be centered about zero frequency (i.e., its power-weighted mean frequency is zero), whereas the spectrum of $V(kT_s)$ is centered on the Doppler frequency f_d. Both n_k and s_k are zero-mean Gaussian processes (i.e., their average amplitude is zero), but s_k is narrowband compared to the receiver bandwidth. It is desirable to have the second spectral moment of s_k narrow compared to the Nyquist interval squared, T_s^{-2}, in order to improve the accuracy of the various estimators. This is because, for a fixed dwell time, the Nyquist interval expands as the number of samples M increases and, furthermore, coherency from sample to sample strengthens; both improve estimator precision. However, other considerations such as multiple trip echoes require sample separation (PRT) large, so the second moments can not be made arbitrarily small compared to T_s^{-2}.

Let the average signal power be S. The magnitude of the normalized correlation function ρ at lag mT_s depends on the sample spacing and σ_v. Thus the autocorrelation function of the process $V(kT_s)$ is

$$R(mT_s) = R_m = E\{V^*(kT_s)V[(k+m)T_s]\} = S\rho(mT_s)e^{j\omega_d mT_s} + N\delta_m, \quad (6.2)$$

where N is defined as the mean white noise power and δ_m is 1 for $m = 0$ and zero otherwise. We can deduce from Eq. (5.18) that when $\rho(mT_s)$ is real the power spectrum is symmetric about $f = 0$. Conversely, the correlation function $\rho(mT_s)$ of the s_k is real if its power spectrum is symmetric about $f = 0$. We shall assume this to be the case. However, the correlation of the $V(kT_s)$ is complex, and its power spectrum $S(f)$ is centered about f_d.

It was explained in Chapter 5 that weather echo signals from regions of uniform reflectivity should have spectra closely resembling a Gaussian function when linear shear and many broadening mechanisms independently contribute to the spectrum shape. Thus in analyzing the statistical properties of various estimators, it is convenient to assume a Gaussian power spectrum,

$$S(v) = \frac{S}{(2\pi)^{1/2}\sigma_v} \exp\left[-(v - \bar{v})^2/2\sigma_v^2\right] + \frac{2NT_s}{\lambda}, \quad (6.3a)$$

whose corresponding autocorrelation at lag mT_s equals

$$R(mT_s) = S \exp\left[-8(\pi\sigma_v mT_s/\lambda)^2\right]e^{-j4\pi\bar{v}mT_s/\lambda} + N\delta_m. \quad (6.3b)$$

The magnitude of the normalized signal correlation is then

$$\rho(mT_s) = \exp\left[-8(\pi\sigma_v mT_s/\lambda)^2\right]. \quad (6.3c)$$

Accurate estimates of three parameters—the power S, mean velocity $\bar{v} = -\lambda f_d/2$, and spectrum width σ_v—are required for meaningful interpretation of Doppler radar data.

6.3 ECHO SAMPLE INTENSITY ESTIMATION

6.3.1 Sample Time Averaging

The sampled echo power is the weighted sum of powers returned from the individual scatterers, which are moving relative to each other. This relative movement produces fluctuation in the samples of echo power S_k. Added to this is the sampled receiver noise power N_k. To obtain a quantitative estimate of the total mean power $\bar{P} = \bar{S} + \bar{N}$, samples P_k must be averaged over a period long enough that the uncertainty in the mean power estimate is reduced to a tolerable level. The mean and variance of the P population are completely described by its probability density.

6.3 Echo Sample Intensity Estimation

6.3.1.1 STATISTICAL PROPERTIES OF ECHO SAMPLES FROM LINEAR AND NONLINEAR RECEIVER–DETECTORS

In the sequel we consider common receiver–detector combinations that act on the complex signal $V(k)$. Detectors in an incoherent radar operate on the IF signals and generate at the output, after filtering, low-frequency envelopes of $V(k)$. Those envelopes are generally nonlinear functions of the magnitude $|V(k)|$, and consequently the output signal Q of a radar receiver can have one of many dependences on the input signal power. Common receiver transfer functions are

(1) square law, in which case the output sample Q_k is proportional to the input power P_k, i.e., $Q_k \propto P_k \propto |V(k)|^2$;
(2) linear amplitude, so that $Q_k \propto P_k^{1/2} \propto |V(k)|$; and
(3) logarithmic, or $Q_k \propto \log P_k \propto \log|V(k)|$.

The linear detector, which rectifies IF signals, has a linear transfer characteristic for only the positive or negative portion of the IF signal and gives zero output for the opposite polarity. It should be evident that any power transfer function can be obtained by operating on the magnitudes $|V(k)|$ at the output of the coherent (phase) demodulator.

In order to improve the accuracy of power estimates, Q_k samples are averaged and the sample mean probability density derived from the probability density of P_k and the receiver transfer function (Marshall and Hitschfeld, 1953). A uniform average,

$$\hat{P} = \frac{1}{M} \sum_{k=0}^{M-1} P_k, \qquad (6.4)$$

usually gives very good estimates of $S + N$. The number of samples, M in (6.4), is obtained at the pulse repetition rate.

Radar reflectivity, from which liquid water content and the rainfall rate can be estimated, is proportional to $\bar{P}$. The $\bar{P}$ values of meteorological interest can easily span a range of 10^8, and often the choice of receiver type hinges on the cost of meeting this large-dynamic-range requirement. The problem is to estimate $\bar{P}$ from sample averages of Q. The estimation is complicated because Q is not linearly related to P (except for the square law receiver). That is, when one operates on the mean output estimates $\hat{Q}$,

$$\hat{Q} = \frac{1}{M} \sum_{k=0}^{M-1} Q_k, \qquad (6.5)$$

with the inverse of the receiver transfer function (i.e., Q versus P) to obtain estimates $\hat{P}$, biases are generated and the uncertainty is larger than if one averages P_k directly. [An estimate is biased if its expected value $E(P)$ differs from the mean of the variable being estimated (i.e., $\bar{P}$).] The output mean for linear and logarithmic receivers is related to $\bar{P}$. $\bar{P}$ estimate bias and standard deviation for the three receivers are given in Fig. 6.2 for an average that contains a number M_I of independent samples (Zrnić, 1975). If the number of independent samples in the average Q is less than about 30, the bias in power estimates $\hat{P}$ becomes dependent on the number of samples averaged, and the

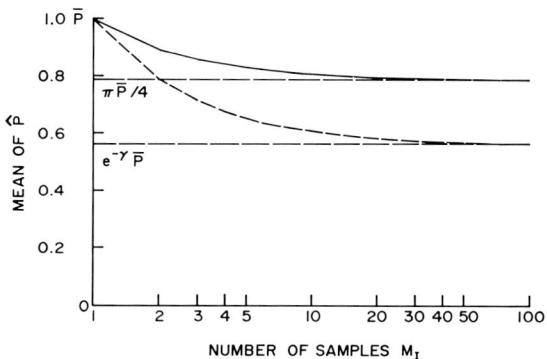

Fig. 6.2a. Expected value of the estimated power $\hat{P}$ from averaged outputs of a linear and a logarithmic receiver for Rayleigh-distributed input amplitudes. The number of independent samples is M_I, and discrete values are connected for visual clarity. The Euler constant $\gamma = 0.577215$. ——, linear; ---, logarithmic.

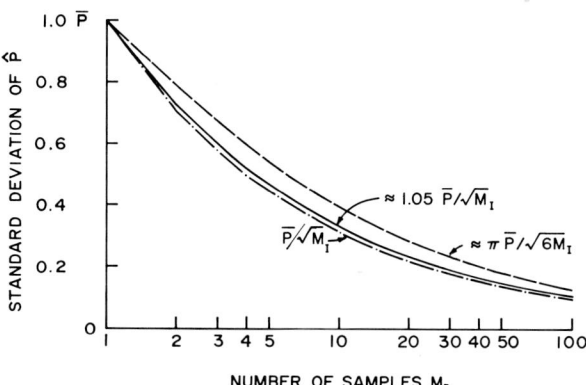

Fig. 6.2b. Power estimate standard deviations referred to the unbiased mean $\bar{P}$ from the output of the square law, a linear, and a logarithmic receiver. Asymptotic large-M_I (>10) formulas are also indicated. —·—·, square law; ——, linear; ---, logarithmic.

6.3 Echo Sample Intensity Estimation

standard deviations of $\hat{P}$ are no longer exactly inversely proportional to the square root of the number of samples (Fig. 6.2b).

Although the uniform average in blocks (6.5) is conceptually simple, it does not provide a continuous update of the output (i.e., an update with each new input sample). One may choose a uniformly weighted running average, which is more complex to implement and requires considerable memory storage, or use a first-order recursive filter (an exponentially weighted running average), for which the estimate is

$$\hat{Q}_k = (1 - b)\hat{Q}_{k-1} + bQ_k, \qquad (6.6)$$

where b is a number between zero and 1. It has been shown that the means and variances of the input power estimates obtained from (6.6) for various receivers (e.g., log or square law) are almost identical to those obtained from a uniform average of M samples if $M = (2 - b)/b$ (Zrnić, 1977a).

6.3.1.2 Number of Equivalent Independent Samples

The total number M of samples obtained from the resolution volume is determined by the PRT and dwell time. However, because considerable correlation may exist from sample to sample, we determine the equivalent number M_I of independent samples in order to calculate the reduction in estimate variance achieved by averaging. The degree of correlation between samples is a function of the radar parameters (e.g., wavelength, PRT, beamwidth, and pulse width) and the meteorological status (e.g., degrees of turbulence and shear) in the resolution volume.

If the estimate $\hat{Q}$ of the true output mean $\bar{Q}$ is derived from an average of M_I uniformly weighted independent samples (i.e., $M = M_I$), the output single sample estimate variance σ_Q^2 is reduced by a factor of $1/M_I$. That is,

$$\sigma_{\hat{Q}}^2 = \sigma_Q^2/M_I. \qquad (6.7)$$

However, if we have M samples with correlation from sample to sample, the estimate variance $\sigma_{\hat{Q}}^2$ does not vary as $1/M$. Instead, for a stationary process and equal-spaced samples, the estimate-variance reduction factor for the M-sample average is given by (Papoulis, 1965)

$$\frac{\sigma_{\hat{Q}}^2}{\sigma_Q^2} = M_I^{-1} = \sum_{m=-(M-1)}^{M-1} \frac{M - |m|}{M^2} \rho_Q(mT_s). \qquad (6.8)$$

The normalized correlation ρ_Q of the receiver's output samples can be expressed in terms of the Doppler spectrum, and the parameters of this spectrum (in a particular spectral width) can be related to atmospheric and radar system parameters. To determine rigorously the autocorrelation of the output Q, we must transform this spectrum by the receiver transfer function (which is nonlinear for the three detectors). For example, for a square law receiver the output correlation ρ_s equals the square of the input correlation ρ (Papoulis, 1965). Correlation functions for the output of the other detectors are more complicated, and the reader is referred to other works for details (Kerr, 1951; Davenport and Root, 1958).

Assuming the input power spectrum to be Gaussian (6.3a) with a corresponding autocorrelation function (6.3c) and zero noise, we may obtain the correlation of the output of a square law receiver,

$$\rho_s(mT_s) = \exp[-(2\pi m\sigma_{vn})^2], \qquad (6.9a)$$

where for convenience

$$\sigma_{vn} \equiv \sigma_v/2v_a = 2\sigma_v T_s/\lambda \qquad (6.9b)$$

is the normalized (to the Nyquist interval) spectrum width, and $v_a = \lambda/4T_s$ is the unambiguous velocity (see Section 3.6).

Combining (6.8) and (6.9a) gives the variance reduction factor for the square law detector:

$$M_I^{-1} = \sum_{m=-(M-1)}^{M-1} \frac{M - |m|}{M^2} \rho_s(mT_s). \qquad (6.10)$$

The variance reduction factors for the other two detector types have been given by Walker et al. (1980). For the linear detector,

$$M_I^{-1} = \sum_{m=-(M-1)}^{M-1} \frac{\pi}{4-\pi} \left(\frac{M-|m|}{M^2}\right) \sum_{n=1}^{\infty} A_n n^{-2} \rho_s^n(mT_s), \qquad (6.11)$$

where

$$A_n = (2n-1)!!/2^n(n-1)!(2n-1) \qquad (6.12)$$

and $(2n-1)!!$ is the consecutive product of odd integers. For the logarithmic detector,

$$M_I^{-1} = \sum_{m=-(M-1)}^{M-1} \frac{6}{\pi^2} \left(\frac{M-|m|}{M^2}\right) \sum_{n=1}^{\infty} n^{-2} \rho_s^n(mT_s). \qquad (6.13)$$

Walker et al. (1980) also give correlation functions for these two receiver–detectors.

The equivalent number M_I of independent samples is plotted versus the product $\sqrt{2}(M-1)\sigma_{vn}$ in Fig. 6.3; Fig. 6.3a compares (6.10), (6.11), and (6.13) for $M = 10$, whereas Figs. 6.3b and 6.3c are plots of (6.10) and (6.13) with M as a parameter. The second label of the abscissa refers to relative shifts of resolution volumes in the azimuth, which are discussed in Section 6.3.1.3. The logarithmic receiver decorrelates the output samples most and thus has the largest number of independent sample M_I (Fig. 6.3). For equal

6.3 Echo Sample Intensity Estimation

M_1, however, the square law detection results in the least error among the three detector types (Fig. 6.2b). With Figs. 6.2 and 6.3 we are in a position to compare the standard deviations of input power estimates from an M-sample average of correlated outputs. Given a product $(M - 1)\sigma_{vn}$, the equivalent number of independent samples M_1 is read off Fig. 6.3 and then entered in Fig. 6.2b to obtain the standard deviation. This comparison favors the square law detector over the other two types. For example, at large M the equivalent number M_1 for a square law detector is $2M\sigma_{vn}\sqrt{\pi}$, whereas for the logarithmic detector one has $M\pi^{5/2}\sigma_{vn}/4$. (These asymptotic solutions are valid for narrow spectral widths, i.e., for $\sigma_{vn} < 0.2$.) Now, asymptotically the standard deviation of the input power estimate (unbiased) from M_1 averaged independent output samples of the logarithmic receiver is $\pi/\sqrt{6} \approx$ 1.28 times larger than that for the square law, and we conclude that correlation decreases that ratio to 1.15.

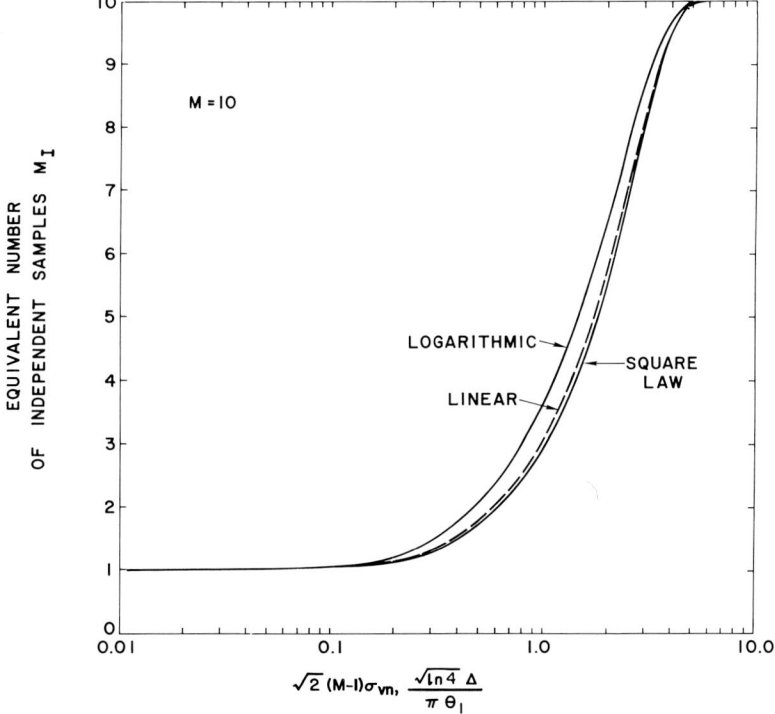

Fig. 6.3a. Equivalent number of independent samples for the three receiver types. The number of samples is 10, and the time or angle averaging parameters of a Gaussian correlation are on the abscissa. (From Walker *et al.*, 1980.)

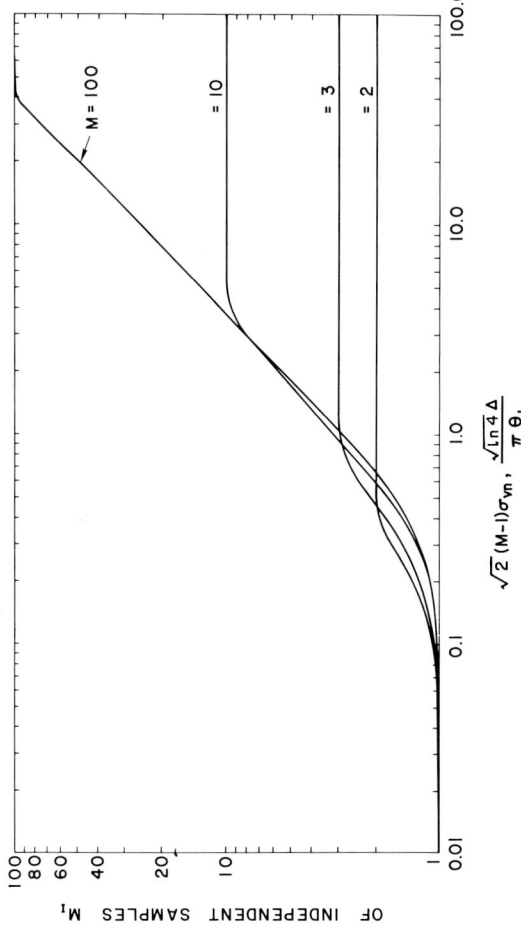

Fig. 6.3b. Equivalent number of independent samples for the square law receiver. A similar graph with slightly more independent samples applies to a linear receiver. (From Walker *et al.*, 1980).

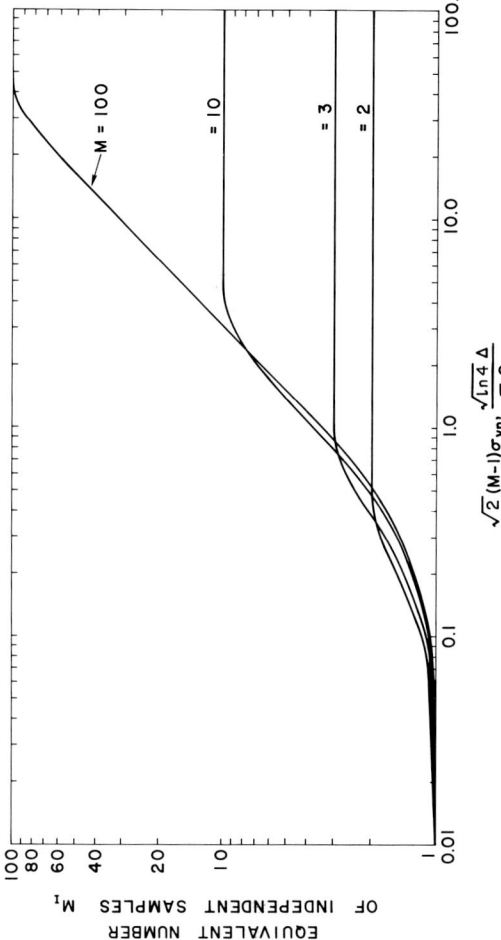

Fig. 6.3c. Equivalent number of independent samples for the logarithmic receiver. (From Walker et al., 1980.)

6.3.1.3 INDEPENDENT SAMPLES DUE TO SHIFTS IN RESOLUTION VOLUME LOCATION

Acquisition of an ensemble sample series while the antenna is rotating results in a continuous alteration in sample volume location. The echo samples for resolution volumes corresponding to two beam positions are correlated if the two radiation patterns overlap and if there is no significant decorrelation from reshuffling of the scatterers' positions. In Appendix C we show that the correlation of echo samples taken at different beam positions equals the correlation of the one-way radiation pattern lagged by the angular spacing. For many types of meteorological measurements the antenna main lobe can be approximated adequately by a Gaussian function. In this case the number of independent samples available with a moving antenna can be found from an analysis similar to that for sample–time averaging. M_1 is also given in Fig. 6.3 for M samples averaged over an angular interval Δ and for an antenna of half-power beamwidth θ_1. The scaling factor $\sqrt{\ln 4}\, \Delta/\pi\theta_1$ comes directly from (C.23) if $\theta_e \approx 0$ and $\Delta = (M - 1)\alpha T_s$.

6.3.2 Range Time Averaging

Estimate variance can be reduced by averaging along the range time τ_s as well as the sample time. Because the range extent of the resolution volume is usually small compared to its size perpendicular to the beam axis, average over a range interval Δr results in a more symmetric spatial resolution volume with little degradation of the spatial resolution. However, range–time averaging the output of a linear or logarithmic receiver introduces a systematic bias of the estimate caused by reflectivity gradients, and this bias limits the maximum Δr useful for averaging (Rogers, 1971). Nevertheless, we have a reasonable latitude in choosing Δr. The incremental spacing between samples multiplied by the number of range samples being averaged gives Δr. The sampling increment is chosen by considering the autocorrelation of the consecutive range samples of the return signal plus receiver noise. Averages containing range-correlated samples have an estimate variance that depends not only on the number of samples, but also on the sample correlation.

Quite often the bandwidth of the receiver may be from about two to three times the reciprocal of the transmitter pulse width τ; there is then significantly less correlation of echo samples spaced a pulse width τ apart along τ_s (see Fig. 4.5). For matched receivers, samples become significantly correlated only when the product $\delta\tau_s B_6$ of range sampling increment and receiver bandwidth is less than about 0.5 (see Section 4.5). This suggests a range sample increment smaller than that based solely on the condition of complete independence, which holds when $\delta\tau_s \geq \tau$ and $B_6 \to \infty$. The reduction in estimate variance in the case of correlated samples is given by Eq. (6.8), where T_s is replaced by the range time increment (see Appendix C) and ρ_Q is now the correlation along the range time axis τ_s.

6.4 MEAN FREQUENCY ESTIMATORS

As for reflectivity estimators, we shall consider only those methods of mean frequency extraction that have been implemented on meteorological Doppler radars. They are based either on the autocovariance or on the DFT of I and Q samples.

6.4.1 Autocovariance Processing: The Pulse Pair Processor

The pulse pair estimator calculates the first two moments of the Doppler spectrum from estimates of the autocovariance function at lag T_s. Using (5.19), we can express $R(T_s)$ in terms of the power-weighted mean Doppler frequency f_d:

$$R(T_s) = e^{j2\pi f_d T_s} \int_{-1/2T_s}^{1/2T_s} S(f) e^{j2\pi T_s (f - f_d)} \, df. \tag{6.14}$$

Note that f_d can be calculated from the argument, $2\pi f_d T_s$, of $R(T_s)$ if the integral in (6.14) is real, which implies that the imaginary part of the integral is zero; i.e.,

$$\varepsilon_s \equiv \int_{-1/2T_s}^{1/2T_s} S(f) \sin[2\pi T_s (f - f_d)] \, df = 0. \tag{6.15}$$

Thus if the spectrum $S(f)$ is symmetric with respect to f_d, the *argument* of $R(T_s)$ yields the mean frequency; otherwise there is a bias error ε related to the magnitude of the sine integral of $S(f)$. Consider the following commonly used definition of the estimate for the power-weighted mean Doppler frequency $\hat{f}_d$:

$$\hat{f}_d = \int_{-1/2T_s}^{1/2T_s} f \hat{S}_n(f) \, df, \tag{6.16}$$

where $\hat{S}_n(f)$ is the normalized estimate of the spectrum of sampled data. It is easily seen that when the spectrum $S(f)$ has spectral components that exceed the Nyquist limits $\pm(2T_s)^{-1}$, these components are undersampled. This leads to aliasing, with the result that $\hat{f}_d$ is a biased estimate of the true mean Doppler f_d. However, some thought reveals that when $S(f)$ is symmetric about f_d but has some frequency components that exceed the Nyquist limit, the argument of $R(T_s)$ still yields f_d (to within an uncertainty equal to an integral multiple of the Nyquist interval). Another way of seeing this is to change variables in (6.15) to get the following instructive formula:

$$\varepsilon_s = \int_{-1/2T_s - f_d}^{1/2T_s - f_d} S(f' + f_d) \sin(2\pi f' T_s) \, df', \tag{6.17}$$

where $S(f' + f_d)$ has its center at the origin $f' = f - f_d = 0$. Because $S(f)$ is periodic with period T_s^{-1}, we can add f_d to the limits of the integral (6.17), making them symmetric about $f' = 0$. Note that $\sin 2\pi(f - f_d)T_s$ has period T_s^{-1} and is odd with respect to f_d. Therefore (6.17) and (6.15) are zero whenever $S(f)$ is symmetric about f_d (see Fig. 6.4).

With reference to Fig. 6.4, we deduce that, even if the spectra are not symmetric, the argument of $R(T_s)$ yields an almost unbiased estimate provided that $S_n(f)$ has its frequencies contained in a band small compared to $(T_s)^{-1}$. In view of the foregoing, we conclude that white noise (because it has a symmetric spectrum) will not bias the autocovariance estimator.

If a sequence of M uniformly spaced pulses are transmitted, the autocorrelation at a sample time lag T_s (for echoes with range delay τ_s) can be estimated from the sum

$$\hat{R}(T_s) = \frac{1}{M} \sum_{m=1}^{M-1} V^*(m)V(m+1), \tag{6.18}$$

so that the mean velocity estimate becomes

$$\hat{v} = -(\lambda/4\pi T_s) \arg \hat{R}(T_s). \tag{6.19}$$

The argument of $\hat{R}(T_s)$ is in radians, and the negative sign signifies that positive Doppler shifts create negative velocities, in accordance with (3.7).

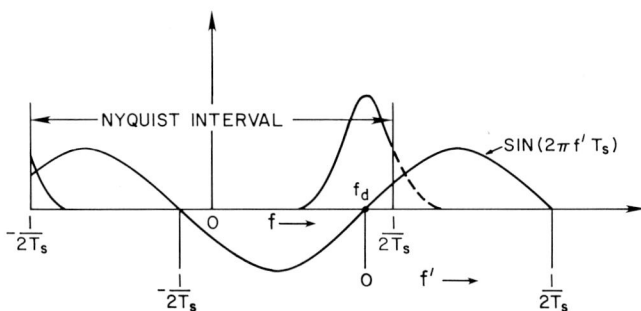

Fig. 6.4. Automatic unbiasing of the estimate of the signal autocovariance argument. The autocovariance estimate of the mean Doppler frequency is unbiased by the frequency components that are undersampled. ———, sampled power spectrum; ---, undersampled spectrum $S(f)$.

6.4 Mean Frequency Estimators

It should be clear that pulses need not form a uniform sequence in order for $R(T_s)$ to be estimated. In fact, pulses can be transmitted in pairs such that the spacing between the pairs is much larger than the intrapair period (Fig. 6.5). In this case $R(T_s)$ is the covariance for the time lag T_s, which is equal to the intrapair period, and the sample estimates $V^*(m)V(m+1)$ have themselves a correlation that depends on the interpair spacing T. Spaced pairs have been proven advantageous in alleviating the contamination of velocity estimates by range-overlaid echoes (see Chapter 7).

Perturbation analysis has been successfully used to derive the variance of the mean velocity (Zrnić, 1977b) of the pulse pair estimator. However, such an analysis is not without flaws: For very narrow spectrum widths or low signal-to-noise ratios, it matches experimental results only if a very large number of samples is used. Briefly, the following two conditions are necessary in order for the perturbation analysis to be valid:

$$2\pi M \sigma_{vn} \gg 1, \qquad (6.20a)$$

$$\rho^2(T_s)M \gg (N/S + 1)^2. \qquad (6.20b)$$

Condition (6.20a) expresses the requirement of a large number of independent samples, and (6.20b) ensures that the argument of $\hat{R}(T_s)$ has a distribution width small compared to 2π. With these assumptions, an expression for the mean frequency variance of correlated but spaced pairs has been derived (Zrnić, 1977b):

$$\text{var}(\hat{v}) = \lambda^2 [32\pi^2 T_s^2 \rho^2(T_s)]^{-1} \{M^{-2}[1 - \rho^2(T_s)] \sum_{m=-(M-1)}^{M-1} \rho^2(mT)(M - |m|)$$
$$+ N^2/MS^2 + (2N/MS)[1 + \rho(2T_s)(1/M - 1)\delta_{T-T_s,0}]\},$$

$$(6.21)$$

where now M is the number of sample pairs, the spacing between pairs is T, and $\delta_{T-T_s,0} = 1$ for $T = T_s$ and zero otherwise. This expression for the variance is perfectly valid when sample pairs are contiguous (i.e., share a common sample when transmitted pulses are uniformly spaced) or when they are independent (which can be achieved by increasing the separation T or by changing the transmitted frequency between pairs). The plot of (6.21) in

Fig. 6.5. Spaced and correlated sample pairs for calculating the mean velocity or spectrum width. The spacing between the pulses of a pair is T_s, and the pair separation is T.

Fig. 6.6 reveals that the variances are not greatly different for independent and contiguous pairs. At narrow widths and low SNR, contiguous pairs have a lower standard deviation of estimates; otherwise the opposite is true.

Superimposed on Fig. 6.6 are results from simulations for contiguous pairs. Several 8192-point periodograms were generated with Gaussian-shaped spectra. The inverse of such periodograms produced long-time records from which chunks of 64 samples were processed. Altogether 1024 chunks were averaged to estimate the mean and variance. Because the number of processed samples is much less than the total number in the sequence, the method effectively introduces a rectangular window. The simulation results at narrow widths and low SNR deviate from the theoretical predictions because only 63 sample pairs were used in each estimate and thus (6.20) was not satisfied. With more samples this deviation decreases.

A good approximation of (6.21), valid when the sum can be replaced with an integral and a Gaussian spectrum is assumed, is

$$\text{var}(\hat{v}) \approx \lambda^2 [32\pi^2 M \rho^2(T_s) T_s^2]^{-1} \{[1 - \rho^2(T_s)] T_s / 2\sigma_{vn} T \sqrt{\pi} \\ + N^2/S^2 + 2(N/S)[1 - \rho(2T_s)\delta_{T-T_s,0}]\}. \quad (6.22\text{a})$$

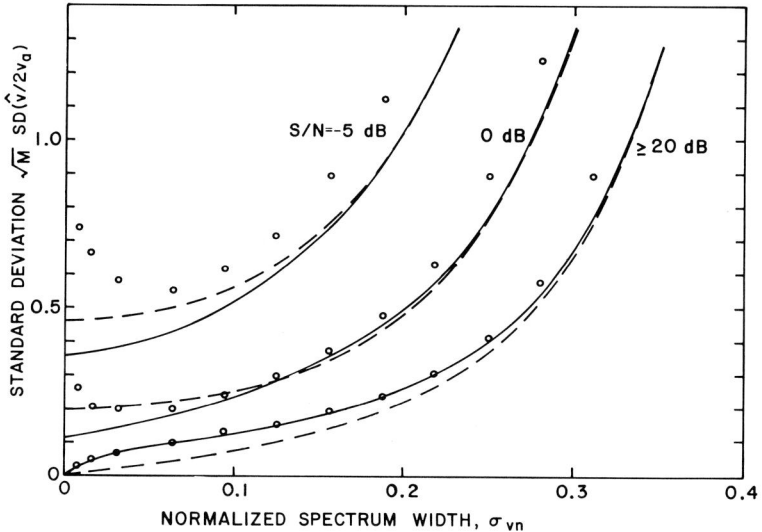

Fig. 6.6. Standard deviations of the mean frequency estimate (autocovariance processing). Spectrum width is normalized to the Nyquist interval $2v_a$. Note a gradual increase between 0 and 0.2 and an exponential rise thereafter. ——, contiguous pairs; ---, independent pairs; ○, simulation.

6.4 Mean Frequency Estimators

When sample pairs are independent, (6.21) reduces to

$$\operatorname{var}(\hat{v}) \approx \lambda^2 \left[32\pi^2 M \rho^2(T_s) T_s^2\right]^{-1} \left[(1 + N/S)^2 - \rho^2(T_s)\right]/\rho^2(T_s). \quad (6.22\text{b})$$

The direct proportionality to ρ^{-2} means that the variance is heavily dependent on σ_{vn}^2 [and hence on the exponent in (6.3c)]. Thus the effect of increasing T_s or σ_v is an exponential growth in the variance (Fig. 6.6).

For large signal-to-noise ratios, narrow spectrum widths, and contiguous pairs, (6.22a) becomes

$$\operatorname{var}(\hat{v}) \approx \sigma_v \lambda / (8 M T_s \sqrt{\pi}). \quad (6.23)$$

6.4.2 Spectral Processing

The advent of the FFT and the subsequent decrease in the cost of the circuits for the required computation make it attractive to use direct spectral methods for mean frequency calculations. Still, the complexity and cost of FFT processors are an order of magnitude larger than those of their autocovariance counterparts. To calculate the mean frequency, one first forms the periodogram (5.22), which is an estimate of the power spectrum that may contain aliases if there are frequency components outside the Nyquist interval. The next step is to obtain a rough mean frequency estimate, k_m/MT_s, where k_m ($-M/2 \leq k_m \leq M/2$) could be the index of the strongest Fourier coefficient. Then the straightforward mean velocity estimate becomes

$$\hat{v} = -\frac{\lambda}{2M} \left\{ \frac{k_m}{T_s} + \frac{1}{\hat{P} T_s} \sum_{k_m - M/2}^{k_m + M/2} (k - k_m) \hat{S}[\operatorname{mod}_M(k)] \right\}, \quad (6.24)$$

where $\hat{P}$ is the total power in the periodogram and $\operatorname{mod}_M(k)$ is the remainder on dividing k by M. This form, in contrast to that given in (6.16), eliminates biases due to aliasing for symmetric spectra. Modifications of (6.24) are used to improve accuracy; among the simplest are the thresholding of spectral powers above the noise level or around k_m. The standard errors of the estimates when spectra are thresholded are listed by Sirmans and Bumgarner (1975). Berger and Groginsky (1973) have derived the variance of a power-weighted mean velocity estimate in which the noise spectral density is subtracted from the periodogram and the summation is replaced with the integral over the Nyquist interval. The variance reads

$$\operatorname{var}(\hat{v}) = \frac{\lambda^2}{4 M T_s S^2} \int_{-1/2T_s}^{1/2T_s} f^2 S^2(f + f_d) \, df. \quad (6.25)$$

If Gaussian signal spectra (6.3a) of narrow width [i.e., $\rho^2(T_s) \simeq 1$] are substituted into (6.25), the variance simplifies to

$$\text{var}(\hat{v}) = \frac{\lambda^2}{4MT_s^2}\left[\frac{\sigma_{vn}}{4\sqrt{\pi}} + 2(\sigma_{vn})^2\frac{N}{S} + \frac{1}{12}\left(\frac{N}{S}\right)^2\right]. \qquad (6.26)$$

It can be shown that the first term of (6.26) is equal to the corresponding term of the first-order expansion in the pulse pair variance (6.22a). However, the other terms in (6.26) are larger than their counterparts in (6.22a), which makes the variance (6.26) larger. Consequently, at small SNR and narrow widths ($\sigma_{vn} < 1/2\pi$) the autocovariance estimator performs better. This is not the case at larger widths. Simulations show that the Fourier-derived mean has a variance that increases much more slowly than the exponential increase associated with the autocovariance method. Another advantage offered by spectral processing is the ease of editing data so that anomalous targets and system malfunctions can be identified and often eliminated.

6.5 ESTIMATORS OF THE SPECTRUM WIDTH

While only coherent radars can provide the mean frequency information, the Doppler spectrum width can also be obtained from incoherent radars (Bello, 1965). One of the early devices, the *R meter*, estimates the width from the rate at which the detected echo envelope [i.e., some function of $|V(k)|$] crosses a threshold (Rutkowski and Fleisher, 1955). Neither the R meter nor any other method based on measurement of the envelope is widely used in radar meteorology. Although the spectrum width of the detected envelope is uniquely related to the Doppler spectrum width, the envelope spectrum is broader, thus degrading the width accuracy more heavily. Nevertheless, methods to extract the width used with coherent radars can also be employed on the signal envelopes from incoherent radars.

6.5.1 Autocovariance Processing

When weather echo spectra closely follow a Gaussian shape, it is natural to estimate the spectrum width from estimates of the autocorrelation coefficient (6.3c). One first needs to estimate $\hat{R}_1 = \hat{R}(T_s)$ and $\hat{S}$, after which the logarithm of their ratio, from (6.3b), will retrieve $\hat{\sigma}_v$:

$$\hat{\sigma}_v = \frac{\lambda}{2\pi T_s\sqrt{2}}\left|\ln\left(\frac{\hat{S}}{|\hat{R}_1|}\right)\right|^{1/2}\text{sgn}\left[\ln\left(\frac{\hat{S}}{|\hat{R}_1|}\right)\right]. \qquad (6.27)$$

The signal power estimate is obtained by subtracting the known noise power

6.5 Estimators of the Spectrum Width

from the average of the squares of the magnitudes:

$$\hat{S} = \frac{1}{M} \sum_{k=0}^{M-1} |V(k)|^2 - N. \qquad (6.28)$$

The autocorrelation estimate $\hat{R}_1$ at lag T_s is given by (6.18). The sgn term in (6.27) warrants some discussion. Since $\hat{S}$ and $|\hat{R}_1|$ are estimates, it is possible at narrow spectrum widths and low SNR for the logarithm in (6.27) to become negative. The sgn term tags these negative cases and allows the assignment of widths to a category of small values.

A related estimator is obtained by expanding the logarithm in (6.27) for values $|\hat{R}_1|$ close to $\hat{S}$, values that occur when $\sigma_{vn} \ll 1$:

$$\hat{\sigma}_v = \frac{\lambda}{2\sqrt{2\pi}T_s} \left|1 - \frac{|\hat{R}_1|}{\hat{S}}\right|^{1/2} \text{sgn}\left(1 - \frac{|\hat{R}_1|}{\hat{S}}\right). \qquad (6.29)$$

At large widths (6.29) has an asymptotic ($M \to \infty$) bias, whereas (6.27) does not (provided the spectrum is Gaussian).

It can be shown that the variances of the spectrum width estimates (6.27) and (6.29) are equal when the spectra are Gaussian and the asymptotic bias is removed from (6.29). Under those conditions, perturbation expansion yields the following expression for the variance (Zrnić, 1977b, 1979a):

$$\text{var}(\hat{\sigma}_v) = \frac{\lambda^2}{128 M \pi^4 \sigma_{vn}^2 \rho^2(T_s) T_s^2} \Big\{ 2[1 - (1 + \delta_{T-T_s,0})\rho^2(T_s)]$$

$$+ \delta_{T-T_s,0}\rho^4(T_s)]N/S + [1 + (1 + \delta_{T-T_s,0})\rho^2(T_s)]N^2/S^2$$

$$+ \rho^2(T_s) \sum_{m=-(M-1)}^{M-1} \{2\rho^2(mT) + \rho^2(mT)\rho^{-2}(T_s)$$

$$+ \rho^2[mT + T_s(1 - \delta_{T-T_s,0})]$$

$$- 4\rho(mT + T_s)\rho(mT)\rho^{-1}(T_s)\}(1 - |m|/M)\Big\}. \qquad (6.30a)$$

With independent pairs (6.30a) simplifies to

$$\text{var}(\hat{\sigma}_v) = \frac{\lambda^2}{128 M \pi^4 \sigma_{vn}^2 \rho^2(T_s) T_s^2} \{[1 - \rho^2(T_s)]^2$$

$$+ 2[1 - \rho^2(T_s)]N/S + [1 + \rho^2(T_s)]N^2/S^2\}. \qquad (6.30b)$$

Because (6.30) is obtained from a perturbation analysis of the expansions around the mean autocovariance $\bar{R}(T_s)$ and the $\bar{\sigma}_{vn}$, it is well to emphasize that at very narrow widths ($\sigma_{vn} < 0.01$) or when the signal-to-noise ratio and the number of pairs are small [e.g., when (6.20) is violated], the results become erroneous. Nevertheless, the results are valid for conditions commonly found in meteorological radars.

Equation (6.30a) for contiguous pairs and Eq. (6.30b) for independent pairs are plotted in Fig. 6.7. There is little difference between the two, and it can be seen that the estimators are good (the standard deviation is small) as long as $0.02 < \sigma_{vn} < 0.2$ and SNR > 5 dB. At larger widths the estimate, like the mean frequency estimate, degrades exponentially. Simulated results (also included in Fig. 6.7) agree well with the theory for a wide range of widths and SNRs, but in regions where the conditions (6.20) are violated, deviations from the plotted curves occur.

Aliasing from undersampling ($\sigma_{vn} > 1/2\pi$) does not bias the estimator. Sample-dependent bias is inherently present, but because it is proportional to M^{-1}, it can be neglected compared to the standard deviation (Zrnić, 1977b). Probably the most serious defect of the estimator is that it is not adaptable to editing, i.e., to identifying errors arising from spurious spectral peaks. At $\sigma_{vn} < (2\pi)^{-1}$, the approximate variances for contiguous pairs are

$$\text{var } \hat{\sigma}_v \approx (3\lambda^2/128\sqrt{\pi}MT_s^2)\sigma_{vn} \quad (6.31)$$

for high SNR and

$$\text{var } \hat{\sigma}_v \approx 3\lambda^2 N^2/128\pi^4 \sigma_{vn}^2 T_s^2 M S^2 \quad (6.32)$$

for low SNR.

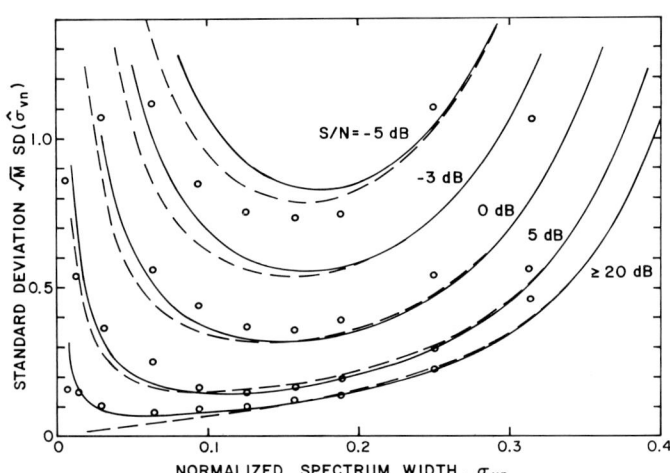

Fig. 6.7. Standard deviation of the width estimate (autocovariance processing). Note the poor quality of the estimator at widths $\sigma_{vn} > (2\pi)^{-1}$ and low SNR. The left portion of the curve cannot be extended toward the origin because the standard deviation there follows a $M^{-1/4}$ law. ——, contiguous pairs; ---, independent pairs; ○, simulation.

6.5.2 Spectral Processing

The following estimate of the width, using spectral estimates obtained from the DFT, avoids bias due to aliasing provided the width is small compared to the Nyquist interval:

$$\hat{\sigma}_v^2 = \frac{\lambda^2}{4\hat{P}T_s^2} \sum_{k_m - M/2}^{k_m + M/2} \left(\frac{k}{M} + \frac{2\hat{v}T_s}{\lambda} \right)^2 \hat{S}[\mathrm{mod}_M(k)]. \tag{6.33}$$

All terms in (6.33) are as in (6.24). Note that the positive sign results from the polarity difference between Doppler frequency and radial velocity conventions. The computed width contains a bias σ_b^2 due to the finite-time-window effect. In general, this bias is difficult to compute exactly. One may argue that both the mean frequency (6.24) and the width (6.33) should be obtained from integrals of the power spectrum (using a continuum of k from $-M/2$ to $M/2$), rather than summations, in order to account exactly for the bias introduced by the window. In practice this would require an enormous amount of computations; furthermore, the sum in (6.33) is close to the integral when spectra are broad compared to the Doppler frequency resolution, $1/MT_s$. For such cases the bias σ_b^2 can be computed exactly because the measured spectrum is equal to a convolution (5.32) of the true spectrum with the lag window transform:

$$\sigma_b^2 = \frac{\lambda^2}{4} \int_{-1/2T_s}^{1/2T_s} f^2 D^2(f)\, df \Big/ \int_{-1/2T_s}^{1/2T_s} D^2(f)\, df, \tag{6.34}$$

where $D(f)$ represents the Fourier transform of the data window (see Table 5.2). Figure 6.8 demonstrates that uniformly weighted data can contribute a significant bias even at $M = 64$ samples.

It should be emphasized that the estimated width (6.33) is biased even at large SNR if spectra have width comparable to the Nyquist interval; in this respect the autocovariance method is superior to Fourier processing. In order to maintain accuracy, however, the systems must be designed to have maximum expected spectrum widths reasonably smaller than T_s^{-1} so that the width estimated by Fourier processing will have negligible bias. For narrow spectra both width estimates have biases that are small but can be an appreciable fraction of the estimate (Zrnić, 1979b; Fig. 6.8).

Next we present a formula for the variance of the width assuming that the integral form of (6.33) is used, $\sigma_{vn} \ll 1$, and the noise spectral density is subtracted (Berger and Groginsky, 1973):

$$\mathrm{var}(\hat{\sigma}_v) = \frac{\lambda^2 T_s}{16M\sigma_{vn}^2 S^2} \int_{-1/2T_s}^{1/2T_s} (f^2 - \sigma_f^2)^2\, S^2(f + f_d)\, df. \tag{6.35}$$

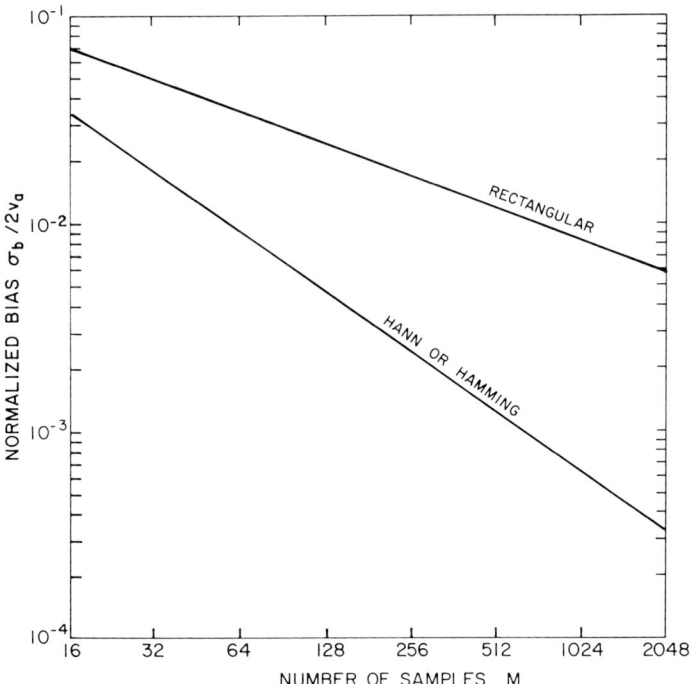

Fig. 6.8. Bias in the width estimate because of the finite time window associated with Fourier spectral processing. The curves are valid provided $\sigma_{vn} \gg M^{-1}$.

In (6.35), $\sigma_f = 2\sigma_v/\lambda$. For narrow Gaussian spectra and white noise this formula becomes

$$\text{var}(\hat{\sigma}_v) = \frac{\lambda^2}{4MT_s^2}\left[\frac{3\sigma_{vn}}{32\sqrt{\pi}} + \sigma_{vn}^2\frac{N}{S} + \left(\frac{1}{320\sigma_{vn}^2} - \frac{1}{24} + \frac{\sigma_{vn}^2}{4}\right)\frac{N^2}{S^2}\right]. \quad (6.36)$$

Unlike the autocovariance estimator of the spectrum width, the variance here does not increase exponentially with width. Note that the first term of (6.36) is the right side of (6.31), meaning that applying the Fourier method is equivalent to the pulse pair method at large SNR. In contrast, at low SNR the pulse pair variance, (6.32), is about three times smaller than the corresponding term of (6.36).

It is significant that mean frequency and width estimators have variances comprising a signal contribution, a noise part, and a cross product of signal and noise. At large SNR (20 dB or so) the signal contribution dominates and the accuracy is solely a function of spectrum width. In such a case, averaging over range and azimuth decreases the error at the expense of resolution.

However, at low SNR the variance is proportional to SNR^{-2}, and a much more efficient method of reducing it is to increase the SNR rather than to average the estimates along range. For distributed targets this can be accomplished by increasing the pulse length and using a matched filter such that the desired range resolution equals the combined weighting function of the filter and the transmitted pulse (Zrnić and Doviak, 1978; Section 4.4).

6.6 MINIMUM VARIANCE BOUNDS

It is characteristic of maximum-likelihood estimators that they provide estimates with minimum variance when a large number of samples M is used. Although these estimators may be quite complex and their algorithms are often unknown, we can nevertheless compute their variance, which establishes a lower bound to be compared with the variance achieved using simpler algorithms. A comprehensive treatment of the Cramer–Rao bounds is that of Zrnić (1979a), from which we have extracted the principal results. The minimum variance for the estimates of mean velocity, assuming a pure sinusoid in white Gaussian noise, is

$$\min \text{var}(\hat{v}) = 3\lambda^2 N / 8 M^3 \pi^2 T_s^2 S. \tag{6.37}$$

For weather echoes of Gaussian-shaped spectra, the following two approximate expressions describe the minimum variance well:

$$\min \text{var}(\hat{v}) = 3\lambda^2 \sigma_{vn}^4 / M T_s^2 (1 - 12\sigma_{vn}^2) \tag{6.38a}$$

in the case of zero noise, and

$$\min \text{var}(\hat{v}) = (\sqrt{\pi} \lambda^2 \sigma_{vn}^3 / M T_s^2)(N^2 / S^2) \tag{6.38b}$$

for large noise.

Discussion of these bounds and comparison where appropriate with the best that the pulse pair algorithm can do follows. We shall constrain the analysis to a constant dwell time $T_d = MT_s$ because (1) this is a parameter that, together with the antenna rotation rate, determines the apparent antenna beamwidth (see Chapter 7) and (2) a comparison of estimators is meaningful if they operate on a time series of constant length. For the *zero noise* case and contiguous samples, from (6.22a), we find the minimum variance in pulse pair estimates of $\bar{v}$ is achieved when

$$T_s(\text{optimal})_{\text{pp}} = 0, \quad N = 0. \tag{6.39a}$$

Thus continuous sampling achieves the largest reduction of the pulse pair velocity variance, which becomes

$$\text{var}(v) = \lambda \sigma_v / 8 T_d \sqrt{\pi}. \tag{6.39b}$$

For this case of continuous sampling the bound (6.38a) is zero, which, of course, can not be achieved with any estimator with a finite dwell time. This is because the assumption of independent samples on which (6.38a) is based is not valid as T_s becomes very small. Therefore, we shall compare the pulse pair estimator to the Cramer–Rao bound for a constant dwell time and with $T_s = T_d/M_I$, where M_I is the equivalent number of independent samples (6.10). For a Gaussian spectrum $M_I = M\sigma_{vn}\sqrt{2\pi}$, so that (6.38a) becomes

$$\min \operatorname{var}(\hat{v}) = 6\lambda\sigma_v/T_d(2\pi)^{3/2}. \qquad (6.40a)$$

The corresponding variance for the pulse pair estimator is found from (6.22a):

$$\operatorname{var}(\hat{v}) = (e^{2\pi} - 1)\lambda\sigma_v/16\pi\sqrt{\pi}\,T_d. \qquad (6.40b)$$

This is 15.7 times larger than the minimum bound (6.40a).

A similar derivation for weak signal power ($N \gg S$) leads to the value of T_s that minimizes the variance of the pulse pair velocity estimates:

$$T_s(\text{optimal})_{pp} = \lambda/4\pi\sigma_v\sqrt{2}, \qquad N \gg S. \qquad (6.41)$$

The reason for a minimum lies in two opposing effects. First, because the pulse pair measures the phase from which the frequency (velocity) is computed, it follows that for any given error in phase, the error in the velocity decreases with an increase in T_s. However, as T_s increases, the correlation between phase samples decreases, which in turn increases the phase errors. The corresponding minimum variance of the pulse pair velocity estimates is

$$\operatorname{var}(\hat{v}) = (\lambda\sigma_v\sqrt{2e}/8T_d\pi)(N^2/S^2), \qquad (6.42)$$

and the Cramer–Rao bound (6.38b) under the appropriate condition (6.41) becomes

$$\min \operatorname{var}(v) = (\lambda\sigma_v\pi^{-3/2}/4T_d)(N^2/S^2). \qquad (6.43)$$

Thus, when signal-to-noise ratio is low, the pulse pair algorithm produces mean velocity estimates with a variance that is only $\sqrt{e\pi/2} \approx 2$ times larger than the best possible.

At a 10-cm wavelength and $\sigma_v = 4\ \text{m s}^{-1}$ we find $T_s(\text{optimal}) \approx 1.4$ ms, which is about 40% larger than the values typically used. Other factors (see Chapter 7) have motivated the use of the shorter pulse repetition times.

Expressions similar to (6.38a) and (6.38b) are available for the minimum variance of the Doppler spectrum width estimate. These are

$$\min \operatorname{var}(\hat{\sigma}_v) = 45\lambda^2\sigma_{vn}^6/MT_s^2 \qquad (6.44a)$$

in the zero noise case, and

$$\min \operatorname{var}(\hat{\sigma}_v) = \sqrt{\pi}\lambda^2\sigma_{vn}^3 N^2/MT_s^2 S^2 \qquad (6.44b)$$

in the case of large noise.

In summary, the Fourier method and autocovariance processing give comparable accuracy in spectral moments. Because the Fourier transform makes use of more information, one would expect it to be always better, but this is the case only in some situations, such as a pure sinusoid in white noise. The optimum processor, then, is a parallel bank of narrowband filters (these can be FFT coefficients), and the mean velocity is associated with the filter having the largest output (Whalen, 1971). Surprisingly, autocovariance (pulse pair) processing of a correlated sequence produces a maximum likelihood estimate of mean velocity if the signal has an exponential correlation function (Zrnić, 1979a). Thus in general the underlying signal statistics dictate the optimum processing, and for a range of signal-to-noise ratios and spectrum widths there is no unique solution. However, it is gratifying that relatively simple algorithms perform very well; they approach and even achieve the minimum variance bound for some class of signals.

6.7 PERFORMANCE ON DATA

Examples of data scattergrams are shown in Figs. 6.9 and 6.10. The Fourier-transform-derived velocity versus pulse pair velocity follows a 45° straight line (Fig. 6.9). The data are taken from a storm, and processing was done with a hard-wired pulse pair calculator. Time series from the same resolution volumes were simultaneously recorded and processed later on a computer. The radar's unambiguous velocity is ± 34 m s^{-1}.

In obtaining the scattergram of widths, time series were recorded and pulse pair [(6.27)] and Fourier widths [(6.33)] were calculated on a computer. The scatter is good in the midrange and has large deviations and a negative bias at low widths, as predicted by the theory [see Fig. 6.7 for deviations and Zrnić (1977b) for bias]. On the average the pulse pair width is larger than the FFT width. This is partly because a 15-dB threshold below the spectral peak was set to suppress noise and spurious peaks in the FFT method, in part because the spectra may not have a Gaussian shape and because of signal quantization.

A typical Doppler spectrum from a thunderstorm is plotted on Fig. 6.11. This spectrum is obtained from a Fourier transform of 64 samples weighted with a von Hann window. The unambiguous velocity in this example is 28.5 m s^{-1}, so the spacing of the spectral coefficients is 0.9 m s^{-1}. Both autocovariance and spectral analysis (with a 15-dB threshhold) were used to obtain the mean velocity and spectrum width. The two moments computed from the autocovariance are the mean velocity $\hat{v} = 14.9$ m s^{-1} and spectrum width $\hat{\sigma}_v = 2.5$ m s^{-1}. The Fourier method yields $\hat{v} = 15.4$ m s^{-1} and $\hat{\sigma}_v = 2$ m s^{-1}. To find the uncertainty in the estimates derived from the autocovariance, one can use Figs. 6.6 and 6.7 or Eqs. (6.23) and (6.31). Since we do not

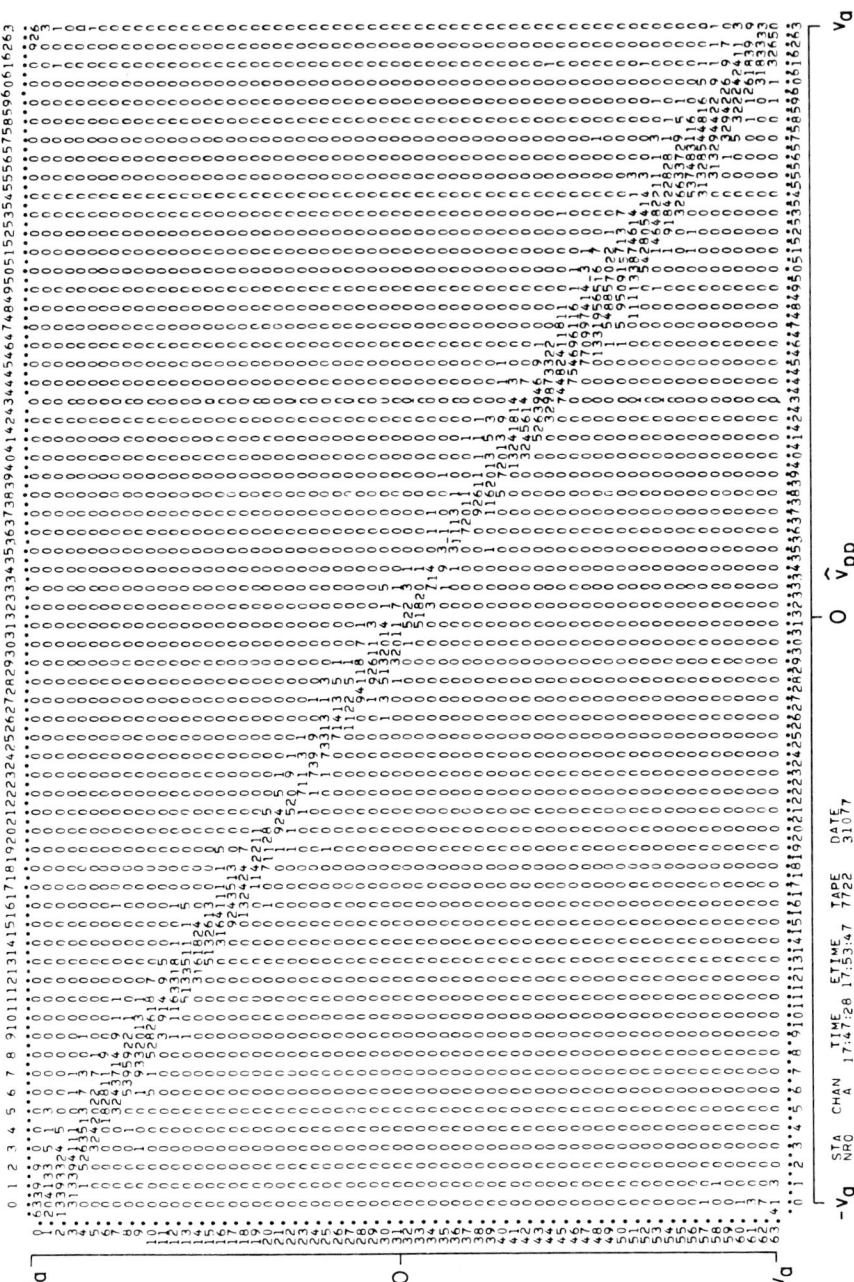

Fig. 6.9. Scattergram of the hard-wired pulse pair velocity $\hat{v}_{pp}$ and power-spectrum-derived velocity $\hat{v}_{FFT}$ for common pulse volumes. Only velocities of echoes with signal-to-noise ratio between 0 and 15 dB are included. The Nyquist velocity v_a is 34 m s^{-1}.

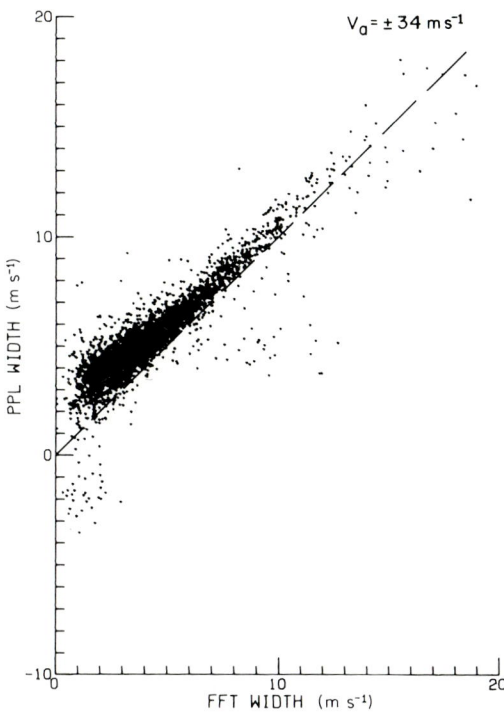

Fig. 6.10. Scattergram of pulse pair versus power spectrum (FFT) widths in meters per second. The Nyquist velocity is 34 ms^{-1}; SNR > 10 dB, and the data are from a storm 100 km away. The L in PPL indicates that (6.27) has been used to estimate the width. The square root nonlinearity of (6.27) bunches the data into the lower and upper groups near zero.

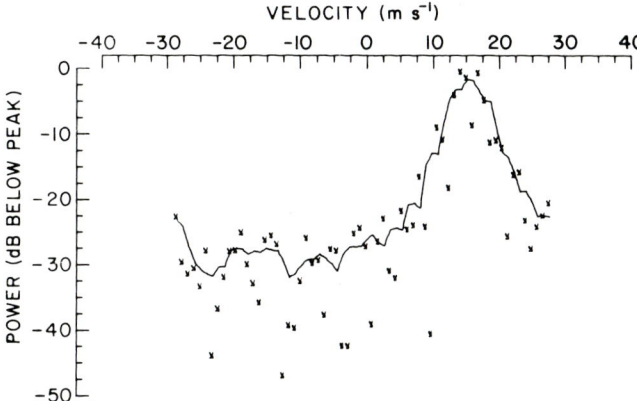

Fig. 6.11. Doppler spectrum from a storm collected at midnight on May 3, 1978. The solid line is a five-point running average of spectral points (×), the data are weighted with the von Hann window, and the signal-to-noise ratio is in decibels. Azimuth: 98.7°; elevation: 2.4°; altitude: 4.180 km; range: 88.752 km; gate 12; SNR: 17.

know the true spectrum width, let us take the average, 2.25 m s^{-1}, of the two measurements. SNR = 17 dB, and one can safely use (6.23) (with $\lambda/4T_s =$ 28.5 m s^{-1} and $M = 64$) to find for $\hat{v}$ a standard deviation of 0.53 m s^{-1}. The same result is obtained for the Fourier-derived mean velocity because at large SNR (6.23) is equivalent to (6.25). The standard deviation of the spectrum width is found from Fig. 6.7. With our values, the standard deviation of $\hat{\sigma}_v$ is 0.57 m s^{-1}. For Fourier-derived widths we use the first term of (6.36) and find 0.32 m s^{-1}. Both of these errors are typical, and the accuracy is adequate for most measurements in storms.

Finally, a display on an oscilloscope of the three moment estimates along a radial demonstrates what is to be expected form uniform precipitation (Fig. 6.12). Two traces (d) and (e), respectively, are nonintegrated and integrated logarithmic video samples. There are 762 range gates spaced 1 μs apart, the pulse repetition period is 768 μs, and the recursive first-order filter (digital integrator) operates on logarithmic samples of power with a time constant of 24 ms for this example, which is equivalent as far as the variance reduction to $M = 49$ uniformly weighted samples. With spectrum widths of 3 m s^{-1} there are about ten independent samples.

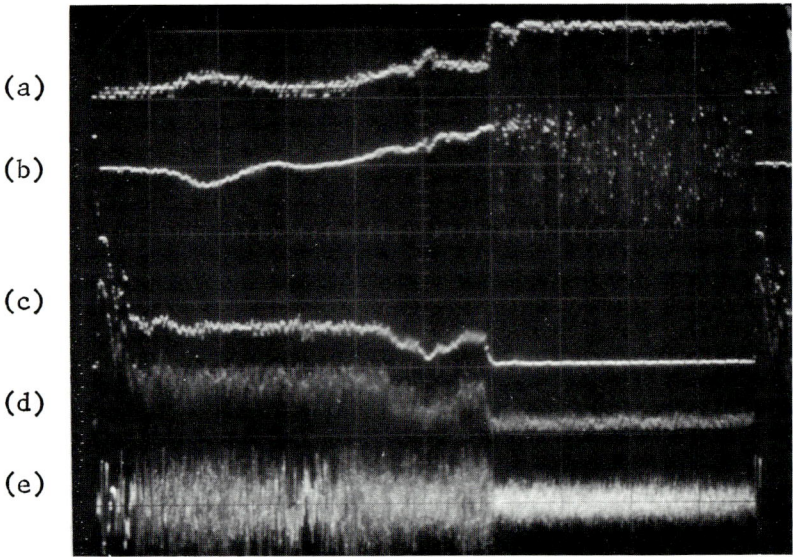

Fig. 6.12. Display as a function of range time (along a radial) of (a) the mean spectrum width (of noise plus signal), calculated from the autocovariance, with a vertical scale of 15 m s^{-1}/div; (b) the mean velocity, also from the autocovariance, and with a scale of 34 m s^{-1}/div; (c) the mean (integrated) power estimates $\hat{P}$ in log units (40 dB/div) estimated with a first-order recursive filter (exponentially weighted integrator); (d) nonintegrated power samples; and (e) one component of the Doppler channel (either I or Q). The horizontal scale is 11.5 km/div.

The mean velocities (b) and spectrum widths (a) are analog representations of values digitally calculated using the autocovariance algorithms (6.19) and (6.29). Data were collected while the antenna was stationary but elevated to 4°. The number of samples for spectrum width and velocity calculations is 64. Fairly uniform rain created echoes that last in range until the beam exceeds the precipitation top (4 km at a range of about 60 km). The continuity of and low spread in the width and velocity estimates up to 60 km are indicative of a large SNR. Note the uniform spread of velocities due to noise in the region of no echoes. Because the width is computed from (6.29) but is noise biased (i.e., N is not subtracted from the power estimate $\hat{P}$, it shows a sharp discontinuity at 60 km and large values at places where there is no echo.

In summary, the first three spectral moments can be routinely obtained at over 1000 contiguous range locations at speeds commensurate with real-time applications (i.e., antenna rotation rate of around a few rpm's). The speed and convenience of velocity and width calculations were improved by the advent of the pulse pair processor and associated digital circuitry. As these circuits become even more accessible, Fourier methods of spectral processing may find wider use in weather radars, especially when signals can be contaminated by spectral artifacts (side lobes, power supply ripple frequencies, etc.). Although spectral processing that uses the classical definition of moments is inferior to the autocovariance method at low SNR and narrow spectrum widths, the estimators based on the autocovariance for broader spectra have an exponential increase in the standard error; this is not the case for the Fourier method, which creates finite errors in both moment estimates and merely develops a bias in the width estimate. More important advantages of the Fourier method are the absence of bias due to nonsymmetric spectra and the feasibility of eliminating anomalous spectral powers.

We emphasize that an estimate of the power, mean velocity, or spectrum width at a single (isolated) range gate has little meteorological meaning; there remain statistical uncertainty and possible contamination from range-overlaid echoes, anomalous targets, ground clutter, etc. Before accepting spectral moment data, one must check them for meteorological consistency over contiguous range and azimuthal locations, or else an examination of the power spectrum should be made.

REFERENCES

Bello, P. A. (1965). On the RMS bandwidth of nonlinearly envelope detected narrow-band Gaussian noise. *IEEE Trans. Inf. Theory* **IT-11**, 236–239.

Berger, T., and Groginsky, H. H. (1973). Estimation of the spectral moments of pulse trains. *Int. Conf. Inf. Theory, 1973.*

Davenport, W. B., and Root, W. L. (1958). "An Introduction to the Theory of Random Signals and Noise." McGraw-Hill, New York.

Kerr, D. W. (1951). "Propagation of Short Radio Waves." McGraw-Hill, New York.
Marshall, J. S., and Hitschfeld, W. (1953). Interpretation of the fluctuating echo from randomly distributed scatterers. *Can. J. Phys.* **31**, Pt. I, 962–995.
Mueller, E. A., and Silha, E. J. (1978). Unique features of the CHILL radar system. *Prepr., Conf. Radar Meteorol., 18th, 1978* pp. 381–382.
Papoulis, A. (1965). "Probability, Random Variables, and Stochastic Processes." McGraw-Hill, New York.
Rogers, R. R. (1971). The effect of variable target reflectivity on weather radar measurements. *Q. J. R. Meteorol. Soc.* **97**, 154–167.
Rutkowski, W., and Fleisher, A. (1955). R-meter: An instrument for measuring gustiness. *MIT Weather Radar Res. Rep.* No. 24.
Sirmans, D., and Bumgarner, W. (1975). Numerical comparison of five mean frequency estimators. *J. Appl. Meteorol.* **14**, 991–1003.
Walker, G. B., Ray, P. S., Zrnić, D., and Doviak, R. (1980). Time, angle, and range averaging of radar echoes from distributed targets. *J. Appl. Meteorol.* **19**, 315–323.
Whalen, A. D. (1971). "Detection of Signals in Noise." Academic Press, New York.
Zrnić, D. S. (1975). Signal-to-noise ratio in the output of nonlinear devices. *IEEE Trans. Inf. Theory* **IT-21**, 662–663.
Zrnić, D. S. (1977a). Mean power estimation with a recursive filter. *IEEE Trans. Aerosp. Electron. Syst.* **AES-13**, 281–289.
Zrnić, D. S. (1977b). Spectral moment estimates from correlated pulse pairs. *IEEE Trans. Aerosp. Electron. Syst.* **AES-13**, 344–354.
Zrnić, D. S. (1979a). Estimation of spectral moments for weather echoes. *IEEE Trans. Geosci. Electron., Spec. Issue Radio Meteorol.* **GE-17**(4), 113–128.
Zrnić, D. S. (1979b). Spectrum width estimates for weather echoes. *IEEE Trans. Aerosp. Electron. Syst.* **AES-15**(5), 613–619.
Zrnić, D. S., and Doviak, R. J. (1978). Matched filter criteria and range weighting for weather radar. *IEEE Trans. Aerosp. Electron. Syst.* **AES-14**, 925–930.

7

Considerations in the Observation of Weather

This chapter examines limitations in pulsed-Doppler radar observations of the atmosphere caused by range-velocity ambiguities, target decorrelation, ground clutter, and antenna side lobes and rotation. Furthermore, because of the statistical uncertainties associated with Doppler spectral moment estimation, sometimes an undesirably long dwell time is required for acceptable measurement accuracy. Various techniques to mitigate these restrictions are described. Finally, we briefly discuss how radar hardware affects measurement accuracy.

7.1 RANGE AMBIGUITIES

To illustrate range ambiguities, Fig. 7.1 shows a conglomerate of thunderstorm cells, to ranges beyond 300 km, as seen by a (WSR-57) radar having a long PRT (6 ms, with an unambiguous range of 900 km). One of the storm cells produced a tornado that was tracked with a Doppler radar located near the WSR-57. The Doppler radar's PRT is significantly shorter in order to have a reasonably large (34 m s^{-1}) unambiguous velocity, and consequently its unambiguous range is only 115 km; Fig. 7.2 shows the range-ambiguous storms as observed by the Doppler radar. Not only are the ranges ambiguous, but because there were storms within as well as beyond 115 km, the tornado-producing cell at 150 km (within the Doppler radar's second-trip $cT_s/2$ interval in this particular case) is partially overlaid with (obscured by) echoes from storms in other trips (first and third $cT_s/2$ intervals). This may make accurate spectral moment estimation difficult.

Doppler velocity and spectrum width measurements are possible only within unobscured regions of this second-trip storm when the Doppler radar, as in this case, is fully coherent from pulse to pulse so that phase information is preserved for all echoes. However, if echoes happen to be overlaid, then

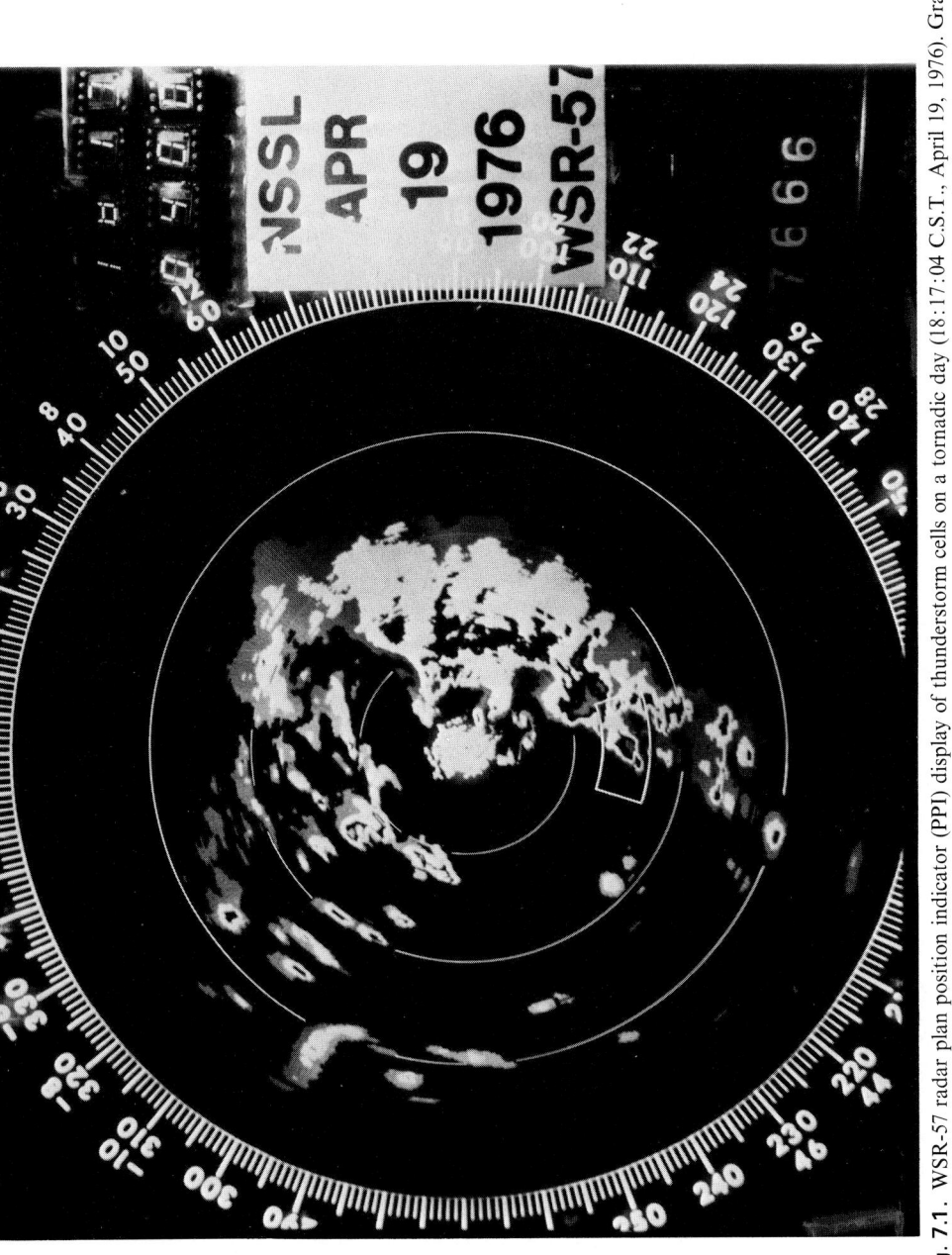

Fig. 7.1. WSR-57 radar plan position indicator (PPI) display of thunderstorm cells on a tornadic day (18:17:04 C.S.T., April 19, 1976). Gray shadings (dim, bright, black, dim, etc.) represent dBZ levels differing by about 10 dBZ starting at 17 dBZ. Range marks are 100 km apart; elevation angle = 0.0°. The unambiguous range is 900 km. The boxed area outlines a tornadic storm cell whose mesocyclone signature was detected in real time by NSSL's Norman (NOR) Doppler radar, which is nearly colocated with the WSR-57.

7.1 Range Ambiguities

moment estimation is still possible with a fully coherent radar for the echo that has significantly more power (≥ 10 dB for velocity and 15 dB for width) than the sum of the other trip echoes being overlaid. If a phase diversity radar is used with appropriate filtering (see Section 7.4.1), then echo power need not be greater than the sum of other echoes in order to obtain reasonably accurate velocity measurements.

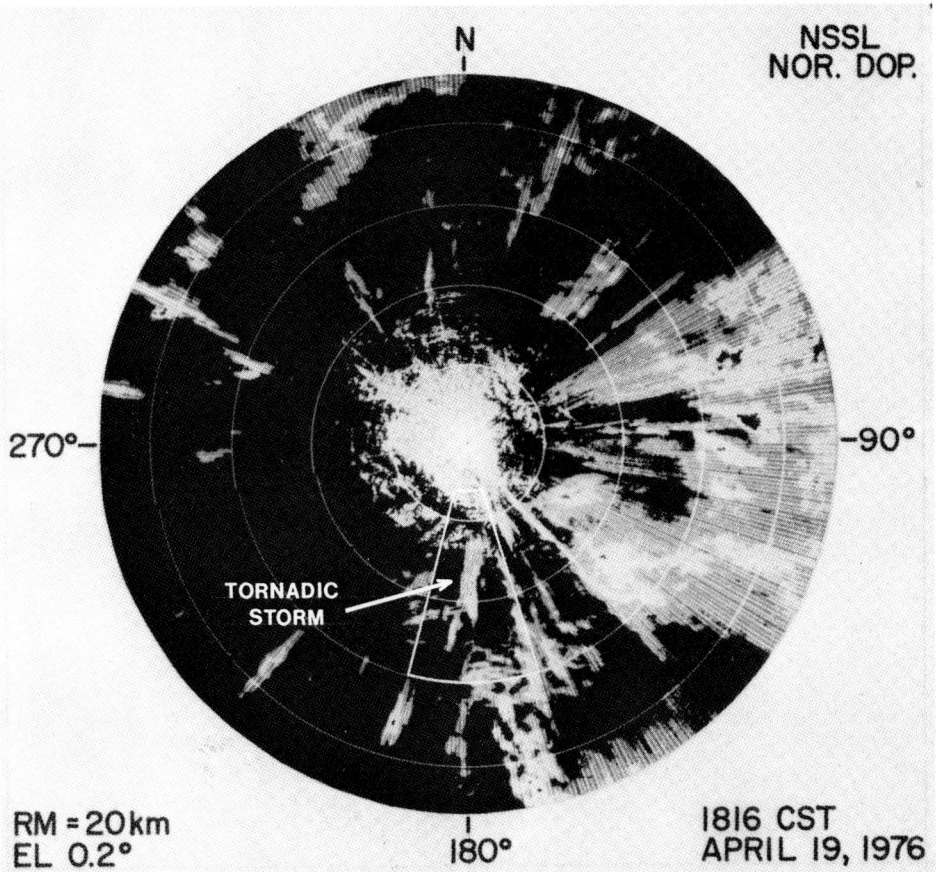

Fig. 7.2. Same storm system as in Fig. 7.1 (18:16:35 C.S.T.), seen with the Doppler radar having 115-km unambiguous range. 10 log Z brightness categories (dim, bright, etc.) start at at 10 dBZ and are incremented in ~ 10-dBZ steps. The 10 log Z scale applies only to first-trip echoes. Some range-overlaid echoes can be recognized by their radially elongated shape. The box outlines the same areas as in Fig. 7.1. Range marks are 20 km apart. Part of the tornadic storm is obscured by ground clutter and a nearby (30–60-km range) first-trip storm.

Range-overlaid storm cells can obscure radar signatures associated with significant meteorological phenomena, such as tornado cyclones and downdrafts, that harbor wind shears hazardous to aircraft near the ground in their descent into or departure from an airport. The characteristics of signatures are related to meteorological phenomena in Chapter 9. Gust fronts produced by strong downdrafts can extend far beyond the precipitation areas, where their effective reflectivity factors could be as low as 0 dBZ and sometimes even less (Table 9.2). Tornado cyclone reflectivities are usually significantly higher (>10 dBZ). Rain evaporation, while intensifying the downdraft, decreases the reflectivity, which can allow highly reflective second-trip storms to obscure observation of significant wind shear.

7.1.1 Probability of Obscuration by a Single Cell

The discussion in this section applies to uncoded transmitted pulses that produce a uniform sequence of echoes used for mean Doppler velocity estimation. Because range-overlaid cells are troublesome only when their echo power exceeds about one-tenth the power P_s associated with a signature echo, we compute the expected overlaid area A_{cn} (from an nth-trip storm cell, $n > 1$) within the first-trip ($n = 1$) radar coverage having echo power larger than $\frac{1}{10}P_s$. In the following development we assume that the signature echo comes from a storm within the first trip, and that second-trip and more distant storms have diameters small compared to r_a and are located at the middle range of each trip. Storm cells of diameter $2r_0$ are assumed to have a symmetric profile of reflectivity $Z = Z_0(1 - r/r_0)^2$. Geometric considerations lead to

$$A_{cn} = \frac{\pi r_0^2 [1 - (2n - 1)(Z_s/20Z_0)^{1/2}]^2}{2n - 1}, \qquad Z_0 \gtrsim \tfrac{1}{20}Z_s(2n - 1)^2$$

$$= 0 \quad \text{otherwise} \tag{7.1a}$$

(Doviak et al., 1978), where Z_s is the reflectivity factor of the first-trip echo (from a resolution volume at the approximate expected range $r_a/2$). If a signature is assumed to be a point representation of a phenomenon and to have equal likelihood of being anywhere within r_a, the probability P_0 that its detection would be obscured by a single overlaid cell is simply the ratio of A_{cn} to πr_a^2. Thus

$$P_0 = \frac{r_0^2[1 - (2n - 1)(Z_s/20Z_0)^{1/2}]^2}{(2n - 1)r_a^2}, \tag{7.1b}$$

Equation (7.1b) is plotted in Fig. 7.3 for the case $Z_s/Z_0 = 0$ and $Z_s = Z_0$ and for obscuring cells in different trip zones. This figure shows P_0 plotted

7.1 Range Ambiguities

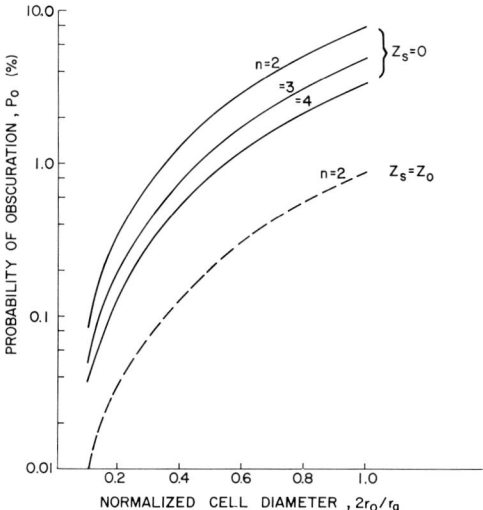

Fig. 7.3. Probability of signature obscuration by a single storm cell: n is the trip zone where the overlaying cell is located, Z_0 its peak reflectivity factor, and Z_s the signature echo reflectivity factor (mm^6 m^{-3}). Signatures are for phenomena (downdraft or mesocylone, etc.) assumed to be located within the first-trip area πr_a^2.

versus the normalized cell diameter $2r_0/r_a$. When $Z_s = 0$, P_0 estimates the percent of multiple-trip echo area that is overlaid onto the unambiguous area πr_a^2. If there are N cells located in the nth-trip zone, the probabilities of obscuration by these cells can be assumed to be the sum of obscuration probabilities for single cells. The conditions $Z_s = 0$ mm^6 m^{-3} and $n = 2$ represent a worse case, giving an upper limit to the percentage of obscuration of the first-trip area by a cell in a second trip. Consider, for example, a storm system in which there are no more than ten second-trip cells having diameters of 20–30 km and $r_a = 100$ km; reference to Fig. 7.3 shows P_0 to be less than 10%. Estimates of cell diameter, trip number, and number of cells from Fig. 7.1 were used to calculate P_0. Echoes with reflectivity larger than the 17 dBZ displayed on the PPI were assumed to cause obscuration. With $r_a = 115$ km the worst case (i.e., $Z_s = 0$) probability of obscuration is calculated to be about 20%, in relatively good agreement with the overlaid area actually seen in Fig. 7.2.

In conclusion, weather Doppler radars having r_a large compared to (i.e., five or more times) storm cell diameters should infrequently experience obscuration of first-trip phenomena by randomly distributed cells. However, as shown in the following section, when cells are organized along a line as is the case for a squall line, P_0 increases significantly.

7.1.2 Obscuration by a Squall Line

We now estimate the probability of obscuring detection of a phenomenon within the first-trip portion of a squall line assuming thunderstorm cells to be uniformly distributed along a straight line, as in Fig. 7.4. Hazardous phenomena such as tornados or gust fronts are assumed to occur with uniform probability throughout the first-trip area A_s of the storms (Fig. 7.5).

Cells are assumed to be spaced d_s apart where $d_s \geq 2r_0$, and the obscuration probability P_{0s} by a squall line is the ratio of the total overlaid echo area (within which echo power is larger than $\frac{1}{10}P_s$) to the area A_s. Because cells in the second-trip zone subtend the largest angle, cells in higher-order zones are contained within this angle (or fall outside A_s, as occurs when $d \to r_a$). Hence their obscuration area is likely to overlap those areas obscured by cells in the second trip. For that reason, and for simplicity, we shall ignore cells beyond the second trip.

Therefore, P_{0s} becomes proportional to the ratio of summed overlaid areas $\sum_k A_{c2}(k)$ to A_s:

$$P_{0s} \propto \sum_k A_{c2}(k)/A_s. \quad (7.2a)$$

Not all $A_{c2}(k)$ obscure A_s, because the farther the squall line is from the radar, the smaller is the portion of $\sum_k A_{c2}(k)$ that overlaps A_s. The area of overlap is assumed to be proportional to

$$L/r_a = \begin{cases} 0, & d \geq 2r_0, \\ (r_0 - 0.5d)/(r_0 + d), & 0 \leq d \leq 2r_0, \\ 1, & -2r_0 \leq d \leq 0, \end{cases} \quad (7.2b)$$

where L/r_a (see Fig. 7.5) is the fraction of second trip echoes that overlay echoes in A_s when the beam is directed to the squall line center at $1.5r_a$. Furthermore, we consider, as in the previous section, that second-trip cells contribute equal overlaid area A_{c2} and number about $2r_a/d_s$. Thus we can combine the foregoing equations to obtain

$$P_{0s} = \pi r_0^2 \left[1 - 3\left(\frac{Z_s}{20Z_0}\right)^{1/2}\right]^2 \left[\frac{2r_a(r_0 - 0.5d)}{3d_s A_s(r_0 + d)}\right] \quad (7.3)$$

for the probability of obscuration by a squall line. Equation (7.3) is plotted in Fig. 7.6 for the worst case (i.e., $Z_s = 0$, $d_s = 2r_0$). We note that obscuration commences when the squall line is at a distance equal to its width ($2r_0$). When $d \geq 2r_0$, second-trip storm echoes do overlay into the first-trip zone, but overlaid areas A_{cn} fall outside A_s and hence should not obscure the detection of phenomena (e.g., mesocyclones) that are within the squall line.

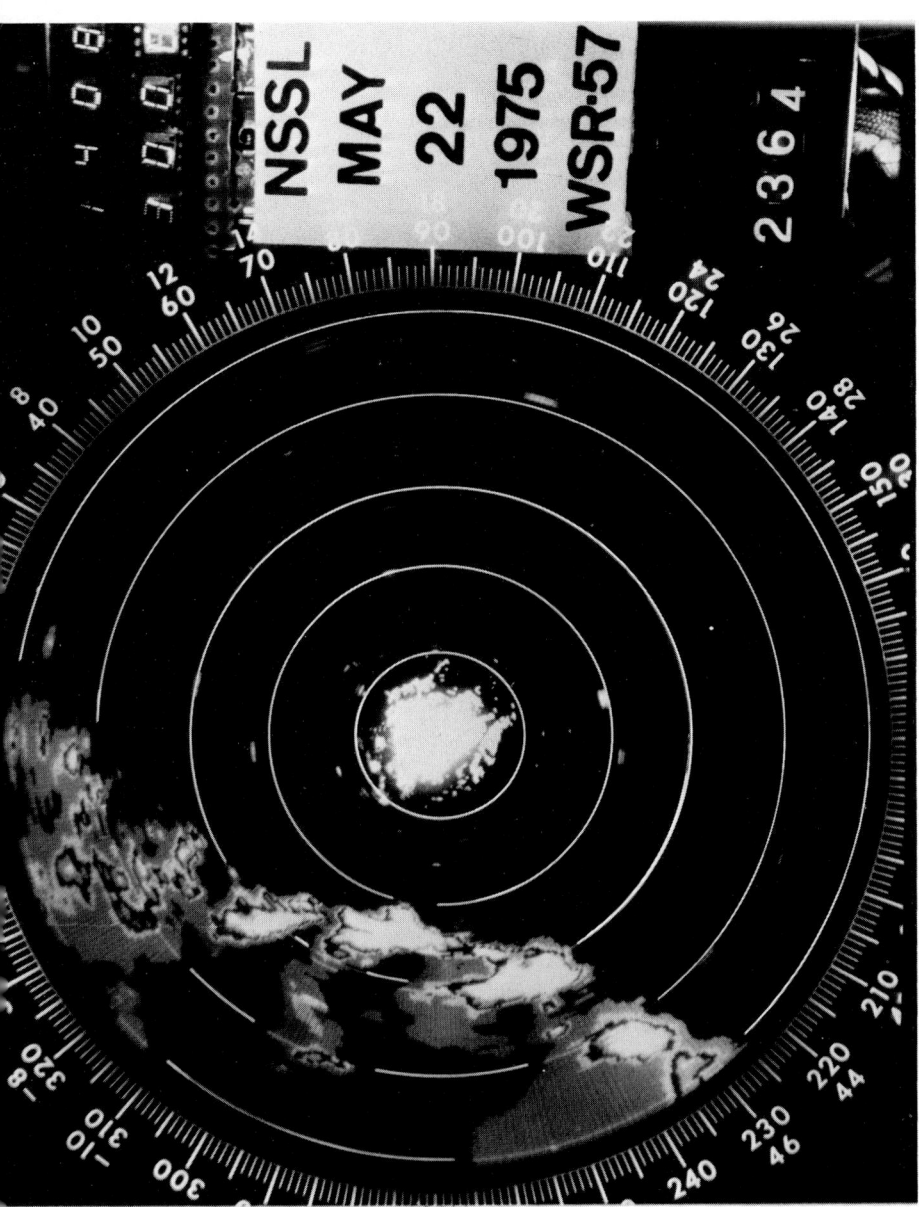

Fig. 7.4. Thunderstorm squall line (14:08:10 C.S.T., May 22, 1975) from a WSR-57 radar system. Range marks are 40 km; the unambiguous range is 900 km.

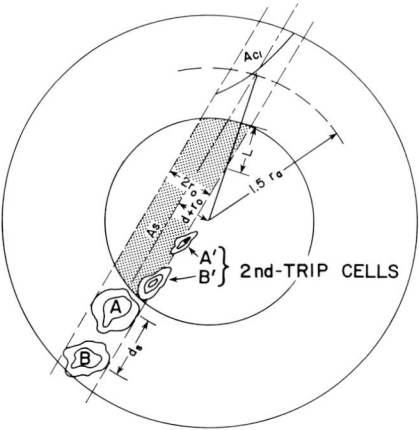

Fig. 7.5. Distribution of overlaid cells in a squall line. A_s is the area (stippled) within which signatures might occur; the squall line distance from the radar is d, and d_s is the cell separation. A_{cl} is the area that, when overlaid into the first-trip zone, overlaps A_s.

However, phenomena such as gust fronts can extend to distances far beyond the boundaries of the cells, and portions or all of these might be obscured by overlaid echoes.

Figure 7.6 also shows that when r_a is much larger than (for instance, two or more times) the squall line width, P_{0s} is relatively independent of r_a. If r_a is merely equal to $2r_0$, there are significant increases in P_{0s}, as can be seen in Fig. 7.6 for the case $2r_0 = 40$ and $r_a = 50$ km. When the squall line is at the radar site, P_{0s} increases to values near 30%, which is nearly independent of parameters. Thus, it appears that little can be done to decrease significantly the probability of obscuration when the squall line is near a radar transmitting uncoded pulses (Section 7.4).

Although we have limited our analyses to the obscuration of phenomena lying within the first-trip area, we can deduce that phenomena located in

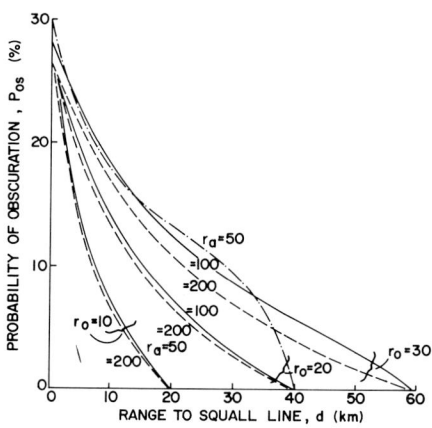

Fig. 7.6. Probability of signature obscuration by a squall line. Signatures are assumed to lie within the first-trip portion of the squall line. r_a is the unambiguous range, $2r_0$ the squall line width, d the distance of the leading edge of the squall line from the radar, and d_s the cell separation. $d_s = 2r_0$. $Z_s = 0$.

7.2. Velocity Ambiguities

the second- (or higher-order) trip zones would experience a larger probability of obscuration. This is because storm cells in the first trip occupy a larger fraction of the first-trip area and hence a much larger fractional area when overlaid onto the higher-order trip zone. Furthermore, first-trip echo power is usually larger because it has an r^{-2} advantage.

7.2 VELOCITY AMBIGUITIES

Target velocities become ambiguous when we cannot distinguish between actual Doppler shifts and aliases that are spaced in frequency by the pulse

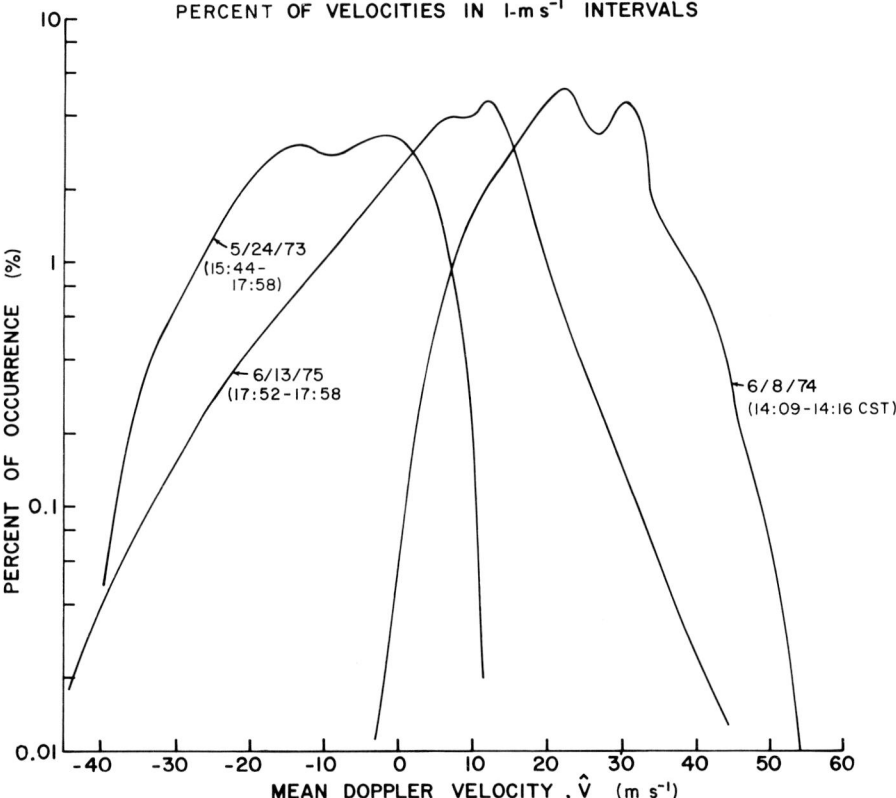

Fig. 7.7. Relative frequency of occurrences of the mean Doppler velocity estimates for three tornadic storms. Data samples are uniformly spaced throughout most of the convective cell. Note the large spread of radial Doppler velocities, which needs to be measured unambiguously. The estimates were obtained from the autocovariance algorithm.

repetition frequency. The range–velocity product

$$r_a v_a = c\lambda/8 \tag{7.4}$$

typifies the ambiguity resolution capabilities of a Doppler radar with uniformly spaced pulses. When both I and Q samples are processed to resolve the sign of the Doppler shift, the unambiguous velocities span the interval $\pm v_a$. The equation shows the advantage of longer wavelengths, but other factors may control this choice. Radar waveform designs formulated to remove ambiguities when targets are discrete and finite in number, or when cross sections do not span a large dynamic range (Deley, 1970), do not work well with weather targets that are distributed quasicontinuously over large spatial regions (tens to hundreds of kilometers) and whose echo strengths can span an 80-dB power range. In Section 7.4 we discuss techniques to resolve many of the ambiguities.

The plots of mean radial velocities shown in Fig. 7.7 illustrate the velocity distributions that can be found in severe storms. Some 20,000 sample points from resolution volumes near ground to about 10 km in altitude are plotted. The centers of the velocity distributions (Fig. 7.7) are displaced relative to one another and to zero, in part because of storm motion. More important than the mean motion or peak radial speeds is the $>50\text{-m s}^{-1}$ spread in velocities for any of these storms. Doppler radars that observe severe storms require an unambiguous velocity span of at least ± 35 m s^{-1} in order to limit velocity aliasing to a few percent or less.

7.3 ECHO COHERENCY

In principle one can choose T_s large enough that no second- or higher-order trip echoes (from storms) will ever be received, but increasing T_s is limited in that echo samples spaced T_s apart must be correlated for accurate Doppler measurements. Correlation exists when

$$\lambda/2T_s \gg \sigma_v, \tag{7.5a}$$

where σ_v is the velocity spectrum width of echoes at range r. Condition (7.5a) merely states that the Doppler width should be much smaller than the Nyquist interval. When T_s is increased, the correlation (6.3c) decreases exponentially, causing the variance in mean Doppler velocity estimates $\hat{v}$ and Doppler width estimates $\hat{\sigma}_v$ to increase exponentially, as can be seen from (6.22) and (6.30). This leads one to consider the inequality

$$\lambda/2T_s \geq 2\pi\sigma_v \tag{7.5b}$$

as necessary for the estimation of Doppler spectral moments. Equality occurs

7.3 Echo Coherency

when the square of the correlation equals e^{-1}, and this condition is chosen as a convenient correlation threshold. When the correlation decreases below this threshold it causes an exponential increase in the variance estimate, as shown in Figs. 6.6 and 6.7. In terms of the unambiguous range,

$$c\lambda/4r_a \geq 2\pi\sigma_v \quad (7.5c)$$

is the condition to maintain echo sample correlation. Requirement (7.5c) places a limit on r_a for a given σ_v and wavelength, whereas (7.4) restricts r_a only if ambiguities due to velocity aliases need to be suppressed by choosing a large v_a. Methods (see Section 7.4) to resolve velocity aliases work provided σ_v is sufficiently small. Thus (7.5c) is a necessary condition to maintain signal sample correlation, and σ_v limits r_a for a chosen λ or vice versa. Note that the spectrum width σ_v is dictated primarily by the weather target, as well as to some extent by the antenna rotation rate and resolution volume size.

To be definite, let equality in (7.5c) specify the maximum width σ_c to which coherency is maintained for given r_a, or, given a spectral width, the maximum r_c (i.e., T_s). Figure 7.8 relates σ_c to r_c with λ as a parameter. It is apparent that, unless spectrum widths are less than a few meters per second, 10-cm or shorter wavelength radars cannot eliminate range-ambiguous echoes (i.e., by having $r_a \geq 500$ km). Data from severe storms (Fig. 10.14) show a median width of 4 m s^{-1}, and about 20% of measured widths are larger than 6 m s^{-1}. Therefore, weather radars ($\lambda \leq 10$ cm) will usually be plagued by ambiguities

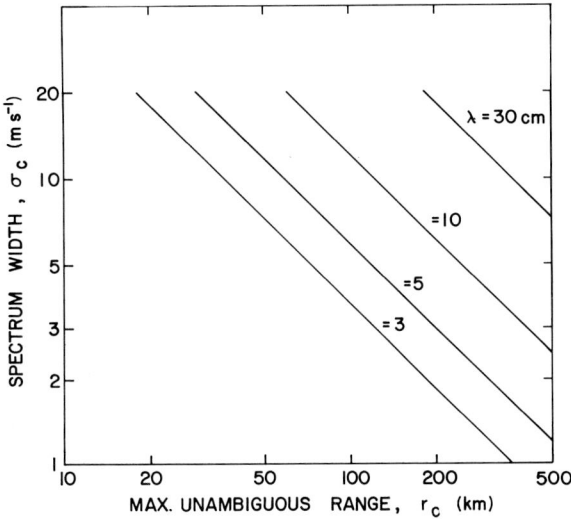

Fig. 7.8. Maximum spectrum width σ_c versus maximum unambiguous range r_c within which echoes are coherent. λ is the radar wavelength.

if Doppler measurements are to be made throughout most of the thunderstorm volume.

7.4 TECHNIQUES TO EXTEND THE UNAMBIGUOUS RANGE AND VELOCITY AND TO REDUCE THE LOSS OF INFORMATION FROM OVERLAID ECHOES

In the absence of practical methods to eliminate simultaneously range and velocity aliases, schemes have been devised that separate range and velocity measurements or minimize the deleterious effects of overlaid echoes. Some promising techniques will now be discussed.

7.4.1 Phase Diversity

Given a radar of wavelength λ, suppose that a suitable v_a is chosen to provide acceptable velocity aliasing. Then r_a is determined (7.4) and, in most cases, is so small that there can be second- and third-trip targets overlaying one another. Let the initial phase ψ (3.26) of the transmitted pulse be random from pulse to pulse, as occurs with transmitters using magnetrons. This random phase can be measured, for instance, by letting a very small fraction of the transmitter pulse leak into the receiver so that its I and Q signals can be measured (Nutten et al., 1979). Thus

$$\psi_k = \tan^{-1}[Q_k(0)/I_k(0)], \tag{7.6}$$

where k denotes the kth transmitted pulse and τ_s is set to zero, indicating that the I and Q are not caused by an echo but by a leakage voltage from the transmitter. It suffices to store the $Q_k(0)$ and $I_k(0)$ for as many PRTs as needed for coherent measurement of velocities in the desired $cT_s/2$ range intervals. The signals received must then have their phases corrected to account for the arbitrary phase of each transmitted pulse.

The echoes from the kth transmitted pulse can have their phases corrected by ψ_k over a period in excess of T_s. Thus, if one wants to make measurements in the first- and second-trip region, ψ_k must be stored for a $2T_s$ period, and phase corrections must be made (Fig. 7.9) in each of two channels, one for the first, the other for the second trip. However, the transmitter reference phase is updated in the first-trip channel immediately after the transmitter is fired, whereas in the second-trip channel the updated phase is lagged by one T_s period.

Echoes outside the selected trip are incoherent (i.e., like white noise because of random phases) when overlaid into the selected trip and thus appear

7.4 Techniques to Extend the Unambiguous Range and Velocity

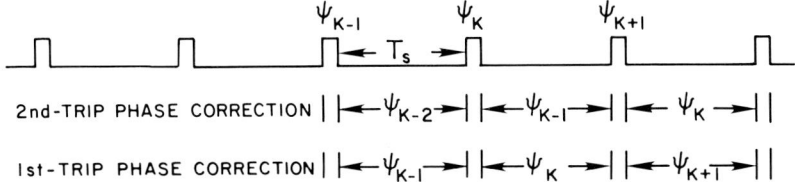

Fig. 7.9. Processing of a random phase pulse train. Transmitted phases ψ_k are random, and in this example the phase correction on echoes is made at different times in two separate receivers, which allows coherent measurement to a range cT_s.

as an increase in noise level. Although the variance of the spectral moment estimates increase because there is a decrease in the effective signal-to-noise ratio (e. g., Fig. 6.6), mean velocity estimates are not biased. To avoid this increase in noise one can adaptively filter out the unwanted signals in a two-step process: (1) in one of two channels (assuming only first- and second-trip echoes are present) the first-trip echoes are coherently summed, whereas in the other the second-trip echoes are coherently summed; (2) the coherent signal in the first (second) channel is then filtered and the residuals coherently summed, from which moment estimates are made for the second- (first-) trip echoes. Because the signal properties change from resolution volume to resolution volume, the filter characteristics should be adaptively adjusted using, for example, maximum-likelihood techniques similar to those described by Waldteufel (1976). Although the procedure would probably give the best overall performance, the complexity of signal processing is much greater than with other techniques.

Our description implies digitally storing the phase, but analog phase storage can be accomplished if two phase-locked oscillators are used to store the initial phase and then to retrieve coherent measurements on first- and second-trip echoes. Each COHO will hold a transmitter phase for two system periods with phase updates for the COHOs offset in time by T_s. The first-trip channel must be serviced by the COHO that had a phase update last; then after each new system period, one COHO has its phase updated, and both COHOS must be switched to alternately service the first- and second-trip receivers. Of course, more phase-locked oscillators are needed if one desires simultaneously to measure velocities in more than two trip regions.

7.4.2 Spaced Pairs with Polarization Coding

The use of spaced pulse pairs of orthogonally polarized samples (Fig. 7.10) for weather radars can reduce the occurrence of those echoes that overlay the desired echo (Doviak and Sirmans, 1973). The $P_t(V)$ and $P_t(H)$ are transmitted powers with vertical and horizontal polarization (electric field). This technique has been successfully adapted to radars measuring ionospheric motions (Woodman and Hagfors, 1969). Also, it has been implemented on weather radars but without orthogonal polarization diversity (Campbell and Strauch, 1976). The first advantage of the method is that, with T_2 sufficiently large, overlay is limited to first- and second-trip targets. Furthermore, overlaid echoes are incoherent, so they do not bias velocity estimates but only decrease the effective signal-to-noise ratio. We can achieve a third advantage when the pulses of a pair are orthogonally polarized because the overlaid echo power is then decreased, possibly by as much as 20 dB. Of course, this is at the expense of a more complicated receiver with two channels and associated microwave circuitry for separating the two orthogonal fields. However, the payoff might be worth the cost because the occurrence of pesky overlaid echoes would be insignificant.

However, a factor that needs to be considered in a polarization diversity radar is the presence of differential phase shift between the two orthogonally polarized signals caused by (1) propagation through rain in which drops are not spherical and (2) differential phase shift upon scattering caused by nonsphericity of the drops within the resolution volume. Because drops are flattened (see Fig. 8.1), signals with horizontal electric fields induce stronger dipole moments and have different phase shifts than those with vertically polarized fields. However, if covariance estimates are made on pairs of pulses in which the sequence of polarization is reversed on each subsequent pair, this bias can be eliminated.

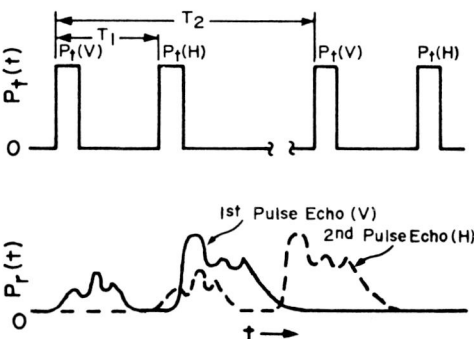

Fig. 7.10. Signal diagram for orthogonally polarized (V,H) samples of a pulse pair; P_t is the transmitter power and P_r the received mean power. Weather-type targets are assumed to produce the echo pattern of P_r.

A disadvantage of spaced pairs is that a longer time is required to collect a sufficient number of sample pairs to reduce the velocity estimate variance to acceptable limits. However, fewer sample pairs are needed to achieve the same measurement accuracy with spaced pairs than with uniformly spaced pulses because spaced sample pairs are less correlated [see Eq. (6.21) and Fig. 6.5]. Furthermore, by changing the carrier frequency between pairs of pulses, one can make pairs completely independent without the necessity of having T_2 larger than T_1, and hence the time to acquire the number of sample pairs needed to achieve a desired precision is reduced appreciably. In this case, the spaced pair technique is similar to the frequency diversity method described in section 7.5.1, but here we may have the advantage of polarization coding to reduce overlaid echo power between first- and second-trip echoes. Another disadvantage of this technique is that, for all practical purposes, ground clutter cannot be canceled.

7.4.3 Staggering the PRT to Increase the Unambiguous Velocity

Staggered PRT belongs to a general class of techniques whereby autocovariance or velocity estimates from two PRT's are suitably combined to effectively increase the composite unambiguous velocity. We shall briefly describe the philosophy behind methods that use two PRT's that may or may not be staggered. In a two-PRT technique, a velocity estimate $\hat{v}_1$ is obtained from echo samples spaced by T_{s1}, whereas a second velocity estimate $\hat{v}_2$ is derived from samples spaced by T_{s2}. For example, uniform spacing T_{s1} can be used during one scan, and the scan can be repeated immediately with the different but uniform sample time spacing T_{s2}. Because $\hat{v}_1, \hat{v}_2$ are associated with different Nyquist intervals, velocity aliasing can cause them to be significantly different (Fig. 7.11), and these differences can be used to resolve the true velocity. However, mean velocity aliases can only be resolved so long as the expected difference $E(\hat{v}_2 - \hat{v}_1)$ remains unambiguous.

Strictly, $E(\hat{v}_2 - \hat{v}_1)$ becomes ambiguous only when the "waveforms" represented by the dotted lines (Fig. 7.11) repeat. This occurs when

$$v_t = mv_{a1} = nv_{a2}, \qquad (7.7a)$$

where m and n are integers, v_t is the true velocity, and v_a is the Nyquist velocity. However, errors in resolving aliases (dealiasing) may occur because

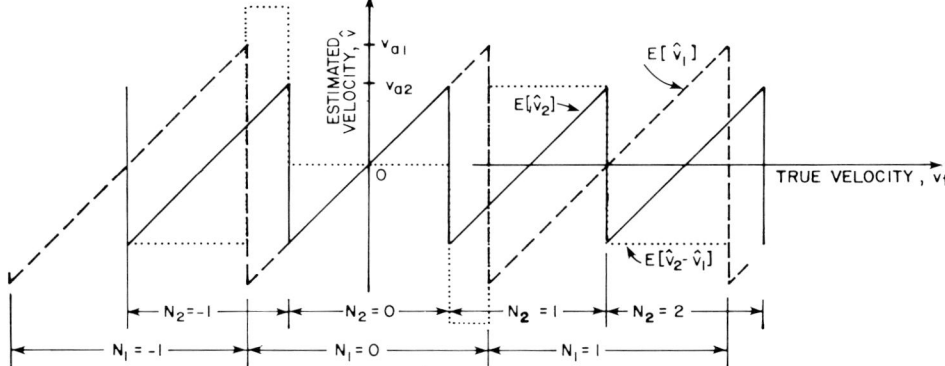

Fig. 7.11. Expected estimated velocities ($\hat{v}_1$, $\hat{v}_2$, and $\hat{v}_2 - \hat{v}_1$) versus the true velocity for $T_{s2}/T_{s1} = 1.5$. N_1, N_2 define ambiguity intervals for samples spaced T_{s1}, T_{s2} apart.

one cannot estimate $E(\hat{v}_2 - \hat{v}_1)$ with zero variance. Sirmans et al. (1976) examine the probability of error in dealiasing and give a comprehensive discussion of the statistical precision of the estimates $\hat{v}_1$, $\hat{v}_2$, and $\hat{v}_2 - \hat{v}_1$.

In the staggered PRT technique (Fig. 7.12) autocovariance estimates $\hat{R}_1$ at lag T_{s1} and $\hat{R}_2$ at lag T_{s2} are combined so that the velocity is obtained from the phase difference of the two:

$$\hat{v} = \frac{\lambda}{4\pi(T_{s2} - T_{s1})} \arg\left(\frac{\hat{R}_1}{\hat{R}_2}\right). \tag{7.7b}$$

To decrease the velocity estimate variance, the covariance estimates $\hat{R}_1$ and $\hat{R}_2$ are an average of M sample-pair covariance estimates $\hat{R}_{1i}$ and $\hat{R}_{2i}$. It can be seen from (7.7b) that the velocity estimate becomes ambiguous when the phase difference, $\arg \hat{R}_1 - \arg \hat{R}_2$, is outside the $-\pi, \pi$ interval. Thus, the "unambiguous" velocity for this staggered scheme is

$$v_m = \pm \frac{\lambda}{4(T_{s2} - T_{s1})}. \tag{7.7c}$$

Equation (7.7c) shows that the smaller the difference between T_{s2} and T_{s1}, the larger is the unambiguous velocity. However, the difference cannot be made too small, because the errors in $\hat{v}$, due to statistical fluctuations in estimates $\hat{R}_1$ and $\hat{R}_2$, are inversely proportional to this difference.

7.4 Techniques to Extend the Unambiguous Range and Velocity

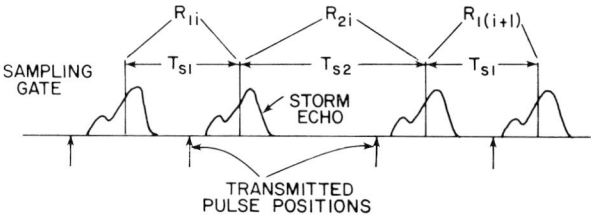

Fig. 7.12. A possible spacing of sampling gates for a staggered PRF system to obtain covariance $\hat{R}_1$, $\hat{R}_2$ estimates at two different lags.

A spectrum width can be estimated from the ratio of the magnitudes:

$$\hat{\sigma}_v = \frac{\lambda}{2\pi\sqrt{2(T_{s2}^2 - T_{s1}^2)}} \left[\ln \left| \frac{\hat{R}_1}{\hat{R}_2} \right| \right]^{1/2}. \tag{7.7d}$$

So we see that staggered PRT offers simple mean-velocity and spectrum-width estimates. Moreover, the width estimate is not biased by white noise.

Before concluding, it is important to review the philosophy employed in selecting staggered or dual PRT parameters. We want v_{a1} and v_{a2} as small (T_{s1} and T_{s2} as large) as possible in order to obtain a large unambiguous range. One must bear in mind that the smallest value of v_{a2} is dictated by the requirement for coherency (7.5c). Spectrum-width data (Fig. 10.11) suggest that v_{a2} probably cannot be much smaller than 16 m s^{-1} if the coherency condition for a large percentage (75%) of severe storm data is to be maintained. Next we need to decide on a value of T_{s1}, which we shall obtain from the desired unambiguous velocity (7.7c). So for a 10-cm radar with an unambiguous velocity $v_{a2} = 16$ m s^{-1} we find $v_{a1} = 24$ m s^{-1} will produce an unambiguous velocity $v_m = \pm 48$ m s^{-1}, a value that should resolve all but the most extreme velocity aliases.

What have we gained with staggered PRF? A single PRF radar having a Nyquist velocity v_a of 48 m s^{-1} has an unambiguous range given by (7.4), which for 10- and 5-cm radars is about 78 and 39 km, respectively. A staggered PFR radar, however, would have an unambiguous range r_{a1} equal to

$$r_{a1} = c\lambda/8v_{a1}, \tag{7.8}$$

which gives a fourfold increase in the unambiguous area over the non-staggered PRF radar, and this is quite significant. Although this can be an important improvement, an issue is whether obscuration is significantly decreased. Studies with the simplified storm model of Section 7.1.1 suggest that the obscuration probability of first-trip echoes is small if $r_a > 130$ km.

Because echo waveforms repeat over the fundamental interval $T_{s1} + T_{s2}$, assume that there are no targets beyond $\frac{1}{2}c(T_{s1} + T_{s2})$. Then an unwanted overlaid signal U is present in only one sample of the T_{s1} or T_{s2} spaced sample pair [e.g., 2, 3 in Fig. 7.13 is an unwanted overlaid echo for the covariance estimate $R(T_{s1})$ of target cluster 2]. Thus U will not contribute coherently to the covariance estimate. To demonstrate this, consider the covariance estimate at lag T_{s1}:

$$\hat{R}(T_{s1}) = [U^*(t) + V^*(t)]V(t + T_{s1}).$$

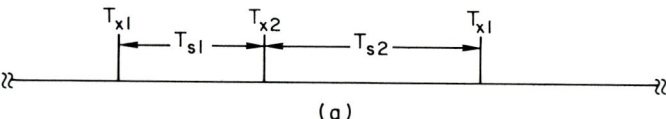

(a)

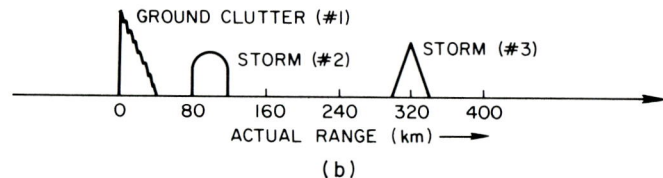

(b)

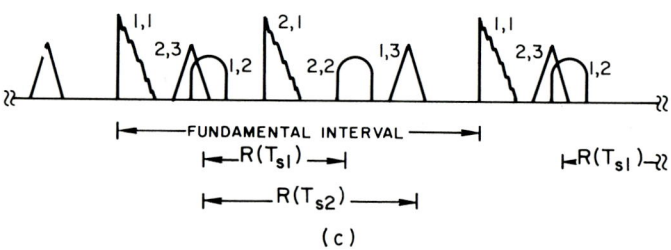

(c)

Fig. 7.13. Example of overlaid echoes in a staggered PRF radar ($v_{a1} = 24$, $v_{a2} = 16$ m s^{-1}). A number pair (e.g., 1,2) identifies the transmitted pulse and its echo cluster (e.g., storm), respectively. (a) Transmitted pulse sequence; (b) actual ranges of target clusters; (c) echo cluster range times.

7.4 Techniques to Extend the Unambiguous Range and Velocity

Its expected value for time-stationary random signals is

$$E[\hat{R}(T_{s1})] = E\{[U^*(it) + V^*(it)]V[i(t + T_{s1})]\}$$
$$= E\{V^*(t)V(t + T_{s1})\}$$

because signal U is uncorrelated with V. Thus overlaid echoes will not bias velocity estimates but will instead increase the standard error of the covariance estimate only if its power is comparable to or larger than the desired signal power.

In summary, a staggered PRF technique increases the unambiguous range, causes overlaid echoes to be incoherent, and increases the unambiguous velocity v_m so that velocity aliasing can be reduced significantly and therefore will not be as great a threat to the accurate interpretation of meteorological data.

7.4.4 Interlaced Sampling

Even though Doppler signatures may be obscured by overlaid multiple-trip echoes, observers should be given a velocity data field in which the range to a datum is unambiguous and velocity values are credible (i.e., are not in error owing to scrambled echoes). This can be accomplished by taking echo reflectivity samples during an interval T_2 sufficiently long to remove, for practical purposes, all overlaid echoes and by having this sampling period interlaced with another whose PRT is short enough to allow coherent measurements for velocity estimates (Fig. 7.14). By interlacing the velocity

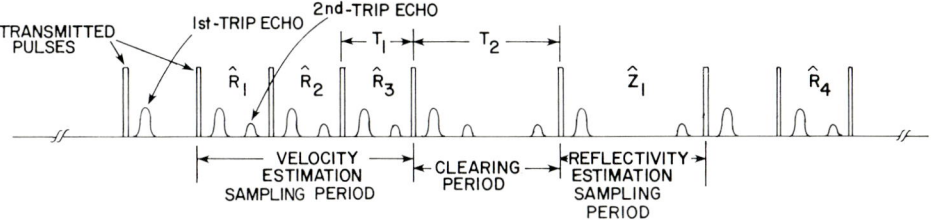

Fig. 7.14. Interlaced sampling technique, where $\hat{R}_1$, $\hat{R}_2$, $\hat{R}_3$, ... are covariance measurements (at equal lags) whose average is used to derive mean Doppler velocity estimates. We depict only first- and second-trip echoes and assume $T_2 = 2T_1$. The clearing period T_2 removes multiple-trip echors from reflectivity estimation in a contiguous T_2 interval.

estimation period MT_1 with one for reflectivity (T_2), we can have nearly colocated resolution volumes for velocity and reflectivity measurements. Figure 7.14 shows one block of samples that contain $M = 3$ covariance estimates and one reflectivity estimate (for each resolution volume). To reduce the velocity and reflectivity estimate variance, we need to average covariance and reflectivity estimates from several (K) blocks.

In order to have all n trip echoes sampled ($n = 2$ in Fig. 7.14) in one T_1 period, sampling should start in the interval nT_1 because the nth multiple-trip echo will not appear until then. Interlaced sampling provides reflectivity data without range ambiguities. This allows us to determine, through comparison of echo powers at range locations separated by $cT_1/2$, those velocity data that are significantly contaminated (obscured) by scrambled multiple-trip echoes and so to eliminate them from observation. Furthermore, we can assign correct ranges to the surviving valid velocity data. Such a dual-sampling system has been in operation at NSSL ($M = 7, K = 8, T_1 = 768$ μs, $T_2 = 4T_1$). Velocity fields displayed in real time are not range ambiguous. The most persistent obscuration to plague the interlaced sampling radar is caused by ground clutter echoes overlaid onto the second trip (i.e., ground clutter seen just beyond r_a) as well as ground clutter within the first trip. Ground clutter obscuration can be lessened with cancelers and by displacing (through changes in T_1) the second-trip ground clutter ring from the storm of interest; otherwise, at low elevation angles there is a 10–15-km range interval wherein the Doppler radar is blinded by clutter (see Color Plates 1a and 1b).

The interlaced mode can accommodate a staggered PRF during the velocity estimation period to allow an increase in both r_a and v_a at the expense of an increased data acquisition time for a given velocity estimate accuracy.

A closely related method of increasing the range to which reflectivity can be resolved unambiguously is to transmit signals at two different frequencies ω_2, ω_1 (each at different PRTs) so that simultaneous reception is possible. The long PRT yields a large unambiguous range for reflectivity estimates, whereas a short PRT is used for velocity estimation. This technique and its signal processing are analogous to the interlaced sampling technique. Its advantage is a reduction of the acquisition time, and it also offers the possibility of better clutter canceler design.

7.4.5 Correcting Aliased Velocities

Velocity aliases can usually be identified, because true velocity fields must be continuous whereas aliasing causes unrealistic gradients (discontinuities) in the measured Doppler field. However, echo-free regions disrupt spatially

7.4 Techniques to Extend the Unambiguous Range and Velocity

continuous velocity measurements, and furthermore, naturally occurring large shears and poor radar resolution make it difficult to dealias velocities in all situations.

However, there is a simple technique that works quite well and has been applied with real-time data. It requires only knowledge of the environmental wind v_e as a function of height. Storms are assumed to perturb the environmental flow, and if $|v_{max} - v_{min}| < 2v_a$ (where v_{max} and v_{min} are the true radial velocity maximum and minimum at that altitude) and wind perturbations grouped about v_e, then velocity aliases can, in principle, be completely resolved.

The estimate of the true Doppler velocity $\hat{v}_t$ is always given by

$$\hat{v}_t = \hat{v} + 2lv_a, \tag{7.9a}$$

where l is a positive or negative integer and $\hat{v}$ is the estimated mean Doppler velocity. Velocity ambiguities are resolved if l can be found for each $\hat{v}$. The parameter l can be estimated for each $\hat{v}$ at altitude h by substituting for $\hat{v}_t$ the radial component v_{er} of the environmental wind and solving for l:

$$\hat{l} = (v_{er} - \hat{v})/2v_a. \tag{7.9b}$$

The distribution of $\hat{l}$s is a mirror image of the distribution of $\hat{v}$s shifted by v_{er}. Figure 7.15 is a sample distribution of $\hat{v}$s and $\hat{l}$s for a tornadic thunderstorm. On this day the environmental wind, averaged from the surface to the tropopause, was $230°/30$ m s^{-1}, giving a 24 m s^{-1} radial component v_{er} of averaged environmental wind for this storm that had about a 15° bearing from the radar. The $\hat{l}$s are tightly clustered near integers that would correctly dealias the vs when perturbations are small compared to $2v_a$. Whenever the true velocity distribution straddles v_a or its aliases, $\hat{l}$s form into two clusters, each group near an integer. If all data in each cluster are assigned the integer nearest the group, the $\hat{v}$s will be correctly dealiased. Researchers have found that when $\lambda = 10$ cm and $r_a < 160$ km it is not necessary to search for $|l|$ larger than unity (Hennington, 1981). That is, rarely do true speeds exceed 70 m s^{-1}. Therefore, they simply make the following assignments:

$$l = \begin{cases} 1 & \text{if } \hat{l} > 0.5, \\ -1 & \text{if } \hat{l} \leq -0.5, \\ 0 & \text{otherwise.} \end{cases} \tag{7.9c}$$

In order to facilitate recognition of mesocyclone circulations, they have selected a v_e, constant in height, that approximates the translational velocity of the vortex and subtracted its radial component from the dealiased velocities given by (7.9a). This procedure symmetrizes the cyclone pattern on the

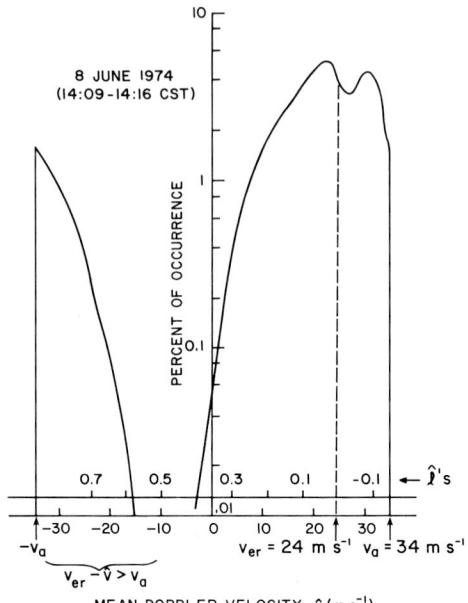

Fig. 7.15. Distribution of $\hat{r}$s and $\hat{l}$s from which decisions for dealiasing are made. Frequency of occurrence is given as a percent of velocities, in 1 m s^{-1} intervals. All aliased velocity estimates between -34 and -10 m s^{-1} can be corrected.

color display [and also on the multimoment display (Section 9.22)], as seen in Color Plates 1c and 1d, and significantly improves the recognition of Doppler velocity signatures (see Chapter 9) of mesocyclones.

The technique we have described is easily implemented on real-time radar computers and works well whenever $2v_a$ is large (e.g., >50 m s^{-1}) compared to the width of the perturbation velocity distribution. It fails most often in regions of strong divergence (e.g., at storm tops) and circulation, where true velocity spreads can be as large as 100 m s^{-1} (see Color Plate 2b). However, there are methods by which the radar's waveform can be designed to increase v_a without decreasing r_a (see Section 7.4.3) so that velocity dealiasing can be accomplished for nearly every storm.

7.5 METHODS TO DECREASE THE ACQUISITION TIME

The averaging time for reflectivity and velocity estimation is dictated by the desired accuracy of the estimates. Considerable savings in averaging

7.5 Methods to Decrease the Acquisition Time

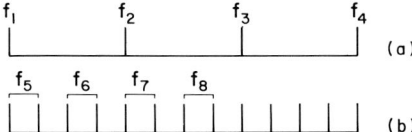

Fig. 7.16. Frequency diversity for (a) reflectivity and (b) velocity measurements. Note that (a) and (b) can be interlaced or transmitted simultaneously.

time can be achieved by judiciously choosing transmitting schemes that increase the equivalent number of independent samples. This can be done either by transmitting several frequencies (simultaneously or consecutively) or by using broadband noiselike signals.

7.5.1 Frequency Diversity

Signals with nonoverlapping spectra are uncorrelated, and such signals, when transmitted, generate on return uncorrelated and thus independent echoes. Because an rf pulse with carrier frequency f has its energy concentrated mainly in the band τ^{-1} around f, a simple way to create uncorrelated signals consists of offsetting the frequencies in successive transmitted pulses by more than the reciprocal of pulse width τ. A train of M such pulses, each with different frequency (Fig. 7.16a) will produce M virtually independent echoes, which, after averaging, will give improved reflectivity estimates for the same dwell time.

Velocity estimates can be obtained if pairs of pulses are transmitted at the same carrier frequency (Fig. 7.16b). Autocovariance-type processing must be used to retrieve the mean velocity or spectrum width, because Fourier analysis is not suitable. A disadvantage of this scheme is that, for all practical purposes, it does not allow ground clutter canceling. Note that, if implemented as in (7.16b), measurement of velocities in the second trip involves a more complicated receiver because, in processing the coherent returns of f_6 from targets in the first trip, one must also process the f_5 returns from targets in the second trip. This necessitates two receiving channels following the mixer (more if there are more trip regions to be examined simultaneously). Filters in the two receivers can eliminate the overlay of echoes from the pair f_5 on the pair f_6, etc., but the second trip echo from the first pulse of f_5 can be overlaid on the first trip from the second pulse of f_5. This is the same as with spaced pairs; i.e., the overlaid echo behaves like noise and is less damaging to the velocity or spectrum width estimates.

7.5.2 Random Signal Transmission

The uncertainty in reflectivity estimates is caused by echoes from different scatterers (within the resolution volume), which add vectorially so that the resultant signal depends on the relative position of the scatterers [the second term in (4.2)]. A noiselike signal with bandwidth B consists of a continuum of frequencies, and echoes are essentially uncorrelated if sampled at spacings larger than B^{-1}. Now, if the bandwidth B is much larger than the reciprocal of pulse duration τ^{-1}, integration along range time τ_s for the duration τ will reduce significantly the fluctuating term of (4.2). For instance, with $B = 40$ MHz the decorrelation range is about 3.75 m, and if $\tau = 1 \mu s$ ($c\tau/2 = 150$ m), there are 40 independent range samples available for averaging. The random signal radar (Fig. 7.17) takes advantage of this concept. It uses a noise source transmitter (a resistor and amplifier with bandwidth B) and a wideband receiver followed by a square law detector and integrator.

We now briefly demonstrate the reduction of variance obtained with the random signal radar and compare it with the variance obtained with the conventional single-frequency radar. The variance of the power estimate $\hat{P}_r$ is the expected value of the square of the second term in (4.2):

$$\operatorname{var}_\rho(P_r) = \tfrac{1}{4} \sum_{i \ne k} \sum_{m \ne n} E(A_i A_k^* A_m A_n^*) E(W_i W_k^* W_m W_n^*)$$
$$\times E\{\exp[j4\pi(r_i - r_k + r_m - r_n)/\lambda]\}$$
$$= \tfrac{1}{4} \sum_{i \ne k} E(|A_i^2||A_k^2|) E(|W_i W_k|^2). \qquad (7.10)$$

Each summation in (7.10) is over two indices, and the quadruple sum reduces to a double sum because the expected value of the third term is zero except when $i = n$ and $k = m$. The subscript ρ indicates that the variance is for the random signal radar, and the weighting function W consists of amplitude and phase fluctuations of the transmitted signal (created by the noise source). When the transmitted pulse contains a single frequency (i.e., when there is no amplitude or phase modulation during τ) and $B_6 \gg \tau^{-1}$, $|W_i|^2 = |W_k|^2 = 1$ for all i, k in the range interval $c\tau/2$. Furthermore, A_i and A_k are uncorrelated when $i \ne k$, so $E(|A_i|^2|A_k|^2) = E(|A_i|^2)E(|A_k|^2)$. Because we deal with a large number of scatterers, the sum of diagonal terms $i = k$ is considerably smaller from the rest. Then the variance of the echo powers from unmodulated transmitted pulses can be approximated by

$$\operatorname{var}(P_r) = \tfrac{1}{4}[\sum_i E(|A_i|^2)]^2, \qquad (7.11)$$

which is the square of the expected value of the first term in (4.2). Thus the variance of P_r is equal to the square of its mean.

The ratio of (7.10) to (7.11) describes the decrease in estimate variance that can be achieved with a random signal radar over that radar transmitting unmodulated pulses. A quantitative comparison is made easier if we assume that the scattering medium is homogeneous. Then $E(|A_i|^2|A_k|^2) = \text{const}$,

7.5 Methods to Decrease the Acquisition Time

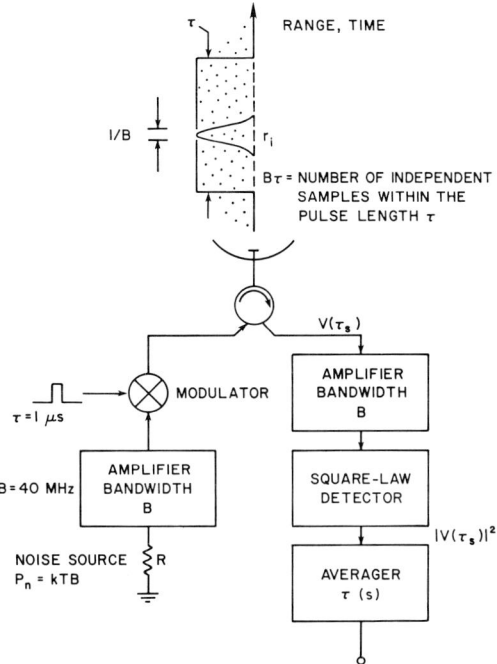

Fig. 7.17. Block diagram of the random signal radar for reflectivity estimation. The noise source is a resistor at a temperature T. (From Krehbiel and Brook, 1979; © 1979 IEEE.)

so the ratio of variances becomes

$$\text{var}_\rho(P_r)/\text{var}(P_r) = \sum_{i \neq k} |W_i W_k|^2 / N_s^2$$

$$= \sum_{l=-N_s}^{N_s} \sum_{i=1}^{N_s - |l|} |W_i W_{i+l}|^2 / N_s^2, \quad (7.12)$$

where N_s is the number of scatterers, and the reasoning leading to (5.24) was used to represent the last sum in (7.12). N_s is proportional to the pulse length τ, and the length $|W_i W_k|^2$ is the correlation magnitude squared, $|R_\rho(\delta\tau_s)|^2$, at $\delta\tau_s = 2(r_k - r_i)/c = 2r_l/c$. Evaluation of the inner sum in (7.12) leads to the following formula:

$$\frac{\text{var}_\rho(P_r)}{\text{var}(P_r)} = \sum_{l=-N_s}^{N_s} \frac{(N_s - |l|)|R_\rho(r_l)|^2}{N_s^2} \approx \frac{1}{\tau} \int_{-\tau}^{\tau} \left(1 - \frac{|\delta\tau_s|}{\tau}\right) |R_\rho(\delta\tau_s)|^2 \, d(\delta\tau_s). \quad (7.13a)$$

The approximation by the integral in (7.13) can be safely made because the distance between scatterers is small compared to the spatial pulse length

$c\tau$. As an example, consider a white noise signal of bandwidth $B \gg \tau^{-1}$ for which the power spectrum is

$$S_\rho(f) = \begin{cases} B^{-1} & \text{for } |f| \leq B/2, \\ 0 & \text{otherwise.} \end{cases} \quad (7.13b)$$

$R_\rho(\delta\tau_s)$ differs significantly from zero only for $\delta\tau_s < B^{-1}$. Hence the term $|\delta\tau_s|/\tau$ in (7.13a) can be neglected, and the integral is approximately

$$\int_{-\tau}^{\tau} |R_\rho(\delta\tau_s)|^2 \, d(\delta\tau_s) = \int_{-B/2}^{B/2} |S_\rho(f)|^2 \, df, \quad (7.14a)$$

where equality follows from Parseval's theorem (Papoulis, 1965). With this the ratio of variances simplifies to

$$\text{var}_\rho(P_r)/\text{var}(P_r) = 1/B\tau, \quad (7.14b)$$

and therefore the ratio of the mean value $\bar{P}_r$ to rms fluctuations increases by $\sqrt{B\tau}$. This is significant because it means that a random signal radar with a bandwidth ten times larger than τ^{-1} can achieve the same precision of reflectivity estimates from one echo as can a monochromatic radar from ten independent echoes. (Remember also that the single-frequency radar may have to transmit many more than ten pulses to obtain ten independent echoes.) This improvement in observation time is achieved at the expense of radar sensitivity if the transmitter average power is not changed. Because the random signal radar has statistically fluctuating transmitter power, one must either make a measurement of the power from pulse to pulse or else τ must be long enough that the average power over τ seconds estimates well the mean power of the noise tube. This is needed in order to avoid errors in reflectivity estimates. A detailed description of such a radar has been given by Krehbiel and Brook (1979).

7.6 EFFECTIVE ANTENNA PATTERN OF A SCANNING RADAR

As explained in Chapter 6, a number of signal samples need to be processed to reduce the uncertainty in estimates of reflectivity, velocity, and spectrum width. When the antenna is stationary, the resolution in the cross beam dimensions is dictated by the antenna beamwidth. However, antenna motion (usually azimuthal rotation) combined with pulse-to-pulse averaging creates an effective broadened beamwidth and shifts the position of the maximum for the apparent antenna pattern. In the following pages the *apparent* antenna pattern is investigated, and a design procedure is presented for selecting the

7.6 Effective Antenna Pattern of a Scanning Radar

rotation rate and the number of samples processed to achieve a desired azimuthal resolution.

The effective antenna pattern will be derived for the output of a square law detector, but the results also apply to derived products such as spectral moments calculated from the power spectrum or from the autocorrelation function, (6.18). For the sake of clarity, we consider the one-dimensional problem and assume time integration (the summation over M) to be continuous. This is a good approximation when the pulse repetition period–antenna rotation rate product ($T_s \alpha$) is much less than the antenna beamwidth. However, it should be understood that time-continuous integration sums echoes from a single range bin or resolution volume that changes its azimuthal position in space as the antenna rotates.

We shall base our development on the procedure followed in Appendix C. Consider the antenna positioned at $\theta_e = 0$, ϕ_0 and viewing a distribution of scatterers [see Eq. (C.9)]. Furthermore, assume that the power $2\sigma_\Omega^2(\phi)$ returned from a unit azimuthal angle at a delay τ_s is a function of azimuth and that the antenna pattern function is product separable, $f(\theta, \phi) = f(\theta)f(\phi)$. The equation relating the pattern and the expected echo power $\bar{P}(\phi_0)$ analogous to (C.13) becomes

$$\bar{P}(\phi_0) = I \int_{-\pi}^{+\pi} 2\sigma_\Omega^2(\phi) f^4(\phi - \phi_0) \, d\phi, \qquad (7.15)$$

where the integral $I = \int f^4(\theta) \, d\theta$ is a constant that we shall ignore. $\bar{P}(\phi_0)$ is available at the output of the detector (Fig. 7.18).

Now let us assume that the antenna turns at a uniform rotation rate α, so that the azimuth position of the beam axis is $\phi_0 = \alpha t$. Hence the expected power $\bar{P}(\phi_0)$ is time dependent if the reflectivity is azimuthally nonuniform. Suppose that an integrator with impulse response $h(t)$ acts on the power samples $P(\phi_0)$ in order to reduce the estimate variance. Strictly speaking, the integrator is acting on instantaneous power samples that are statistically fluctuating [such as that given by (C.11)], not on the expected time-continuous function $\bar{P}(\phi_0)$. In this discussion, however, we are interested in the mean power $\bar{P}(\phi_0)$ and need to have the expected value after the integrator. Integration and expectation are commutative, and therefore the expected power $\bar{P}_0(\phi_0)$ at the integrator output is obtained by convolving (7.15) with

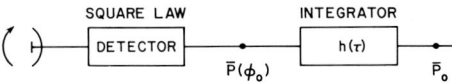

Fig. 7.18. Schematic of the power estimation that affects the azimuthal position and width of the resolution volume.

the integrator's impulse response $h(\tau)$:

$$\bar{P}_0(t) \sim \int_{-\infty}^{t} h(t - \tau) \int_{-\infty}^{\infty} 2\sigma_\Omega^2(\phi) f^4(\phi - \alpha\tau) \, d\phi \, d\tau. \quad (7.16)$$

Note that the infinite limits of integration in azimuth simplify calculations and can be safely used because the two-way pattern $f^4(\phi)$ has significant values only over a very small angular span. Interchanging the order of integration produces the following formula:

$$\bar{P}_0(t) = \int_{-\infty}^{+\infty} 2\sigma_\Omega^2(\phi) \left[\int_{-\infty}^{t} h(t - \tau) f^4(\phi - \alpha\tau) \, d\tau \right] d\phi. \quad (7.17)$$

Because (7.17) is similar in form to (7.15), we are prompted to define a two-way apparent antenna pattern,

$$f_a^4(\phi - \phi_0) = \int_{-\infty}^{t = \phi_0/\alpha} h(t - \tau) f^4(\phi - \alpha\tau) \, d\tau, \quad (7.18)$$

for the scanning antenna. $\bar{P}_0(t) = \bar{P}(\phi_0)$ given by (7.15) applies strictly for a stationary antenna.

It can also be shown that $f_a(\phi - \phi_0) \to f(\phi - \phi_0)$ when $\alpha \to 0$. The equivalence in the mean signifies that a motionless antenna with pattern f_a^4 sees, on the average, the same reflectivity factor as a scanning antenna whose rate is α and pattern is f^4. We caution the reader that the maximum of $f_a(\phi - \phi_0)$ does not occur necessarily at $\phi = \phi_0$, as we shall demonstrate by the example to be given shortly. A discussion concerning the apparent pattern when detectors are logarithmic or linear is provided by Zrnić and Doviak (1976).

Two parameters of interest for the apparent pattern are (1) the apparent displacement of the antenna axis from the end of integration and (2) the apparent beamwidth. Let us examine the finite-time block integration because it can be used for reflectivity estimation and because both the autocovariance and spectral processing employ it. Its impulse response is given by

$$h(\tau) = \begin{cases} 1/MT_s & \text{for } 0 \leq \tau \leq MT_s, \\ 0 & \text{otherwise.} \end{cases} \quad (7.19)$$

Some thought (see Fig. 7.19) reveals that the maximum of (7.17), and hence the apparent pattern maximum for an assumed symmetric pattern, occurs when (7.19) is centered with respect to f^4 in the integral (7.18). This means that the apparent pattern maximum occurs at

$$\phi_a(\text{maximum}) = \phi_0 - \tfrac{1}{2}\alpha MT_s. \quad (7.20)$$

7.6 Effective Antenna Pattern of a Scanning Radar

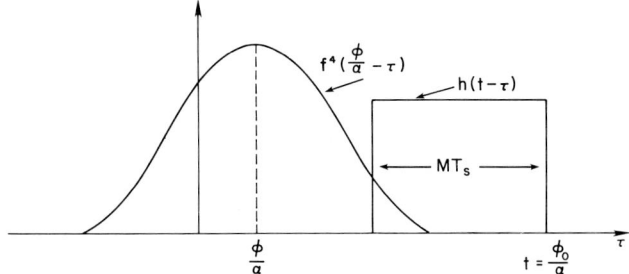

Fig. 7.19. The evaluation of the integral in (7.18) leading to the apparent pattern. For symmetric patterns, the integration is a convolution, and the apparent pattern maximum occurs when the two curves are centered with respect to each other.

Equation (7.20) shows that the azimuth of targets that contribute most (assuming spatially uniform reflectivity) to the integrated output lags behind the beam position at the time the integration is complete.

The one-way half-power apparent beamwidth ϕ_a is found by solving for the one-quarter-power points of the two-way apparent pattern, (7.18). Assuming a Gaussian radiation pattern, [(C.16)] and substituting it and (7.19) into (7.18), we obtain

$$\text{erf}(2\sqrt{\ln 4}\,\phi/\theta_1) - \text{erf}[2\sqrt{\ln 4}(\phi - \alpha M T_s)/\theta_1] - \tfrac{1}{2}\text{erf}(\sqrt{\ln 4}\,\alpha M T_s/\theta_1) = 0, \quad (7.21)$$

which has two solutions ϕ_1 and ϕ_2 that determine the apparent pattern width $\phi_a = \phi_2 - \phi_1$.

A graph of the normalized apparent beamwidth versus the product of rotation rate with integration time (Fig. 7.20) demonstrates that there is appreciable increase in ϕ_a after $\alpha M T_s/\theta_1$ reaches 1. We can use Fig. 7.20 to advantage in choosing values for radar data acquisition parameters (α, M, etc.). Assume, for example, that we know the desired resolution ϕ_a, which, given θ_1, determines the ordinate value in Fig. 7.20 and hence a value for $\alpha M T_s/\theta_1$. Now, for a chosen T_s (which is dictated by echo coherency and the desired unambiguous range and velocity), the product αM is found. Note here that M is the number of pulses in a contiguous equispaced train. If the transmission is more complicated (such as interlaced pulses), then $M T_s$ needs to be replaced with the appropriate dwell time. The desired accuracy of spectral moments dictates M, which then leads to the rotation rate α. If this rate is too slow, either the accuracy requirement must be relaxed or a different signal design and processing scheme should be attempted.

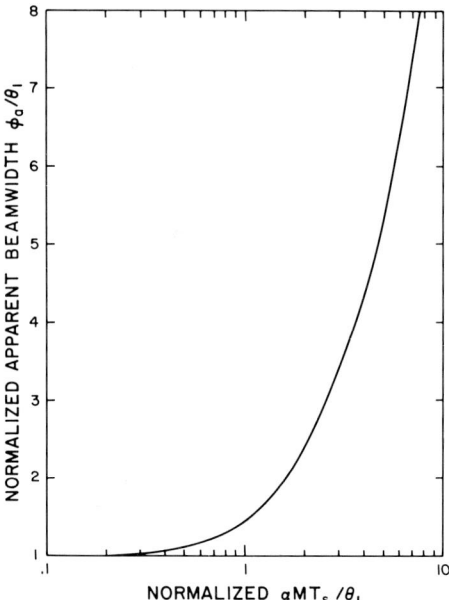

Fig. 7.20. One-way normalized apparent beamwidth versus the normalized rotation rate for an assumed Gaussian antenna pattern with a one-way 3-dB beamwidth equal to θ_1.

7.7 ANTENNA SIDE LOBES

Targets seen though antenna side lobes are detrimental since they can bias the reflectivity, velocity, and spectrum width estimates. Sometimes they may totally obscure weather echoes. Ground clutter is an example of a return that usually enters the receiver through side lobes, but, as we shall see in Section 7.8, this echo power can be reduced considerably by proper filtering since its mean velocity is always zero. This is not the case with moving clutter targets such as aircraft, birds, and automobiles. Also, in the presence of intense precipitation cores, the radar measurements several degrees away from cores may be contaminated with echoes received through side lobes.

The main lobe shape and width, as well as the near side-lobe levels, are dictated by the wavelength, the antenna reflector size, and the tapering of its illumination. More distant side lobes are caused by imperfection in the antenna and blockage of radiation caused by the feed horn, its supports (Fig. 3.3), radome, and so on.

Figure 7.21 shows a contoured portion of a measured antenna gain pattern extending in azimuth from $-16.6°$ to $+13.6°$ and in elevation angle from

7.7 Antenna Side Lobes

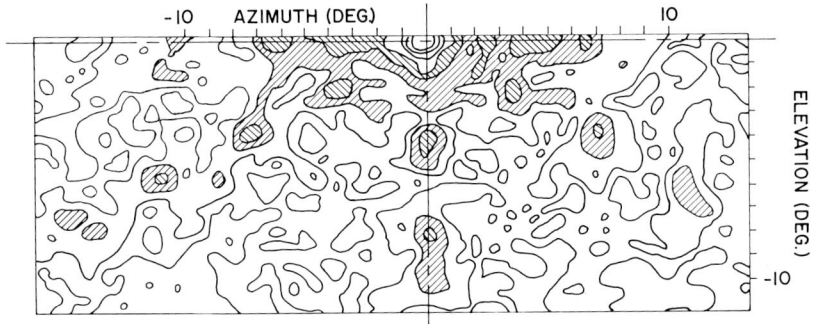

Fig. 7.21. Antenna gain pattern of NSSL's Norman Doppler radar (two way) on 17 October 1974. The data were smoothed using a Shuman nine-point formula. Lines of equal power below maximum are spaced by 12 dB. ▨, 60–72 dB; ▩, 48–60 dB. [From Waldteufel (1976).]

−11.6° to 0.6° with respect to the main axis. The diagram accuracy is thought to be, on average, ±1.5 dB in magnitude and ±0.1° in position, and the overall picture seems satisfactory in that there is excellent continuity throughout the pattern, with a consistent decrease away from the main lobe. The diagram is far from annular: The highest side lobes have an angular extent broader than the main lobe and are distributed along the principal axes (azimuth and elevation) and at ±30° from the line $\theta_e = 0°$. Near the main beam the first species prevails, whereas away from it the opposite tends to happen. We shall not dwell on this pattern structure, but there is little doubt that the four-legged structure supporting the horn that illuminates the reflector and scattering from radome ribs are instrumental in building this particular secondary pattern. From Fig. 7.21 it seems reasonable to admit a circular main lobe with an angular diameter of 2°, at roughly 40 dB below the two-way pattern peak. Therefore, near side lobes, contamination becomes significant only if reflectivity gradients exceed 40 dB deg^{-1}.

To estimate echo intensity received through side lobes the two-way antenna pattern has to be cross correlated with the reflectivity η pattern at a given range from the radar. Because this η map is itself provided by the radar with an angular resolution corresponding to, at best, the antenna main beam (i.e., about 1°), it is convenient to integrate the pattern over each square degree and then normalize it to the main-lobe peak. An example of such a two-way pattern (Fig. 7.22) shows that peaks 46 dB below the integrated main lobe can be found at 4°. Thus, if the main beam is 4° away from a strong reflectivity core and pointing at a location with 40 dB less reflectivity (which is not so uncommon), a considerable error in the data fields may result (Fig. 7.23). Estimate error in $\hat{Z}$ would be at least 1 dB, and if $\hat{v}$ is computed from the autocovariance, it will have increased uncertainty and bias errors.

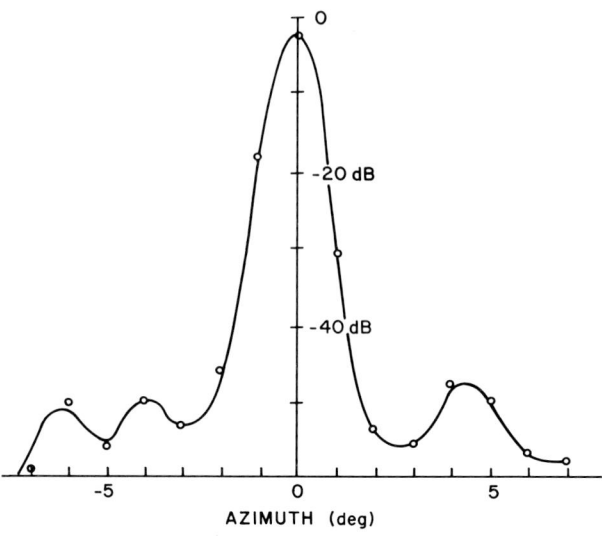

Fig. 7.22. Relative two-way antenna gain pattern integrated over $1° \times 1°$. The 0-dB level is the peak of the nonintegrated pattern. ○, data points (Norman Radar Observatory antenna).

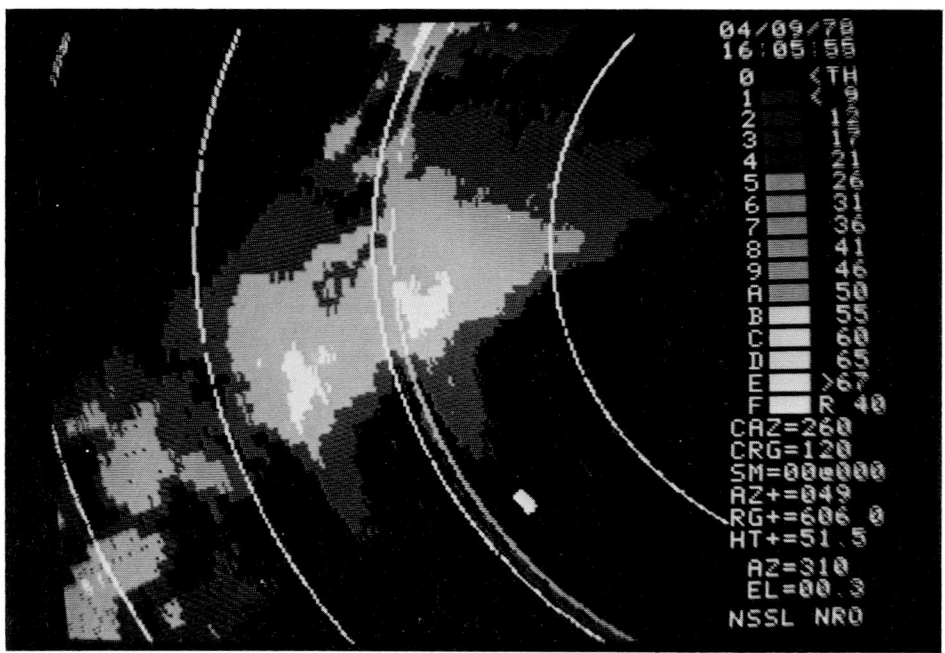

Fig. 7.23. Examples of echoes received through side lobes. The reflectivity factor spans a range from -5 to nearly 65 dBZ. The faintest gray areas are from -5 to 21 dBZ. The two-way antenna gain at angles as far as $12°$ away from the main lobe axis can be as high as -60 dB below the maximum antenna gain for the antenna used here. Thus the echo levels that stretch southward from the reflectivity maximum can be echoes (i.e., Fig. 7.21) seen through antenna side lobes, so reflectivity and velocity fields in these regions are questionable. Range marks are 40 km apart, and the inner range arc is at 80 km.

Because power enters through *all* side lobes, the integrated contribution must be considered. For the pattern on Fig. 7.21 the ratio of power in the main lobe to power in the side lobes was found to be 31.5 dB. This simply means that, with uniform reflectivity, the effective signal-to-noise ratio (i.e., main lobe to side lobe echoes) can be as low as 31.5 dB. If the shear of radial velocities were linear, there would be no bias in the mean velocity, but there would be an increased spectrum width, otherwise biases would be created. The exact assessment of side lobe influence on spectral moments requires knowledge of the reflectivity and velocity fields (Waldteufel, 1976).

There are no simple solutions to the side lobe problem. Shrouding the antenna helps, and so does the elimination of radomes (or the use of better ones). Radar meteorologists must be aware that data may have contaminations through side lobes.

7.8 GROUND CLUTTER AND ITS SUPPRESSION

Returns from the ground occur whenever the transmitted energy hits the ground by way of main beam or side lobes. At lowest elevation angles, when portions of main beam intercept the ground, the clutter echo is strongest. On the PPI scope of a radar with poor resolution, clutter appears as a large bright area centered in the middle (e.g., Figs. 7.1 and 7.2). However, when reflectivity is mapped (Fig. 7.24) with a radar having a very narrow transmitted pulse (0.3 μs), a wideband receiver (10 MHz), and an antenna with a narrow beamwidth (0.8°), the clutter pattern clearly shows recognizable targets. The data in Fig. 7.24 were collected with a narrow range weighting function (about 50 m), producing very high-range resolution of targets at the NSSL radar site in Norman, Oklahoma. The returns depicting roads are most probably caused by trees and posts along the roads. Other diffuse returns are from clusters of homes and forested areas. The bright arc to the southwest at range 9 km is from a ridge that borders the southern boundary of the Canadian River valley. The bright round dot in the center represents transmitter pulse leakage and returns from the immediate vicinity of radar. Strong returns from the southeast are from buildings.

In dealing with ground clutter, it is customary to define the target reflectivity η_c as the cross section per unit area (Section 4.3 defines volume reflectivity):

$$\eta_c = (\Delta A)^{-1} \sum_{\Delta A} \sigma_{bi}, \qquad (7.22)$$

where, analogously to (4.10), σ_{bi} is a cross section of scatterers located in the area ΔA. Since the scatterers are spread over an area rather than a

7 Considerations in the Observation of Weather

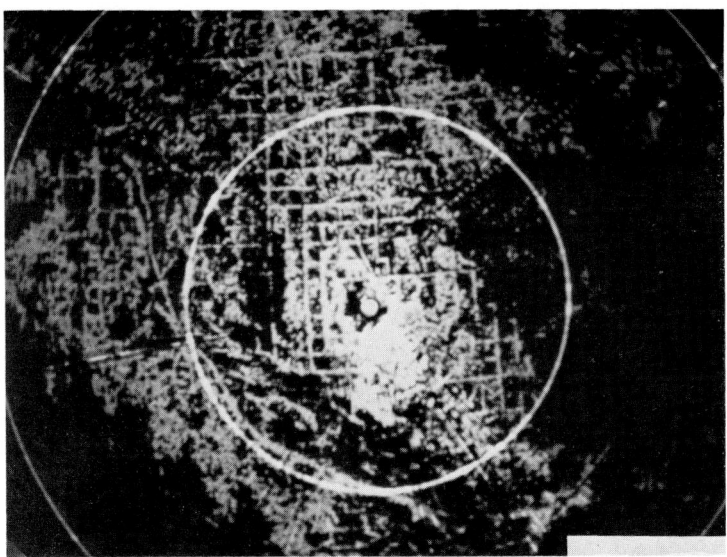

Fig. 7.24. Ground echoes. The transmitter pulse is 0.3 μs long, the elevation angle is 0.6°, the beamwidth is 0.8°, and the range marks are 5 nautical miles (about 9 km) apart. The receiver bandwidth is 10 MHz. The radar is located in Norman, Oklahoma.

volume, the radar equation takes a slightly different form:

$$\bar{P}(r_0) = \frac{P_t g^2 \lambda^2 l^2 \eta_c}{(4\pi r_0)^3} \int_0^\infty \frac{|W(r_0 - r)|^2 \, dr}{\sin \gamma} \int_0^\pi f^4(\theta, \phi) \, d\phi', \qquad (7.23)$$

where all the terms are as in (4.12) with exception of η_c and γ, which is the angle between the normal to the earth and the line connecting the antenna to a point where clutter is calculated (Fig. 7.25). Note the r^{-3} dependence of the received power in (7.23). This is a result of the linear increase in area with r and the r^4 decrease in power scattered by ground targets. In order to integrate the pattern, θ and ϕ must be expressed in terms of ϕ', h, and r.

Details about ground clutter, in particular cross sections per unit areas of various terrains, can be found in books by Long (1975) and Nathanson (1969). Here we stress that echo power has a strong dependence on terrain type and the incident angle of electromagnetic waves. Moreover, because the foliage and water on the ground change, considerable variation with seasons is not uncommon.

As a rule, ground clutter echoes have long correlation times, which means that their spectrum widths are very narrow. In addition, their mean velocity is zero and thus can easily be recognized in the Doppler spectrum (Fig. 7.26). Therefore, when spectral processing of Doppler signals is employed, removal

7.8 Ground Clutter and Its Suppression

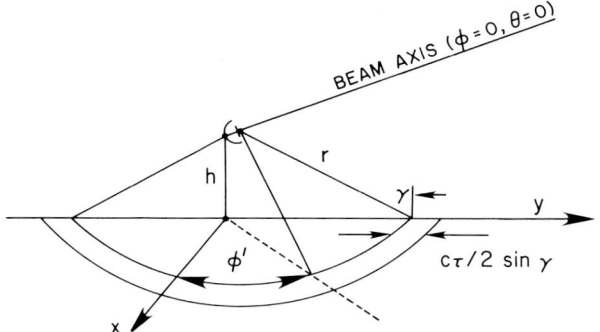

Fig. 7.25. Radar antenna at a height h above ground pointing at ϕ_0, θ_0 receives ground clutter from a range $r = c\tau_s/2$ at range time τ_s. The circular patch is illuminated by various portions of the side lobes, and the pattern has to be integrated accordingly.

of the dc line and one or two adjacent spectral coefficients prior to Doppler moment estimation usually restores weather spectrum moment estimates if saturation has not occurred somewhere in the receiver chain.

Recursive filters operating on the digitized I, Q video signals with a notch at dc have been used to suppress ground returns. Groginsky and Glover (1980) present an elliptic filter with three poles that has a 50-dB notch, a 1-dB ripple in the passband, and a notch width between 2% and 4% of the

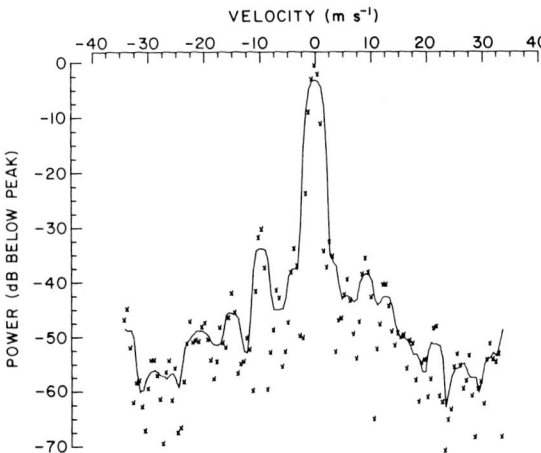

Fig. 7.26. Doppler spectrum of ground clutter for an antenna scanning in azimuth at rate $\alpha = 10°\ \text{s}^{-1}$. The width is $0.45\ \text{m s}^{-1}$; ×s are spectral powers, and the solid line is a five-point running average. Time: 15:01:42; azimuth: 241.5°; elevation: 0.5°; altitude: 0.015 km; range: 5.315 km; gate: 01; SNR: 100. von Hann window.

Nyquist interval. Such sharp filters have long-lasting transients and therefore may not be the best choice if automatic gain control (AGC) circuits are employed. AGC circuits usually change the gain in steps, which causes step changes of the I, Q signals, which in turn can cause transients (ringing) at the output of recursive filters. The transients can be controlled with more sophisticated processing. Zrnić et al. (1982) have examined the performance of an initialized filter whereby the first few samples of incoming signals are used to set the memory elements to anticipated steady state values. They conclude that the performance with a small number of samples is degraded by 10 dB for power estimation and by 20 dB for velocity as compared to the performance with a long uninterrupted sequence. To remove ground clutter effectively, the dwell time (over which the filtering is done) should be about equal to the reciprocal of the clutter spectrum width.

In order to use ground clutter cancelers, the echoes must not saturate the receiver chain, and one must have a good idea of the clutter spectrum. An intuitively satisfying hypothesis for stationary antennas is to model the I or Q clutter voltage as a dc line (i.e., delta function) plus a narrowband fluctuation in ground target reflectivity. The strong dc is due to rigid objects (buildings, posts, tree trunks, etc.), while the fluctuating part is due to the motion of vegetation, wires, etc. When the antenna is rotating, the fluctuating component increases according to Eq. (5.61) because the targets receive changing illumination, so that even totally rigid objects produce a time-varying signal.

The wind vibrates trees and other objects, and when the displacements are a significant fraction of rf wavelength, they cause fluctuations in the returned echo very similar to that produced by the reshuffling of hydrometeors. Typical values of the clutter spectrum width range from 0.02 to 1 m s^{-1}, the lower value corresponding to no wind, the upper being measured at 30 m s^{-1} (Nathanson, 1969). Ground clutter is a problem that has not been solved completely, and only recently have weather radar engineers turned their attention to it.

7.9 SPECTRAL ARTIFACTS

Imperfection in the receiver chain and signal processing affect to various degrees the output signal spectrum. First, let us examine receiver nonlinearities that distort the in-phase and quadrature components for the case in which receiver noise N can be neglected. Assume that a nonlinear transfer function g distorts both components equally so that the output signal is

$$V_0(t) = g[I(t)] + jg[Q(t)]. \tag{7.24}$$

7.9 Spectral Artifacts

If $g(V)$ is an odd function, then $V_0(t)$ has an autocorrelation (Davenport and Root, 1958, p. 292)

$$R_{V0}(\tau) = 2 \sum_{k=1}^{\infty} \frac{h_{0k}^2}{k!} \{\operatorname{Re}^k[\tfrac{1}{2}R(\tau)] + j\operatorname{Im}^k[\tfrac{1}{2}R(\tau)]\}. \tag{7.25}$$

$R(\tau)$ is the correlation of the input signal, and k takes odd values. The coefficients h_{0k} are

$$h_{0k} = \frac{1}{\pi j} \int_c f_+(s) \exp(\tfrac{1}{2}\sigma^2 s^2) s^k \, ds, \tag{7.26}$$

where $f_+(s)$ is the Laplace transform of $g(x)$ for $x > 0$, σ^2 is the power of I or Q, and the integration contour is along the $s = j\omega$ axis. We know that odd powers of time-varying quantities generate odd harmonics. Therefore, the power spectrum of the output [i.e., the Fourier transform of $R_{V0}(\tau)$] will contain odd harmonics. It is less obvious but nevertheless true that the harmonics in our particular case alternate with positive and negative frequencies so that at $k = 1, 5, 9, \ldots$ they have one sign, at $k = 3, 7, \ldots$ the other. With sampled signals the higher-frequency components introduced by nonlinearity are aliased into the Nyquist interval unless some kind of filtering is used. From (7.25) and (7.26) it may be seen that the power in the undistorted portion of the output spectrum is $h_{01}^2 R(0)$. The noise power N_g generated by the nonlinearity g is

$$N_g = \sum_{k \neq 1} \frac{h_{0k}^2 R^k(0)}{k! 2^{k-1}}. \tag{7.27}$$

It is simpler to calculate this noise from the total power at the output than to evaluate all the coefficients h_{0k}:

$$N_g = \frac{2}{\sqrt{2\pi}\,\sigma} \int_{-\infty}^{\infty} g^2(I) \exp\left(-\frac{I^2}{2\sigma^2}\right) dI - 2h_{01}^2 \sigma^2. \tag{7.28}$$

The exponential is the probability density of the input I component [see Eq. (4.5)], so the first term is the expected value of the output power. In (7.28) only the I component is used because its statistics equal those of Q. The factor of 2 accounts for the power in both components, and the exponent is due to the Gaussian distribution of I. The ratio of the signal S to distortion noise N_g at the output of the nonlinear device can therefore be defined as

$$S/N_g = h_{01}^2 \sigma^2 / \{E[g^2(I)] - h_{01}^2 \sigma^2\}. \tag{7.29}$$

This expression will be used to calculate the spectral degradation due to the transfer characteristic of an A/D converter.

7.9.1 Quantization and Saturation Noise

Saturation may occur in the rf or IF portion of the receiver, in the video amplifiers, or in the A/D converter. Saturation and quantization noise of the A/D converter will now be analyzed. In the following discussion $b\sigma$ is the A/D converter's clipping level (Fig. 7.27), n is the number of bits, including sign, $q = 2b\sigma/(2^n - 1)$ is the quantization step size, and 2^n is the number of converter levels. Because the A/D has odd transfer function, the preceding method is suitable to calculate the coefficients h_{0k}:

$$f_+(s) = \frac{2}{s}\left(\frac{1}{2} + \sum_{l=1}^{J} e^{-lqs}\right), \qquad J = 2^{n-1} - 1. \tag{7.30}$$

Contour integration along the imaginary axis yields h_{01}.

$$h_{01} = \frac{2\sqrt{2}}{\sqrt{\pi}\,\sigma}\left[\frac{1}{2} + \sum_{l=1}^{J} \exp\left(-\frac{q^2 l^2}{2\sigma^2}\right)\right]. \tag{7.31}$$

Now h_{01} and N_g from (7.28) are inserted in (7.29) to obtain the signal-to-distortion ratio S/N_g at the A/D output. If the number of A/D bits is fixed, the plot (Fig. 7.28) shows that the signal-to-noise ratio is maximized at a unique value of b. At larger clipping levels, quantization noise is dominant, whereas at lower levels it is the distortion noise produced by saturation. Normally an automatic gain control circuit controls the gain setting prior to the A/D converter (Fig. 6.1). If it is set so that $b \approx 4$ (with $n = 10$ bits), an optimum results.

When a narrowband Gaussian signal is passed through an A/D converter, odd "harmonics" are generated that have alternately positive and negative frequencies. Owing to sampling, the harmonics fold back into the Nyquist interval. In the case of saturation by receiver nonlinearity (e.g., in the rf or IF amplifiers), this folding can be reduced with filtering prior to A/D conversion. However, in the A/D converter, folding happens during conversion, and hence the quantization and saturation effects cannot be filtered.

Zeoli (1971) defines distortion as a rise in the spectral skirts of a signal. Comparison of such distortion (for a Gaussian spectrum) due to rf (or IF) saturation with those due to A/D saturation demonstrates that rf (or IF) saturation has, on the average, less distortion. Therefore, to make saturation less troublesome, it is better to have it occur in the chain before the A/D converter and to design the receiver so that A/D saturation cannot occur.

Figure 7.29 illustrates how saturation of the A/D converter affects a narrowband Gaussian spectrum. In (7.29a) is a spectrum obtained from a weighted (von Hann) time series. The mean velocity of this echo is 11 m s^{-1}, the spectrum width is 2.6 m s^{-1}; and no significant distortions are present

7.9 Spectral Artifacts

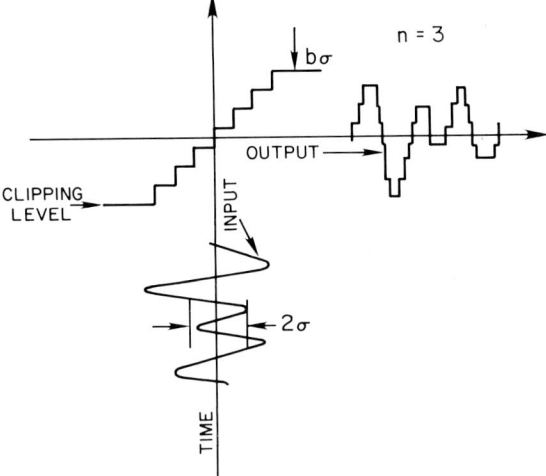

Fig. 7.27. The transfer characteristic of a three-bit A/D converter, with the input signal and its rms value σ indicated.

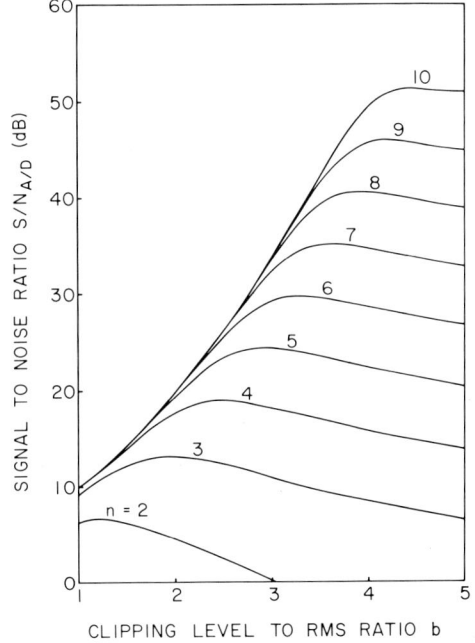

Fig. 7.28. Distortion at the output of an A/D converter. The number of bits n is indicated.

(i.e., harmonics have powers less than 30 dB below the peak). Next, this time series was clipped to cause saturation at the level $b = 1$. The periodogram presented in (7.31b) contains odd alternating harmonics: the third is at -33 m s^{-1}, the fifth is at 55 m s^{-1}, and the seventh is buried in the noise. Higher harmonics with smaller amplitudes are aliased and cannot be seen.

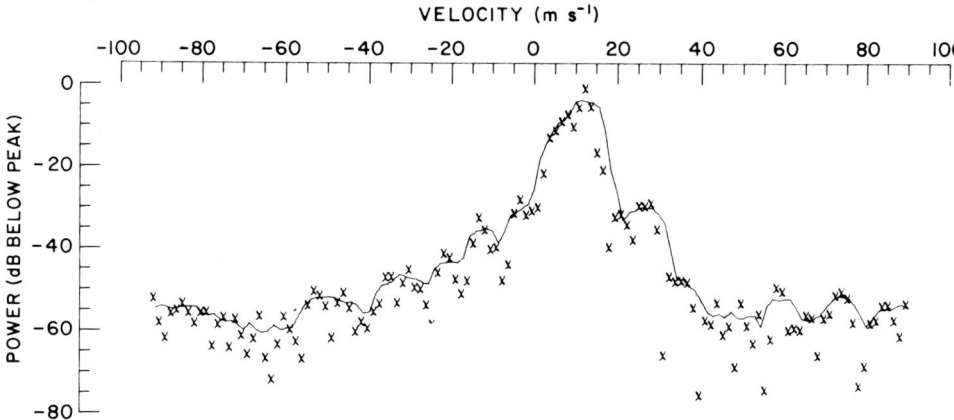

Fig. 7.29a. Doppler spectrum of a weather signal. The unambiguous velocity is 91 m s^{-1}, and the time series (von Hann weighted) contains 128 samples; $n = 10$, $b \simeq 4$. Date: 20 May 1977; time: 18:53:52; azimuth: 3.7°; elevation: 3.1°; altitude: 1.880 km; range: 33.559 km; gate: 01; SNR: 39.

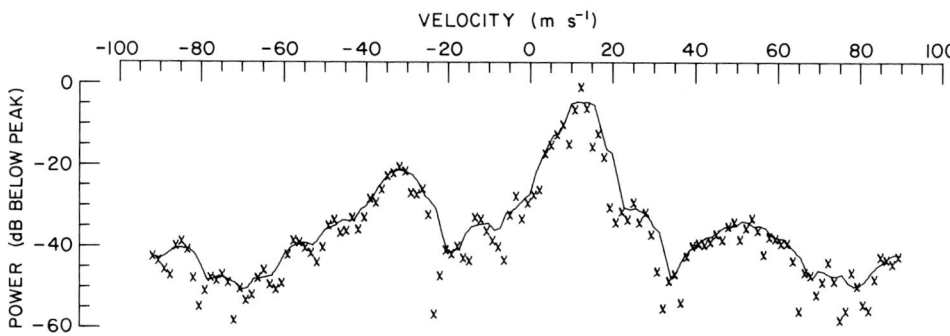

Fig. 7.29b. The time series that produced (a) is clipped, von Hann weighted, and then Fourier analyzed; $n = 8$, $b = 1$. Date: 20 May 1977; time: 18:53:52; azimuth: 3.7°; elevation: 3.1°; altitude: 1.880 km; range: 33.559 km; gate: 01; SNR: 26.

7.9 Spectral Artifacts

7.9.2 Amplitude and Phase Imbalance

Imbalances in the gain and nonquadrature phase shift between the I and Q signal channels create an image spectrum having frequencies symmetric to the actual spectrum. This effect can be seen by considering the Fourier expansion of an arbitrary signal. When the channels are perfectly matched,

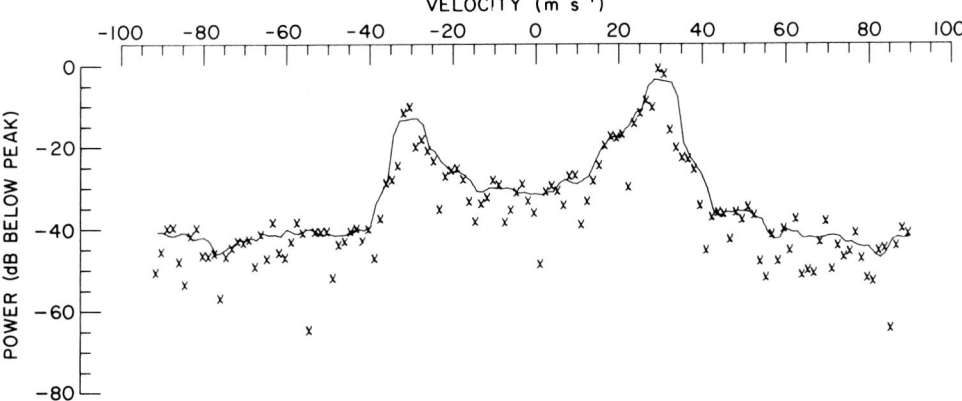

Fig. 7.30a. The image in this spectrum was created by multiplying one component of the complex signal by $K = 0.5$. Date: 20 May 1977; time: 18:53:47; azimuth: 8.9°; elevation: 3.2°; altitude: 2.082 km; range: 35.957 km; gate: 05; SNR: 25. von Hann weighting.

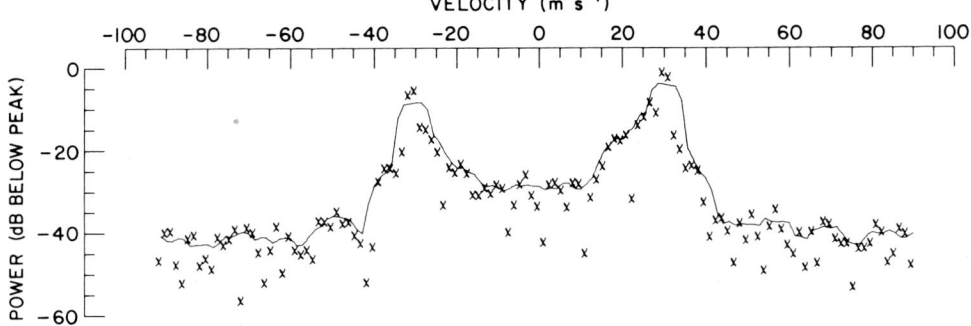

Fig. 7.30b. Imbalance of 60° in phase created this spectrum, but $K = 1$. This imbalance is for spectral components at 30 m s^{-1} and was accomplished by shifting the Q component of the signal one T_s interval with respect to I. Date: 20 May 1977; time: 18:53:47; azimuth: 8.9°; elevation: 3.2°; altitude: 2.082 km; range: 35.957 km; gate: 05; SNR: 25; von Hann weighting.

each term from the Fourier series expansion contributes to the time series an amount

$$A_i e^{j(\omega_i t + \theta_i)}, \tag{7.32a}$$

where A_i is the magnitude and θ_i the phase of the ith coefficient. An imbalance $K = G_I/G_Q$ in the gains G_I, G_Q of the I, Q channels, and a differential phase shift Δ in the Q channel gives

$$A_i K \cos(\omega_i t + \theta_i) + j A_i \sin(\omega_i t + \theta_i \pm \Delta)$$
$$= \tfrac{1}{2} A_i (K + e^{\pm j\Delta}) e^{j(\omega_i t + \theta_i)} + \tfrac{1}{2} A_i (K - e^{\mp j\Delta}) e^{-j(\omega_i t + \theta_i)} \tag{7.32b}$$

The first term on the right side of (7.32b) at frequency ω_i is (7.32a) modified in amplitude and phase, while the second term at $-\omega_i$ is the image.

An amplitude imbalance of 0.5 is responsible for the image spectrum on Fig. 7.30a. The image peak at -30 m s^{-1} is 10 dB below the signal spectrum. The time series that was amplitude imbalanced to create the spectrum in Fig. 7.30b was phase imbalanced by shifting (in time) the I or Q sample by one pulse repetition period T_s. A phase imbalance of 60° (for the mean velocity 30 m s^{-1}) generates the distorted spectrum shown in Fig. 7.30b.

Image suppression L (Fig. 7.31), in decibels, is defined as the power ratio of the two terms in (7.32b):

$$L = 10 \log |(K + e^{\pm j\Delta})/(K - e^{\mp j\Delta})|^2. \tag{7.33}$$

Usually the phase demodulator (Fig. 6.1) introduces imbalances both in phase and amplitude, whereas the video amplifiers may have different gains but normally do not introduce differential phase shifts. In order to maintain 40 dB of image suppression, the amplitude imbalance must be less than 2% and phase imbalance less than 1.2° (Fig. 7.31).

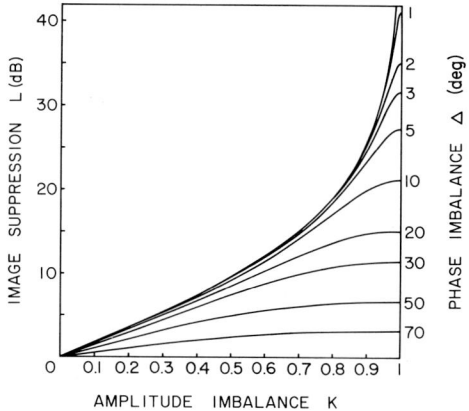

Fig. 7.31. Image suppression L versus amplitude and phase imbalance.

7.9.3 Phase Jitter

The relative phase between transmitted pulses, as well as those phase shifts between echoes sequentially traversing the receiver, must be kept stable (coherent, fixed, or known) in order to make possible precise Doppler measurements. We must accept the phase jitter produced by target motion or propagation path changes over which we have no control, but additional phase jitter may originate in the transmitter chain, the local oscillator, or in the second mixer (COHO). This phase jitter also causes phase deviations in the complex video signal. For example, ψ in Eq. (3.29) fluctuates in an unknown and, most probably, random manner. Aside from phase shifts produced by the target [i.e., by changes in r in (3.29)], the phase deviation at the mth sample is $\psi'_m = \psi_m - \psi_0$, where ψ_0 is the mean phase. Assuming a joint probability density $p(\psi'_m, \psi'_{m+k})$ of composite phases at times mT_s, $(m+k)T_s$, the following equation for the autocorrelation $R_j(kT_s)$ of echoes with radar-induced jitter is obtained:

$$R_j(kT_s) = E[V^*(mT_s)e^{-j\psi'_m}V(mT_s + kT_s)e^{j\psi'_{m+k}}]$$
$$= R(kT_s)E[e^{j(\psi'_{m+k} - \psi'_m)}]$$
$$= R(kT_s)\int_{-\pi}^{\pi}\int_{-\pi}^{\pi} p(\psi'_m, \psi'_{m+k})e^{j(\psi'_{m+k} - \psi'_m)}\,d\psi'_m\,d\psi'_{m+k}. \quad (7.34)$$

The integrals multiplying the signal autocorrelation $R(kT_s)$ are Fourier transforms of the jitter probability density function. This probability density is a normalized power spectrum of the jittery transmitted signal. Typically, the power spectral density, for most transmitters, has a narrow peak on a pedestal composed of harmonics. In this case, the peak's width determines the frequency (velocity) resolution, and the pedestal's height reduces proportionately the dynamic range over which the signal spectrum can be observed. Only short-term phase instabilities (with variations that occur during the dwell time) affect the spectral moment estimates. Phase changes due to slow frequency drifts do not significantly broaden the spectrum, but they may eventually shift the signal frequencies out of the receiver's passband.

7.10 DETECTION OF WEAKLY SCATTERING WEATHER TARGETS

Pulsed radar transmitters have limited peak power and thus the detection of weak scatterers is also limited. Average power cannot be increased (to improve detection) by increasing τ without compromising resolution or by decreasing r_a because this may cause additional multiple-trip targets. However, the proper encoding of transmitted signals can increase the average power and the detection of weak scatterers while neither increasing peak power or PRF nor degrading range resolution. Of course, various compromises must be made, and the purpose of this section is to discuss the trade-offs between transmitted power and resolution (i.e., range weighting function).

As discussed in Chapters 2 and 3, the peak transmitter power P_t of a pulsed-Doppler radar is the average power over that cycle of the rf that gives maximum value. The average transmitter power P_{av} is an average of the power over the pulse repetition period. Consider rectangular pulses for which P_t is constant over the pulse length τ. The average power P_{av} then is just

$$P_{av} = P_t \tau / T_s, \tag{7.35}$$

and the ratio $\tau/T_s = d$, often expressed as a percentage, is referred to as the *duty factor*. Now, for the simple modulation schemes we have considered so far, d is to a large extent dictated by the weather targets. Unambiguous range and echo coherency require a few milliseconds for T_s, whereas the pulse duration τ determines the range resolution. For most meteorological applications, d has a value around 0.001. As can be seen from the weather radar equation, the SNR, and hence the minimum target cross section that can be detected, depend directly on P_t. The P_t required to detect rain showers is usually less than several hundred kilowatts, which is easily handled (at wavelengths 10 cm and longer) by waveguides and microwave components. Thus, the average power P_{av} generated by microwave tubes is rather small [1 kW is typical for what is considered a moderate-power radar and equals the power emitted at infrared wavelengths (Fig. 2.1) by a household iron].

However, clear-air refractive index fluctuations have target cross sections that can be several orders of magnitude smaller than rain and require considerably larger P_t for detection. For efficient transmitter tubes, the removal of heat losses is not terribly difficult, so that P_{av} can be increased an order of magnitude or more to allow a corresponding increase in P_t. But if P_t is close to the transmission line breakdown point, it cannot be further increased, and the only way to increase P_{av} is by lengthening τ. (In Chapter 11 it is shown [Eq. (11.154)] that echo power detectability depends on P_{av}.) This degrades the range resolution unless more sophisticated signal coding techniques are employed.

7.10.1 Pulse Compression

To increase the SNR without exceeding the peak permissible power and to improve the range resolution for a given pulse width, designers of radars have invented pulse compression. Here we shall describe this concept by way

7.10 Detection of Weakly Scattering Weather Targets

of an example and discuss its implications for weather radars. To begin with, consider an rf pulse of length τ_p that consists of segments τ seconds long. Suppose that the rf phase of each segment can be independently controlled. A method often used is to change the phases by 180° in a prescribed manner. In Fig. 7.32a such a phase-change sequence is denoted $c(n\tau)$, where $+1$ signifies a certain transmitter phase and -1 a 180° phase shift. A replica of the transmitted signal is stored, and a cross correlation with the returned echoes is performed as part of the detection process. The output of a correlation receiver when a point target at a delay τ_s has created the echo is shown in Fig. 7.32b. We note that the signal correlation is concentrated around the peak, which in this example is four times larger than the side lobes. It is important to realize that the output peak $R(0)$ is not the power of the incoming signal, because the correlation was performed with a stored replica rather than with the signal itself. Incidentally, had we transmitted a pulse of length τ with the same peak power, on correlation we would have obtained exactly a symmetrical triangular response having amplitude 1 and base width 2τ. Instead, with a pulse four times longer we have obtained a return that is four times larger at $\tau = 0$, and the resolution (which we define as the width of the main peak) has remained as good as with the short pulse. In this example, the compression ratio $(\tau_p/\tau = \tau_p B)$, also called the time–bandwidth product, characterizes the gain and improvement in resolution achievable with pulse compression.

We note in passing that the effective increase in signal power over the single pulse of length τ is proportional to the code length squared, or M^2 (the signal voltage increases as M), which for the example of Fig. 7.32 is 16. Although this is an impressive gain, what really matters is the signal-to-noise ratio. The total noise power is M times larger as it is obtained by summing that many uncorrelated noise samples. Thus the SNR is improved by a factor of M.

A schematic of the correlator for the code of Fig. 7.32 consists of delay elements, a sign reversal, and an adder (Fig. 7.33a). The device operates in a

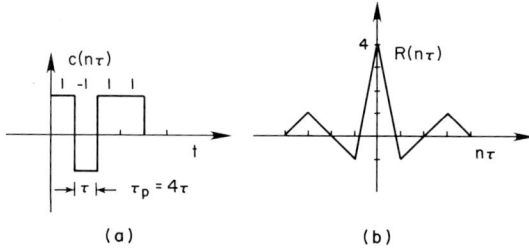

Fig. 7.32. (a) A four-element two-phase code and (b) its autocorrelation function.

pipeline mode so that at every τ interval it generates an output that corresponds to an input that occurred 4τ earlier. Two correlators are needed—one for I and the other for Q—so a possible complex output with noise appears, as in Fig. 7.33b. The price paid for the increased SNR and good resolution is a more complicated system and undesirable range side lobes.

Why are the side lobes undesirable? If we recall the explanation of the range weighting function (see Section 4.1), we immediately recognize that the weighting function $|W(r)|^2$ is precisely the output of the correlator $|R(n\tau)|^2$, which here has the role of a receiver filter. Thus, in the presence of strong reflectivity gradients, weak scatterers weighted with the main peak can be completely masked by powerful targets at the side lobes. To make matters worse, short codes with 25 dB or more of main lobe/side lobe ratio do not exist. For this reason and because storms generally provide ample return power, conventional weather radars do not use pulse compression. The picture changes, however, if one considers faint clear-air echoes from fluctuations in the refractive index, as found throughout the troposphere. Very high-frequency (40–60 MHz) radars use signal designs closely related to pulse compression. These are discussed next.

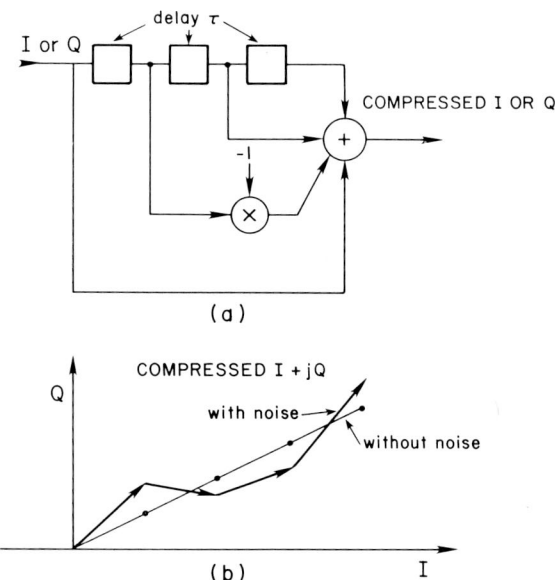

Fig. 7.33. A correlator (decoder) for the code on Fig. 7.32. Each delay element introduces a delay of τ. With the indicated delays and a reversal of phase of the second element, an in-phase addition of the complex I, Q signal occurs. Because of noise, the compressed $I + jQ$ deviates somewhat from the ideal case.

7.10.2 Complementary Codes

These remarkable codes, discovered by Golay (1961), have the following property:

$$R_1(n\tau) + R_2(n\tau) = \begin{cases} 2M & \text{if } n = 0, \\ 0 & \text{otherwise,} \end{cases} \qquad (7.36)$$

where $R_1(n\tau)$ is the autocorrelation function of code $c_1(n\tau)$ and $R_2(n\tau)$ is the autocorrelation of code $c_2(n\tau)$. The elements of $c_1(n\tau)$ can have one of the two values ± 1:

$$R_1(n\tau) = \sum_{k=0}^{M-n-1} c_1(k\tau) c_1(k\tau + n\tau). \qquad (7.37)$$

Both codes are of equal length M, which means they have M segments of duration τ where the phases can be controlled in 180° increments. Figure 7.34 illustrates a four-segment complementary code pair.

A radar operating with complementary codes transmits and receives alternately the two codes c_1, c_2 (Fig. 7.34). The resulting autocorrelation

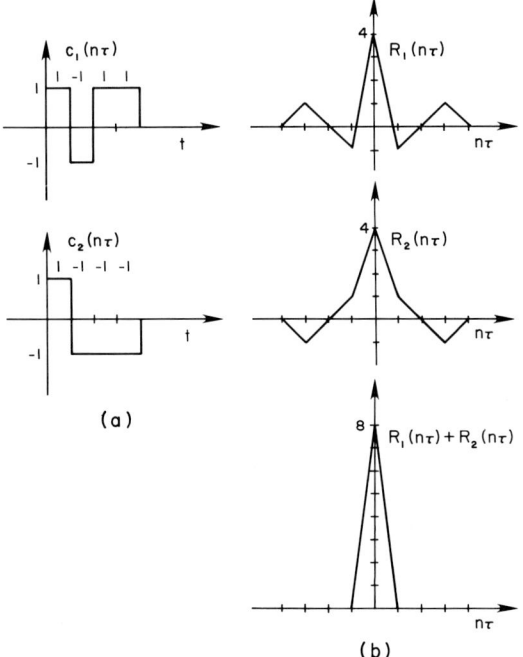

Fig. 7.34. (a) Two complementary codes with four elements each and (b) their autocorrelation functions.

functions at the output of the correlation receivers are the range weights given to the scatterers by each code–correlator combination. Those range weights $[R_1(n\tau, \tau_s)$ and $R_2(n\tau, \tau_s)]$ are summed to produce a weighting function with a single peak and no side lobes. In order for the side lobes to be truly zero, the echoes must be totally correlated within the time period of one code switch T_s, and the Doppler shift must be small compared to the Nyquist interval. Because the decorrelation time of clear-air echoes at very high frequency (VHF) is a few seconds and $T_s \leq 1$ ms complementary codes have been applied at those frequencies (Schmidt *et al.*, 1979).

The correlation can be done at IF or on the complex I, Q signals. The spectral moments are found by processing the (complex) autocorrelation sum in the same manner as for conventional I, Q voltages (Chapter 6).

In summary, complementary codes of length M provide (1) a compression ratio $2M$ and (2) a range weighting function without side lobes. These nice properties hold only when the echo decorrelation time is much longer than the switching time between codes and when the Doppler shifts are small. Note also that, as far as the Doppler processing of echoes is concerned, the effective pulse repetition time is $2T_s$. Overlaid echoes due to second-trip targets have a weighting function equal to the sum of the cross correlations between the two codes.

7.10.3 FM cw Doppler Radar

Frequency-modulated continuous-wave (FM cw) transmission is another coding scheme whereby large SNR can be achieved by having a high duty cycle without excessive peak transmitter power. A block diagram of an FM cw radar is shown in Fig. 7.35. The radar requires two antennas—one for transmission, the other for reception—and it transmits a continuous signal whose frequency is linearly swept with period T_s (Fig. 7.36). Analytically,

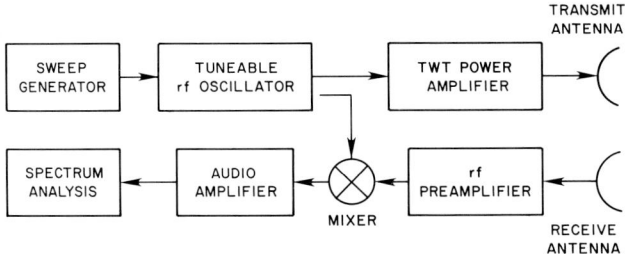

Fig. 7.35. Block diagram of an FM cw radar. (Courtesy of Strauch, Wave Propagation Laboratory.)

7.10 Detection of Weakly Scattering Weather Targets

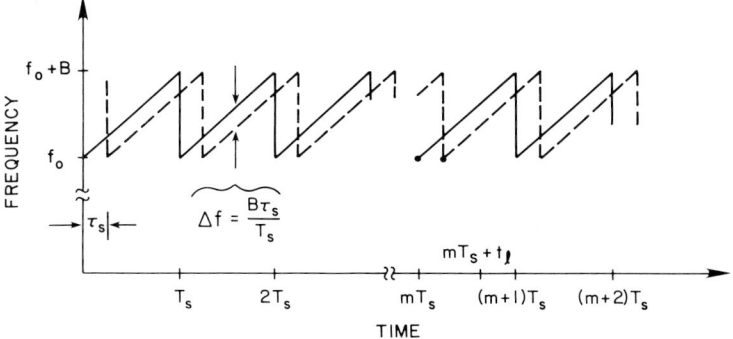

Fig. 7.36. Transmitted (——) and received (---) frequrncy variations. The target is assumed to be stationary. (Courtesy of Strauch, Wave Propagation Laboratory.)

the sinusoidal dependence of the transmitted signal $V_t(t)$ can be expressed

$$V_t(t) = \cos\{2\pi[f_0 + (B/2T_s)(t - mT_s)](t - mT_s)\}, \quad mT_s \le t < (m+1)T_s, \tag{7.38}$$

where m is an integer. In (7.38) use was made of the following relationship between the transmitter's instantaneous phase ψ_t and the frequency:

$$f(t) = \frac{1}{2\pi}\frac{d\psi_t}{dt}, \tag{7.39a}$$

or

$$\psi_t = 2\pi \int_0^{t_l} [f_0 + (B/T_s)t_l]\, dt_l, \tag{7.39b}$$

where

$$t_l = t - mT_s \tag{7.39c}$$

is the time relative to the start of the transmitter's mth sweep. A stationary target at range $r = c\tau_s/2$ returns a signal that is mixed with the one being transmitted (Fig. 7.38) to produce a difference frequency:

$$\Delta f = 2rB/cT_s. \tag{7.40a}$$

Thus Δf is a measure of the target range, and the range resolution is determined by the resolution with which one can measure Δf This is, of course, equal to the reciprocal of the available time $(T_s - \tau_s)$, which for near targets is almost T_s. Therefore, short-range targets have a range resolution of

$$\Delta r = c/2B. \tag{7.40b}$$

Most often B is tens of megahertz and the range resolution is a few meters. The decisive advantage of the FM cw radar is that the range resolution can be increased simply by spanning a larger frequency band B without changing the average transmitted power.

The received signal (the dashed line in Fig. 7.36) is both delayed in time and Doppler shifted. For purposes of illustration, let us examine a point target at range

$$r(t) = r_0 + vt = r_0 + mT_s v + t_l v, \tag{7.41}$$

where r_0 is the range when $t = 0$. The echo $V_r(t)$ from this discrete target replicates the transmitted signal with a delay $2r(t)/c$. Therefore

$$V_r(t) = V_t[t - 2r(t)/c]. \tag{7.42}$$

This received signal is mixed with the transmitted one (Fig. 7.35) to produce a low-frequency signal whose phase ψ_l is the difference between the transmitted phase ψ_t and the received phase ψ_r. From (7.38) and (7.42),

$$\psi_t = 2\pi[f_0 t_l + \tfrac{1}{2}(B/T_s)t_l^2], \tag{7.43a}$$

$$\psi_r = 2\pi\{f_0[t_l - 2r(t)/c] + \tfrac{1}{2}(B/T_s)[t_l - 2r(t)/c]^2\}, \tag{7.43b}$$

so that the phase difference during the time interval $T_s - \tau_s$ becomes

$$\psi_l = 2\pi\{2f_0 r(t)/c + \tfrac{1}{2}(B/T_s)[4t_l r(t)/c - 4r^2(t)/c^2]\}. \tag{7.44}$$

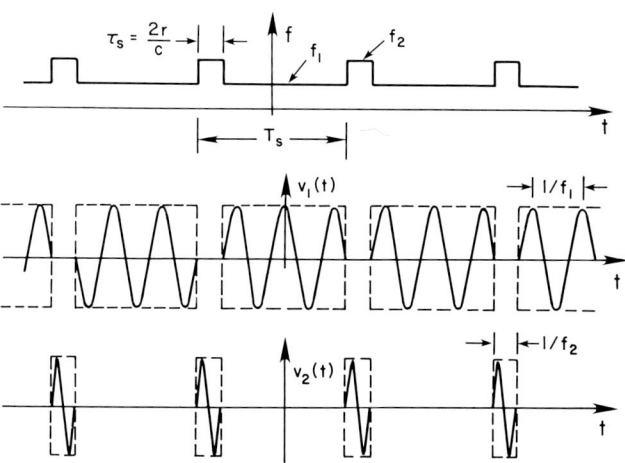

Fig. 7.37. Frequency and amplitude plots of the received signal after it has been mixed down. (Courtesy of Strauch, Wave Propagation Laboratory.)

7.10 Detection of Weakly Scattering Weather Targets

Now the difference frequency Δf_l is found from the derivative of (7.44):

$$\Delta f_l = \frac{1}{2\pi}\frac{d\psi_l}{dt_l} = \frac{2f_0 v}{c} + \frac{2Br_0}{cT_s} + \frac{2Bmv}{c} + \frac{4Bvt_l}{T_s c} - \frac{4Bv}{T_s c^2}(r_0 + mT_s v + t_l v). \tag{7.45}$$

For a practical FM cw radar only the first three terms in (7.45) are significant. The first,

$$2f_0 v/c = 2v/\lambda, \tag{7.46a}$$

is the Doppler shift; the second,

$$2Br_0/cT_s, \tag{7.46b}$$

represents the initial range (r_0) term [see (7.40a)]; and the third,

$$2Bmv/c = (2B/cT_s)(vmT_s), \tag{7.46c}$$

is due to the accumulated range up to time mT_s. It should be evident that mixing during the time interval from 0 to τ_s will produce a different frequency than the one during the interval from τ_s to T_s. This is illustrated in Fig. 7.37. Appropriate filtering or signal sampling eliminates such undesirable higher frequencies. Consistent with the approximation modifying (7.45), we can write the phase difference, from (7.44),

$$\psi_l = -\frac{4\pi r_0}{\lambda} - \frac{4\pi m T_s v}{\lambda} - 2\pi \Delta f_m t_l, \tag{7.47}$$

where

$$\Delta f_m = \frac{2v}{\lambda} + \frac{2Br_0}{cT_s} + \frac{2v}{c}Bm.$$

Shortly we shall demonstrate that the change of phase, from sweep to sweep, given by the term $4\pi m T_s v/\lambda$ enables one to perform the Doppler measurement. Consider an L-sample representation of the signal obtained during one sweep at the mixer output. An L-point discrete Fourier transform of these samples can locate the frequency

$$2v/\lambda + 2Br_0/cT_s, \tag{7.48}$$

but there is no way that the range and Doppler terms can be separated from this single measurement. Since the signal is real, the DFT generates L coefficients, one-half of which are conjugates of the other half. Thus, there must be $L/2$ distinct coefficients, corresponding to as many frequency locations. Those coefficients $A(f)$ are found from the discrete Fourier transform:

$$A(f) = \sum_{l=0}^{L-1} \cos \psi_l e^{-j2\pi f lT_s/L}$$

$$= \frac{\sin[2\pi(f - \Delta f_m)T_s]}{2\sin[2\pi(f - \Delta f_m)T_s/L]} \exp\left[j\left(\phi_0 - \frac{4\pi mT_s v}{\lambda}\right)\right] e^{j\phi_1}$$

$$+ \frac{\sin[2\pi(f + \Delta f_m)T_s]}{2\sin[2\pi(f + \Delta f_m)T_s/L]} \exp\left[-j\left(\phi_0 - \frac{4\pi mT_s v}{\lambda}\right)\right] e^{j\phi_2}, \quad (7.49)$$

where $t_l = lT_s/L$, $\phi_0 = -4\pi r_0/\lambda$,

$$\phi_1 = \pi[(f - \Delta f_m)T_s - (f - \Delta f_m)T_s/L],$$

$$\phi_2 = \pi[(f + \Delta f_m)T_s - (f + \Delta f_m)T_s/L].$$

To measure the reflectivity alone, the power spectra [i.e., $|A(f)|^2$ from several sweeps are averaged. The frequency locations are ranges that may be shifted by velocity, but with a zenith-pointing radar in clear air, where the radial velocity is usually negligible, one can call them the range (or height). This is not the case when velocities are high, because if $2v/\lambda$ exceeds $2Br_0/cT_s$, then the target appears in adjacent $2Br_0/cT_s$ intervals.

To measure both range and velocity, the complex FFT coefficients are calculated on a number M of consecutive sweeps. Frequencies f at which $A(f)$ is evaluated are chosen to be multiples of the Nyquist frequency T_s^{-1}; i.e., $f = k/T_s$, where k is an integer that generates L coefficients $A(k)$, two for each frequency location. For a frequency k/T_s near Δf_m, the first $(\sin x)/\sin(x/L)$ term in (7.49) is dominant while the second (its mirror image) is heavily attenuated. Furthermore, the only term that changes appreciably from sweep to sweep, i.e., with m, is the exponent $4\pi mT_s v/\lambda$. The other m-dependent term ($2vBm/c$ in Δf_m) is much smaller because typically $f_0 = c/\lambda \sim 10^9$, whereas $B \sim 10^7$. Therefore, M discrete Fourier transforms performed over M consecutive sweeps generate a matrix of coefficients $A(k, m)$ (k denotes the range location and m the sweep number). In view of the preceding discussion, these coefficients can be represented

$$A(k, m) = C(k)e^{-j4\pi mT_s v/\lambda}, \quad (7.50)$$

where for the kth range location [$k = fT_s$ in (7.49)] $C(k)$ is essentially constant. Now, to determine the Doppler shift associated with each frequency location, it suffices to perform $L/2$ additional Fourier transforms of length

7.10 Detection of Weakly Scattering Weather Targets

M over consecutive sweeps. This leads to

$$S(n) = \left| C(k) \sum_{m=0}^{M-1} \exp\left[-j2\pi\left(\frac{2vmT_s}{\lambda} + \frac{mn}{M}\right)\right] \right|^2$$

$$= |C(k)|^2 \left| \frac{\sin[2\pi(2vT_sM/\lambda + n)]}{\sin[2\pi(2vT_s/\lambda + n/M)]} \right|^2, \qquad (7.51)$$

where $S(n)$ is the power spectrum equivalent to the Doppler spectrum of the pulsed radar and has the same unambiguous velocity $\pm\lambda/4T_s$. The rectangular window transform (see Chapter 5) is maximized for $v = -\lambda n/2T_sM$. [Alternatively, instead of the Fourier transform of the M-point complex samples, autocovariance processing may be applied to (7.50) for a constant k.] To summarize, each of M discrete Fourier transforms of length L generates a matrix $[A(k, m)]$ of complex coefficients. The $L/2$ columns (with index k) correspond to fixed range locations, and the M rows correspond to sweeps. A second set of transforms, one over each column, yields the Doppler frequency spectra for the corresponding range locations. This linear processing, derived for a point target, is applicable to weather echoes as well because returns from many point targets linearly combine to generate the composite signal.

An alternative procedure to retrieve the Doppler and range information from the FM cw radar requires a single (long) ML-point discrete Fourier transform over the M sweeps. To show the result, let us perform a DFT on the mth sweep and sum the result over m. First, t_l in (7.47) must be replaced with $lT_s/L - mT_s$ (where $l = 0, 1, \ldots, L, L+1, \ldots$ and $l \geq Lm$). With this substitution, (7.47) becomes

$$\psi_1 = \phi_0 + 2\pi(2Br_0/cT_s + 2vBm/c)mT_s - 2\pi\Delta f_m lT_s/L. \qquad (7.52)$$

Now the long DFT, $D(k)$, can be written

$$D(k) = \sum_{m=0}^{M-1} \sum_{l=mL}^{(m+1)L-1} \frac{e^{j\psi_1} + e^{-j\psi_1}}{2} e^{-j2\pi lk/ML}. \qquad (7.53)$$

It suffices to perform the summations over $e^{-j\psi_1}$, which corresponds to positive frequencies; the term $e^{j\psi_1}$ is for negative frequencies. The sum over l yields

$$D(k) = \frac{e^{-j\phi_0}}{2} \sum_{m=0}^{M-1} \exp\left[-j2\pi m\left(\frac{k}{M} - \frac{2v}{\lambda}T_s\right)\right]$$

$$\times \frac{1 - \exp[-j2\pi(k/M - \Delta f_m T_s)]}{1 - \exp[-j2\pi(k/M - \Delta f_m T_s)/L]}. \qquad (7.54)$$

174 7 Considerations in the Observation of Weather

As in deriving (7.51), one can neglect the changes in Δf_m due to m, so only the exponents in (7.54) need be summed. This done, the magnitude squared, $|D(k)|^2$, reduces to

$$|D(k)|^2 = \frac{1}{2}\left|\frac{\sin[\pi(k/M - 2vT_s/\lambda)M]}{\sin[\pi(k/M - 2vT_s/\lambda)]} \frac{\sin[\pi(k/M - \Delta f_m T_s)]}{\sin[\pi(k/M - \Delta f_m T_s)/L]}\right|^2. \quad (7.55)$$

For the moment, assume that the Doppler shift is 0 and let $f = k/MT_s$. The second term is a slowly varying envelope of the form $\sin(\pi f T_s)/\sin(\pi f T_s/L)$ and is plotted in Fig. 7.38 for a point target at range

$$r_0 = (cT_s/2B)(2/T_s) = c/B. \quad (7.56)$$

On the other hand, the first term, $\sin(\pi M f T_s)/\sin(\pi f T_s)$, varies rapidly and repeats at T_s^{-1} intervals. The product of the two terms isolates the point target, as shown in Fig. 7.38. With a Doppler shift, the rapidly varying term is displaced from the envelope and the envelope is also shifted, but only slightly (Fig. 7.39). Figure 7.40 is a pictorial description of the processing. The frequency axis is divided into M segments, the first and last of which are $2/T_s$ long and have $L/2$ spectral coefficients; the rest are T_s^{-1} long with L coefficients. Zero Doppler shift corresponds to the center of each segment.

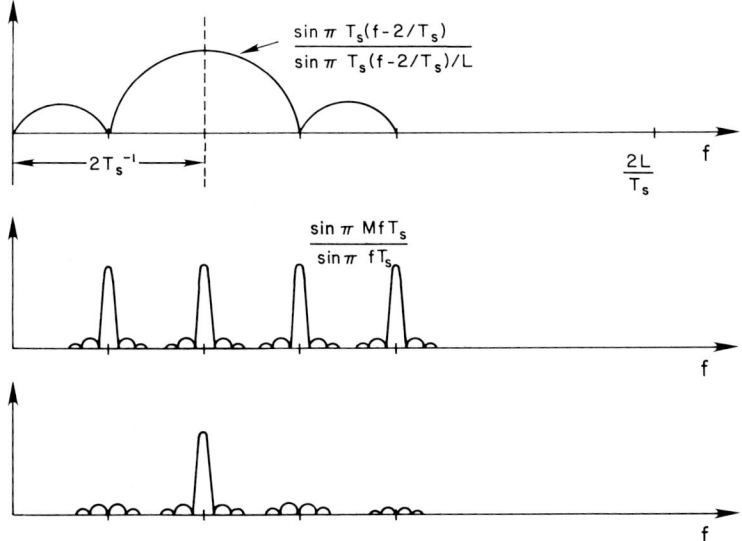

Fig. 7.38. Spectrum shape of a point target with zero Doppler shift that would be obtained from a discrete Fourier transform along consecutive range time samples. The upper curve is the slowly varying envelope, the middle the fast one. The product of the two, at bottom, is the spectral shape.

7.10 Detection of Weakly Scattering Weather Targets

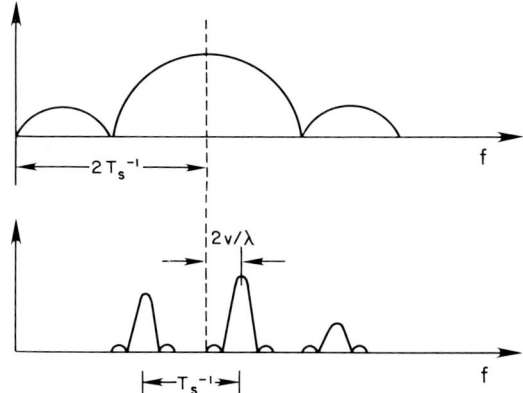

Fig. 7.39. Spectral shape of a point target with a Doppler shift (bottom), shown relative to the center of the slowly varying envelope.

The distributed targets appear in a correct range interval as long as their velocity does not exceed the unambiguous velocity $\lambda/4T_s$. Otherwise, they fold into adjacent range intervals.

Linear sawtooth modulation of the microwave frequency gives the FM cw radar characteristics analogous to the pulsed-Doppler radar. The reciprocal of B is analogous to an effective pulse length for the pulse radar, whereas T_s is equivalent to the pulse repetition time.

Because the peak-to-average power ratio of the FM cw radar is unity, we immediately infer that the range resolution can be increased without changing the effective per-pulse transmitted energy. However, with pulsed radar a decrease in τ reduces the transmitted pulse energy (assuming constant peak power). Therefore, *per-pulse* SNR is decreased in proportion to the square of pulse width for a pulsed radar, whereas the equivalent SNR of the FM cw radar decreases linearly with the effective pulse width (i.e., B^{-1}).

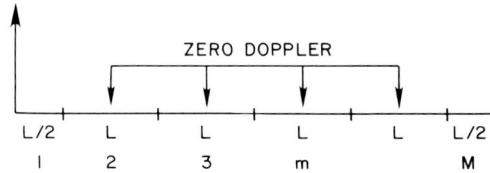

Fig. 7.40. Frequency axis and its relation to the unambiguous range and Doppler for a long discrete Fourier transform. Range intervals are numbered from 1 to M, and zero Doppler shift is indicated. Note that only positive Doppler shifts are measured unambiguously in the first range interval, and only negative shifts are measured in the last.

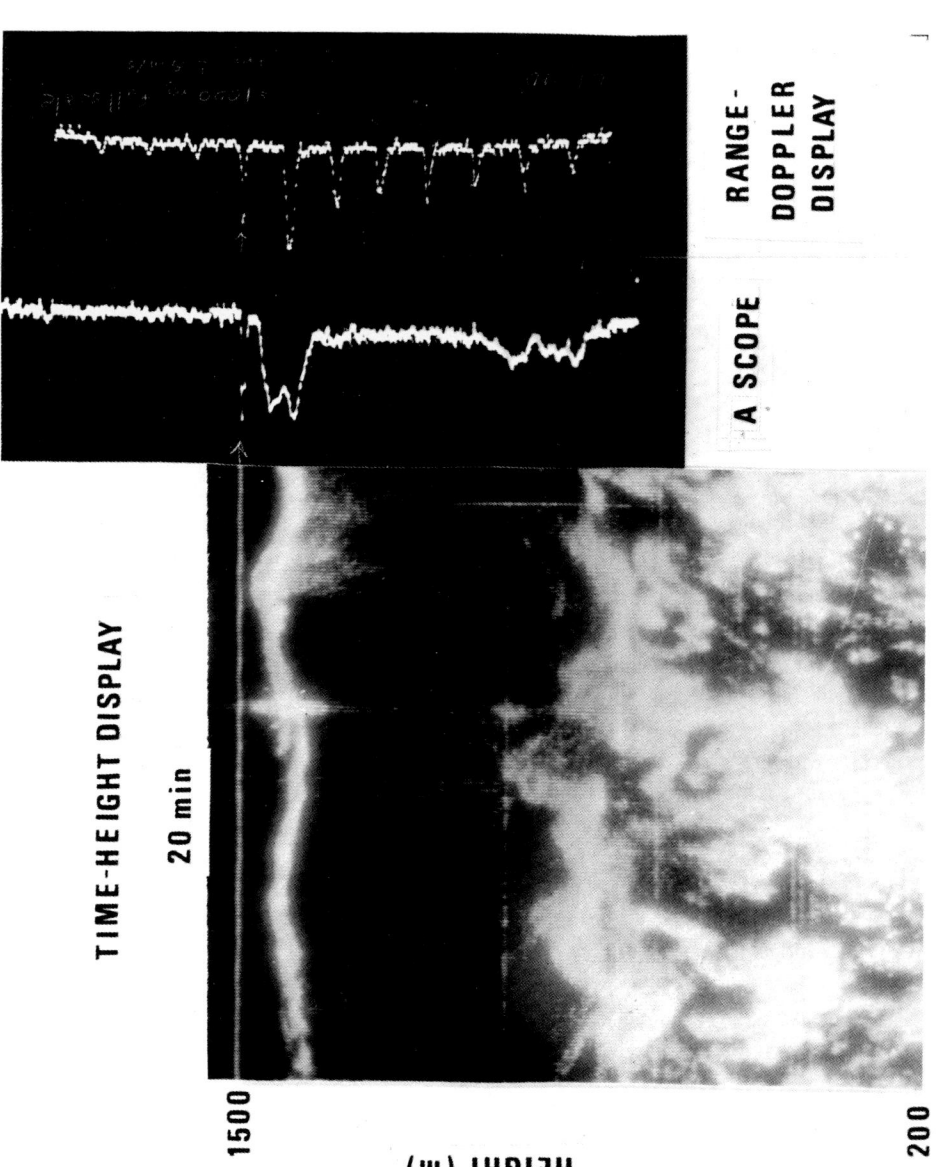

Fig. 7.41. FM cw radar display of echo intensity versus range time (A-scope) and range-Doppler display (right side) and a time–height display (left side) of echo intensity. (From Gossard *et al.*, 1978, copyrighted by the American Geophysical Union.)

This unquestionable advantage of the FM cw vanishes when considering Doppler modes and coherent echoes. In this case there is (ideally) no difference in the two types of radars if the average power is the same. The pulsed radar may even prove advantageous. A slight drawback of the FM cw radar is that for equal-signal bandwidth it is difficult to achieve the same signal-to-noise ratio as with the pulsed radar. This is because frequencies are swept during mixing, and the noise above and below the reference frequency (i.e., both sideband noises) are superimposed on the signal. With a preselector filter the pulsed-Doppler radar easily eliminates one sideband of noise that the simple (homodyne) FM cw radar described here cannot.

The following parameters (Strauch, 1976) are typical of an FM cw radar: Carrier frequency, 2.8 GHz; frequency deviation, $B = 10$ MHz; sawtooth period, $T_s = 1$ ms. Besides the paper by Strauch (1976), a report by Barrick (1973) contains a comprehensive discussion of signal processing. A paper by Chadwick and Strauch (1979) demonstrates that an FM cw Doppler receiver is equivalent to a correlation receiver.

An example of data output from a FM cw radar shows a finely resolved feature in the planetary boundary layer (Fig. 7.41). The intensity of backscatter is shown at left versus height for a vertically pointing antenna. The range resolution is about 3 m. The backscatter is due to refractive index fluctuations, and the returned power is proportional to the structure parameter C_n^2 (see Chapter 11). As in Fig. 7.40, the range axis is also a velocity axis, for which the Doppler spectra are shown on the right in 87-m height increments. The velocity span within a height increment is 10 m s^{-1}.

REFERENCES

Barrick, D. E. (1973). FM/CW radar signals and digital processing. *NOAA Tech Rep.* **ERL 283-WPL 26**, 1–22.

Campbell, W. C., and Strauch, R. C. (1976). Meteorological Doppler radar with double pulse transmission. *Prepr., Conf. Radar Meteorol., 17th, 1976* pp. 42–44.

Chadwick, R. B., and Strauch, R. G. (1979). Processing of FM-CW Doppler radar signals from distributed targets. *IEEE Trans. Aerosp. Electron. Syst.* **AES-15**(1), 185–188.

Davenport, W. B., and Root, W. L. (1958). "An Introduction to the Theory of Random Signals and Noise." *McGraw-Hill, New York.*

Deley, G. W. (1970). Waveform design. *In* "Radar Handbook" (M. I. Skolnik, ed.), pp. 3-1–3-47. McGraw-Hill, New York.

Doviak, R. J., and Sirmans, D. (1973). Doppler radar with polarization diversity. *J. Atmos. Sci.* **30**, 737–738.

Doviak, R. J., Sirmans, D., Zrnić, D., and Walker, G. B. (1978). Considerations for pulse-Doppler radar observations of severe thunderstorms. *J. Appl. Meteorol.* **17**, 189–205.

Golay, M. J. E. (1961). Complementary series. *IRE Trans. Inf. Theory*, **IT-7**, 82–87.

Gossard, E. E., Chadwick, R. B., Moran, K. P., Strauch, R. G., Morrison, G. E., and Campbell,

W. C. (1978). Observation of winds in the clear air using an FM-CW Doppler radar. *Radio Sci.* **13,** 285–289.

Groginsky, H. L., and Glover, K. M. (1980). Weather radar canceller design. *Prepr., Conf. Radar Meteorol., 19th, 1980* pp. 192–198.

Hennington, L. (1981). Reducing the effects of Doppler radar ambiguities. *J. Appl. Meteorol.* **20,** 1543–1546.

Krehbiel, P. R., and Brook, M. (1979). A broad-band noise technique for fast-scanning radar observations of clouds and clutter targets. *IEEE Trans. Geosci. Electron., Spec. Issue Radio Meteorol.* **GE-17**(4), 196–204.

Long, M. W. (1975). "Radar Reflectivity of Land and Sea." Heath, Indianapolis, Indiana.

Nathanson, F. E. (1969). "Radar Design Principles." McGraw-Hill, New York.

Nutten, B., Amayenc, P., Chong, M., Hauser, D., Rouse, F., and Testud, J. (1979). The Ronsard radars: A versatile C-band dual Doppler facility. *IEEE Trans. Geosci. Electron.* **GE-17**(4), 281–287.

Papoulis, A. (1965). "Probability, Random Variables, and Stochastic Processes." McGraw-Hill, New York.

Schmidt, G., Ruster, R., and Cezchowsky, P. (1979). Complementary code and digital filtering for detection of weak VHF radar signals from the mesosphere. *IEEE Trans. Geosci. Electron.*, **GE-17**(4), 154–161.

Sirmans, D., Zrnić, D., and Bumgarner, W. (1976). Estimation of maximum unambiguous Doppler velocity by use of two sampling rates. *Prepr., Conf. Radar Meteorol., 17th, 1976* pp. 23–28.

Strauch, R. G. (1976). Theory and application of the FM-CW Doppler radar. Ph.D. Dissertation, Dept. of Electrical Engineering, University of Colorado, Boulder.

Waldteufel, P. (1976). An analysis of weather spectra variance in a tornadic storm. *NOAA Tech. Memo.* **ERL NSSL-76,** 1–80.

Woodman, R. F., and Hagfors, T. (1969). Methods for the measurement of vertical ionospheric motions near the magnetic equator by incoherent scattering. *J. Geophys. Res. Space Phys.* **75,** 1205–1212.

Zeoli, G. W. (1971). IF versus video limiting for two-channel coherent signal processors. *IEEE Trans. In. Theory* **IT-17,** 579–587.

Zrnić, D. S., and Doviak, R. J. (1976). Effective antenna pattern of scanning radars. *IEEE Trans. Aerosp. Electron. Syst.* **AES-12**(5), 551–555.

Zrnić, D. S., Hamidi, S., and Zahrai, A. (1983). Considerations for the design of ground clutter cancelers for weather radar. Final Rep. No. FAA-RD-82-68. Prepared for U.S. Dept. of Transportation. FAA Systems Res. and Develop. Serv., Washington, D.C.

8

Rain Measurements

Accurate estimates from radar measurements of the rainfall rate R or of the precipitation liquid water content M require detailed knowledge of the drop size distribution $N(D)$. By comparing R and the reflectivity factor Z, with both computed from $N(D)$ as measured by disdrometers (Richards and Crozier, 1981), it can be seen that different distributions giving the same Z can cause rainfall rates to differ by as much as a factor of 4.

Although radar techniques have practical limitations and their accuracy in rainfall rate estimation is highly suspect, they have the decided advantage because they can survey vast areas and make millions of measurements in minutes. Radars (e.g., dual polarization or Doppler) capable of measuring more than one parameter (e.g., vertical and horizontal reflectivities or a spectrum of terminal velocities) in each resolution volume offer improved estimates of critically important parameters of the drop size distributions so that high-resolution measurement of the spatial distribution of rainwater can be made. Multiple-parameter radars in combination with satellites (Atlas *et al.*, 1982), rain gauges, and other instruments may give the sought-for accuracy in rainfall estimates together with the ease of data collection inherent in remote sensing, without needlessly compromising spatial and temporal resolution.

To characterize accurately the relations between water density and cloud dynamics with good spatial resolution and to sense reliably the threat of unusual but significant events such as flash floods, efforts have been and still are under way to improve the accuracy with which radars can estimate rainfall.

8.1 DROP SIZE DISTRIBUTIONS

The distribution of drop sizes is of central importance in determining the reflectivity factor Z, liquid water content M, and rainfall rate R. The drop size distribution is the volume density of drops per unit drop diameter and therefore has units of m^{-4}. Cloud droplets are usually formed by water vapor

Fig. 8.1. Typical shape of large drops falling at terminal velocity w_t. The diameter D of the drops, from left to right, is 8.00, 7.35, 5.80, 5.30, 3.45, and 2.70 mm, corresponding to the following values of w_t: 9.2, 9.2, 9.17, 9.13, 8.46, and 7.70 m s^{-1}. (From Pruppacher and Beard, 1970.)

8.1 Drop Size Distributions

condensing on particulates which serve as condensation nuclei. Supersaturation (humidity of $>100\%$) is required to condense water vapor in pure air, but condensation is too slow a process to produce precipitating drops (rain) within the lifetime of clouds. The coalescence of colliding drops is required for rapid growth of raindrops.

Drops of diameter $D < 0.35$ mm are essentially spherical, and drops up to 1 mm in diameter have a shape well approximated by an oblate spheroid (see Fig. 8.1; Pruppacher and Pitter, 1971). Larger drops have a progressively flattened and then concave base ($D \gtrsim 4$ mm), the larger deformed drops having a high probability of breaking up into smaller fragments. The processes of coalescence and breakup determine the size distribution of raindrops. Srivastava (1974) gives a review of the theoretical models used to determine size distributions and compares theory with observations. We present here a brief description of drop formation and give some results from drop growth models as an approach to the size distribution of cloud particles and raindrops. For a thorough treatment, the reader is referred to texts on cloud physics, such as that by Mason (1971).

8.1.1 Cloud Drop Distributions

Consider a droplet of water in an environment of water vapor at a pressure such that equilibrium is achieved between the rate of evaporation of the water molecules in the drop and the rate of condensation of the molecules surrounding the drop. The vapor pressure required to achieve this equilibrium depends exponentially on the inverse diameter D^{-1} of the drop—smaller droplets have much higher vapor pressure. Air above a plane surface ($D \to \infty$) of water has a humidity $H = 100\%$ if equilibrium is established. Thus a droplet introduced into a barely saturated ($H = 100\%$) atmosphere will evaporate; it cannot persist, much less grow, unless the environment is supersaturated ($H > 100\%$) by an amount equal to the vapor pressure of the droplet.

The droplet diameter at which equilibrium is achieved is a critical diameter because larger drops continue to grow and smaller drops evaporate as long as the vapor pressure is maintained. In pure air, water vapor will not condense until the air becomes supersaturated by several hundred percent. In this highly saturated state, condensation does occur on ions produced by passing cosmic particles. The Wilson cloud chamber, used to map the trajectories of ionizing nuclear particles, is based on this fact.

A pure environment rarely exists in the atmosphere because of the myriad of aerosols, both natural and man-made, that serve as centers of condensation (i.e., condensation nuclei). These condensation nuclei require considerably lower vapor pressure to achieve equilibrium than water drops of the same diameter, and hence they prevent great supersaturation from occurring in the natural atmosphere.

Not all condensation nuclei are solid particles. Some are hygroscopic liquids, such as nitric and sulfuric acid. Some have the property of condensing water vapor at pressure less than saturation (i.e., $H = 100\%$). As the droplet grows on the aerosol, be it solid or liquid, the increased drop volume shields the nuclei's influence, and the drop looks more like pure water. The vapor pressure required for continued growth then increases and the drop reaches an equilibrium size. If $H < 100\%$, droplets never achieve sufficient sizes to form a visible cloud (i.e., drop diameters $\gtrsim 5$ μm, where scattering of light is practically nonselective with respect to color), although they may contribute to haze which causes a moderate diminution of visibility. However, if drops reach the critical diameter and $H \gtrsim 100\%$, then clouds form, and cloud droplets may continue to grow to precipitation size. Nevertheless many drops compete in the reservoir of vapor and will decrease saturation, so the growth of droplets is restricted.

Now let us consider a rising parcel of moist air. Drops immediately begin to form about condensation nuclei when the air parcel, lifted by convection, is brought below its dew point temperature (i.e., the temperature at which

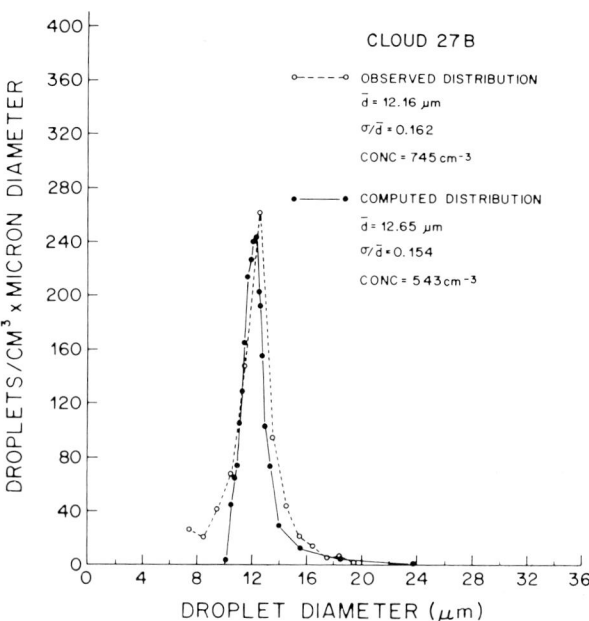

Fig. 8.2. Comparison of computed and observed drop size spectra. Condensation alone is considered. (After Fitzgerald, 1972.)

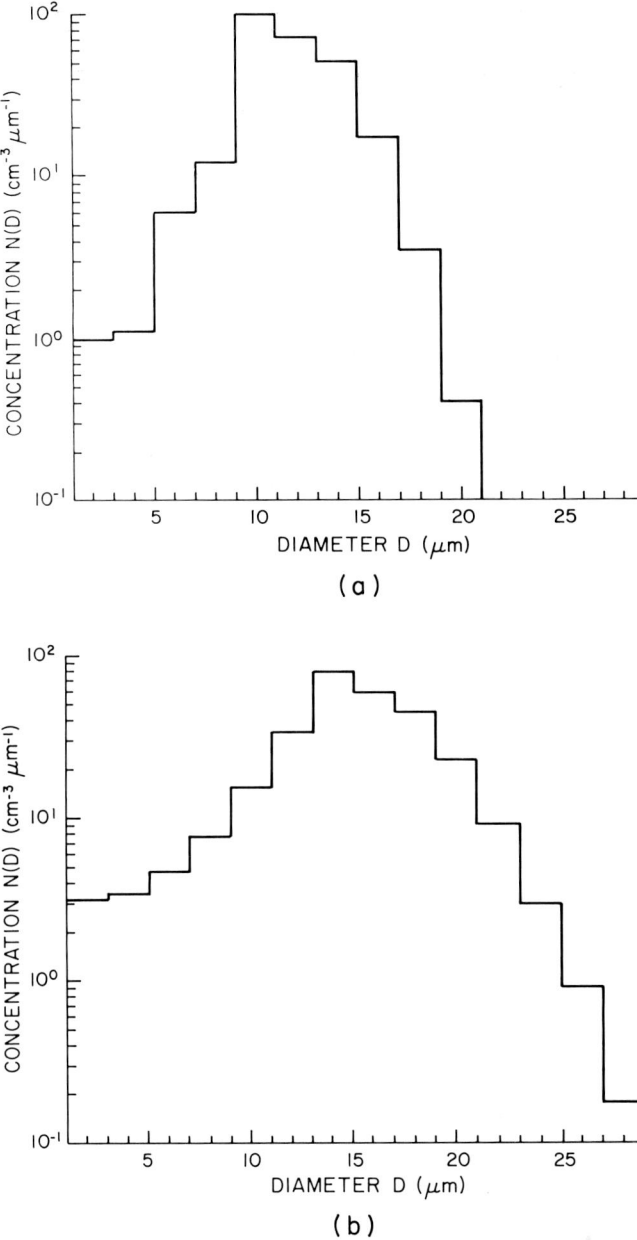

Fig. 8.3. Observed cloud drop size spectra. (a) The drop spectrum in fair weather cumulus taken in weak updraft (2–3 m s^{-1}) at 4.7 km (temperature $-2°$C) about 200 m above the cloud base. The water content is 0.55 g m^{-3}, and the drop concentration is 523 cm^{-3}. (b) Drop spectrum in growing cumulus congestus taken at 7.5 km ($-18°$C) about 3.2 km above the cloud base. The measured updraft was 15 m s^{-1}, the liquid water content 1.19 g m^{-3}, and the drop concentration 582 cm^{-3}. (Courtesy of Dr. T. E. Dye, NCAR.)

water vapor begins to condense). More efficient nuclei allow drops to grow more rapidly, and when moisture is made available by the updraft at the rate equal to the rate at which it is being condensed on the nuclei, the concentration of cloud drops equals the concentration of nuclei activated. The air is supersaturated at this point, but subsequently supersaturation decreases rather rapidly to a quasi-steady state in which the rate of condensation is very nearly balanced by the rate at which moisture in excess of saturation is made available by the updraft. Because the rate of drop diameter increase is inversely proportional to D (Mason, 1971, p. 122), the size distribution narrows as drops grow. Figure 8.2 shows a computed size distribution for droplets near the base of a cloud and a comparative observation.

However, drop size observations higher in the cloud do not have such narrow distributions. Some sample drop spectra in fair weather cumulus and cumulus congestus are given in Fig. 8.3, which shows larger drops in the congestus cloud. Both measurements were made using a forward scattering spectrometer probe on the NCAR/NOAA Explorer sailplane (Knight et al., 1982). The depicted cloud drop distribution can be used to estimate the reflectivity factor Z and the cloud liquid water, as discussed in Section 8.3.

8.1.2 Raindrop Size Distributions

Water drops commonly do not coalesce upon collision but rebound. However, when slightly electrified, drops do unite on collision. Thus, small drops are lost and bigger ones created while the number density decreases. The theoretical framework for drop size growth by stochastic coalescence is given by Telford (1955). The growth of big drops and the loss of small ones are balanced by drops breaking up when they reach large sizes. The probability of break-up has been observed to increase exponentially with drop diameter (Komabayasi et al., 1964). For example, the lifetime of a 7.5-mm drop is observed to be 10 s.

Srivastava (1971) has determined the steady state size distribution for raindrops generated by their breakup and coalescence. He shows that the steady state raindrop size distribution is independent of the assumed initial distribution (which could be the cloud droplet distribution) but depends on the liquid water content. Calculations by Srivastava (1967) show that exponential distributions are quasi-stable with respect to coalescense in that they change rather slowly, whereas narrow distributions tend rather rapidly toward the exponential shape. Figure 8.4 shows, for different rainfall rates, stationary raindrop size spectra that evolved from assumed initially exponential cloud drop distributions. As rain rates R increased, the stationary distribution was achieved faster: With $R = 1050$ mm hr^{-1}, the stationary distribution was attained in ~ 300 sec. Also plotted in Fig. 8.4 are an exponential rain drop spectra using the Marshall–Palmer (1948) parameters.

8.1 Drop Size Distributions

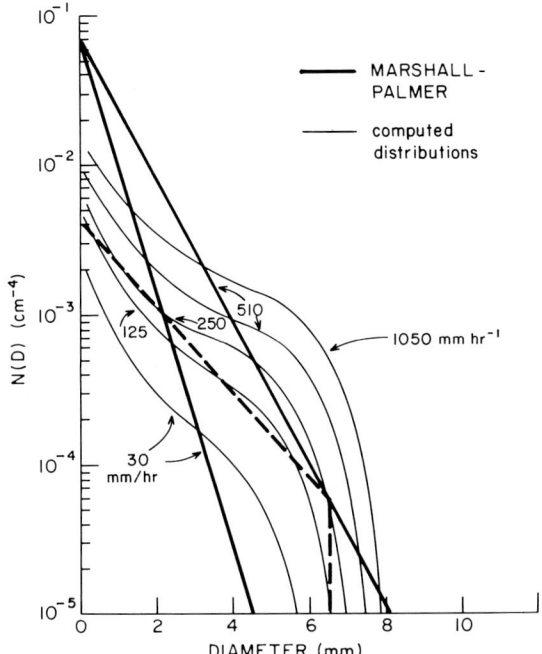

Fig. 8.4. Steady state raindrop size spectra assuming coalescence and the spontaneous breakup of large drops. The Marshall–Palmer (1948) exponential distributions for two rain rates are drawn for comparison. The dashed line is a subjectively fitted truncated exponential distribution for $R = 125$ mm hr^{-1}. (After Srivastava, 1971.)

An exponential distribution of raindrop sizes had been observed by Marshall and Palmer (1948), who used a filter paper to measure directly the density of drop diameters at the surface, and Sekhon and Srivastava (1971) inferred from vertically pointed Doppler radar data that spectra were exponential at several altitudes below the melting layer.

It is interesting to note that, for the Marshall–Palmer (M–P) data (Fig. 8.5), the drop size distribution $N(D)$ has a tendency to flatten at small drop diameters. This is consistent with the observations of Laws and Parsons (1943), whose data (not presented here) also show a rapid increase in $N(D)$ for even smaller drops, producing a shape similar to the theoretical ones in Fig. 8.4. The similarity in shape suggests that the natural processes forming raindrops are taken into account in the theoretical model, but the differences in magnitude of $N(D)$ might be explained if steady state conditions were not met during observation. Airborne measurements of $N(D)$ reported by Carbone and Nelson (1978) for penetrations into growing and mature storms are in reasonable agreement with the model results of Srivastava.

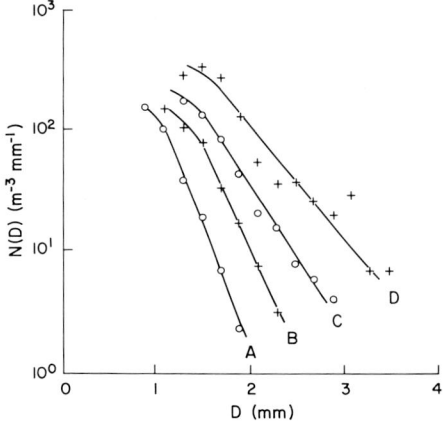

Fig. 8.5. Raindrop size distribution versus drop diameter. (Data from Marshall and Palmer, 1948.) Curves ABCD are for rainfall rates of 1.0, 2.8, 6.3, and 23.0 mm hr^{-1}, respectively.

The parameters of the exponential size distribution that fit the Marshall–Palmer data are†

$$N(D) = N_0 \exp(-\Lambda D), \qquad (8.1a)$$

$$\Lambda = 4.1 R^{-0.21} \quad \text{mm}^{-1} \qquad (8.1b)$$

$$N_0 = 8 \times 10^3 \quad \text{m}^{-3} \text{mm}^{-1}, \qquad (8.1c)$$

where R is the rainfall rate in millimeters per hour. This distribution is widely used to compute rainfall rates from reflectivity factor measurements, although actual drop size spectra differ vastly depending on geographical location, type of rain storm, season, and region within the storm.

Comparison of the M–P raindrop spectra with the theoretical stationary spectra (Fig. 8.4) for the middle drop size range shows that (1) the stationary distributions are flatter than the M–P ones and (2) the stationary distributions are roughly parallel to each other. Pasqualucci's (1982) measurements with a radar ($\lambda = 8.6$ mm) confirm that the rate of decrease for exponential $N(D)$ approaches a constant when the liquid water content exceeds 1 g m^{-3}, but the number density of liquid drops per unit diameter begins to grow. Although the stationary spectra appear quite different from the M–P exponential spectra, they could be fitted with truncated exponential functions in the drop diameter range (1–7 mm) that contributes most to R and Z (see, for example, Figs. 8.7 and 8.9) so that simple analytic formula may still be used. In most observations, $N(D)$ is well represented by an untruncated

---

† In the remaining formulas of this chapter, $N(D)$ and other quantities are in mks units, but in the figures and in discussion conventional units are often used.

exponential function (Waldvogel, 1974). In general N_0 and Λ, as well as the maximum drop diameter, are dependent on R. Indeed, many observations suggest that both N_0 and Λ are highly variable (see, for example, Figs. 8.20c and 8.20d and Waldvogel, 1974).

There are other functions containing two parameters that might better fit the $N(D)$ observed (Cataneo and Stout, 1968; Austin and Geotis, 1979) and hence would be better for estimating $N(D)$ parameters from radar measurements. Although functions containing three or more parameters can better fit a wider variety of observed $N(D)$, more independent measurements may be required to specify these parameters (see Section 8.5.3).

The theoretical distribution, using experimental data for the model parameters, demonstrates that raindrops have an upper diameter limit at about 8 mm. However, larger drops are created by the melting of giant snowflakes or graupel, and actual distributions can be markedly different than the theoretical.

8.2 TERMINAL VELOCITY OF DROPS

Gunn and Kinzer (1949) made precise measurements of the terminal velocity w_t of water droplets in stagnant air at sea level. These data are commonly used in calculating the rainfall rate, on the ground, from $N(D)$ and in deriving the Doppler velocity power spectra for vertically pointed radars. Figure 8.6 shows the terminal velocity for drops in the diameter range 0.27–5.76 mm. Drops of diameter larger than about 6 mm were found to be unstable and to break into smaller ones. A useful formula for terminal velocity is (Atlas *et al.*, 1973)

$$w_t(D) = 9.65 - 10.3 \exp(-600D) \quad \text{m s}^{-1}, \qquad (8.2)$$

where D is the diameter in meters; if D lies between 6×10^{-4} and 5.8×10^{-3} m, there is less than 2% error from the measured values of Gunn and Kinzer. However, we shall find it useful to have a power law fit to $w_t(D)$ data. Atlas and Ulbrich (1977) show that

$$w_t(D) \approx 386.6 D^{0.67} \qquad (8.3)$$

fits the data of Gunn and Kinzer in the diameter range 5×10^{-4} m $< D < 5 \times 10^{-3}$ m. An exponent of 0.5 in (8.3) would be expected from theory, but this is correct only for drops of the very smallest diameter (<0.3 mm), drops that remain spherical. At the larger drop diameters found in moderate-to-heavy rainfall, drops are distorted and a theoretical solution is difficult, so we resort to experimental data and use empirical formulas. Equation (8.3) is also plotted in Fig. 8.6 for comparison.

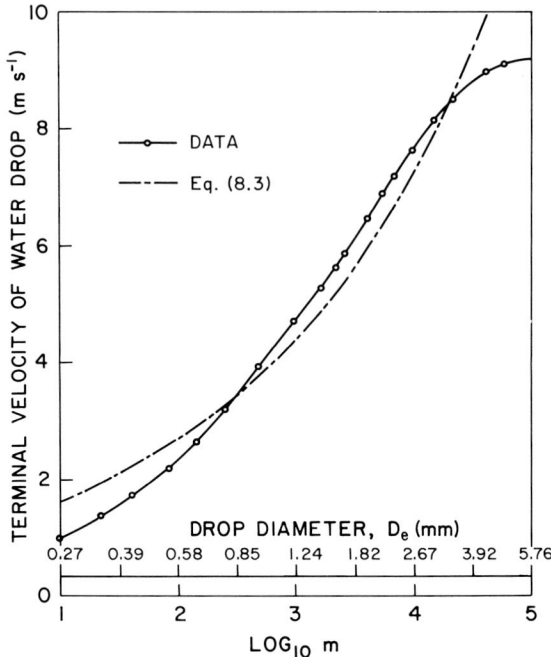

Fig. 8.6. Terminal velocity (solid line) of distilled water droplets in stagnant air at 76-cm-Hg pressure, 20° Centigrade, and 50% relative humidity, as function of the mass m (in micrograms) or the equivalent spherical diameter D_e. (Data from Gunn and Kinzer, 1949.)

Drops fall faster in rarified air, so $w_t(D)$ needs to be adjusted if the terminal velocity is required at higher altitudes. Foote and duToit (1969) have examined data on terminal velocities of drops falling in a partially evacuated tube. They deduce that dependence of the terminal velocity $w_t(D, \rho)$ on the air density can be approximated by

$$w_t(D, \rho) = w_t(D)(\rho_0/\rho)^{0.4}, \tag{8.4}$$

where ρ_0 is the density of air at a pressure of 760 mm Hg and a temperature of 20°C.

8.3 RAINFALL RATE, REFLECTIVITY, ATTENUATION, AND LIQUID WATER CONTENT

In Chapter 4 we showed that the average or expected power of echo samples is proportional to the reflectivity factor Z if drop diameters are small compared to the wavelength. In this section we present the assumptions

8.3 Rainfall Rate, Reflectivity, Attenuation, and Liquid Water Content

that need to be made in order to relate the radar-measurable parameters Z and attenuation rate K to the more interesting meteorological variables, the liquid water content M of the cloud and the rainfall rate R.

8.3.1 Liquid Water Content

An important parameter in rain production is the efficiency with which the water vapor is converted into cloud water. Thus radar measurements of the distribution of M throughout the cloud are often desired. Jet aircraft engines ingesting water may "flame out" if the liquid water concentration is large. The liquid water mass density contribution by drops having diameter between D and $D + dD$ is

$$dM(D) = (\pi/6)D^3 \rho_w N(D)\, dD,$$

where ρ_w is water density of the drop (10^3 kg m^{-3}). The cloud's water density is then

$$M = \int dM(D) = (\pi \rho_w/6) \int_0^\infty D^3 N(D)\, dD. \tag{8.5}$$

We shall find the following formula quite useful when we consider exponential drop size distributions:

$$\int_0^\infty x^{\nu-1} e^{-\mu x}\, dx = (1/\mu^\nu)\Gamma(\nu), \qquad \mathrm{Re}\,\mu > 0, \quad \mathrm{Re}\,\nu > 0, \tag{8.6}$$

where $\Gamma(\nu)$ is the gamma function [if ν is an integer, n, $\Gamma(\nu) = (n - 1)!$]. Values of $\Gamma(\nu)$ for noninteger ν are given in mathematical tables (e.g., Abramovitz and Stegun, 1964). If we assume an untruncated exponential drop diameter distribution [Eq. (8.1a)], we can easily solve the integral (8.5) using (8.6) to obtain a cloud water density of

$$M = \pi \rho_w N_0 / \Lambda^4. \tag{8.7}$$

The diameter of a drop such that half the water is contained in larger drops is defined as the *median volume diameter* D_0. Thus, D_0 is the solution of the equation

$$(\pi \rho_w/6) \int_0^{D_0} D^3 N(D)\, dD = M/2. \tag{8.8}$$

This can be solved for D_0, giving

$$D_0 = 3.67/\Lambda. \tag{8.9}$$

The median drop diameter is used frequently instead of Λ because D_0 is an easily identified physical attribute of the cloud. The exponential drop distribution can then be written

$$N(D) = N_0 e^{-3.67 D/D_0}. \tag{8.10}$$

Equation (8.1a) may be a good approximation in many situations even if N_0 and Λ are not limited to those values specified by Marshall and Palmer. In general, N_0 and Λ are unknowns and cannot be determined from the measurement of the single parameter Z. Waldvogel (1974) finds significant variability in N_0 and Λ as determined from disdrometer measurements. Pasqualucci (1978) deduced drop size distributions from the Dopper spectra of vertical velocities that also show considerable variability in N_0 and Λ. Section (8.5) discusses methods whereby radar can be used to measure N_0 and Λ and hence M.

8.3.2 Reflectivity Factor Z

Z has been defined as the sixth power of the drop diameter summed over all drops in a unit volume [Eq. (4.27b)]. The reflectivity factor is thus given in terms of the drop size distribution by

$$Z = \int_0^\infty N(D) D^6 \, dD. \tag{8.11}$$

It is instructive to plot, as in Fig. 8.7, the integrand of (8.11) to show the relative weights that different drop diameters give to Z. In Fig. 8.7 Z is expressed in the commonly used logarithmic units; that is $\mathrm{dBZ} = 10 \log Z$ ($\mathrm{mm}^6/\mathrm{m}^3$). For sake of illustration, we used the exponential drop diameter distribution with parameters deduced by Marshall and Palmer.

Even though the smallest drops are most numerous, the sixth power of D causes the fewer larger diameter drops to be the most important contributors to Z. Therefore, even if the exponential distribution poorly estimates the actual drop size density at small drop diameters, this should not produce much error in Z provided that Λ and N_0 accurately describe the distribution for diameters that strongly contribute to Z. Also shown in Fig. 8.7 is the median volume drop diameter for each distribution. Note that the reflectivity factor increases by two orders of magnitude but that the diameters that contribute most to Z increase by no more than a factor of 2. For diameter distributions that can be expressed by (8.1a), we derive the relation

$$Z = N_0 (6!)(D_0/3.67)^7 = N_0 (6!) \Lambda^{-7} \tag{8.12}$$

between the reflectivity factor and the two parameters defining the distribution.

8.3 Rainfall Rate, Reflectivity, Attenuation, and Liquid Water Content

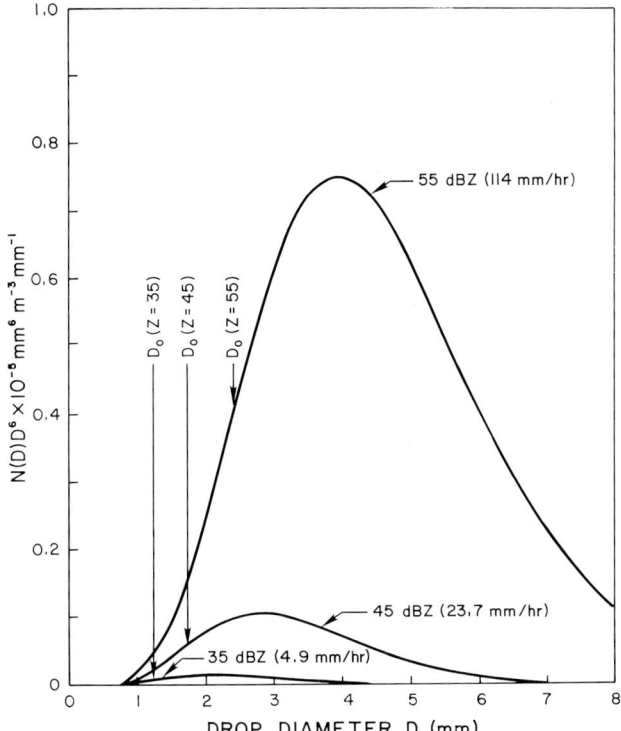

Fig. 8.7. Reflectivity integrand $D^6 N(D)$ versus drop diameter for three echo intensities (35, 45, and 55 dBZ). $R = (4.1 D_0/3.67)^{4.76}$ mm hr^{-1}; $N_0 = 8 \times 10^3$ m^{-3} mm^{-1}; and D_0 (mm) is the median volume drop diameter.

8.3.2.1 Truncated Size Spectra

Equation (8.12) was derived under the assumption that drop diameters extend to infinitely large values, which is unrealistic. The estimate of Z is not significantly in error if the actual drop size spectrum is well approximated by an exponential function and integration is made assuming all sizes are present, but large errors can result for spectra that are flatter (i.e., have a smaller Λ) and that are not truncated at large diameters. Consider the exponential drop size distribution truncated at $D = D_{\max}$. Then it can be shown that reflectivity factor is given by

$$Z = N_0 \Lambda^{-7} \gamma(7, a), \tag{8.13a}$$

where $\gamma(7, a)$ is the incomplete gamma function (Gradshteyn and Ryzhik, 1965) and

$$a = \Lambda D_{\max}. \tag{8.13b}$$

The truncated drop size distribution is

$$N(D) = \begin{cases} N_0 e^{-\Lambda D}, & D < D_{max}, \\ 0, & D_{max} < D. \end{cases} \qquad (8.14)$$

Now consider a truncated spectrum that fits the stationary drop size spectra for $R = 125$ mm hr^{-1}, shown in Fig. 8.4. $N_0 = 4 \times 10^2$ m^{-3} mm^{-1}, $\Lambda = 0.644$ mm^{-1}, and $D_{max} = 6.5$ mm produce a truncated spectrum that fits well this stationary distribution. Using these values to compute $\gamma(7, a)$, we can show that (8.12) overestimates Z by almost an order of magnitude relative to the more accurate Eq. (8.13a).

8.3.2.2 Median Volume Diameter for Truncated Spectra

D_0 in (8.10) is the median volume diameter only as $D_{max} \to \infty$. The median volume diameter D_{0t} for a truncated distribution is obtained by solving the transcendental equation

$$\gamma(4, \Lambda D_{0t}) = \tfrac{1}{2}\gamma(4, \Lambda D_{max}) \qquad (8.15)$$

for ΛD_{0t} given ΛD_{max}. Equation (8.15) is obtained by solving (8.8) for D_{0t} when the upper limit of (8.5) is D_{max}. The solution of (8.15) is given in Fig. 8.8. As ΛD_{max} becomes small, $N(D)$ approaches a uniform distribution over the interval $0 \leq D \leq D_{max}$, and D_{0t}/D_{max} approaches the asymptotic limit

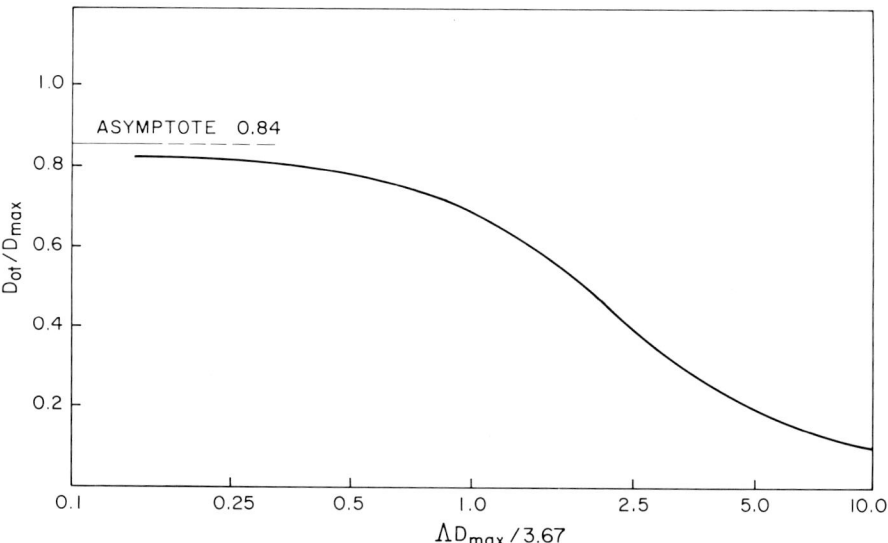

Fig. 8.8. Median volume diameter D_{0t} versus normalized exponential drop size parameter ΛD_{max} for a truncated exponential drop size distribution.

8.3 Rainfall Rate, Reflectivity, Attenuation, and Liquid Water Content

0.84. [One should be careful to distinguish the median volume diameter D_{0t} and the median diameter of $N(D)$, which is $0.5D_{max}$ in this limit.] At the limit as ΛD_{max} gets large, the distribution approaches an exponential one, and $D_{0t} \simeq D_0$ when $\Lambda D_{max} \geq 10$.

8.3.3 Rainfall Rate

When we know the drop diameter spectrum, we can directly compute the rainfall rate in still air using $w_t(D)$ from (8.2). The number of drops $N_A(D)$ of diameter D between D and $D + dD$ impacting upon an area A per unit time dt is

$$N_A(D)/dt = N(D)Aw_t(D)\,dD. \tag{8.16}$$

Each drop has a liquid water mass $dm(D)$:

$$dm(D) = (\pi/6)D^3\rho_w.$$

Therefore the total mass of water per unit area and time is

$$M/A\,dt = \int_0^\infty N_A(D)\,dm(D)\Big/A\,dt = (\rho_w\pi/6)\int_0^\infty D^3 N(D)w_t(D)\,dD. \tag{8.17}$$

The rainfall rate R is usually measured as depth of water per unit time, so

$$R = M/\rho_w A\,dt = (\pi/6)\int_0^\infty D^3 N(D)w_t(D)\,dD. \tag{8.18}$$

We can derive a relation between R and the two parameters of the exponential distribution of drop diameters. Substitution of (8.1a) and (8.2) into (8.18) gives, after integration, the following formula for R:

$$R = \frac{\pi N_0}{\Lambda^4}\left[9.65 - \frac{10.3}{(1 + 600/\Lambda)^4}\right]\text{ m s}^{-1}, \tag{8.19}$$

where mks units are implied throughout. To convert to the more commonly used units of millimeters per hour we need to multiply (8.19) by the factor 3.6×10^6.

It is instructive to find the drop diameters that contribute most to the rain. We have plotted in Fig. 8.9 the rainfall rate integrand of (8.18) assuming an exponential drop size distribution. We can see that, even for the highest rain rate shown, most of the contribution to rain comes from drops having diameters less than 6 mm. However, the rain rate for Fig. 8.9 was computed using the M–P parameters. If the spectra are flatter (i.e., Λ is smaller), as some observations show, then larger-diameter drops contribute significantly to R and (8.19) is then not accurate. In this case one needs to integrate a

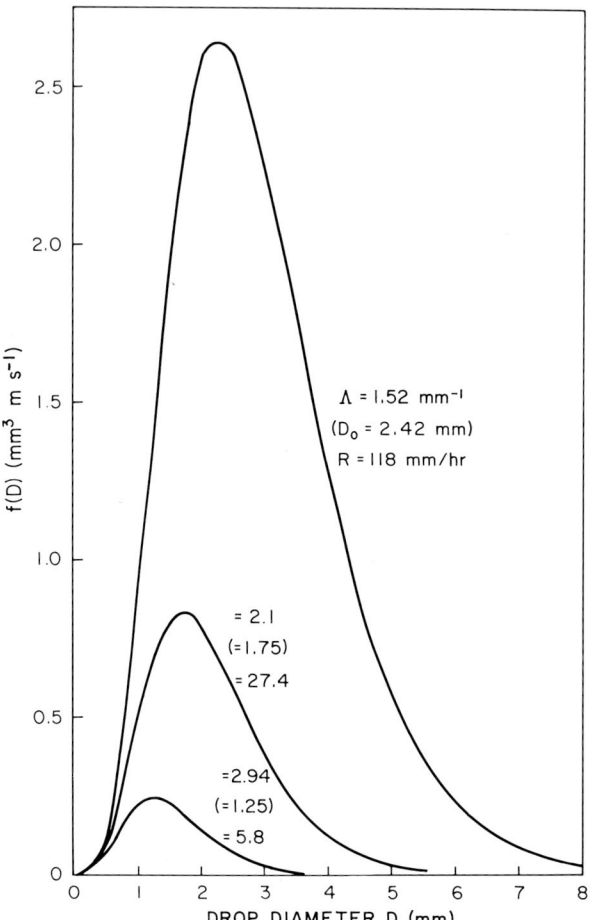

Fig. 8.9. The integrand of the rainfall rate equation (8.18) showing which drops contribute most, assuming an exponential drop size distribution for three different rainfall rates. The rainfall rate integrand is $(\pi N_0/6) f(D) \times 10^{-9}$; $f(D) = (9.65 - 10.3 e^{-0.6D}) \times D^3 e^{-\Lambda D}$.

truncated size spectra to derive a formula for R that depends on three parameters, N_0, Λ, and D_{max}.

In deriving (8.19) we have assumed (8.2) to be valid for all drops. Obviously when $D \to 0$ we obtain unrealistic values for w_t. However, Fig. 8.9 shows that drops smaller than 0.5 mm contribute very little to the rainfall rates when R is significant (i.e., $\gtrsim 5$ mm hr^{-1}), and thus the error has little practical effect on our results. Of course, it would be best to integrate (8.18) numerically using the exact data for $w_t(D)$.

8.3 Rainfall Rate, Reflectivity, Attenuation, and Liquid Water Content

We note that if the Λ and N_0 from the Marshall–Palmer relations (8.1b) and (8.1c) are substituted into (8.19) we obtain a value of R somewhat higher than that implied by (8.1b). This difference arises because (8.1b) is an empirical relation based on measured R for each observed $N(D)$ distribution, and the Marshall–Palmer measurements extend over only a limited drop diameter interval (1 mm $< D <$ 3.5 mm), within which $N(D)$ is nearly exponential. Thus the actual distribution can significantly depart from the assumed exponential distribution for both large and small drops. This departure from the actual $N(D)$ could account for the small differences in R.

8.3.4 Attenuation Rate K

We now discuss the relation between the drop size spectra and the attenuation rate. In Chapter 3 (Section 3.3) we derived the attenuation rate K:

$$K = 4.34 \times 10^3 \frac{1}{\Delta V} \sum_n (\sigma_{an} + \sigma_{sn}) \quad \text{dB km}^{-1}, \tag{8.20}$$

where ΔV is a unit volume (in cubic meters), and the absorption cross section σ_{an} and total scattering cross section σ_{sn} (in square meters) of the nth scatterer are summed for all scatterers in ΔV. The sum $\sigma_t = \sigma_a + \sigma_s$ is called the *extinction cross section*, or sometimes the *attenuation cross section*. Because there is a close connection between the energy loss and the amplitude of the forward-scattered wave, attenuation need not be computed by explicitly evaluating σ_a and σ_s. It can be shown that for linearly polarized radiation the rate at which incident energy is lost in scattering and absorption is proportional to the amplitude component of the forward-scattered wave having the same polarization as the incident one (Born and Wolf, 1964, p. 658). For relatively small drop-diameter/wavelength ratios, the leading term in the series solution formulated by Mie (1980) for the sphere scattering problem gives the following approximation for σ_a and σ_s (see Chapter 3);

$$\sigma_a = (\pi^2 D^3/\lambda) \, \text{Im}(-K_w) \tag{8.21a}$$

and

$$\sigma_s = (2\pi^5 D^6/3\lambda^4)|K_w|^2. \tag{8.21b}$$

It can be shown that $\sigma_s < \sigma_a$ when the first terms of Mie's solution (i.e., the Rayleigh approximation) are compared. Because of this, one might be tempted to use σ_a for attenuation estimation at wavelengths $\lambda \geq 10$ cm, for which all raindrops satisfy the Rayleigh condition $D \leq \lambda/16$. However, we must be cautious because the complete solution to the scattering problem shows that there is a significant contribution to absorption from the remaining terms of the series (even at $\lambda = 10$ cm although $D \leq \lambda/16$ for all raindrops). This is illustrated by plotting, as in Fig. 8.10, the ratio of the complete solution σ_t (Mie) to σ_a given by (8.21a). The upper curve is this ratio at $\lambda = 10$ cm, and it is quite evident that the approximation (8.21a) is acceptable only for diameters less than about 2 mm. Thus the Rayleigh approximation for attenuation is in error for moderate to heavy rains, even at $\lambda = 10$ cm. Nevertheless, the one-way attenuation for $\lambda = 10$ cm is usually less than 1 dB through convective cells of high rain rate (see Fig. 3.4).

In order to determine the attenuation of rain at microwave wavelengths, we must retain higher-order terms in the Mie solution for σ_t. Examples of $\sigma_t(D)$ for four wavelengths are given in Fig. 8.11. The range of drop diameters for each wavelength is shown at the top of the figure. The solutions are for spherical water drops, and because drops begin to flatten when $D \gtrsim 0.3$ mm, the solutions are approximate.

We note that as the drop diameter becomes large compared to the wavelength, the extinction coefficient approaches an asymptotic value of 2.0. This result appears somewhat paradoxical, because with objects large compared to wavelength one would have expected the geometric optics approximation

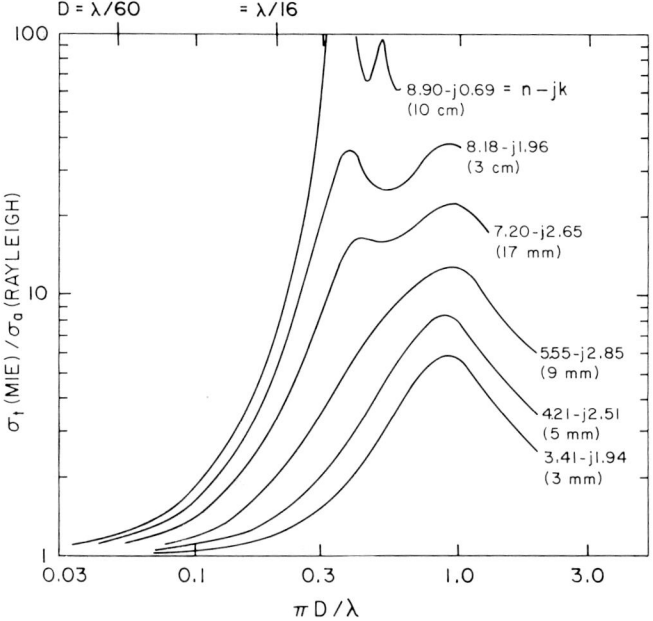

Fig. 8.10. Ratio of the Mie attenuation cross section to the Rayleigh absorption cross section for a water sphere as a function of $\pi D/\lambda$. ($T = 18°C$.) (Data from Gunn and East, 1954.) The pair (n,k) represents the complex refractive index of water (see Section 3.2).

8.3 Rainfall Rate, Reflectivity, Attenuation, and Liquid Water Content

to apply so that the sum of scatter and absorption cross section would be equal to $\pi D^2/4$. The explanation of this apparent contradiction is that, no matter how large the scatterer, there is always a narrow region (the neighborhood of the edge of the geometric shadow) where the geometric optics approximation does not hold. For further discussion, the reader is referred to the article by Sinclair (1947).

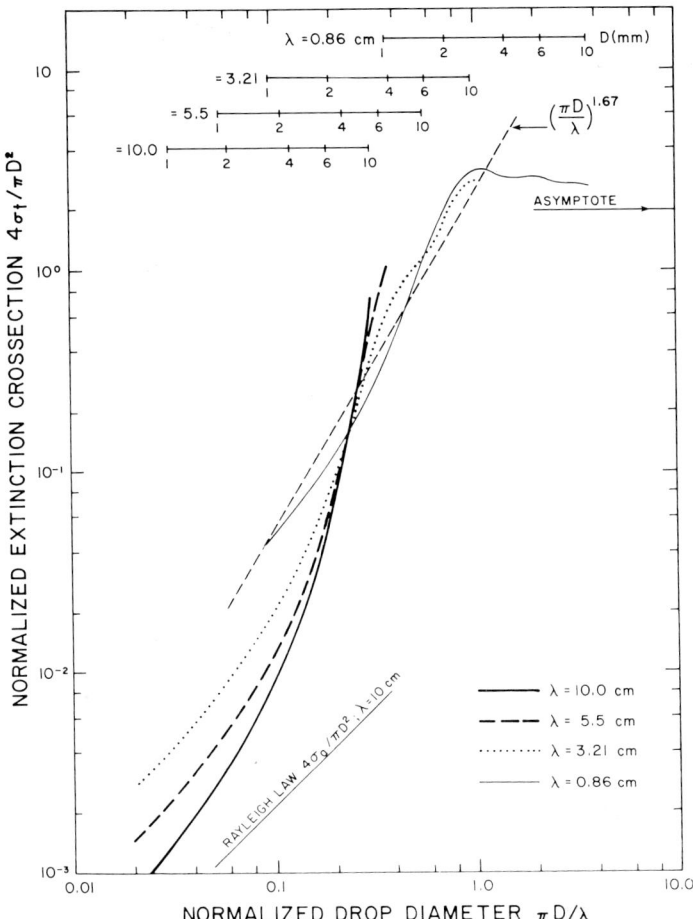

Fig. 8.11. The normalized extinction cross section versus normalized drop diameter for spherical water drops at a temperature of 0°C and at four wavelengths. (Data from Herman *et al.*, 1961.) The solid straight line is the asymptotic absorption cross section in the Rayleigh limit. The dashed straight line shows the power law for normalized cross section that fits the data at $\lambda = 0.86$ cm (see Section 8.4.2).

Given the extinction cross section $\sigma_t(D)$, the attenuation rate is determined if $N(D)$ is known. Expressing the sum (8.20) as an integral in terms of $N(D)$ yields

$$K = 4.34 \times 10^3 \int_0^\infty N(D)\sigma_t(D)\,dD$$

$$= K(\text{absorption}) + K(\text{scatter}). \tag{8.22}$$

Figure 8.12 shows the relative contribution of scatter and absorption to total attenuation. At centimeter wavelengths, absorption loss dominates for all rain rates, but when wavelength is less than 1 cm, scatter can dominate

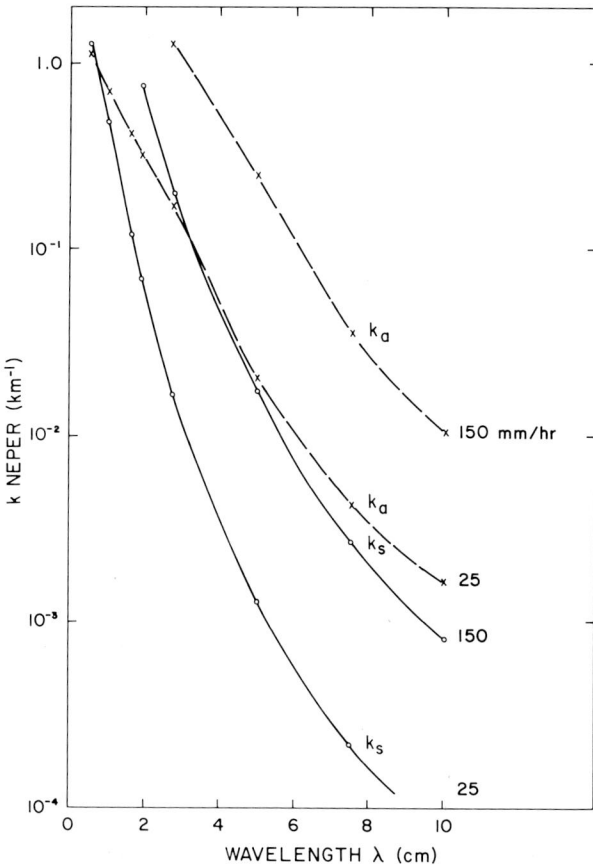

Fig. 8.12. Absorption k_a and scattering k_s coefficients vs wavelength and rain rate at 20°C. (Data from Setzer, 1970.)

absorption at low rain rates. There is no easy solution to (8.22), so investigators have resorted to numerical methods using specified drop size distributions. Burrows and Attwood (1949) have used the Laws and Parsons (1943) drop size data to compute attenuation at various wavelengths and temperature. The attenuation curves shown in Fig. 3.4 assume this drop size spectra. Waldteufel (1973) calculated K for M–P drop spectra and rainfall rates of 1, 10, 100 mm hr^{-1}. He numerically calculated values of $\sigma_t(D)$ using the Mie equations for temperatures of 0°, 10°, 18°, and 40°C.

8.4 SINGLE-PARAMETER MEASUREMENT TO ESTIMATE THE RAINFALL RATE

In this section we discuss methods whereby remote measurement of one parameter, such as the reflectivity factor Z or attenuation rate K, is used to estimate the rainfall rate.

8.4.1 R, Z Relations

Remote measurement of R has considerable practical interest. For many years radar meteorologists have attempted to find a useful formula that relates R to the radar measured parameter Z. Unfortunately, there is no universal relationship connecting these parameters, although it is common experience that larger rainfall rates produce more intense echoes. Examining Eq. (8.19), we see that, for the exponential drop size distribution, we need to measure two parameters in order to obtain R. Moreover, a real drop size distribution requires an indefinite number of parameters to characterize it, and thus the radar-determined value of Z alone cannot provide a unique measurement of R.

There has been considerable effort to establish whether some of the drop size distribution parameters might be known for a given type of rain (stratiform, thunderstorm, etc.). Measurements of drop size distributions around the globe under different climatic conditions have been made, and Battan (1973) lists no fewer than 69 different R, Z relations. Even for rain conditions that were supposedly the same (stratiform), Atlas and Chmela (1957) show considerable variability in the R, Z relations. For example, their data show that the same reflectivity factor Z could be associated with either $R = 33$ mm hr^{-1} or $R = 11$ mm hr^{-1}, a possible 300% error depending on which measured drop size distribution is used. Because these R, Z relations were obtained from actual drop size measurement, they should be accurate for each of the rain events. Even though both rains were classified as stratiform, they differed significantly in size distribution.

It is quite difficult to calibrate radars to within a decibel, and there could be a systematic bias in the radar measured reflectivity. Some of this error can be compensated by choosing an appropriate R, Z relation. According to Cain and Smith (1976), the relation $Z = 1.55R^{1.88}$ removes any pervasive bias in the radar estimate rainfall (RER) in North Dakota, whereas in Miami, Florida, Woodley et al. (1975) reported that the relation $Z = 300R^{1.4}$ worked better. We should recognize that even if the actual drop size distribution were on the average the same at two different locations, errors in radar calibration could lead meteorologists to develop different R, Z relations appropriate to each region. Because neither R, Z relation holds in an absolute sense—another radar at the same location could find a different R, Z relation—the radars need to be reliably calibrated. Other sources of error are (1) horizontal winds that cause ground level droplet deposition to be displaced from the elevated radar resolution cell location (radars rarely detect precipitation close to the ground except for very near ranges in flat country), (2) attenuation from atmospheric gases, rain, and wetted radome (see Fig. 8.25 for an example of atmospheric attenuation effects), (3) reflectivity enhancement in the melting layer (Hill et al., 1981), (4) incomplete beam filling, (5) evaporation (Hardy, 1963), (6) beam blockage (Harju and Puhakka, 1980), (7) rain rate gradients (Zawadzki, 1981), and (8) vertical air motion, which can induce R, Z errors by changing the vertical rain water flux and by changing the drop size distribution in vertically inhomogeneous situations (Carbone and Nelson, 1978).

If the vertical air velocity w is subtracted from w_t [in (8.2), for example], the result is substituted into (8.18), and this equation is then integrated for an assumed exponential drop size distribution, the rainfall rate equation obtained is similar to (8.19) except that w must be subtracted from 9.65 m s^{-1}. From this result it is easy to determine that a 2-m s^{-1} updraft causes a 50% overestimate and a 2-m s^{-1} downdraft a 25% underestimate, both relatively independent of R. At larger w the errors grow rapidly and become more dependent on R. Furthermore, even if the spatial average of w should vanish, the error in the rate of total water fall (and hence total accumulated water) may not, as can be seen from the following general formula for the error in water mass per unit time:

$$\text{error} = \iint w(\mathbf{r})M(\mathbf{r})\,dA,$$

where A is the area over which accumulated water is to be computed (this is usually a catchment area). In general, the integral does not vanish, and usually significant rains occur in regions of downdrafts (negative w) so that total water mass rates are easily underestimated.

8.4 Single-Parameter Measurement to Estimate the Rainfall Rate

In spite of the superfluity of R, Z relations, many of them do not differ greatly at rainfall rates between 20 and 200 mm hr^{-1}. (For further discussion, see Fig. 8.15 and Section 8.5.1.) For stratiform rain the relation

$$Z = 200R^{1.6}, \qquad (8.23\text{a})$$

referred to as the Marshall–Palmer formula (Marshall et al., 1955), with R in mm hr^{-1} and Z in mm^6 m^{-3}, has proven quite useful, although there are notable exceptions (e.g., Jorgensen and Willis, 1982; Cataneo, 1969). Laws and Parsons drop size spectral measurements give

$$Z = 400R^{1.4}. \qquad (8.23\text{b})$$

More recently Joss and Waldvogel (1970) have used

$$Z = 300R^{1.5} \qquad (8.23\text{c})$$

and showed a 42% standard deviation between radar- and disdrometer-measured daily rainfall accumulations for 47 days of rain events throughout 1967. When they considered those 25 days in which rain accumulation was larger than 10 mm, the standard deviation was reduced to 28%. The use of three different R, Z relations (one for drizzle, one for widespread rain, and one for thunderstorms) doubled the accuracy of radar measured amounts of precipitation (standard deviation of $\approx 13\%$), but these accurate estimates are valid for daily accumulations, not the rain rate. On the other hand, Richards and Crozier (1981) conclude that for southern Ontario, Canada, the improvement in accuracy of radar measurements of precipitation from separating the observations into different rainfall types is insignificant.

Although the use of (8.23a), (8.23b), or (8.23c) may produce large dispersion from the actual rain rate measured by a gauge, the accuracy of rainfall measurements can be greatly improved by averaging in space or time. Leber et al. (1961) used (8.23a) to obtain hourly averages of the 10-cm radar rate during extremely heavy rain and integrated these hourly averages to produce a 24-hr isohyet map that compared very well with the accumulated rainfall obtained from a dense network of rain gauges. More recently, Brandes and Sirmans (1976) show that (8.23a) with a well-calibrated radar produced an agreement in the area accumulation of rain to within an average factor of 1.05 with the estimates from gauges for 11 storm days in central Oklahoma. However, deviations of about 2.5 were found in R using (8.23a) for storms on different days if gauge averages are assumed correct. The reasonably good agreement between the *area integrated rainfall* from radar and gauges demonstrates that the M–P drop size distribution may be adequate for storms when the space or time fluctuations in $N(D)$ are averaged out.

To characterize accurately the relations between water and cloud dynamics with greater resolution, and because some of the more unusual rains may produce meteorologically more important events such as floods, efforts are under way to improve the accuracy by which radar can estimate rainfall. Some of these techniques are attenuation methods (Section 8.4.2), the dual wavelength method (Section 8.5.2), the dual polarization method (Section 8.5.3), and rain gauge–radar combinations (Section 8.5.4). In the dual measurement techniques two parameters of an assumed drop size distribution are measured.

8.4.2 Attenuation Method

Communication engineers and radar meteorologists have observed for a wide range of rainfall rates and rain types a consistent relationship between microwave attenuation and the rainfall rate measured with rain gauges along the path of communication. Thus, an accurate measure of attenuation might be used to obtain reasonable estimates of R without prior information concerning $N(D)$. Furthermore, the observations show that K and R appear to be uniquely related by a power law for a wide range of rainfall rates and rain types, being nearly linearly related at 1-cm wavelength. We shall illuminate the underlying cause of this unique property of the attenuation rate that could make its measurement attractive for remote estimation of R.

It can be shown that if $\sigma_t(D)$ is well approximated by a power law dependence on D in the range of diameters that contribute significantly to K, then the liquid water content M and K are also related by a power law, a result that is consistent with many experiments. Moreover, we shall now demonstrate that the power law approximation,

$$\sigma_t(D) \approx CD^n, \tag{8.24}$$

for the normalized extinction cross sections shown in Fig. 8.11, leads to a rainfall rate as a function of attenuation rate that can be independent of the drop size distribution! Atlas and Ulbrich (1974) first illustrated that the power law dependence of the rainfall rate on the microwave attenuation implies the effective power law dependence given in (8.24).

Now let us examine again the rainfall rate formula (8.18). In order for R to be expressed as an integral of a power law function of D, we are motivated to use the approximation (8.3) for the terminal velocity w_t. Using (8.3) in (8.18) yields

$$R = [\pi(386.6)/6] \int_0^\infty D^{3.67} N(D) \, dD. \tag{8.25}$$

8.4 Single-Parameter Measurement to Estimate the Rainfall Rate

Substitution of (8.24) into (8.22) gives

$$K = 4.34 \times 10^3 C \int_0^\infty N(D) D^n \, dD \quad \text{dB km}^{-1}. \tag{8.26}$$

Comparison of (8.25) and (8.26) immediately shows that if the power exponent $n = 3.67$, then R and K are linearly related and, moreover, independent of drop size distribution.

Atlas and Ulbrich (1974) fitted Waldteufel's (1973) numerically computed K, R values with a power law in the R interval of $1-100$ mm hr^{-1}. From these fitted curves they have obtained values of C and n for a wide range of wavelengths (0.1–10 cm) and temperature (0°–40°C). These data (Fig. 8.13) show that $n = 3.67$ for $\lambda = 0.86$ cm. Furthermore, at this wavelength, C and n are essentially independent of temperature over the range 0°–18°C and are altered only negligibly as the temperature rises to 40°C. For comparison, we have plotted on Fig. 8.11 a line (thinly dashed) having the slope $n = 3.67$ (1.67 for the *normalized* extinction cross section) to show how well it fits the $\sigma_t(D)$ data at $\lambda = 0.86$ cm in the diameter range (0.5–5 mm) of drops that contribute significantly to R (see Fig. 8.9).

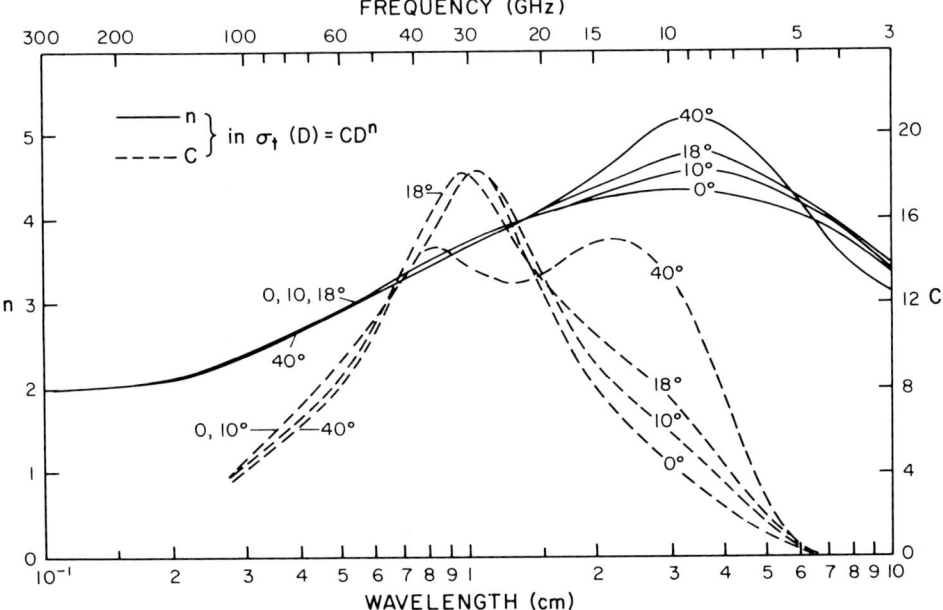

Fig. 8.13. Curves of effective C and n in $\sigma_t = CD^n$ versus wavelength and temperature. (From Atlas and Ulbrich, 1974.) σ_t is in squared centimeters and D in centimeters.

Any wavelength near 1 cm would similarly show that attenuation and rainfall rate relations are relatively independent of the drop size distribution. This conclusion is quickly accepted if one refers to Fig. 8.14, which plots attenuation rate versus rainfall rate for four wavelengths. The circles are K and R values numerically computed by Atlas and Ulbrich (1977) using

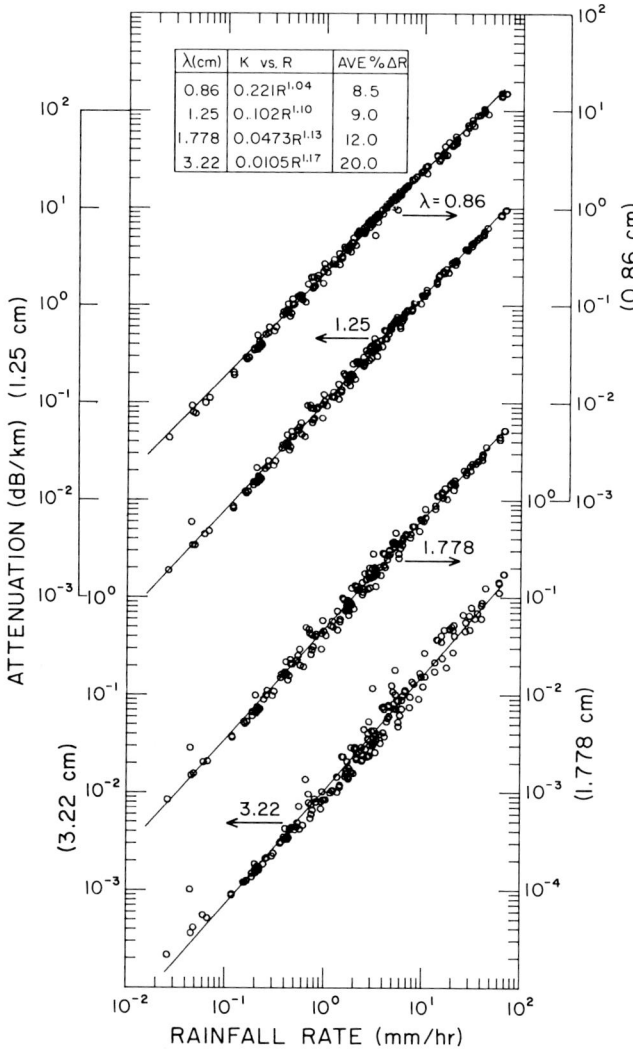

Fig. 8.14. Scattergrams and regression line of K versus R at indicated wavelengths. ($T = 10°C$.) The inset table gives the regression equations and average errors in R (see the text). (Courtesy of C. W. Ulbrich, Clemson University.)

measured drop size distributions (204 of them for three days) and the exact attenuation cross sections for each drop diameter at $T = 10°C$. The numerical values of K versus R are from the best fit to the data of a regression equation $\log K = \alpha + \beta \log R$. The regression equations and the average percentage deviation of R are shown in the inset table of Fig. 8.14. We note how closely packed the data are for $\lambda = 0.86$ cm in spite of the large number of different drop spectra used. We do notice at $\lambda = 3.22$ cm a larger scatter, attributed to differences in drop size spectra. When in situ measurements of the rain rate with rain gauges having excellent spatial (45 m) and temporal (20 s) resolution were made close (2 m) to the microwave path, remarkably good agreement with theory was obtained for both thunderstorm and frontal rain (Norbury and White, 1972).

Even though accurate measurement of K can be related to R if short wavelengths are used, there are practical difficulties in acquiring the data at high rain rates over large areas. For example, the strong attenuation at these short wavelengths imposes limitations on the magnitude and range over which attenuation can be measured. Atlas and Ulbrich (1977) show at $\lambda = 1.25$ cm that the maximum R that can be measured using the two-way radar method is 20 mm hr^{-1} if this rate extends over a range of 30 km. One can trade off range for higher rainfall rate measurements, but in any case the areal coverage is severely limited. There would be a significant increase in areal coverage if one-way measurements could be made. However, this implies separated transmitters and receivers, increasing the cost. Furthermore, a 1-m s^{-1} updraft can cause path attenuation many decibels larger than surface-based measurements of R would suggest (Atlas et al., 1982a). Another practical obstacle to radar measurements of attenuation is a need for a reflector of known cross section at the end of the path where measurements are being made.

The attenuation method assumes the beam to be sufficiently narrow that it is uniformly filled, although the pulse can experience different levels of attenuation (due to cells along the path having different rain rates) as it propagates along the beam. In this case the measured attenuation is path averaged, and the deduced rainfall rate is also the average along the path, because R and K are linearly related (at $\lambda \approx 0.88$ cm).

8.5 DUAL PARAMETER MEASUREMENT TO ESTIMATE THE RAINFALL RATE

We now turn our attention to methods whereby measurement of two parameters can be used to deduce other parameters. For example, measurement of the two remotely sensed parameters, the attenuation rate and reflectivity, can in principle be used to infer the rainfall rate, liquid water content, or any other parameter derived from a two-parameter drop size distribution. Atlas and Chmela (1957) were first to develop a rain parameter diagram that permits one to obtain the rain parameters R, M, D_0, and Z from any pair of the set. Atlas and Ulbrich (1974) developed a more general and accurate rain parameter diagram, which added the microwave attenuation K and optical extinction rate to the other four parameters, thus permitting one to deduce all six parameters from measurement of any pair. The underlying principle is that measurement of any two variables can be used to specify completely a two-parameter drop size distribution. Ulbrich and Atlas (1978) constructed rain parameter diagrams for several wavelengths (0.86, 1.25, 1.78, and 3.22 cm) and temperatures ($-10°$, $0°$, $+10$, and $+20°C$) assuming an exponential drop size distribution and showed how the diagrams can be applied to any size distribution.

We focus our attention on three techniques whereby two variables can be measured: (1) the dual wavelength method, in which the reflectivity factor Z and attenuation rate K are remotely measured; (2) the dual polarization method, in which reflectivity is measured at two different polarizations, with the resulting differential reflectivity shown to be related to Λ; and (3) the rain gauge–radar method, in which in situ point measurements of R and radar measurements of Z are combined to obtain a better assessment of rain over areas between gauges.

8.5.1 Rain Parameter Diagram

Before we proceed, it is well to examine a rain parameter diagram (Fig. 8.15). Although Fig. 8.15 is only for $\lambda = 3.22$ cm and $T = 10°C$, it does illustrate the interrelationship among K, Z, R, and D_0 for an assumed exponential drop size distribution. For example, a measurement of $K = 10^{-1}$ dB km^{-1} and $Z = 2 \times 10^3$ mm^6 m^{-3} gives a rainfall rate estimate slightly larger than 10 mm hr^{-1} and D_0 slightly smaller than 1 mm. The heavy solid lines in the figure are approximate rain parameter curves,

$$K = 8.61 \times 10^{-4} Z^{0.405} R^{0.595} \quad \text{dB km}^{-1}. \tag{8.27}$$

where Z is in mm^6 m^{-3} and R in mm hr^{-1}. Equation (8.27) estimates R given the remotely measurable parameters Z, K and produces values that are in reasonable agreement with the more exact solution of Atlas and Ulbrich. The agreement is good in those regions (shaded) within which most measured drop size distributions fall and at high rainfall rates.

Obviously, rainfall rates less than about 10 mm hr^{-1} require path lengths in excess of several kilometers in order to have measurable attenuation (e.g., ≈ 1 dB). Thus to be useful at lower rainfall rates, a short and more attenuating wavelength would be required. We might estimate K over longer

8.5 Dual Parameter Measurement to Estimate the Rainfall Rate

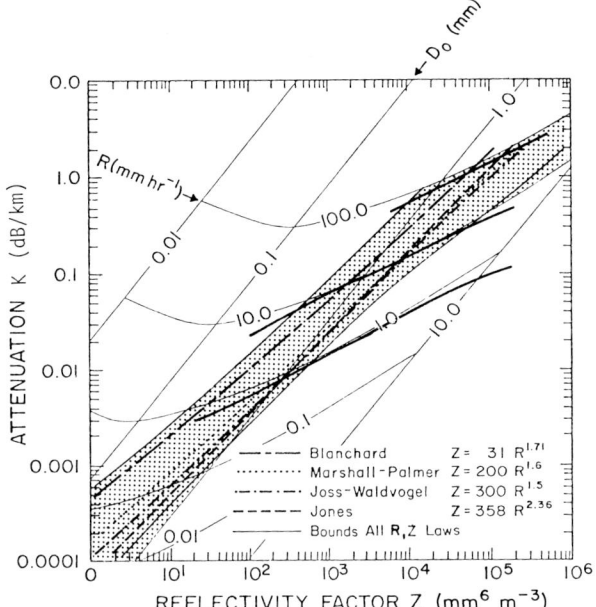

Fig. 8.15. The rain parameter diagram for exponential drop size spectra. The shaded area contains the 69 Z, R relations of Battan (1973). The heavy solid lines are approximate rain parameter curves given by Eq. (8.27). $\lambda = 3.22$; $T = 10°C$. (From Atlas and Ulbrich, 1974.)

paths in order to have a measurable value of attenuation and use the path averaged values $\bar{Z}$, $\bar{K}$ to estimate $\bar{R}$. We often are interested in areal averages of R rather than point values, so long path averages are often not objectionable.

Several commonly used empirical R, Z relations are plotted in Fig. 8.15. These are due to Marshall and Palmer (1948) [Eq. (8.23a)], Blanchard (1953), Jones (1956), and (8.23b), which is due to Joss and Waldvogel (1970). Except for Blanchard's curve, which is applicable to warm orographic rain, the curves are very close to each other. Indicated by the shaded area of Fig. 8.15 are the bounds for the 69 R, Z relations surveyed by Battan (1973). All these are obtained from measured drop size distributions. Although most of them fall close to the M–P curve, they do spread over a tenfold range in R at constant Z, indicating the possible range of error if Z alone is measured and nothing else is known about the drop size distribution.

We again make note, from Fig. 8.15, that K changes relatively little compared to Z for fixed R. In other words, for wavelengths λ as large as 3 cm, K is a measure of R that is relatively insensitive to changes in drop size spectra, as we noted in Section 8.4.2.

8.5.2 Dual Wavelength Method

In Section 8.4.2 we showed that, at wavelengths around 1 cm, attenuation alone can be used to measure R. However, because of the practical limitations that direct attenuation measurements impose, investigators have turned to dual wavelength radars in an attempt to measure K remotely. The dual wavelength technique does not limit attenuation measurements to λ near 1 cm; rather it allows remote measurement of Z and K, from which one can determine two parameters of the drop size spectra.

Alternatively, we can measure the effective reflectivity factor at two wavelengths. One involves Rayleigh scattering, the other a wavelength sufficiently short that substantial Mie scattering (i.e., $D > \lambda/16$) occurs for the larger drops that comprise the scatter domain. From the differences in measured reflectivity factors, one can deduce drop size parameters (Wexler and Atlas, 1963).

Hail can have a significant effect on measurements with a dual wavelength radar; Eccles and Atlas (1973) have proposed a hail detection technique that exploits this effect when spherical hail, wet or dry, has a diameter larger than 8 mm. The technique is based on the range dependence of the decibel ratio of the reflectivities η. Rinehart and Tuttle (1982) have examined the effects of antenna mismatch and have concluded that previous studies that used data collected with dual wavelength radar systems may need to be reexamined. Unfortunately, after many years of work the potential of dual wavelength radar for hail detection is still uncertain.

Although we have several techniques of estimating rainfall rates, we shall confine our attention to one promising dual wavelength method—the independent measurement of Z and K. Eccles (1979) used attenuation measured remotely with a dual wavelength radar but computed R from a modified R, K relation of Atlas and Ulbrich (1974). The dual wavelength radar routinely observed severe convective phenomena in northeastern Colorado in 1972–1974, and the precipitation rate computed from the attenuation was found to be more accurate than that computed from Z.

Although the dual wavelength technique applies to any two-parameter $N(D)$ we shall assume an exponential drop size spectrum in order to show how the parameters N_0 and Λ can be determined from measurement of the reflectivity factor Z and attenuation rate K. We shall arrive at two simultaneous equations containing N_0 and Λ along with measured samples of backscattered power at two wavelengths, one (λ_l) long and nonattenuating, the other (λ_s) short and attenuating. One equation represents the attenuation at λ_s; the other the reflectivity factor at λ_l. We select a nonattenuating wavelength (λ_l) in order to simplify the problem, although two attenuating wavelengths can be used. That is, the attenuation values measured at the two

8.5 Dual Parameter Measurement to Estimate the Rainfall Rate

attenuating wavelengths can be used to determine N_0 and Λ, as can Z and K (Goldhirsh and Katz, 1974). Although the measurement of two attenuation rates eliminates the need to calibrate the radar for absolute power, thus disposing of a major source of error in reflectivity measurements, uniform reflectivity must exist along the path over which K is measured.

We assume the two radar beams are matched in resolution volume size. The mean power backscattered at the two wavelengths are, at range $r = r_1$ [see Chapters 3 and 4, Eqs. (3.13b) and (4.29)]

$$P_l = \frac{C_l Z}{r_1^2}, \tag{8.28a}$$

$$P_s = (C_s Z_s / r_1^2) \exp\left(-2 \int_0^{r_1} k_s \, dr\right), \tag{8.28b}$$

where Z is the reflectivity factor at λ_l and Z_s is the effective reflectivity factor at λ_s (see Section 4.3.3). Because λ_s is small, drop diameters may be larger than the Rayleigh limit $\lambda/16$, so Z_s might not equal Z. The short wavelength attenuation rate k_s [Eq. (8.22)] is

$$k_s = \int N(D) \sigma_t(D) \, dD \quad \text{m}^{-1}. \tag{8.29}$$

Now, at range r_2 we have two equations of identical form except that Z and Z_s may be different. Taking the natural logarithm of the ratio of (8.28b) at two ranges r_1 and r_2, we obtain

$$\ln\left[\frac{P_s(r_1) Z_s(r_2) r_1^2}{P_s(r_2) Z_s(r_1) r_2^2}\right] = 2 \int_{r_1}^{r_2} k_s \, dr, \tag{8.30}$$

in which $Z_s(r_2)$ and $Z_s(r_1)$ are both unknown. However, from measurements of Z at λ_l we might deduce Z_s. For example, it can be shown that for R between 1 and 100 mm hr^{-1} the reflectivity factor Z_s approximates Z well for wavelengths as short as 3 cm (Wexler and Atlas, 1963). We shall assume this condition, so that using (8.28a) we can arrive at the following expression for the path-integrated attenuation rate:

$$2\bar{k}_s = \frac{2}{r_2 - r_1} \int_{r_1}^{r_2} k_s \, dr = \frac{1}{r_2 - r_1} \ln\left[\frac{P_s(r_1) P_l(r_2)}{P_s(r_2) P_l(r_1)}\right]. \tag{8.31}$$

The right side is obtained directly from measurements of the echo power at the two λs and two ranges.

Assuming single scatter is dominant, $\bar{K}$, $\bar{Z}$, and $\bar{R}$ are invariant with respect to the placement of drops along the path (i.e., the drop populations at any one range can be permuted with the population at any other range).

Apart from the weighting given by the beam pattern function, the measured reflectivity factor Z_e and path averaged quantities $\bar{Z}$ and $\bar{R}$ do not depend on how the scatterers are distributed across the beam. This is not the case for path-averaged attenuation rates, because rearrangements of drops across the beam can cause a large variation in the average power flux along the beam. For example, if K is nonuniform across the beam as well as along it, the rearrangement of drops within the beam can cause the average power flux across the beam (after attenuation) to change even though $\bar{Z}$ and $\bar{R}$ remain the same. Thus $\bar{R}$ estimated from $\bar{K}$ is dependent on the distribution of drop size density across the beam. On the other hand, reflectivity and differential reflectivity in the dual polarization technique (see Section 8.5.3) do not depend as strongly on these effects.

Because path-averaged $\bar{Z}$, $\bar{K}$, and $\bar{R}$ are not dependent on how drops are distributed along the averaging length $(r_2 - r_1)$, we might expect that the path-averaged $\bar{R}$ could be retrieved from $\bar{Z}$ and $\bar{K}$. However, if $N(D)$ varies from resolution volume to resolution volume (see Section 4.3.2) even though it is, for example, exponential, it can be shown that the use of path averaged Z, K in a rain parameter diagram does not, in general, give $\bar{R}$. Nevertheless, at a wavelength of 0.86 cm, R is independent of $N(D)$, so that the path average $\bar{K}$ alone determines $\bar{R}$. At other wavelengths we need to assume that the path average $\bar{N}(D)$ is exponential (or some other two-parameter distribution) in order to use the rain parameter diagrams.

Assume $N(D)$ is uniform across the beam and $\bar{N}(D)$ to be an exponential drop size distribution within the volume of the beam along the path $r_2 - r_1$. Using these assumptions we can determine the path-averaged k_s,

$$\bar{k}_s = N_0 \int e^{-\Lambda D} \sigma_t(D) \, dD \quad (\text{m}^{-1}). \tag{8.32a}$$

For precision (8.32a) needs to be numerically evaluated because $\sigma_t(D)$ is not an analytic function. However, as we have already discussed, it can be approximated by a power law in the range of drop diameters that contribute significantly to the rainfall rate (i.e., rates of 1–100 mm hr^{-1}). Thus, substituting (8.24) into (8.32a) and integrating, we obtain

$$\bar{k}_s = \Gamma(n + 1) C' N_0 \Lambda^{-(n+1)}, \tag{8.32b}$$

where $C' = 10^{2n-4} C$, and Λ is in inverse meters. The factor 10^{2n-4} converts data for C in Fig. 8.13 from the centimeter to the meter units used in the equations. Likewise the path-averaged reflectivity factor is obtained from (8.12):

$$\bar{Z} = N_0 (6!) \Lambda^{-7}. \tag{8.33}$$

We thus have two equations and two unknowns. Let us change (8.32b) to

8.5 Dual Parameter Measurement to Estimate the Rainfall Rate

units of decibels per kilometer and divide it by (8.33) to eliminate N_0:

$$\bar{K}_s/\bar{Z} = [4.34 \times 10^3 \Gamma(n+1)C'/(6!)]\Lambda^{6-n} \quad \text{dB km}^{-1} \text{ m}^{-3}. \quad (8.34)$$

Thus Λ (in inverse meters) is determined directly from the ratio of $\bar{K}_s$ to $\bar{Z}$. $\bar{Z}/N_0$ from (8.33) and the ratio (8.34) are plotted in Fig. 8.16 for two attenuating wavelengths $\lambda_s = 5$ and 3 cm at $T = 18°C$ using C and n values obtained from Fig. 8.13. The rainfall rate scale plotted along the abscissa is applicable only for M–P drop size distribution in which $N_0 = 8 \times 10^3$ m^{-3} mm^{-1}.

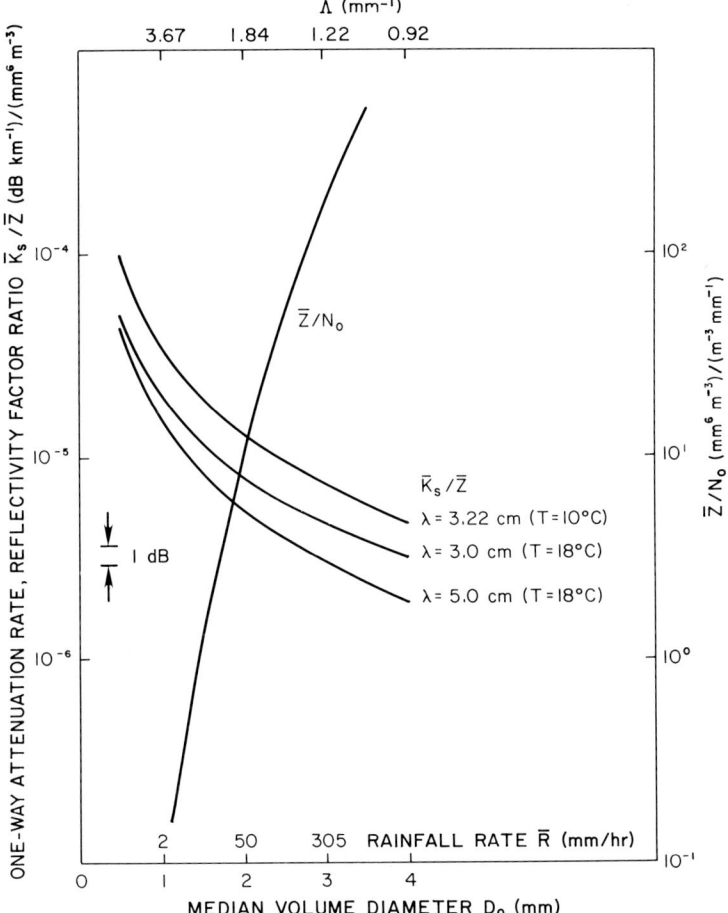

Fig. 8.16. Ratios of the path-integrated attenuation rate reflectivity factor K_s/Z and normalized reflectivity factor $\bar{Z}/N_0$ versus the median volume drop diameter. Rainfall rate R scale (mm hr^{-1}) is only for the path-averaged M–P value of $N_0 = 8 \times 10^3$ m^{-3} mm^{-1}.

To show how Fig. 8.16 is used, let us assume that $\bar{K}_s$, averaged over a 10-km path, is 2 dB km^{-1} at $\lambda = 3.0$ cm and that the path-averaged reflectivity $\bar{Z}$ (at $\lambda = 10$ cm) is 2×10^5 mm^6 m^{-3} (53 dBZ). From the ratio $\bar{K}_s/\bar{Z}$ we find $D_0 = 1.7$ mm. At this D_0 we go down the graph to intersect the curve $\bar{Z}/N_0$ at 3 from which we deduce $N_0 = 6.77 \times 10^4$ m^{-3} mm^{-1}. The values of N_0 and $\Lambda = 3.67/D_0$ are inserted into (8.19) (after converting to mks units) to obtain a path-averaged rainfall rate $\bar{R} = 204$ mm hr^{-1}. However, note that for the same Λ the M–P computed rainfall rate [using, e.g., Eq. (8.1b)] underestimates R at 21 mm hr^{-1}. Using the same values of $\bar{K}_s$ and $\bar{Z}$ for $\lambda = 3.22$ cm we estimate $\bar{R} = 80$ mm hr^{-1}, in reasonable agreement with Fig. 8.15 and consistent with Fig. 8.14, whereas the M–P computed rainfall rate now overestimates $\bar{R}$ at 120 mm hr^{-1}.

We observe that $\bar{K}_s/\bar{Z}$ changes by a factor of 5 over the range of significant (2–300 mm hr^{-1}) path-averaged rainfall rates $\bar{R}$. This is not a very large span considering that errors in $\bar{Z}$ and $\bar{K}_s$ less than 1 dB (a factor of 1.26) are difficult to achieve and that this error level causes considerable error in $\bar{R}$. However, the rainfall rate from Fig. 8.16 is computed using the M–P value for N_0, which may overestimate R at small values of Λ. The difficulties inherent in dual wavelength measurements are pointed out by Eccles and Mueller (1973), who propose a technique to solve the problem caused by the statistical fluctuations of the two signals. A more detailed discussion of error and radar requirements is given by Goldhirsh (1975), who shows that R-estimate errors are significantly less if $\lambda_s = 1$ cm.

Note that at $\lambda \approx 1$ cm the measurement of $\bar{K}_s$ alone estimates $\bar{R}$ as accurately as $\bar{K}_s$ is estimated (see Fig. 8.14). Although N_0 and Λ need to be estimated using both $\bar{Z}$ and $\bar{K}_s$, $\bar{Z}$ does not add more information to the estimate of $\bar{R}$. Z measured at λ_l can be used to estimate the reflectivity Z_s at λ_s so that $\bar{K}_s$ can be computed. However, severe attenuation of 1-cm waves in intense precipitation limits the rain area that can be surveyed. It is important to note here that we are focusing our attention on path-averaged values, and these can often be estimated with greater accuracy than point (i.e., resolution volume) estimates of Z and K_s. Still, the radar needs to be well calibrated, and a dual wavelength radar might well be supplemented with a smaller number of rain gauges for purposes of calibration.

8.5.3 Dual Polarization

Dual polarization, which also attempts to estimate two parameters of the drop size distribution, relies on the echo intensity of two orthogonally polarized waves. Transmitted pulses have alternate horizontal and vertical polarization, and echoes received in the respective principal plane of polarization (same as transmitted) are processed. However, echoes with both

8.5 Dual Parameter Measurement to Estimate the Rainfall Rate

senses of polarization are usually simultaneously present, although the principal polarization is typically stronger.

The basis for the dual polarization scheme is the observation that drops are not spherical but have an oblate spheroidal shape (see Fig. 8.1). In still air, with weak static and vertically oriented electric fields, the drops fall with their minor axis vertical and have eccentricities that depend only on the diameter of the equivalent sphere. Thus, for raindrops satisfying the Rayleigh conditions for scatter, not only do we expect larger power echoes for horizontally polarized waves, but moreover it can be shown that the ratio of reflectivities Z_H/Z_V for horizontally and vertically polarized waves is a direct measure of Λ. However, drops falling in turbulent air or experiencing collisions may oscillate, causing the average raindrop to have a major-axis/minor-axis ratio b/a nearer to unity than that observed for single drops falling in still air (Jameson and Beard, 1982). Thus there could be an uncertainty in the relation between Λ and the ratio Z_H/Z_V. Nevertheless, Seliga et al. (1982) show that dual polarization data should still produce an improvement in radar estimates of R or M. Furthermore, if oscillations result from collisions, then Z_H/Z_V is a function of both N_0 and Λ, a condition that complicates the retrieval of the distribution parameters but does not significantly deteriorate the quantitative applicability of the polarization technique. Although falling drops may generate a spectrum of shapes, let us consider the effect that a single drop falling in still air has on the reflectivity factor.

The axis ratio b/a is related to the diameter D_e of an equivalent volume spherical raindrop as (Green, 1975)

$$D_e = 2\{(T_s/g\rho_w)[(b/a)^2 - 2(b/a)^{1/3} + 1](b/a)^{1/3}\}^{1/2}, \tag{8.35}$$

where $T_s = 72.75 \times 10^{-3}$ J m^{-2} is the surface tension of water, g is the acceleration due to gravitaty, and ρ_w is the water density. Equation (8.35) is plotted in Fig. 8.17, which shows that when $D_e > 1$ mm drops become flattened and a ratio of width to thickness of nearly 2:1 is attained for drops approaching the breakup limit of 8 mm. The diameter D_e of the sphere having a volume $\pi D_e^3/6$ that equals the volume $4\pi b^2 a/3$ of the ellipsoid is chosen as a convenient parameter not only because it uniquely specifies the water drop shape but because it is also the parameter used in relating the terminal velocity to drop size (Fig. 8.6).

We shall now use a simplified theory attributed to Gans (1912) and employed by Seliga and Bringi (1976) to determine the backscatter cross section of oblate spheroids for incident waves of either horizontal or vertical polarization. Gans's work is essentially an extension of the Rayleigh theory for spheres applied to the case of oblate and prolate spheroids. According to Gans's theory, the components of dipole moments aligned along the axes

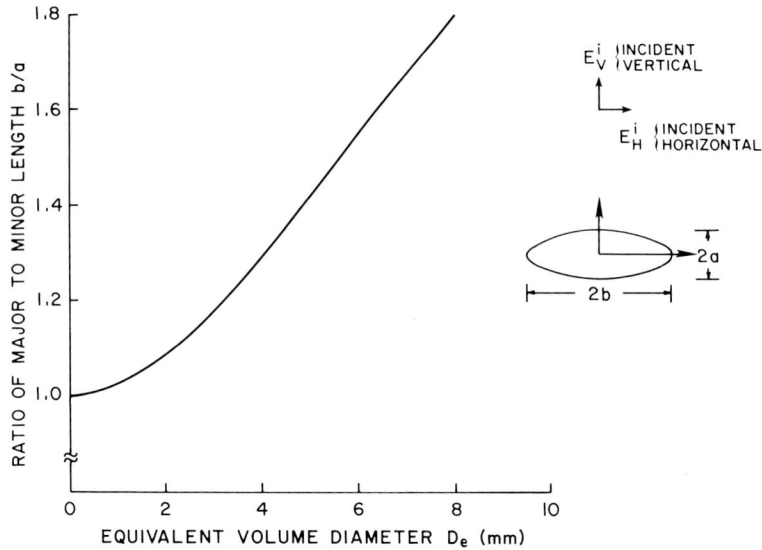

Fig. 8.17. The ratio of the major axis length to minor axis length (b/a) as a function of D_e given by Eq. (8.35).

of the ellipsoid are independently excited by the components of the incident electric field along each axis. As with scattering from a sphere, the dipoles that produce the radiation or scattering have moments not only proportional to the incident electric field strength, volume, and dielectric properties of the scatterer, but also to factors that depend on the shape and orientation.

8.5.3.1 BACKSCATTER CROSS SECTION OF AN OBLATE SPHEROID

The backscatter cross section for horizontal or vertical polarization obtained from the results of Gans (1912) is

$$\sigma_{H,V} = \frac{16\pi^7 D_e^6}{9\lambda^4} \left| \frac{m^2 - 1}{4\pi + (m^2 - 1)P_{H,V}} \right|^2, \qquad (8.36)$$

where m is the complex refractive index and $P_{H,V}$ are geometric factors:

$$P_V = (4\pi/e^2)\{1 - [(1 - e^2)/e^2]^{1/2} \sin^{-1} e\} = 4\pi - 2P_H, \qquad (8.37)$$

where e is the eccentricity of the ellipsoid,

$$e^2 = 1 - (a/b)^2. \qquad (8.38)$$

It can easily be shown that when $a/b \to 1$, $P_H = P_V \to 4\pi/3$ and $\sigma_H \to \sigma_V$, both of which then equal the backscatter cross section given in Eq. (3.6).

8.5 Dual Parameter Measurement to Estimate the Rainfall Rate

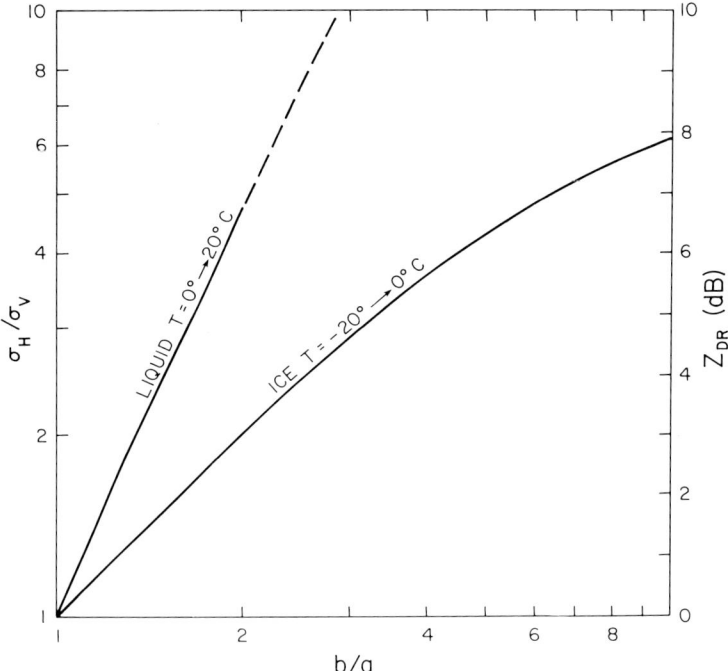

Fig. 8.18. The ratio of backscatter cross sections σ_H/σ_V for oblate spheriodal scatterer as a function of its axis ratio. The dashed portion of the line for a liquid is at ratios where drops are unstable; however, it can be used for water-coated ice.

The ratio of σ_H/σ_V for oblate spheroids (Fig. 8.18) indicates marked differences in the σ_H, σ_V backscatter cross sections for water in the range of b/a expected for liquid drops. This ratio would also be the ratio of reflectivities η_H/η_V for drops of uniform size. The logarithm of the reflectivity ratio defines the differential reflectivity:

$$Z_{DR} = 10 \log(\eta_H/\eta_V). \qquad (8.39)$$

The reflectivity, a function of the drop size distribution, can be expressed in terms of D_e:

$$\eta_{H,V} = \int \sigma_{H,V}(D_e) N(D_e) \, dD_e. \qquad (8.40)$$

We shall assume a truncated exponential drop size distribution, so

$$\eta_{H,V} = \frac{16 N_0 \pi^7}{9 \lambda^4} \int_0^{D_{max}} D_e^6 \left| \frac{m^2 - 1}{4\pi + (m^2 - 1) P_{H,V}} \right|^2 e^{-\Lambda D_e} \, dD_e. \qquad (8.41)$$

We substitute (8.41) into (8.39) to obtain

$$Z_{DR} = 10 \log \left[\int_0^{D_{max}} D_e^6 S_H(m, b/a)(e^{-\Lambda D_e}) \, dD_e \right/$$

$$\left. \int_0^{D_{max}} D_e^6 S_V(m, b/a)(e^{-\Lambda D_e}) \, dD_e \right], \qquad (8.42)$$

where

$$S_{H,V}(m, b/a) = \left| \frac{m^2 - 1}{4\pi + (m^2 - 1)P_{H,V}(D_e)} \right|^2 \qquad (8.43)$$

gives the dependence of Z_{DR} on the shape as well as the temperature and wavelength (through the complex refractive index m). The parameter $P_{H,V}(D_e)$ is a function of D_e through Eqs. (8.35), (8.37), and (8.38). We note that (8.42) is independent of N_0, and thus we can directly determine Λ of the truncated drop size distribution if D_{max} is known.

Seliga and Bringi (1976) have evaluated (8.42) for $D_{max} = 10$ mm and $T = 20°$C. Their results are plotted in Fig. 8.19, where for comparison the more recent Z_{DR} values are plotted using an exact theoretical formulation evaluated by Al-Khatib et al. (1979). We see that the simplified theory agrees well with the more exact formulation for $\Lambda \geq 2.0$ mm^{-1}. The large difference for $\Lambda < 2.0$ mm^{-1} is mostly caused by the difference in D_{max}. When both limits are equal, there is relatively little difference in the entire indicated range of Λ (Seliga and Bringi, 1978).

Once we obtain Λ from Z_{DR} measurements, we need to determine N_0 in order completely to specify the drop size distribution under the assumption that it is exponential. N_0 can be determined from measurements of η_H or η_V. The reflectivity factors $Z_{H,V}$ in terms of $\eta_{H,V}$,

$$Z_{H,V} = (\lambda^4/\pi^5)|K_w|^2 \eta_{H,V}, \qquad (8.44a)$$

have been evaluated by Al-Khatib et al. (1979) using the exact formulation for scattering. We also plot Z_H in Fig. 8.19. Both parameters of the drop size distribution, and hence the rainfall rate, can now be computed from Z_H and Z_{DR} measurements. However, if D_{max} is an independent variable, then D_{max} needs to be estimated; otherwise, we can have different R for the same measured Z_H and Z_{DR}. Scientists at the Air Force Geophysics Laboratory using disdrometer data have found that for rain the maximum drop diameters generally adhered to the relationship (Plank, 1977)

$$D_{max}\Lambda = 7.5. \qquad (8.44b)$$

Ulbrich and Atlas (1982) point out that, when $\Lambda D_{max} > 9.2$, Z_{DR} is not sensitive to the truncation of an exponential distribution but is significantly

8.5 Dual Parameter Measurement to Estimate the Rainfall Rate

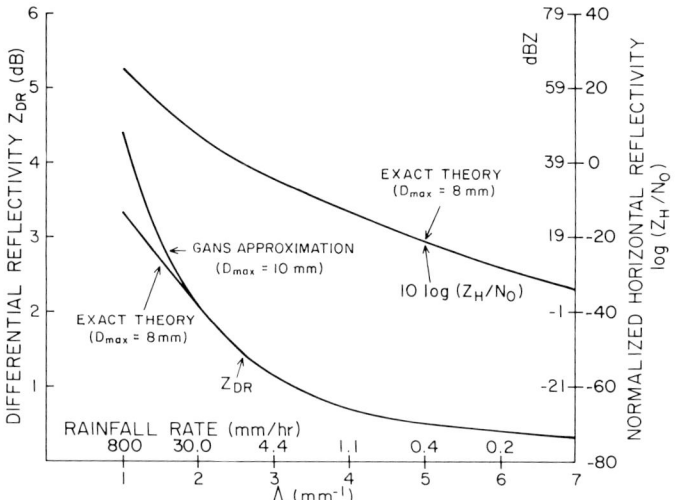

Fig. 8.19. Variations of Z_{DR} and the normalized horizontal reflectivity $10 \log(Z_H/N_0)$ with Λ. R and dBZ scale for $N_0 = 8 \times 10^3$ m^{-3} mm^{-1}, and Z_H is in mm^6 m^{-3}.

affected by a change in the shape of the distribution, a result consistent with the observations of Goddard and Cherry (1982). For reference, we have shown in Fig. 8.19 the rainfall rate along the abscissa and reflectivity factor (in dBZ) along the ordinate using the Marshall–Palmer N_0 value.

8.5.3.2 Sample Results

We show here some sample results using the dual polarization technique. The results plotted in Fig. 8.20 are for times when the radar elevation angle was low enough to ensure that the radar beam was well below the freezing level, so the scatterers were most-likely to be raindrops.

Figures 8.20a and 8.20b give the measured values of Z_H and Z_{DR} along a radial through what appear to be two prominent cells in Oklahoma on May 2, 1979. The reflectivity factor extends up to 54 dBZ, and Z_{DR} is always positive and lies mostly within the expected range predicted by the theoretical values given in Fig. 8.19.

That values of Z_{DR} are larger than 3.5 dB suggests targets other than raindrops or some measurement-related problem (Peterson *et al.*, 1981). Melting ice crystals can have a large ratio of horizontal to vertical dimensions and thus could appear to radar as being similar to liquid drops having large b/a. For example, when one-tenth of the radius of a 2-mm ice sphere melts, the backscatter cross section is nearly the value for an all-water drop of the same radius (Battan, 1973). Water–ice particles having b/a ratios of about

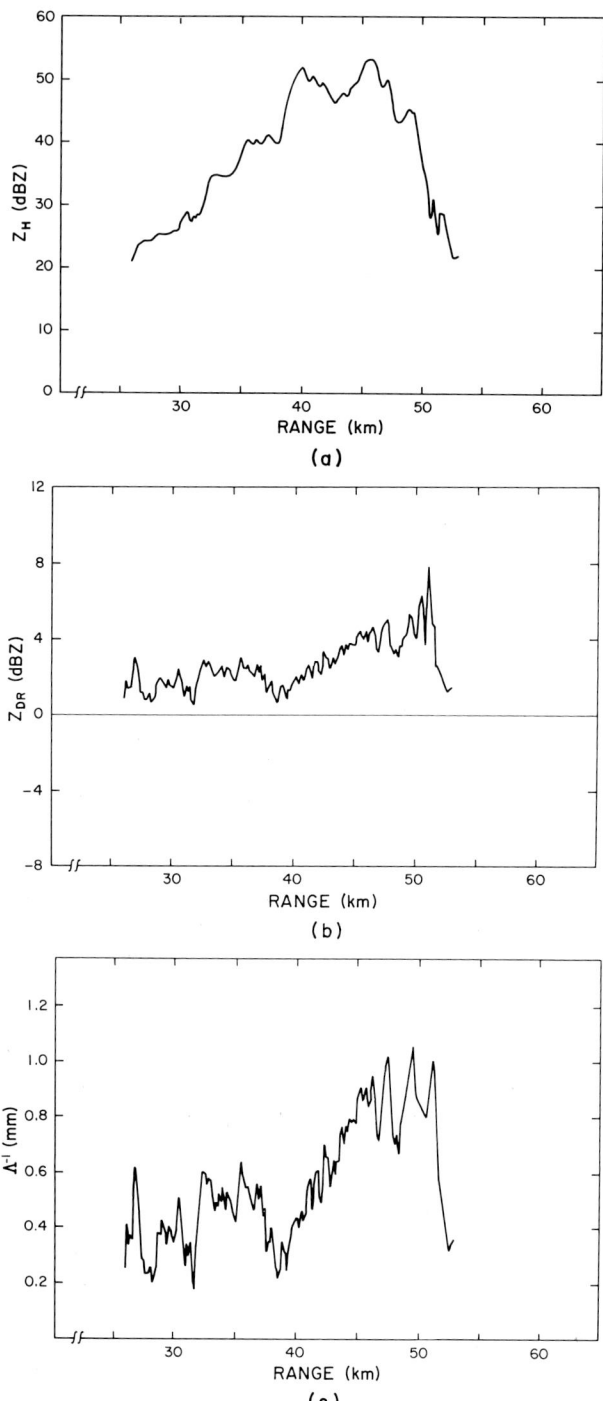

Fig. 8.20. (a) Z_H, (b) Z_{DR}, (c) Λ^{-1}, (d) N_0, and (e) R versus the range for an elevation angle of $0.8°$ at time $23:62:04$. ———, Z_{DR}; ---, $R = (Z/200)^{1/1.6}$. (Courtesy of Dr. T. A. Seliza, Ohio State University.)

8.5 Dual Parameter Measurement to Estimate the Rainfall Rate

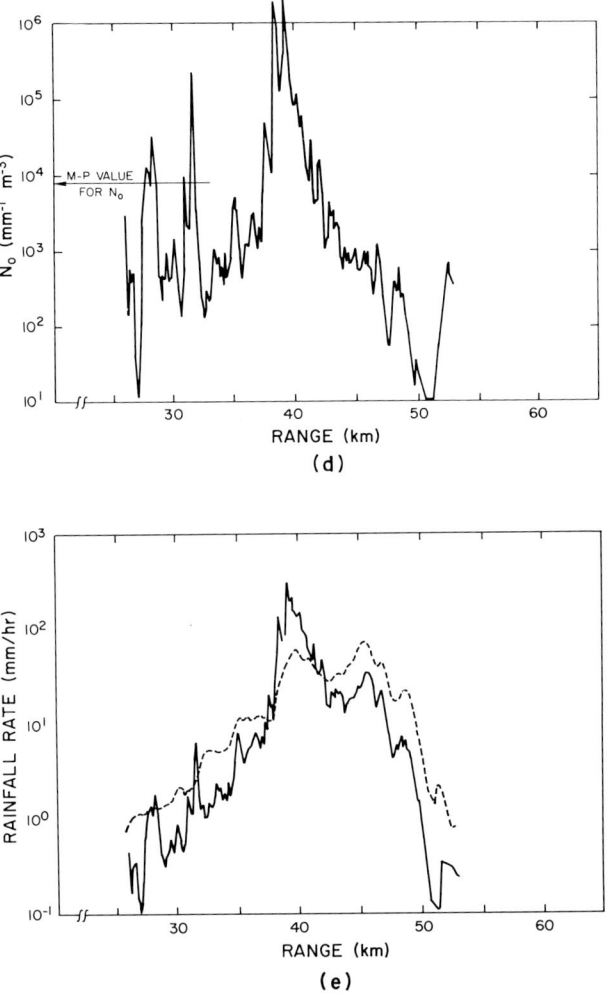

Fig. 8.20. (continued).

2.3 can then be expected to produce Z_{DR} as large as 8 dB (Fig. 8.18). Aggregates of ice crystals of up to 22 mm in maximum dimension were the most frequently observed particle form in Oklahoma thunderstorms (Heymsfield and Hjelmfelt, 1982).

Zrnić et al. (1982) have theoretically examined the effects of the storm's electric field and drop charge on the shape of drops, concluding that the electric fields commonly observed in clouds are not strong enough to cause a significant change in shape but that the presence of drop charge enhances the natural oblateness of the drop, possibly explaining some of the Z_{DR} values that exceed the predicted limit. It remains for simultaneous measurement by radar and in situ sensors to confirm the type of hydrometeors causing these large values of Z_{DR}.

The above-mentioned considerations indicate that significant error (i.e., >1 dB) would be made if reflectivity factor values were deduced from singly polarized radar measurements. For example, vertically (horizontally) polarized measurements of reflectivity in rain tend to underestimate (overestimate) the equivalent spherical drop reflectivity. However, Z_H estimates are closer than Z_V to Z for the equivalent sphere, even for oscillating drops (Dissanayake *et al.*, 1982).

Figures 8.20c and 8.20d show the values of Λ^{-1} and N_0 for the same data set. Of particular importance is the large variability of N_0 measured with this technique. The constant M–P value of N_0 can be seen to be a poor approximation to the highly variable values of N_0 computed from Z_H and Z_{DR}. Figure 8.20e shows the rainfall rate computed from the N_0 and Λ values. Even though N_0 and Λ have highly irregular spatial dependence, the rainfall rate is quite systematic and suggests that one cell produces significantly more rain than the other. Also plotted in Fig. 8.20e is the rainfall rate computed from the Z, R relation $Z = 200R^{1.6}$. This (and any other Z, R relation) shows that, as expected, both cells generate similar rainfall rates, although the Z_{DR} technique suggests almost an order of magnitude difference.

A nearly direct comparison of Z_H, Z_{DR} as measured by a fast-switching dual polarization radar having a small resolution volume (30-m diam by 300-m length) situated 200 m above a disdrometer with Z_H, Z_{DR} computed from disdrometer data shows excellent correlation (Goddard and Cherry, 1982). Radar-measured Z_{DR} was on the average 0.1 dB less than the value calculated from $N(D)$ measured with the disdrometer, and radar-measured Z_H was 1.5 dB higher. These recent results have significantly less bias in Z_{DR} than earlier measurements indicated (Goddard *et al.*, 1982). Even so, this small error in Z_{DR} translates to an error of as much as 25% in the rainfall rate. In order to explain this bias, Goddard and Cherry (1982) suggest that smaller raindrops that are distorted by oscillations are, on the average, more nearly spherical than was observed by Pruppacher and Beard (1970) under laboratory conditions. Indeed, observations of raindrop shapes near the ground (Jameson and Beard, 1982) confirm that, for observed equivalent spherical diameters between 2 and 4 mm, drops are on the average more nearly spherical than suggested by the Pruppacher–Beard observations. Alternatively, the large difference in sample volume size, the 200-m separation in height, a nonuniformity in rain characteristics, vertical air motion, and

8.5 Dual Parameter Measurement to Estimate the Rainfall Rate

size sorting due to shear may account for some of the discrepancy between the radar and disdrometer measurements. Recent airborne in situ measurements of drop shape confirm that on the average the shape is in good agreement with the observations of Pruppacher and Beard, suggesting that environmental conditions controlling shape were different near the surface than aloft (Bringi et al., 1982).

More observations using accurately controlled conditions (e.g., smaller resolution volume and in situ measurements of drop shape and size, as well as shear) will be required before one can pinpoint the cause of discrepancies. Nevertheless, these early results demonstrate the richness of information produced by polarization-sensitive radars and their potential in the study of the microphysics of storms.

There is at least one distinct advantage of the dual polarization technique over the dual wavelength method previously discussed. The dual polarization technique measures the rainfall rate averaged over the resolution volume and does not require the rainfall to be uniformly distributed within this volume. This is not the case for the dual wavelength method, in which the measured value of the attenuation depends strongly on the distribution of the rain across the beam. The dual polarization technique estimates the rainfall rate averaged over the resolution volume and thus gives a measure of the spatial variability in rainfall from volume to volume. This may be useful, especially when radar measurements are combined with rain gauge data, as discussed in the Section 8.5.4.

Another benefit of the dual polarization technique is its capability to differentiate at times between the ice phase and the liquid phase of water. For instance, in the high-reflectivity column (55 dBZ) at 35 km (Fig. 8.21a), water drops below 2 km are responsible for the large (>2 dB) Z_{DR} (Fig. 8.21b). Above 2 km, Z_{DR} is only 0.13 dB, most likely because of randomly oriented ice hydrometeors. The sharp change in Z_{DR} occurs near the 0°C isotherm and marks the transition between ice particles and water. As mentioned earlier, when 25% (10% of the radius) of a spherical ice mass melts, it appears to the radar as an apparent water hydrometeor of the same diameter. Assuming this holds for flattened ice crystal structures, a large increase in Z_{DR} due to change in phase occurs when the ice is partially melted (Fig. 8.18). The reflectivity factor Z often increases (as ice melts) many times more (10–100) than could be caused by phase change alone (4.7 times). These large reflectivity increases are attributed to the coalescence of ice crystals, which begins above the melting layer. Humphries and Barge (1979) have simultaneously observed with dual wavelength (5 and 10 cm) and polarization diversity radar (10 cm) the reflectivity and circular depolarization ratio (CDR) in stratiform precipitation. Their measurements suggest

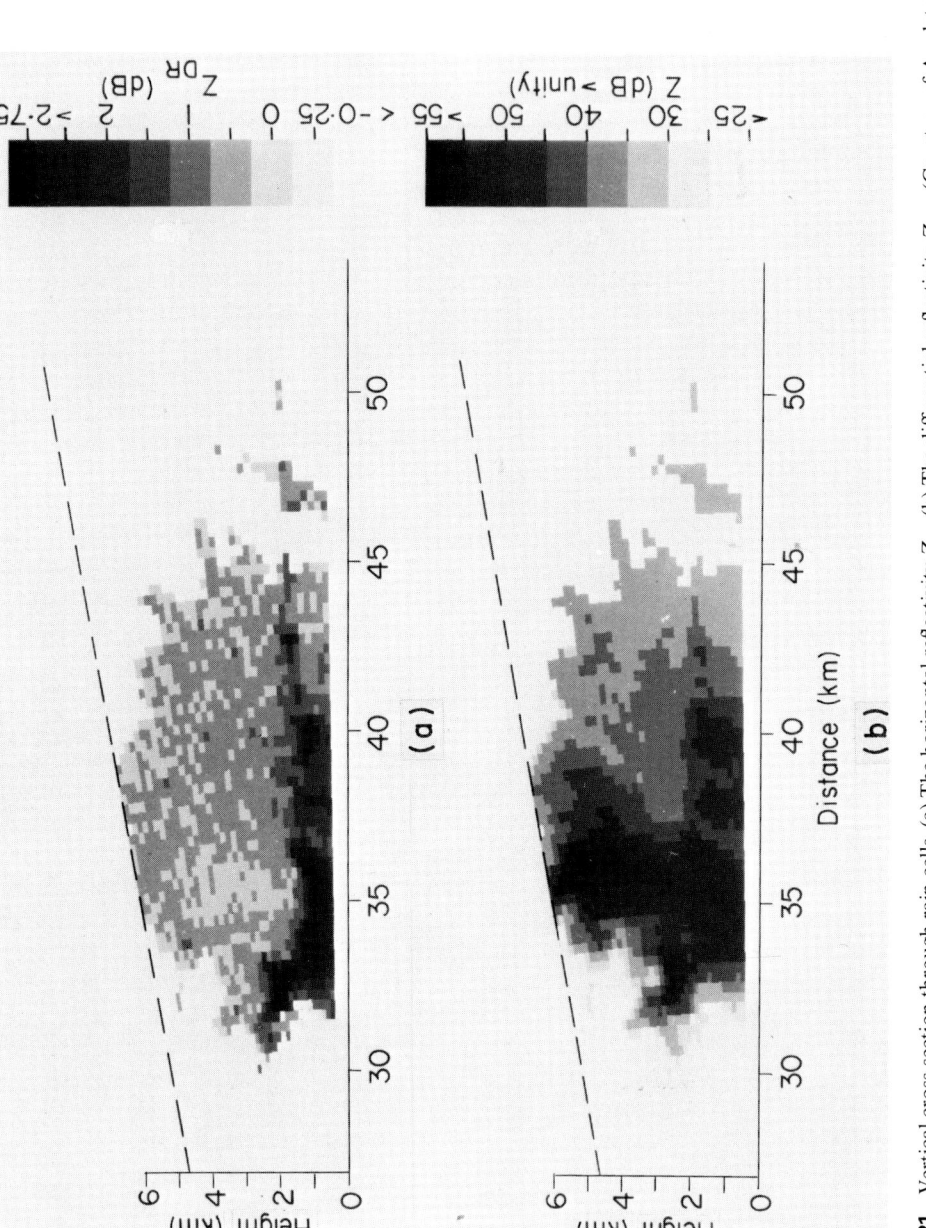

Fig. 8.21. Vertical cross section through rain cells. (a) The horizontal reflectivity Z_H. (b) The differential reflectivity Z_{DR}. (Courtesy of Appleton Laboratory, Slough, England.)

8.5 Dual Parameter Measurement to Estimate the Rainfall Rate

the presence of water-coated non-Rayleigh scatterers from about 200 m above the 0°C level to 500 m below it, and they attribute this to snowflakes coalescing in the presence of water drops at temperatures below freezing just above the 0°C height. Because Z depends on the sixth power of hydrometeor size (for Rayleigh conditions), it is easily accepted that explosive increases in Z occur as crystals coalesce and melt. The increase in fall speed as ice melts and the inevitable break-up of large melting crystals lead to decreases in reflectivity below the melting layer, which forms an enhanced region of high reflectivity, called the *bright band*. In Fig. 8.21a the bright band appears at a range of 43 km. In the bright-band region Z_{DR} is also enhanced, as seen in Fig. 8.21b, because it is there that wet hydrometeors are the most oblate.

It is evident in Fig. 8.21 that the peak of Z_{DR} in the bright band occurs a little below the peak of reflectivity, a result consistent with that of Humphries and Barge (1979), who noted that for circularly polarized radars the maximum CDR is located below the peak of reflectivity. These observations suggest that hydrometeors falling through a melting layer are flattest below the height of maximum size. This would result if large quasi-spherical ice crystals begin to melt (producing enhanced Z) before they are broken and flattened by the airstream. Table 8.1, from Hall *et al.* (1980), relates qualitatively the various types of hydrometeors to the Z and Z_{DR} values.

TABLE 8.1 Expected Characteristics of Z and Z_{DR} at 10-cm Wavelength for Various Hydrometeor Types[a]

Hydrometeor type	Z	Z_{DR}	Comments
Rain	High	High	Includes large oblate drops
Drizzle, cloud, or fog	Low	Low	Small spherical drops of water and/or small ice particles
Dry snow flakes	Medium–low	Medium–low	Large horizontally oriented low-density aggregates
Sleet/wet snow	High	High	Large oblate horizontally oriented particles
Wet graupel	High	Negative	Large conical vertically oriented particles
Wet hail	High	Variable	Large particles; seldom spheres
Dry hail or other high-density ice particles	Medium	Low	

[a] From Hall *et al.* (1980).

8.5.4 Rain Gauge and Radar

The most direct way to measure the rainfall rate is to use a rain gauge—a catchment that measures the depth of water per unit time. Although tipping buckets and weighing gauges are commonly used, they are subject to significant errors caused by wind (Neff, 1977). However, rain gauges, although quite accurate, only measure rainfall at a point. There is usually little interest in point rainfall measurements except to determine the relative accuracies of various gauges. More often, interest lies in the accurate estimation of rainfall in a unit time averaged over large catchment areas, and these estimates are usually expressed in millimeters of water depth. The areal averages find application in hydrology and flash flood forecasting.

Because there may be large errors in the rain depth at any one gauge representing the areal average, hydrologists have resorted to a network of rain gauges and to radar to improve areal average rainfall estimates. There is no doubt that a sufficiently dense network of gauges can measure rainfall better than a radar. In fact, gauge measurements are accepted as the standard against which other measurement techniques are compared. Yet no matter how accurate gauges may be for point measurements (typically 5%–10%), their accuracy for areal averages is a function of the gauge density and the spatial variability of rainfall. We discuss this accuracy in Section 8.5.4.1.

Although the accuracy of the radar-measured rainfall rate is highly suspect, radar has the decided advantage of being able to survey remotely vast areas and to make millions of measurements in minutes. The cost of a gauge network to match these capabilities in spatial continuity and rate of data sent to a central location would be prohibitive. Therefore, meteorologists have combined radar and rain gauge data to take advantage of the best of each—the accuracy of gauge data and the spatial coverage of radar data. The improvement, using radar–gauge combinations to estimate the areal average rain is discussed in Section 8.5.4.2. The combination adjusts the error-prone radar measurements.

8.5.4.1 GAUGE ACCURACIES FOR AREAL AVERAGE RAINFALL DEPTH

Because there is no easy way to establish a true areal average of the rainfall depth, the rain gauge network accuracy is determined relative to the most dense network that can be economically deployed at the time. Huff (1970) used a 49-gauge network in Illinois for the years 1955–1964, and these gauges were deployed over a 100-km^2 area. In 1971 and 1973 Woodley *et al.* (1975) deployed 220 gauges over a 510-km^2 area in Florida. More recently, Hildebrand *et al.* (1979) analyzed gauge accuracies for a network of 109 gauges on a 1500-km^2 area in Montana during the summer of 1976.

8.5 Dual Parameter Measurement to Estimate the Rainfall Rate

The sampling error E, in millimeters of water, represents the absolute difference between the best estimate of the true mean P of the precipitation depth (mm) obtained from the maximum density of gauges and the sample mean precipitation P_g calculated from the gauge amounts for a less-dense network of gauges. This is a measure of how accurately a network of gauges estimates areal averages of rain depth. Silverman *et al.* (1981) have made a study of the accuracy of the best estimate of "true" rainfall and have found that the maximum gauge density used in the above-mentioned experiments was sufficiently high for practical purposes. In his study of 10 years of rain gauge data, Huff found the error E to be a function of mean precipitation P, mean area per gauge or gauge area A_g (km^2), area of the network A_n (km^2), and duration T (hr) of the storm. The regression equation

$$\log E = -0.5 + 0.68 \log P + 0.94 \log A_g - 0.01 \log T - 0.75 \log A_n \quad (8.45)$$

fits the data quite well for 1955–1959, a period in which there was a relatively large percentage (33%) of air mass storms with considerable spatial variability. The data for 1960–1964 has 23% air mass showers and consequently a somewhat smaller error (Huff, 1970). Higher spatial variability suggests a larger natural dispersion of rainfall and hence a larger error for any given network. The regression equation (8.45), plotted in Fig. 8.22, demonstrates that error increases with increasing mean precipitation depth, although the percent error, $100\ E/P$, is seen to decrease. As the gauge density increases, A_g decreases and the error decreases. Also, plotted in Fig. 8.22 are data from Woodley *et al.* (1975) for Florida, where the predominant rain was from air mass showers (see Fig. 8.23), which may explain why errors were larger than in the Illinois data, which were dominated by frontal storms, squall lines, and low-center storms.

Nevertheless, Fig. 8.22 shows that, for those regions where air mass thunderstorms may contribute appreciably to the rainfall, in order to have an accuracy in areal average rainfall of 30% for precipitation greater than 10 mm, a gauge area of 143 km^2 should be adequate. However, even this spacing may be too dense for an economical measurement of rain over large areas.

Also plotted in Fig. 8.22 are data from Montana, as reported by Hildebrand *et al.* (1979). They combined several data sets to give a percentage error independent of P, but we have arbitrarily plotted this percentage error at $P = 10$ mm. Shown parenthetically next to the Montana data is the mean area per gauge. Based on this Montana data, we require a gauge area of 114 km^2 to have an accuracy of 30%, a somewhat larger gauge density (smaller A_g) than required in Florida or Illinois for the same accuracy.

Although the Florida and Illinois data represent 6-hr shower durations, the time dependence for Illinois is quite weak [see Eq. (8.45)], and in Florida

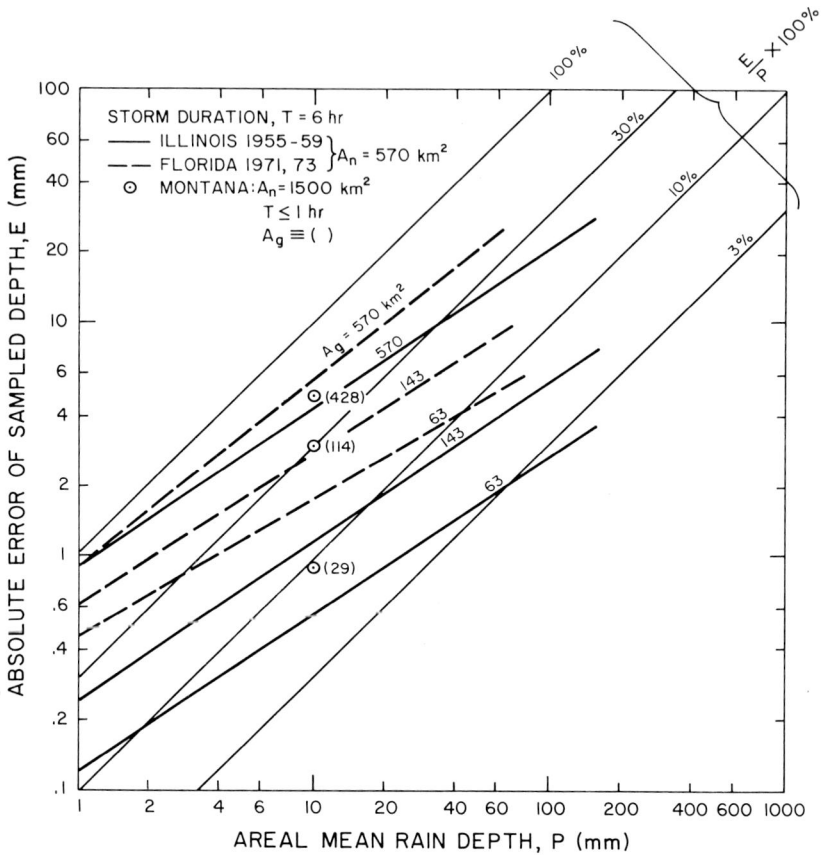

Fig. 8.22. Absolute difference of the mean areal average rain depth, computed for a thined network of gauges having indicated area per gauge A_g, and the depth computed for the full dense network of area A_n. The Illinois data are from Eq. (8.45), the Florida data from Woodley *et al.* (1975), and the Montana data from Hildebrand *et al.* (1979).

most of the rain occurred within a 6-hr period, although individual showers most probably had shorter lifetimes. The data accuracy exhibited in Fig. 8.22 could serve as a guide for determining gauge deployment in the absence of radar. Hildebrand *et al.* (1979) suggest that radar used in conjunction with gauges as outlined in Section 8.5.4.2 improves gauge estimates when gauge areas are larger than about 300 km².

8.5.4.2 RADAR–RAIN GAUGE COMBINATIONS

Before we can confidently accept radar estimates of rainfall, we should be aware of the phenomena that can cause variance from gauge estimates. If

8.5 Dual Parameter Measurement to Estimate the Rainfall Rate

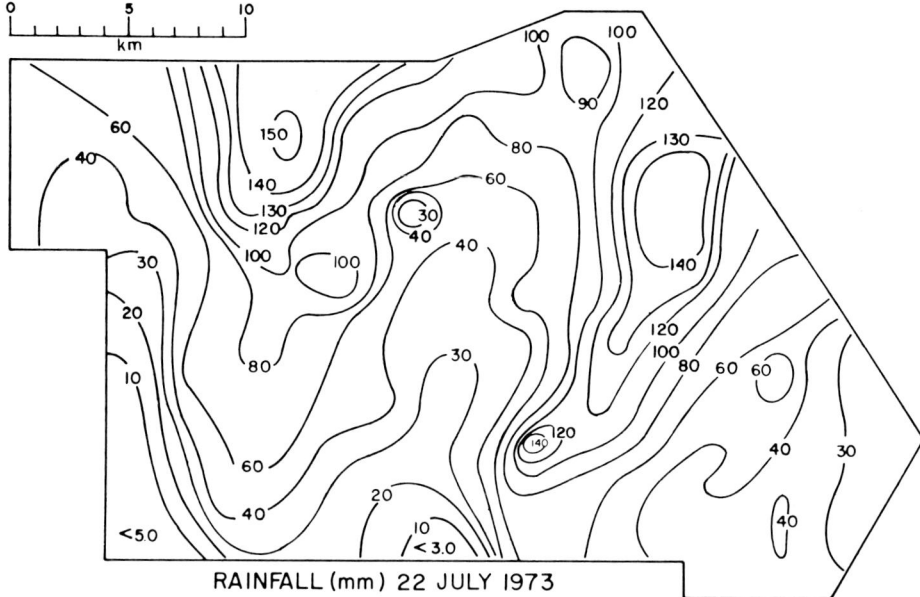

Fig. 8.23. Isohyetal analyses for a Florida air mass shower showing a strong spatial variability of rainfall depth. Contours are depth of water, in millimeters, for 22 July 1973. (Data from Woodley et al., 1975.)

only reflectivity factor measurements are available, one needs to choose an appropriate R, Z relation. As we saw in Fig. 8.20, the parameters N_0 and Λ of an assumed exponential drop size distribution vary considerably from point to point, so one may also expect the R, Z relation to vary. On the other hand, for areal averages of rainfall, there is suggestive evidence that N_0 and Λ may be appropriately chosen to produce an R, Z relation that in the mean predicts the average rainfall measured by a network of gauges (see Fig. 8.24 and the discussion in Section 8.4.1).

Figure 8.25 is a comparison of gauge- and radar-estimated water depths ($\bar{G}/\bar{R}$) at 65 gauge locations for a shower measured by the National Severe Storms Laboratory's WSR-57 radar located in Central Oklahoma. Atmospheric gas attenuation accounted for most of the correction seen in Fig. 8.25b. Although for this day the gauge-estimated rainfall (GER) was 1.38 times higher than that estimated by radar (RER), the average ratio $\bar{G}/\bar{R}$ for 14 rain events is 1.05 when the relation $Z = 200R^{1.6}$ is used (Brandes and Sirmans, 1976). Because daily radar calibrations can be in error, and because different R, Z relations might be appropriate for different days as well as for

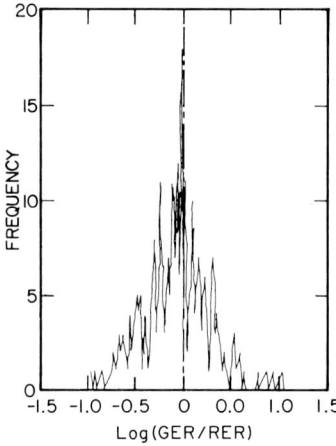

Fig. 8.24. Distribution of the logarithmic ratio of gauge-estimated rainfall to radar-estimated rainfall (from Cain and Smith, 1976), for 300 gauge–hour events in North Dakota during 1972.

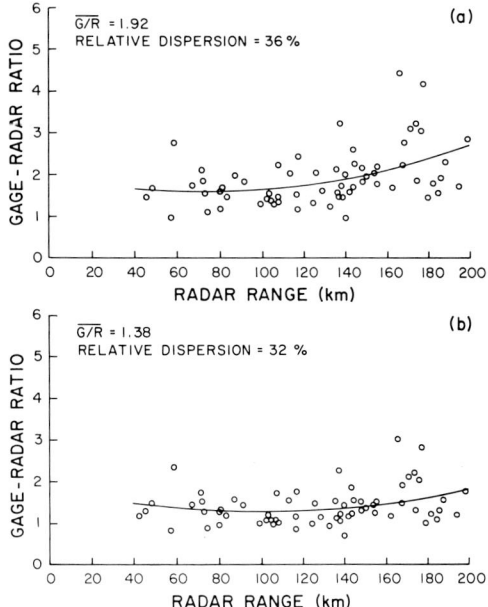

Fig. 8.25. Range distribution of G/R ratios using radar measurements that were (a) uncorrected and (b) corrected for atmospheric absorption, rainfall attenuation, and biases due to reflectivity gradients (see the text). Each point is a radar–gauge comparison, at a gauge location, of the estimated rainfall rate integrated in time (i.e., the depth of water) over 24 hr. The relative dispersion is the standard deviation expressed as a percentage of the mean. (From Brandes and Sirmans, 1976.)

8.5 Dual Parameter Measurement to Estimate the Rainfall Rate

different locations surveyed by the same radar, it becomes mandatory to adjust radar rain estimates in accordance with the in situ gauge measurements.

Brandes (1975) has suggested a technique whereby gauges can be used to adjust the RER. This methodology minimizes the impact of choosing an inappropriate R, Z relation and the need for a nearly perfect radar calibration. The major steps in Brandes's technique are as follows:

(1) Radar-estimated rainfall rates are integrated in time for the selected period for each resolution volume using (8.23a).

(2) G/R ratios are calculated at each gauge having at least 2.5 mm of rain, using the radar data from within a fixed radius about the gauge. The radius is chosen to be small with respect to the gauge spacing and compared to the scale of the precipitation.

(3) The G/R data are then used to determine field-of-adjustment factors, which are applied to the RER field to generate a first-guess corrected radar rainfall field.

(4) Differences between corrected radar-measured rainfall and actual gauge estimates are then used to determine a refinement in the adjustment factors so that the final corrected RER will agree with the gauge data at gauge locations.

Hildebrand *et al.* (1979) have employed this methodology with minor modifications to investigate the characteristics of combined gauge–radar rainfall data. They compare the results with the rainfall distribution estimated by a dense network of gauges (Fig. 8.26). The two results are very close, as might be expected, because Hildebrand *et al.* used all the gauge data to adjust the RER. We should not expect such good agreement if few gauges are used.

An objective of the gauge–radar rainfall estimation technique is to combine sparse gauge and dense radar rainfall data to produce a rainfall estimate with the point accuracy of gauges and the spatial resolution and coverage of radar. The technique would work quite well if the radar measured the spatial variability in R, but in fact it only measures the variability in Z (see Fig. 8.20e). Furthermore, the technique assumes that the gauge measures an R that fills the resolution volume at the individual gauge location and that rain alone is responsible for Z, and vertical air motion is neglected. The capability of the radar to measure variations in R has not been adequately verified and can be expected to be sufficient only if the drop size distribution does not change appreciably from storm cell to storm cell or within a cell. However, if the radar measures Z_{DR} and Z_H or Z_V, it should map the spatial variations in R better. A few sparsely sprinkled rain gauges would then be helpful to remove the pervasive bias due to radar calibration error.

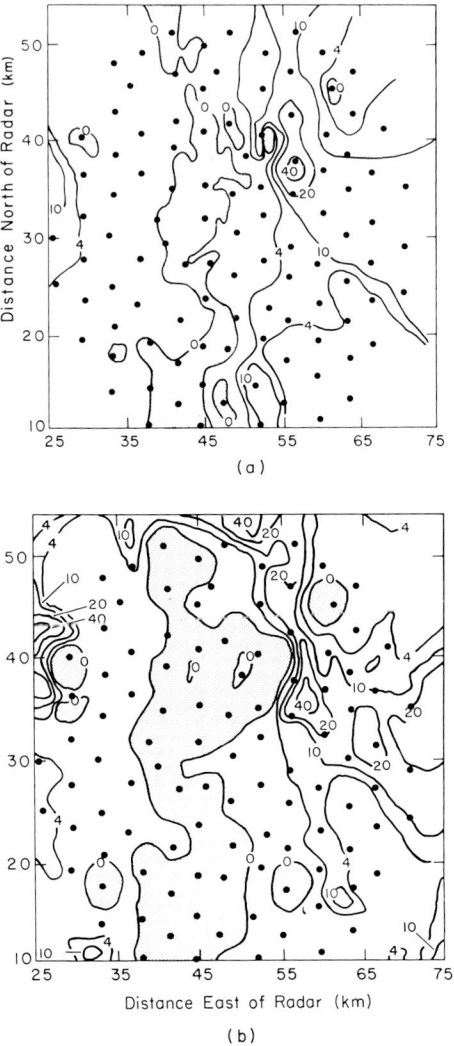

Fig. 8.26. Comparison of (a) the gauge-measured rainfall distribution (the depth of water, in millimeters, for a 1-hr period) with (b) a radar-estimated rainfall adjusted by gauge comparisons and taking evaporation into account. (From Hildebrand *et al.*, 1979.)

Even if the radar poorly measures the actual variation in the rain rate (as could occur if Z alone were measured), there may still be value in the gauge–radar combination if the horizontal scales of $N(D)$ are large compared to the gauge spacings. In the studies of Wilson (1970) and Brandes (1975), the gauge–radar mean rainfall estimates were more accurate than the estimates using only gauges for large-area (29,000 km²), low-gauge-density (no more than one gauge per 700 km²) and long-duration rainfall cases. Brandes (1975) shows that radar-measured rainfall corrected by gauge data improved the accuracy from 24% for measurements by gauge alone to 14% for combined radar–gauge measurements with a gauge density of one gauge per 1600 km². A recent summary of improvement in radar-measured rainfall, corrected by gauge data for denser spacings and smaller areas, is given by Wilson and Brandes (1979, Table 3).

When gauge spacings are large compared to the scales of $N(D)$, then the error in using radar–rain gauge combinations can be large if the radar measures the spatial variability of Z alone. However, if it also measures Z_{DR}, then radar should, in principal, directly measure the spatial variability of R, and few gauges would be required to remove bias errors.

8.6 DISTRIBUTION OF HYDROMETEORS FROM DOPPLER SPECTRA

Measurements of the Doppler spectrum or its moments may be employed to make inferences concerning the drop size distributions. In a sense, this can be considered to be a multiparameter method, since several simultaneous measurements are independent. As a matter of fact, the first application that required the entire Doppler spectrum (as opposed to its moments) was determining the drop size distribution (Boyenval, 1960).

In essence, the method relies on the existence of a unique relationship between the terminal speed and diameter of drops falling in still air (Fig. 8.6). In the presence of up- or downdrafts of velocity $\bar{w}$, uniform within the resolution volume, the relationship between the normalized Doppler spectrum and the drop size distribution is

$$S_n(\bar{w} - w_t)\, dw_t = \sigma(D)N(D)\, dD/\eta, \qquad (8.46)$$

where it is assumed that the resolution volume is much larger than the average spacing between drops for all diameters. Furthermore, S_n must be free from all other spectrum broadening contributions (see Section 5.3). The relation $\sigma(D)/\eta = D^6/Z$, from (3.6) and (4.28), can be inserted in (8.46) to obtain

$$S_n(\bar{w} - w_t)\, dw_t = D^6 N(D)\, dD/Z. \qquad (8.47)$$

Thus, if the vertical speed of air is known, measurement of the reflectivity factor and the Doppler spectrum suffices to determine $N(D)$ with no need for additional assumptions regarding its functional form.

The most serious obstacle to drop size measurements from the Doppler spectra is the need for precise determination of the vertical airspeed. Rogers (1964) was the first to use a relationship between the mean terminal velocity and reflectivity for an assumed exponential drop size distribution. Battan (1973) proposed a lower-bound method that involves assumptions on the minimum size of the detectable hydrometeors and their terminal velocity.

Atlas et al. (1973) used (8.47) as the starting point for a detailed analysis of the properties of Doppler spectra for vertical viewing. They calculated theoretical spectra and spectral moments when $N(D)$ is exponential and $w_t(D)$ is approximated by various analytical expressions. For a power law dependence $w_t = CD^\alpha$, the reflectivity-weighted mean terminal speed is

$$\overline{w_t} = 720^{(7-\alpha)/7}\Gamma(7+\alpha)C(Z/N_0)^{\alpha/7}, \tag{8.48}$$

which must be added to the mean Doppler velocity $\bar{v}$ in order to determine the airspeed $\bar{w}$. Joss and Dyer (1972) point out that errors in $\bar{w}$ of 1 m s^{-1} could lead to roughly a threefold change in the liquid water content.

Recently, Hauser and Amayenc (1981) proposed a least squares fit of the three parameters N_0, Λ, and $\bar{w}$ of the model Doppler spectrum to the observed spectra. Their results exhibit a large variability in the parameters deduced (the vertical air velocity, liquid water content, rainfall rate, N_0, and Λ) on finely spaced scales of the order of 200 m. The general coherence of the results leads to the conclusion that the method is well suited to the study of stratiform precipitation when the hypothesis of exponential drop size distributions holds and when spectral broadening due to turbulence is negligible.

If an independent estimate of $\bar{w}$ were possible, the drop size distribution could be readily obtained from the Dopper spectrum. Such a possibility is offered when lightning occurs in the resolution volume (Section 9.5.6), because the ionized channel acts as a perfect tracer. This is seen in the Doppler spectrum as a distinct peak (Fig. 9.40) located at the vertical speed of air. An example of $N(D)$ for hail obtained from the Doppler spectrum of Fig. 9.40 is plotted in Fig. 8.27. The speed of the updraft was estimated to be 11 m s^{-1}, and a terminal velocity for spherical hail at 7 km above ground is assumed to follow the theoretical relation

$$w_t^2 = \tfrac{4}{3}(\rho_h/\rho)(gD/C_d) \quad \text{m}^2 \text{ s}^{-2}, \tag{8.49}$$

where the hail density $\rho_h = 900$ kg m^{-3}, the air density $\rho = 0.63$ kg m^{-3} (at 450 mbar), and the drag coefficient C_d is 0.4, so that $w_t = 216D^{1/2}$. An estimate of the turbulent broadening was made from the positions of the lightning peak in several consecutive spectra. The peak had an excursion of 4 m s^{-1}, and we therefore assume the turbulent broadening to be $\tfrac{4}{3}$ m s^{-1}.

8.7 Summary

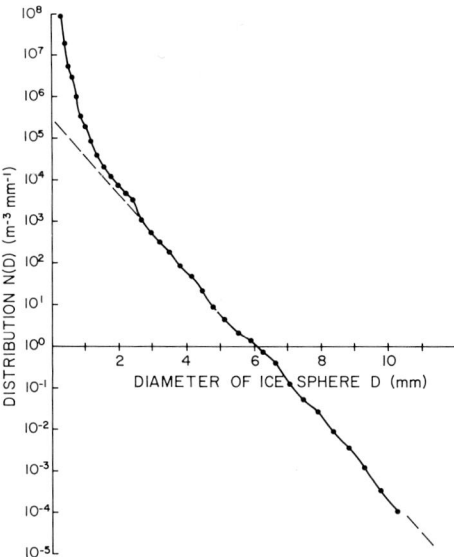

Fig. 8.27. The size distribution of ice spheres, obtained from the Doppler spectrum (Fig. 9.40) of a vertically pointed radar.

Spectral skirts have been reduced accordingly, and Eq. (8.47) has been applied over a 30-dB dynamic range of the spectrum. The equivalent reflectivity factor (here 55 dBZ) was scaled with the ratio of the refractive indexes of hail to water. Evident in Fig. 8.27 is the huge (10^{13}) span of the distribution and the remarkably good fit to an exponential model, $4 \times 10^5 \exp(-2D)$ m^{-3} mm^{-1}, for sizes larger than 1 mm. With these values, the calculated equivalent reflectivity factor (8.12) is 55.5 dBZ. In this particular experiment the resolution volume was about 10^8 m^3, and we conclude from Fig. 8.27 that, even though the probability of hailstones between 9.5 and 10.5 mm is 10^{-4} m^{-3}, there are on the average 10^4 such stones in the resolution volume.

8.7 SUMMARY

Several techniques for rainfall rate measurement have been reviewed. It is concluded that, after many years of development of techniques, there is still no satisfactory proven method for accurately estimating the rainfall rate or liquid water content when high spatial and temporal resolution is required. Radars in combination with satellites (Atlas *et al.*, 1982), rain gauges, and other instruments may give the sought-for accuracy of rainfall estimate with the ease of data collection inherent in remote sensing. Dual parameter measurement techniques appear to offer the best solution for rainfall rate estimation if $N(D)$ is a significant error source and high resolution is required. There is a need to evaluate the relative importance of error sources other than uncertainty in $N(D)$. However, uncertainties in the drop size distribution and in the Z, R relation might not significantly affect estimates of rainfall averaged over a large area and over long durations if bias in the radar-estimated rainfall is removed by comparisons with a few rain gauges (Hudlow and Arkell, 1978).

Srivastava's (1971) results show that, when $N(D)$ is controlled by coalescence and drop break-up and is in steady state, $N(D)$ is a function of the liquid water density only. A one-parameter radar measurement, for example, of Z, should then suffice to estimate R. However, a large variability between R and Z is usually found, even when using R and Z computed from drop size distributions that have been precisely measured by disdrometers, suggesting that steady state has not been achieved. Observed cloud drop size distributions are often exponential, and they evolve as drops grow into the steady state rain spectra shown in Fig. 8.4. Although these spectra are not exponential, they can be fitted by truncated exponential functions. We therefore see that as rain spectra are evolving, an exponential function can describe accurately a continuously changing $N(D)$, but, in contrast to the case of fixed N_0 (as in the M–P distribution), N_0 and Λ must both be measured. For low rain rates, a two-parameter radar measurement should satisfactorily describe the drop size distribution, but as steady state is approached at large R, it will be important at least to truncate the spectrum, because large drops significantly contribute to R. We then can no longer describe $N(D)$ by two parameters unless there is some deterministic relation between Λ, N_0, and D_{max} at large R, as suggested by Plank (1977).

When gauge spacings are large compared to the scale of variations in $N(D)$, the errors can be large in using radar–rain gauge combinations if radar alone is used to measure the spatial variability of Z. However, if radar measures Z_{DR} as well, then it should provide a better measure of the spatial variability of R, and few gauges are then required to remove bias errors. Based on preliminary results from dual polarization measurements, there appears to be the potential for significant improvement in short-time rainfall estimates at small spatial scales.

REFERENCES

Abramowitz, M., and Stegun, I. A. (1964). "Handbook of Mathematical Functions," Natl. Bur. Stand., Appl. Math. Ser. 55, 2nd printing, Supt. Doc. U.S. Govt. Printing Office, Washington, D.C.

References

Al-Khatib, H. H., Seliga, T. A., and Bringi, V. N. (1979). "Differential Reflectivity and its Use in the Radar Measurement of Rainfall," Atmos. Sci. Prog. Rep. No. AS-S-106. Ohio State University, Columbus.

Atlas, D., and Chmela, A. C. (1957). Physical-synoptic variations of drop size parameters. *Proc. Weather Radar Conf., 6th, 1957* pp. 21–30.

Atlas, D., and Ulbrich, C. W. (1974). The physical basis for attenuation-rainfall relationships and the measurement of rainfall parameters by combined attenuation and radar methods. *J. Rech. Atmos.* **8**(1–2), 275–298.

Atlas, D., and Ulbrich, C. W. (1977). Path- and area-integrated rainfall measurement by microwave attenuation in the 1–3 cm band. *J. Appl. Meteorol.* **16**, 1322–1331.

Atlas, D., Srivastava, R. C., and Sekhon, R. S. (1973). Doppler radar characteristics of precipitation at vertical incidence. *Rev. Geophys. Space Phys.* **2**, 1–35.

Atlas, D., Meneghini, R., and Moore, R. K. (1982a). The outlook for precipitation measurements from space. *Atmos.-Ocean* **20**(1), 50–60.

Atlas, D., Ulbrich, C. W., and Meneghini, R. (1982b). The multi-parameter remote measurement of rainfall. NASA Tech. Memo 83971, Goddard Space Flight Center, Greenbelt, Maryland.

Austin, P. M., and Geotis, S. G. (1979). Raindrop sizes and related parameters for GATE. *J. Appl. Meteorol.* **18**, 569–575.

Battan, L. J. (1973). "Radar Observation of the Atmosphere." Univ. of Chicago Press, Chicago, Illinois.

Boyenval, E. H. (1960). Echoes from precipitation using pulsed Doppler radar. *Prepr., Weather Radar Conf., 8th, 1960* pp. 57–64.

Blanchard, D. C. (1953). Raindrop size distribution in Hawaiian rains. *J. Meteorol.* **10**, 457–473.

Born, M., and Wolf, E. (1964). "Principles of Optics" 2nd ed. Macmillan, New York.

Brandes, E. (1975). Optimizing rainfall estimates with the aid of radar. *J. Appl. Meteorol.* **14**, 1339–1345.

Brandes, E., and Sirmans, D. (1976). Convective rainfall estimation by radar: Experimental results and proposed operational analysis technique. *Prepr., Conf. Hydro-Meteorol., 1976* pp. 54–59.

Bringi, V. N., Seliga, T. A., and Cooper, W. A. (1982). Analysis of aircraft spectra and differential reflectivity (Z_{DR}) radar measurements during the cooperative precipitation experiment. Personal communication.

Burrows, C. R., and Attwood, S. S. (1949). "Radio Wave Propagation," p. 219. Academic Press, New York.

Cain, D. E., and Smith, P. L., Jr. (1976). Operational adjustment of radar estimated rainfall with rain gage data: A statistical evaluation. *Prepr., Radar Meteorol. Conf., 17th, 1976* pp. 533–538.

Carbone, R. E., and Nelson, L. D. (1978). The evolution of raindrop spectra in warm-based convective storms as observed and numerically modeled. *J. Atmos. Sci.* **35**, 2302–2314.

Cataneo, R. (1969). A method for estimating rainfall rate-radar reflectivity relationships. *J. Meteorol.* **8**, 815–819.

Cataneo, R., and Stout, G. E. (1968). Raindrop-size distributions in humid continental climates, and associated rainfall rate-radar reflectivity relationships. *J. Appl. Meteorol.* **7**, 901–907.

Dissanayake, A. W., Chandra, M., and Watson, P.A. (1982). "Theoretical Prediction of Differential Reflectivity Using Dual Polarization Radar for Various Types of Hydrometeors." Proc., URSI Comm. F., Open Symp. Multiple-Parameter Radar Meas. Precipitation, pp. 20–23. Rutherford Appleton Lab. Chilton, Didcot Oxfordshire, U.K.

Eccles, P. J. (1979). Comparison of remote measurements by single- and dual-wavelength meteorological radars. *IEEE Trans. Geosci. Electron.* **GE-17**, 205–218.

Eccles, P. J., and Atlas, D. (1973). A dual-wavelength radar hail detector. *J. Appl. Meteorol.* **12,** 847–856.

Eccles, P. J., and Mueller, E. A. (1973). X-band attenuation and liquid water content estimation by a dual-wavelength radar. *J. Appl. Meteorol.* **10,** 1252–1259.

Fitzgerald, J. W. (1972). "A Study of the Initial Phase of Cloud Droplet Growth by Condensation: Comparison between Theory and Observation," Cloud Phys. Lab. Tech. Note, 44, p. 144. University of Chicago, Chicago, Illinois.

Foote, G. B., and duToit, P. S. (1969). Terminal velocity of raindrops aloft. *J. Appl. Meteorol.* **8,** 249–253.

Gans, R. (1912). Uber die Form ultramikroskopischer Goldteilchen. *Ann. Phys. (Leipzig)* [4] **37,** 881–900.

Goddard, J. W. F., and Cherry, S. M. (1982). The Ability of Dual-polarization Radar (Co-polar Linear) to Predict Rainfall Rate and Microwave Attenuation," Proc. URSI Comm. F, Open Symp. Multiple-Parameter Radar Meas. Precipitation, pp. 161–167. Rutherford Appleton Lab. Chilton, Didcot Oxfordshire, U.K.

Goddard, J. W. F., Cherry, S. M., and Bringi, V. N. (1982). Comparison of dual-polarization radar measurements of rain with ground-based disdrometer measurements. *J. Appl. Meteorol.* **21,** 252–256.

Goldhirsh, J. (1975). Improved error analysis of raindrop spectra, rain rate, and liquid water content using multiple wavelength radars. *IEEE Trans. Antennas Propag.* **AP-23,** 718–720.

Goldhirsh, J., and Katz, I. (1974). Estimation of raindrop size distribution using multiple wavelength radar systems. *Radio Sci.* **9**(4), 439–446.

Gradshteyn, I. S., and Ryshik, I. M. (1965). "Table of Integrals, Series, and Products." Academic Press, New York

Green, A. W. (1975). An approximation for the shapes of large raindrops. *J. Appl. Meteorol,* **14,** 1578–1583.

Gunn, K. L. S., and East, T. W. R. (1954). The microwave properties of precipitation particles. *Q. J. R. Meteorol. Soc.* **80,** 522–545.

Gunn, R., and Kinzer, G. D. (1949). The terminal velocity of fall for water droplets in stagnant air. *J. Meteorol.* **6,** 243–248.

Hall, M. P. M., Cherry, S. M., Goddard, J. W. F., and Kennedy, G. R. (1980. Raindrop sizes and rainfall rate measured by dual-polarization radar. *Nature (London),* **285,** 195–198.

Hardy, K. R. (1963). The development of raindrop-size distributions and implications related to the physics of precipitation. *J. Atmos. Sci.* **20,** 299–312.

Harju, A. E., and Puhakka, T. M. (1980). A method of correcting quantitative radar measurements for partial beam blocking. *Prep., Radar Meteorol. Conf., 19th, 1980* pp. 234–239.

Hauser, D., and Amayenc, P. (1981). A new method for deducing hydrometeor-size distribution and vertical air motions from Doppler radar measurements at vertical incidence. *J. Appl. Meteorol.* **20,** 547–555.

Herman, B. M., Browning, S. R., and Battan, L. J. (1961). "Tables of the Radar Cross Sections of Water Spheres," Tech. Rep. No. 9. University of Arizona, Inst. Atmos. Phys., Tucson.

Heymsfield, A. J., and Hjelmfelt, M. (1982). Microphysical characteristics of Oklahoma thunderstorms. *Prepr., Conf. Severe Local Storms, 12th, 1982* pp. 20–23.

Hildebrand, P. H., Towery, N., and Snell, M. R. (1979). Measurement of convective mean rainfall over small areas using high-density rain gages and radar. *J. Appl. Meteorol.* **18,** 1316–1326.

Hill, F. F., Browning, K. A., and Bader, M. J. (1981). Radar and raingauge observations of orographic rain over South Wales. *Q. J. R. Meteorol. Soc.* **107,** 643–670.

Hudlow, M. O., and Arkell, R. E. (1978). Effect of temporal and spatial sampling errors and Z-R variability on accuracy of GATE radar rainfall estimates. *Prepr., Radar Meteorol. Conf., 18th, 1978* pp. 342–349.

Huff, F. A. (1970). Sampling errors in measurement of mean precipitation. *J. Appl. Meteorol.* **9,** 35–44.

Humphries, R. G., and Barge, B. L. (1979). Polarization and dual-wavelength radar observations of the bright band. *IEEE Trans. Geosci. Electron.* **GE-17,** 190–195.

Jameson, A. R., and Beard, K. U. (1982). Raindrop axial ratios. *J. Appl. Meteorol.* **21,** 257–259.

Jones, P. M. A. (1956). "Rainfall Drop-size Distributions and Radar Reflectivity," Res. Rep. No. 6. Meteorol. Lab., Illinois State Water Surv. Urbana.

Jorgensen, D. P., and Willis, P. T. (1982). A Z-R relationship for hurricanes. *J. Appl. Meteorol.* **21,** 356–366.

Joss, J., and Dyer, R. (1972). Large errors involved in deducing drop-size distributions from Doppler radar data due to vertical air motion. *Prepr., Radar Meteorol. Conf., 15th, 1972* pp. 179–180.

Joss, J., and Waldvogel, A. (1970). A method to improve the accuracy of radar measured amounts of precipitation. *Prepr., Radar Meteorol. Conf., 14th, 1970* pp. 237–238.

Knight, C. A., Cooper, W. A., Breed, D. W., Paluch, I. R., Smith, P. L., and Vali, G. (1982). *In* "Hailstorms of the Central High Plains" (Charles A. Knight and Patrick Squires, eds), Vol. 1, Chap. 7, pp. 151–194. Colorado Assoc. Univ. Press, Boulder, Colorado.

Komabayasi, M., Gonda, T., and Isono, K. (1964). Lifetime of water drops before breaking and size distribution of fragment droplets. *J. Meteorol. Soc. Jpn.* **42,** 330–340.

Laws, J. O., and Parsons, D. A. (1943). The relationship of raindrop size to intensity. *Trans. Am. Geophys. Union* **24,** 452–460.

Leber, G. W., Merrit, C. J., and Robertson, J. P. (1961). WSR-57 analysis of heavy rains. *Proc. Weather Radar Conf., 9th, 1961* pp. 102–105.

Marshall, J. S., and Palmer, W. (1948). The distribution of raindrops with size. *J. Meteorol.* **5,** 165–166.

Marshall, J. S., Hitschfeld, W., and Gunn, K. L. S. (1955). Advances in radar weather. *Adv. Geophys.* **2,** 1–56.

Mason, B. J. (1971). "The Physics of Clouds." Oxford Univ. Press (Clarendon), London and New York.

Mie, G. (1980). Contribution to the optics of suspended media, specifically colloidal metal suspensions. *Ann. Phys. (Leipzig)* [4] **25,** 377–445.

Neff, E. L. (1977). How much rain does a rain gage gage? *J. Hydrol.* **35,** 213–220.

Norbury, J. R., and White, W. J. K. (1972). Microwave attenuation at 35.8 GHz due to rainfall. *Electron. Lett.* **8,** 91–92.

Pasqualucci, F. (1978). Radar observations of drop-size distributions and vertical air velocity in a squall line storm. *Prepr., Conf. Radar Meteorol., 18th, 1978* pp. 121–128.

Pasqualucci, F. (1982). "Drop-size Distribution Measurements in Convective Storms with a Vertically Pointing 35-GHz Doppler Radar." Proc. URSI Comm. F, Open Symp. Multiple-Parameter Radar Meas. Precipitation, pp. 139–144. Rutherford Appleton Lab. Chilton, Didcot Oxfordshire, U.K.

Peterson, J. R., Seliga, T. A., and Bringi, V. N. (1981). "A Study of a SESAME 1979 Squall Line Using the Differential Reflectivity Radar Technique," Atmos. Sci. Prog. Rep. No. AS-S-112. Ohio State University, Columbus.

Plank, V. G. (1977). "Hydrometeor Data and Analytical-theoretical Investigations Pertaining to the SAMS Rain Erosion Program of the 1972–73 Season at Wallops Island, Virginia,"

AFGL/SAMS Rep. No. 5, AFGL-TR-77-0149. Air Force Geophys. Lab., Hanscom AFB, Massachusetts.

Pruppacher, H. R., and Beard, K. V. (1970). A wind tunnel investigation of the internal circulation and shape of water drops falling at terminal velocity in air. *Q. J. R. Meteorol. Soc.* **96**, 247–256.

Pruppacher, H. R., and Pitter, R. L. (1971). A semi-empirical determination of the shape of cloud and raindrops. *J. Atmos. Sci.* **28**, 86–94.

Richards, W. G., and Crozier, C. L. (1981). "Precipitation Measurement with C-Band Weather Radar in Southern Ontario," Int. Rep. No. APRB 112, p. 35. Cloud Phys. Res. Div., Atmos. Environ. Serv. Downsnow, Ontario, Canada.

Rinehart, R. E., and Tuttle, J. D. (1982). "Dual-wavelength Processing—The Effects of Mismatched Antenna Beam Patterns." Proc. URSI Comm. F, Open Symp. Multiple-Parameter Radar Meas. Precipitation, pp. 83–88. Rutherford Appleton Lab. Chilton, Didcot Oxfordshire, U.K.

Rogers, R. R. (1964). An extension of the Z-R relation for Doppler radar. *Proc. World Conf. Radar Meteorol.*, 1964 pp. 158–161.

Sekhon, R. S., and Srivastava, R. C. (1971). Doppler radar observations of drop size distributions in a thunderstorm. *J. Atmos. Sci.* **28**(6), 983–994.

Seliga, T. A., and Bringi, V. N. (1976). Potential use of radar differential reflectivity measurements at orthogonal polarizations for measuring precipitation. *J. Appl. Meteorol.* **15**, 69–76.

Seliga, T. A., and Bringi, V. N. (1978). Differential reflectivity and differential phase shift: Applications in radar meteorology. *Radio Sci.* **13**, 271–275.

Seliga, T. A., Ayden, K., and Bringi, V. N. (1982). "Behavior of the Differential Reflectivity and Circular Depolarization Ratio Radar Signals and Related Propagation Effects in Rainfall," Proc. URSI Comm. F, Open Symp. Multiple-Parameter Radar Meas. Precipitation, pp. 35–42. Rutherford Appleton Lab. Chilton, Didcot Oxfordshire, U.K.

Setzer, D. (1970). Computed transmission through rain at microwave and visible frequencies. *Bell Syst. Tech. J.* **49**, 1873–1892.

Silverman, B. A., Rogers, L. K., and Dahl, D. (1981). On the sampling variance of rain gage networks. *J. Appl. Meteorol.* **20**, 1468–1478.

Sinclair, D. (1947). Light scattering by spherical particles. *J. Opt. Soc. Am.* **37**(6), 475–480.

Srivastava, R. C. (1967). On the role of coalescence between raindrops in shaping their size distribution. *J. Atmos. Sci.* **24**, 287–292.

Srivastava, R. C. (1971). Size distribution of raindrops generated by their breakup and coalescence. *J. Atmos. Sci.* **28**, 410–415.

Srivastava, R. C. (1974). The cloud physics of particle size distributions, a review. *J. Rech. Atmos.* **8**, 23–39.

Telford, J. W. (1955). A new aspect of coalescence theory. *J. Meteorol.* **12**, 436–444.

Ulbrich, C. W., and Atlas, D. (1978). The rain parameter diagram: Methods and applications. *J. Geophys. Res.* **83**, No. C3, 1319–1325.

Ulbrich, C. W., and Atlas, D. (1982). "Assessment of the Contribution of Differential Polarization to Improved Rainfall Measurements." Proc. URSI Comm. F, Open Symp. Multiple-Parameter Radar Meas. Precipitation, pp. 1–8. Rutherford Appleton Lab. Chilton, Didcot Oxfordshire, U.K.

Waldteufel, P. (1973). Attenuation des ondes hyperfréquence par la pluie: Une mise au point. *Ann. Tellecommun.* **28**, 255–272.

Waldvogel, A. (1974). The N_0 jump of raindrop spectra. *J. Atmos. Sci.* **31**, 1067–1078.

Wexler, R., and Atlas, D. (1963). Radar reflectivity and attenuation of rain. *J. Appl. Meteorol.* **2**, 276–280.

References

Wilson, J. W. (1970). Integration of radar and gage data for improved rainfall measurement. *J. Appl. Meteorol.* **9,** 489–497.

Wilson, J. W., and Brandes, E. A. (1979). A radar measurement of rainfall—A summary. *Bull. Am. Meteorol. Soc.* **60,** 1048–1058.

Woodley, W. L., Olsen, A. R., Herndon, A., and Wiggert, V. (1975). Comparison of gage and radar methods of convective rain measurement. *J. Appl. Meteorol.* **14,** 909–928.

Zawadzki, I. (1981). The quantitative interpretation of weather radar measurements. *Prepr., Conf. Radar Meteorol., 20th, 1981* pp. 586–587.

Zrnić, D. S., Doviak, R. J., Mahapatra, P. R. (1982). "The Effect of Charge and Electric Field on the Shape of Raindrops," Proc. URSI Comm. F, Open Symp. Multiple-Parameter Radar Meas. Precipitation. pp. 31–39. Rutherford Appleton Lab. Chilton, Didcot Oxfordshire, U.K.

9

Observations of Winds, Storms, and Related Phenomena

The great utility of centimeter-wavelength pulsed-Doppler radar in storm observation derives from its capability to map the reflectivity η and mean radial velocity $\bar{v}$ inside the storm's shield of clouds. If single-beam radars are used, a three-dimensional picture of a storm more than a few kilometers away takes about 2–5 min of data collection time, not only because of antenna rotation limitations, but also because a large number of echoes from each resolution volume needs to be processed in order to reduce the statistical uncertainty in the η and $\bar{v}$ estimates (see Chapter 6). Although the storm can change significantly during this period, leading to distortion of the radar image of the reflectivity and velocity fields, highly significant achievements have been made in depicting the structure and evolution of the thunderstorm.

The meteorologically interesting variables, however, are not η nor $\bar{v}$ but parameters such as the rainfall rate (on the ground) and wind. Doppler radar most often measures the radial speed of hydrometeors, not air, and in certain situations, such as vertically directed beams, these speeds can differ significantly from the radial component of wind. Targets such as water drops have small mass and quickly respond to horizontal wind forces to trace faithfully the wind. Stackpole (1961) has shown that, for the energy spectrum of wind scales following a power of $-\frac{5}{3}$ law (see Chapter 10) to at least 500 m, more than 90% of the rms wind fluctuations are acquired by the drops if their diameter is less than 3 mm. For radar beams at low elevation angles, target terminal velocities (i.e., the steady state vertical velocities relative to the air) give negligible bias error in the radial wind component. At high elevation angles these velocities need to be estimated.

Incoherent radars map η and not wind, but if the radar's resolution volume is sufficiently small and reflectivity estimates have small variance, these radars can track small-scale reflectivity structures, and if the structure drifts with the volume of air, they can map the vector wind (Crane, 1979; Reinhart, 1979). However, terminal velocity and evolution (growth) of the Z field can cause errors in wind estimation. On the other hand, Doppler radar measures—practically instantaneously—velocities in each resolution volume and hence can provide the true radial velocity field with better resolution than obtained by tracking reflectivity.

9.1 THUNDERSTORM STRUCTURE

Although there are authors who comprehensively examine the structure and characteristics of thunderstorms (e.g., Kessler, 1982, has three volumes of chapters by many prominent authors), a brief description will now be given in order to help interpret the weather radar observations shown in this and other chapters. We focus our attention on the severe thunderstorm because it produces damaging wind and hail, and also because it is often part of a larger weather system such as a squall line or hurricane.

Clouds give a good picture of the thunderstorm, and, even though most of the storm's violent interior is obscured from visual observation, the cloud patterns delineate much of the storm's internal motion. The cloud structure of a severe thunderstorm is diagramed in Fig. 9.1a as it might appear to the unaided eye. Even when clouds, precipitation, trees, and buildings do not obstruct one's view, a composite of phenomena such as that shown could not be observed at one time because the scales are so vastly different. Even a thunderstorm photograph (Fig. 9.1b) that shows nearly the entire cloud envelope does not display all the phenomena present in this storm. The diagrammatic view sketched in Fig. 9.1a is toward the northwest. The main storm tower contains the largest and strongest updrafts, although there is a multitude of smaller-scale turbulent vertical motions. Broad zones of warm, dry downdrafts of weak (<5 m s^{-1}) intensity have been observed in the clear air surrounding the upper parts of a severe storm (Sinclair, 1973; Fritsch, 1975), and sometimes the small-scale turrets that form the envelope of the cauliflower cloud structure can be seen descending on the upwind side of the storm. However, the strongest downdrafts are usually encountered within the regions where precipitation exerts drag and evaporatively cools the air, locally increasing its density.

If the convective instability and environmental wind shear are weak, light-to-moderate showers and mild up- and downdrafts are usually all that are contained in the storm, and middle-level (3–6 km) air reaching the ground is not evaporatively cooled very much. Updrafts are usually short-lived, because intensifying precipitation causes drag that slows the upward flow, eventually turning it downward. In this case, precipitation falls to the ground through the air column in which it was formed.

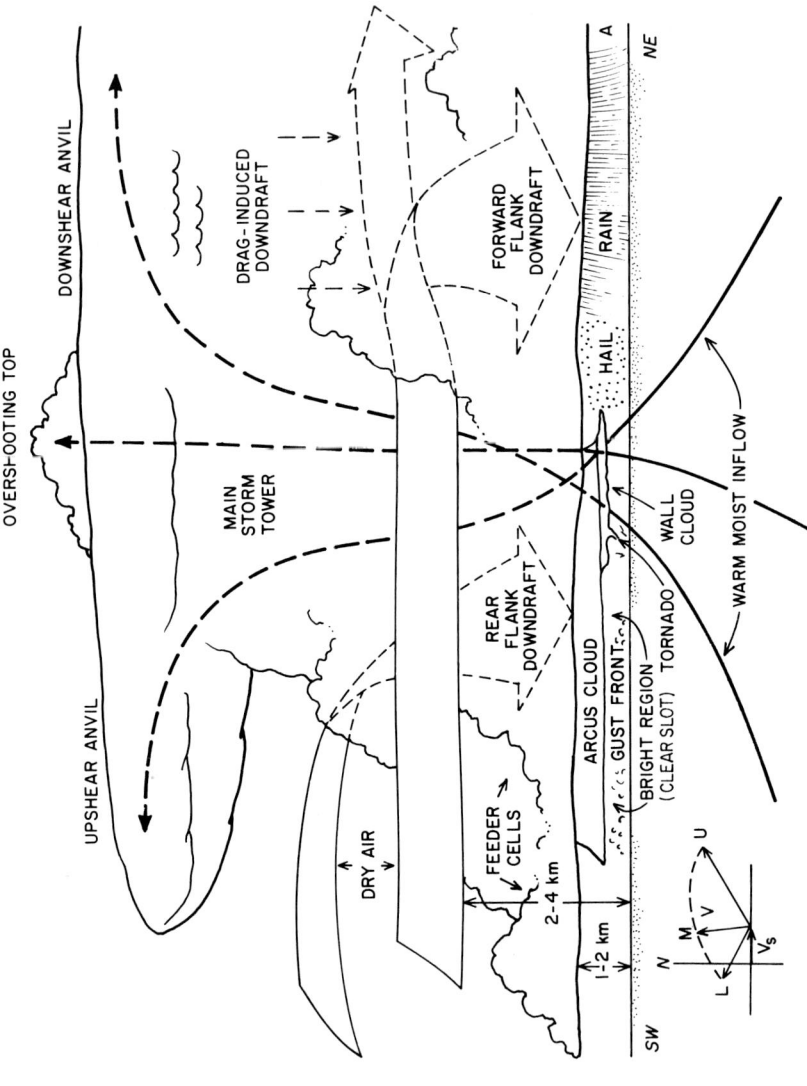

Fig. 9.1a. Diagram of a tornadic supercell storm as viewed from the southeast. The sketch is not to scale. Dashed lines are air trajectories inside the cloud. Environmental wind vectors at low (L), middle (M), and upper (U) levels of the atmosphere are drawn at the lower left. The storm motion vector is v_s.

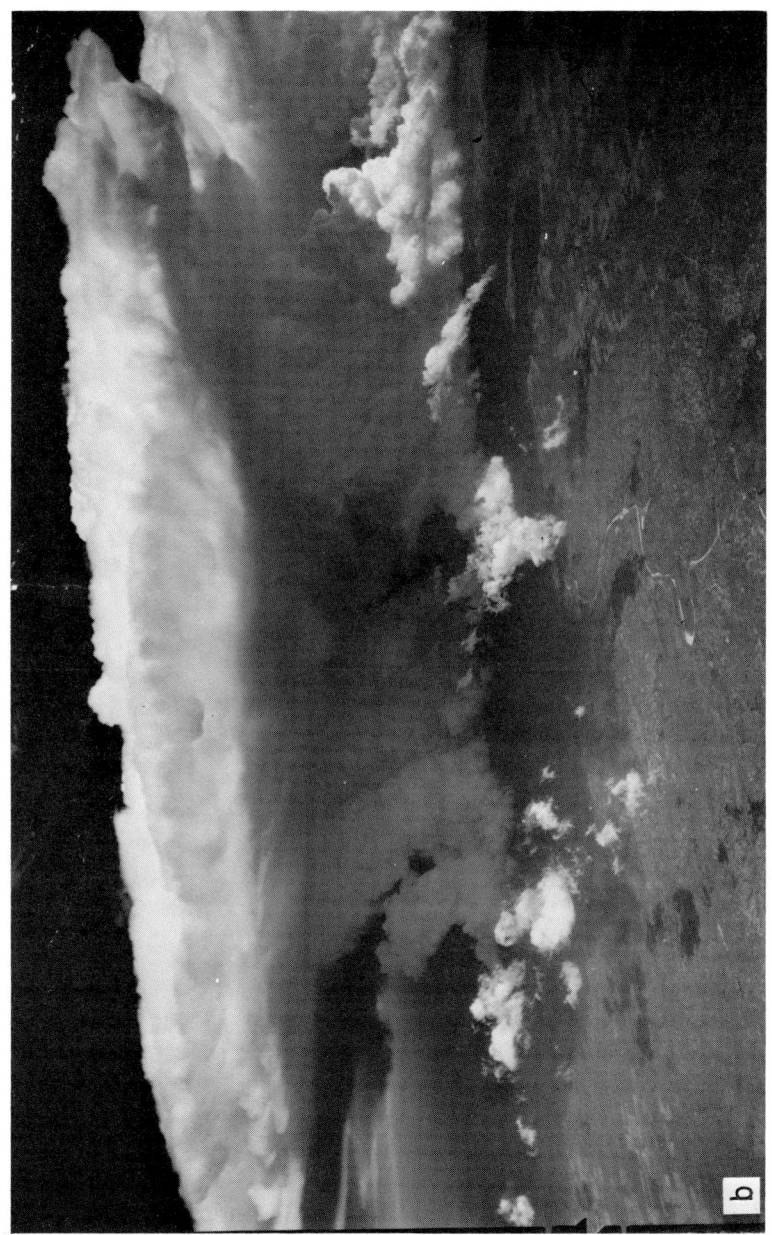

Fig. 9.1b. A severe thunderstorm as seen from an aircraft. The view is toward the upshear anvil. Three distinct updraft currents are seen in this multicell severe storm. (Courtesy of J. Lee, NSSL.)

Fig. 91c. Photograph toward the rear flank downdraft location, showing the wall and arcus clouds. (Photo by H. Bluestein, University of Oklahoma, taken during NSSL's Thunderstorm Intercept Project.)

9.1 Thunderstorm Structure

Fig. 9.1d. Southeast view of a bubble of updraft air overshooting the anvil of clouds. A feeder cell is to the right (southwest) of the main storm. (Courtesy of H. Bluestein, University of Oklahoma.)

However, when there is strong shear of the environmental wind and the instability is considerable, faster updrafts may be created. The shear tilts the updrafts so that the precipitation falls out rather than through the updraft, with a resultant increase in the net buoyancy. The interaction of the updrafts with the sheared environmental wind organizes the flow so that intrinsically cold middle-level air is cooled sensibly by evaporating precipitation and, in descending rapidly, serves as an important additional source of energy for the storm. Cold low-level outflow also initiates new convection and may cause continuous lifting of the potentially unstable boundary layer air. In that case the storm may be long-lived (several hours) and have a quasi-continuous flow of up- and downdrafts. Ordinary thunderstorms can be considered closed systems in the sense that the storm air does not sufficiently interact with the environment to maintain a continuous quasi-steady flow of ascending and descending air. On the other hand, the severe thunderstorm called a supercell, sketched in Fig. 9.1a, is an open system hypothesized to organize a steady flow and to propagate either to the left or right of the mean tropospheric wind with a velocity that allows it a steady interaction with the large-scale environment of the storm (Browning, 1982). In either of the two storm types, air rising and descending does not follow a closed circuit, because the air that goes up does not return and the air that comes down does not immediately go up again; the thunderstorm produces an interchange between the surface and the upper air.

In the middle latitudes of the North American Great Plains, the updraft is fed by boundary layer air, usually entering the storm from the east to southeast in a coordinate system translating at a velocity v_s equal to that of the storm (Fig. 9.1a). Storm motion is usually northeast, although supercells move more easterly and even southeast. Within the updraft, condensation takes place, and above the 0°C altitude freezing of rain droplets can commence, releasing more latent heat, which further increases the buoyancy of the ascending air. At levels where temperatures decrease to below $-30°C$, the proportion of ice increases (Anthes et al., 1982). The rapidly ascending air is associated with a low-pressure region beneath the main tower, where a wall cloud is formed from saturated subcloud air that has been cooled by expansion (Fig. 9.1c). It is here where the most intense updraft below the cloud base is located and where the most intense tornados are found. Usually a wall cloud rotates cyclonically and is part of a larger mesocyclone, which extends up through the main storm tower. Increasing rotational velocity is accompanied by lowering pressure.

The intensity of the updraft can be so strong that its momentum can carry air into regions of large negative buoyancy in the lower stratosphere, and overshooting cloud tops can be observed (Fig. 9.1d). Above the tropopause height the air is stably stratified, and a strong negative buoyancy force acts on the updraft, forcing it back down to the tropopause, where it spreads into a relatively thin layer (which may be more than 3000 m thick, however; see Color Plate 2c), known as the anvil cloud. The outflow speed near the cloud top is so high that the storm air can move upstream against the strong upper level winds and form the upshear portion of the anvil. Although the upwind anvil may extend only several kilometers from the main storm tower (Fig. 9.1b), the downwind portion can be visible for many hundreds of kilometers as a plume of cloud carried off by the environmental winds. The air in the updraft column is usually found to be rotating cyclonically.

The downdraft contains large amounts of dry environmental air that is turbulently mixed with the updraft air containing precipitation. As dry air approaches the storm from the southwest to west, some of it is deflected around the updraft tower, which acts partially like an obstacle. Two distinct zones of downdrafts are depicted on the surface map in Fig. 9.2a: (1) the forward-flank downdraft (FFD) and (2) the rear-flank downdraft (RFD) (Lemon and Doswell, 1979).

Precipitation falling downstream from the updraft column may drag air to contribute to the FFD. This downdraft, abetted by evaporation, is depicted in Fig. 9.1a by downward-directed dashed lines that emanate high in the

9.1 Thunderstorm Structure

storm beneath the downwind anvil. Because the updraft acts as an obstacle to the environmental flow, a pressure deficit can be created near the downstream side of the main updraft tower. Thus pressure gradient forces may divert dry air into the precipitation zone, where it mixes with the forward flank downdraft. An example of a downdraft embedded in rainy air is shown in Fig. 9.2b, where precipitation near the surface is deflected outward as depicted in Fig. 9.1a as region A.

The rear flank downdraft (RFD) is hypothesized by Lemon and Doswell (1979) to begin at the upper levels of the storm, where an upwind stagnation point and a high-pressure region develop that may initiate the downward flow. Air trajectory analysis for numerically modeled storms as well as

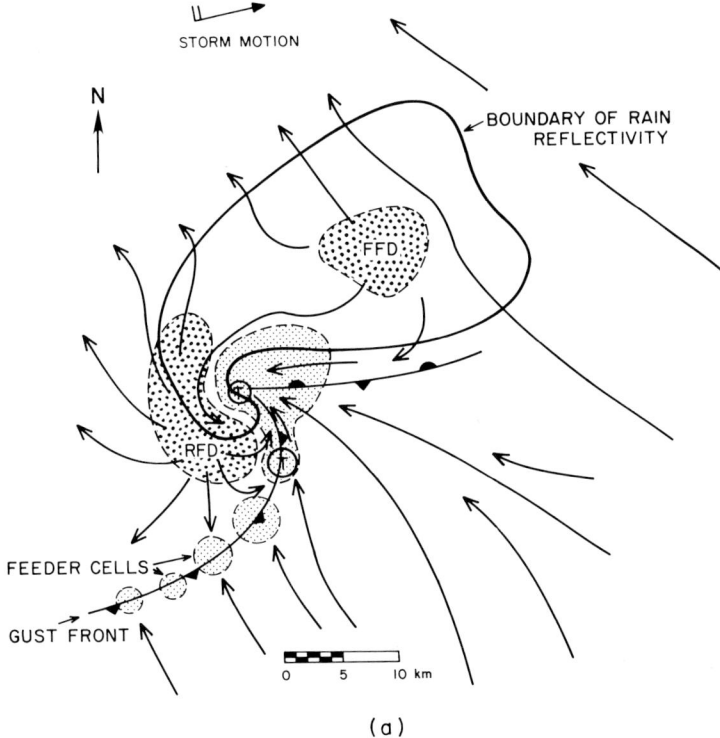

Fig. 9.2a. Plan view of surface weather, showing locations of the forward-flank and rear-flank downdrafts. The solid line outlines the rainy area usually mapped by radar. Note the hooklike feature. Downdraft and updraft regions are delineated by coarse and fine stippling, respectively. T designates the most likely location where tornados form. [Adapted from Lemon and Doswell (1979); courtesy of R. Davies-Jones.]

Fig. 9.2b. View of a rain shaft, showing deflection of precipitation due to downdraft air being turned outward at the ground. (Photo by H. Bluestein, University of Oklahoma, taken from NASA's CV990 aircraft.)

Fig. 9.2c. Photograph toward the rear-flank downdraft, which evaporates cloud particles and thus allows sunlight to penetrate to the surface. Strong outflow of the RFD lifts the moist boundary layer air that forms the arcus cloud seen. (Courtesy of H. Bluestein, University of Oklahoma.)

Doppler data do not exhibit environmental air descending much more than a kilometer or two (Klemp *et al.*, 1981). However, the moisture and temperature characteristics of downdraft air measured at the surface suggest that the air sometimes originates at the middle levels of the troposphere (Barnes, 1978).

When the relative humidity of the downdraft is less than 100%, cloud particles and the smallest raindrops evaporate quickly, leaving behind a nearly cloud-free column through which sunlight can be seen as a bright region at the surface of the rear flank of the storm (Figs. 9.1a and 9.2c). The downdraft builds a pool of air on the ground that is more dense than the surrounding air and often carries with it the horizontal momentum of the lower- to middle-level wind. The air then quickly spreads outward forming a gust front whose presence may be marked by dust clouds and, extending beyond the front, an arcus cloud (Fig. 9.1a). The arcus cloud forms when water vapor in the moist environmental air, lifted by the outward flow, condenses, creating a shelflike cloud that can extend several kilometers beyond the surface location of the gust front. New feeder cells develop along the gust front, and these drift with the southwesterly winds into the main tower enforcing the updrafts.

Many flow characteristics implicit in this broad description are elaborated in the Doppler radar estimated wind and reflectivity fields presented throughout this chapter.

9.2 OBSERVATIONS WITH TWO DOPPLER RADARS

A single Doppler radar maps a field of velocities that are directed toward (or away from) the radar. A second Doppler radar spaced far from the first produces a field of different radial velocities because the true velocities are projected on different radials. The two radial velocity fields can be vectorially synthesized to retrieve the two-dimensional velocities in the plane containing the radials. It is customary to accomplish the synthesis on common grid points to which radar data are interpolated. Mean radial velocities, assigned a location at the center in each of the radar's resolution volumes surrounding a grid point, are not measured simultaneously but are separated in time up to the few minutes required for each radar to scan the storm volume. Furthermore, the respective resolution volumes are usually quite different in size and orientation. Nevertheless, useful estimates of wind can be made on scales of air motion large compared to the largest resolution volume dimension if the velocity field is nearly stationary over the period required for data collection. Stationarity is probably the most severely violated of the many assumptions required in order to derive vector wind fields from dual Doppler data.

9.2.1 Reconstruction of Wind Fields

Wind field determination from two Doppler radar data is greatly simplified if the synthesis is performed in cylindrical coordinates with axis, chosen to be the line connecting the two radars. That is, radial velocities at data points (centers of resolution volumes) are interpolated to nearby grid points on planes having a common axis (the COPLAN technique) (Lhermitte, 1970). Cartesian wind components can be derived from these synthesized cylindrical components. Although one could solve directly for Cartesian wind components, this necessitates solving an inhomogeneous hyperbolic partial differential equation to derive the vertical wind (Armijo, 1969).

The cylindrical coordinate system is illustrated in Fig. 9.3. The mean Doppler velocity needs to be corrected for the scatterers' reflectivity-weighted mean (i.e., average over the resolution volume) terminal velocity $\bar{w}_t$. Thus the estimate of the radial component of air motion is

$$v_{1,2} = v'_{1,2} + \bar{w}_t \sin \theta_{e1,2}, \tag{9.1}$$

where $v'_{1,2}$ are the mean Doppler target velocities measured by radars 1, 2 at data points and $\bar{w}_t$ is positive. To estimate $\bar{w}_t$ for each resolution volume, one could use the empirical expression (Atlas et al., 1973)

$$\hat{w}_t = 2.65 Z^{0.114}(\rho_0/\rho)^{0.4} \quad \text{m s}^{-1}, \tag{9.2}$$

where the parenthetical term is a correction [see Eq. (8.4)] to account for the height-dependent air density ρ. This relation represents to within a standard error of 1 m s^{-1} the experimental data of Joss and Waldvogel (1970) over a large range of Z (i.e., $1 \leq Z \leq 10^5$ mm^6 m^{-3}) and drop size distributions for regions of liquid water, but large errors (up to several meters per second) in $\bar{w}_t$ estimates can be caused by using in regions of hail a $\hat{w}_t(Z)$ relation appropriate for liquid water. Usually there is little or no information to identify these regions uniquely, and errors in vertical wind w can result. However, it has been shown for typical arrangements of storms relative to the two radar

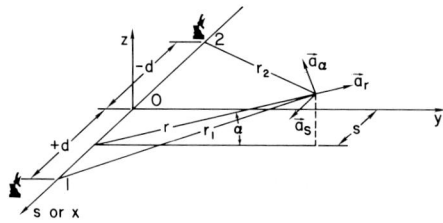

Fig. 9.3. Cylindrical coordinate system used for dual radar data analysis. The radars are located at the points 1 and 2, and $\mathbf{a}_r$, $\mathbf{a}_s$, $\mathbf{a}_z$ are the unit normals defining the direction of the three orthogonal velocity components. The cylinder axis is along the line connecting the radars, and r is the range from the axis to the data point.

9.2 Observations with Two Doppler Radars

placement that the error in the vertical air velocity w caused by errors in $\hat{w}_t$ is significantly smaller than errors in the estimate $\hat{w}_t$ (Doviak et al., 1976).

The estimated radial velocities $v_{1,2}$ of the air can be interpolated to uniformly spaced grid points in planes at an angle α to the horizontal surface containing the baseline. Interpolation filters the data and reduces the variance. Several interpolation schemes are possible, and a particularly simple and effective one employs the Cressman function (Cressman, 1959) W_i to weight data at points inside a volume centered on a grid point from which distance to the ith datum is D_i:

$$W_i = \begin{cases} (R_i^2 - D_i^2)/(R_i^2 + D_i^2) & \text{for } D_i \leq R_i, \\ 0 & \text{for } D_i > R_i. \end{cases} \quad (9.3)$$

R_i is the influence radius that determines the size of the interpolation volume. The shape of this volume, usually selected to be a sphere, is dictated by the functional dependence of R_i on the direction of a datum from the grid point. The cylindrical wind components in the r, s plane are related to $\bar{v}_1, \bar{v}_2$ as

$$w_r = [(s + d)r_1\bar{v}_1 - (s - d)r_2\bar{v}_2]/2dr, \quad (9.4a)$$

$$w_s = (r_2\bar{v}_2 - r_1\bar{v}_1)/2d, \quad (9.4b)$$

where $\bar{v}_{1,2}$ are the interpolated Doppler velocities of air.

The wind component w_α normal to the plane is obtained by solving the continuity equation in cylindrical coordinates:

$$\frac{1}{r}\frac{\partial}{\partial r}(r\rho w_r) + \frac{1}{r}\frac{\partial}{\partial \alpha}(\rho w_\alpha) + \frac{\partial}{\partial s}(\rho w_s) = 0, \quad (9.5)$$

the boundary condition being that $w = 0$ at the ground. One can integrate (9.5) starting from the ground or from the storm top, where the vertical velocity can be estimated to be zero. In either case, it can be shown that the estimate variance of w_α increases monotonically with increasing α (upward integration) or with decreasing α (downward integration). However, because density increases downward, smaller w_α variance results if one integrates from the storm top to the surface (Bohne and Srivastava, 1975). Information about vertical velocity at both the top and bottom of the troposphere can be used to reduce the errors in the vertical velocity estimates (O'Brien, 1970; Ziegler, 1978). The mass density ρ is given by

$$\rho = \rho_0 \exp(-gMr \sin \alpha/RT), \quad (9.6)$$

where g is the gravitational constant (9.8 m s^{-1}), M the mean molecular weight of air (29 g mol^{-1}), T the absolute temperature (K), and R the universal gas constant (8.314 J mol^{-1} K^{-1}). Appropriate values of ρ_0 and T can be obtained from surface measurements and upper air soundings of the environment within which the storm develops. The Cartesian components u, v, w of

Storm motion
20.0 m s⁻¹
Distance north (km)
Distance east (km)
A
B
(a)
Storm motion component in plane
40.0 m s⁻¹
Height (km AGL)
East 49.7 40.9 37.0 33.1 29.3 25.4 21.6
North 36.4 41.0 45.6 50.2 54.6 59.4 64.0
Distance along AB (km)
(b)

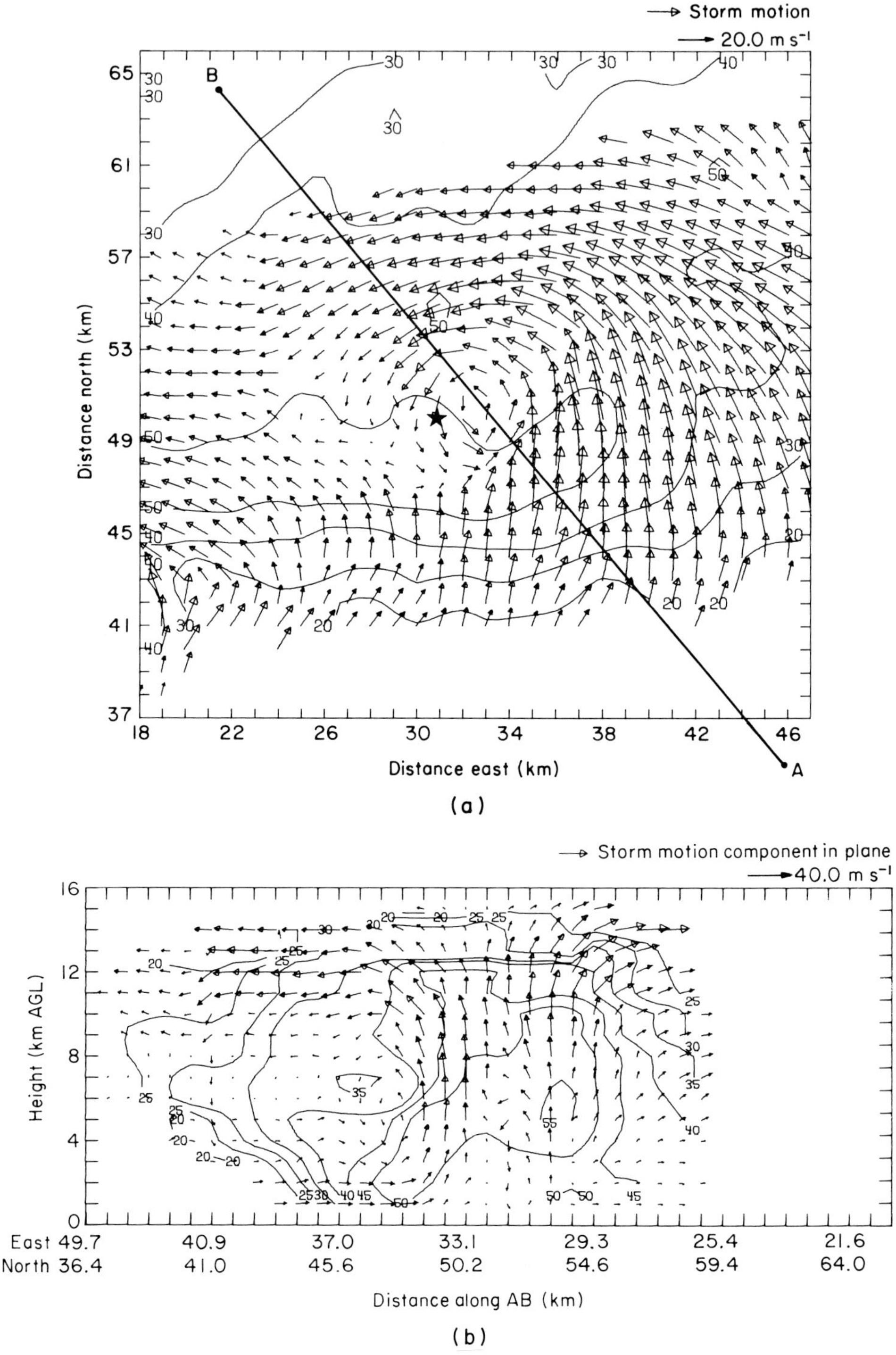

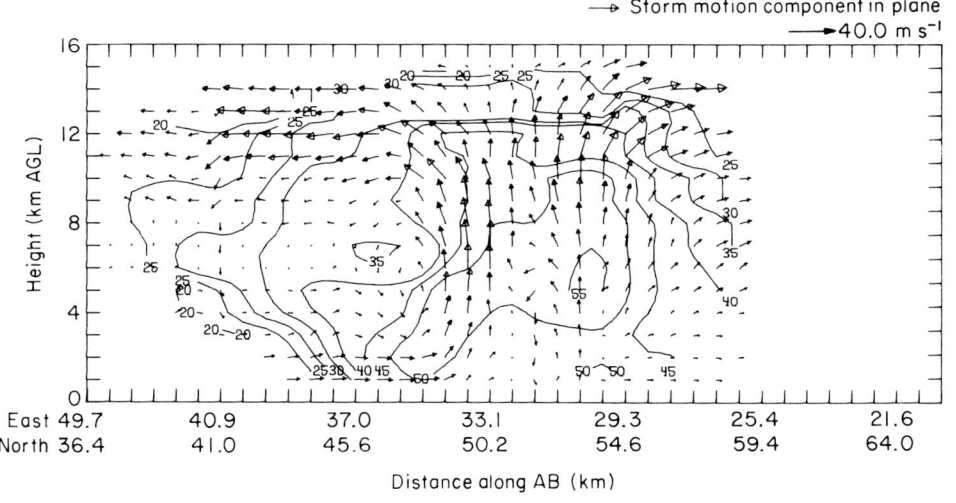

9.2 Observations with Two Doppler Radars

wind can then easily be obtained from w_r, w_s, w_α. Several techniques used to compute the vertical velocity are compared by Ray et al. (1980).

9.2.2 Observation of Tornadic Storms

Doppler radar observations of two tornadic storms will be briefly discussed to illustrate some of the general storm features depicted in Figs. 9.1 and 9.2.

Figure 9.4a shows a horizontal cross section of a tornadic storm that formed in central Oklahoma on 2 May 1979. The three components of wind were obtained from radial velocity measurements made by two radars spaced 70 km apart, a reflectivity–terminal velocity relation, and integration of the mass continuity equation using suitable boundary conditions at the ground and the cloud top. Satellite photographs with superimposed reflectivity contours at 8 km and horizontal wind mapped at 11.9 km by the dual Doppler radars appear as Figs. 9.4c and 9.4d, respectively. The large circulation of air seen in Fig. 9.4a is a thunderstorm cyclone. Because thunderstorm cyclones are much smaller than extratropical or tropical cyclones, including hurricanes, these circulations are called *mesocyclones*. In all the cases observed, large tornados have been preceded by mesocyclones. Usually pressure decreases significantly within the interior of a mesocyclone, and when cyclonic circulation reaches the surface, moist boundary layer air entering the cyclone condenses and forms a wall cloud, making the circulation visible (Fig. 9.1c). Figure 9.4e shows a tornado embedded in the larger mesocyclone, which is also visible. Mescocyclones have diameters ranging from a few kilometers to ~10 km and can extend in height from near the ground to near the tropopause. Tornados are smaller-scale vortices embedded within the mesocyclone, and if their centers do not coincide (as in Fig. 9.4a) the tornado circulates around the mesocyclone center, forming a trochoidal path on the ground as the mesocyclone and tornado are translated by the moving storm. Note that the small-scale tornado vortex is not resolved in Fig. 9.4a.

The inflow of air from the southeast in the first 2 km above ground level (AGL), and updrafts as large as 40 m s^{-1} can be seen in the vertical cross

---

Fig. 9.4a,b. (a) Contours of reflectivity factor (in dBZ) and the storm relative wind field in a horizontal cross section at 2 km above ground level and (b) a vertical cross section of a tornadic thunderstorm on 2 May 1979, at 16:58 C.S.T. Winds are in the coordinate system moving with the storm. Storm motion speed and direction are shown. Line AB in (a) is the location of the vertical cross section. Distances are from a radar at Roman Nose State Park, Oklahoma (Alberty et al., 1979). The star pinpoints the tornado location in (a). The arrow of indicated velocity at the top right of each plot gives the speed which is proportional to the arrow's length. (From E. Brandes and B. Johnson, NSSL, personal communication.)

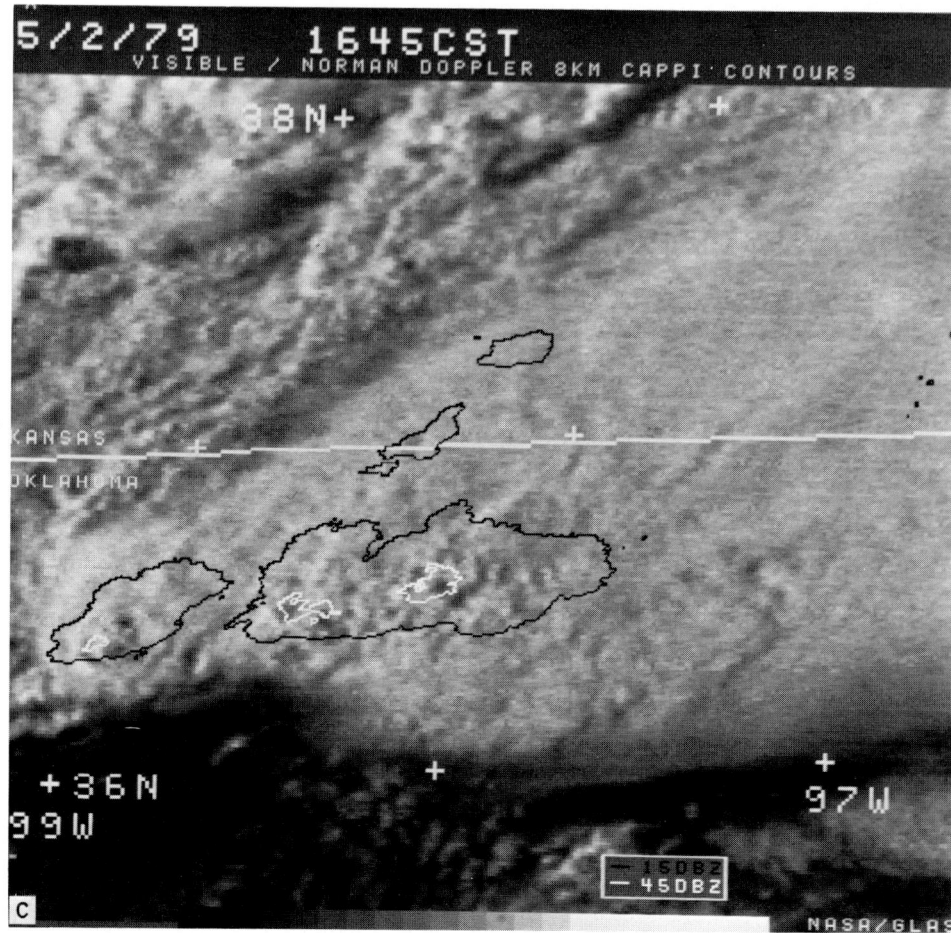

Fig. 9.4c. Satellite photograph of the anvil of clouds over the 2 May 1979 storm in northwest Oklahoma at 16:45 C.S.T. The Norman Doppler radar reflectivity factor contours of 45 (white) and 15 (black) dBZ at the 8-km altitude are superimposed. (Courtesy of Dr. G. Heymsfield, GLAS, NASA.) The fields of reflectivity and the spectrum width of this storm are presented as Figs. 9.24 and 9.25. The corresponding Doppler velocity fields are presented in Color Plates 2a and 2b.

9.2 Observations with Two Doppler Radars

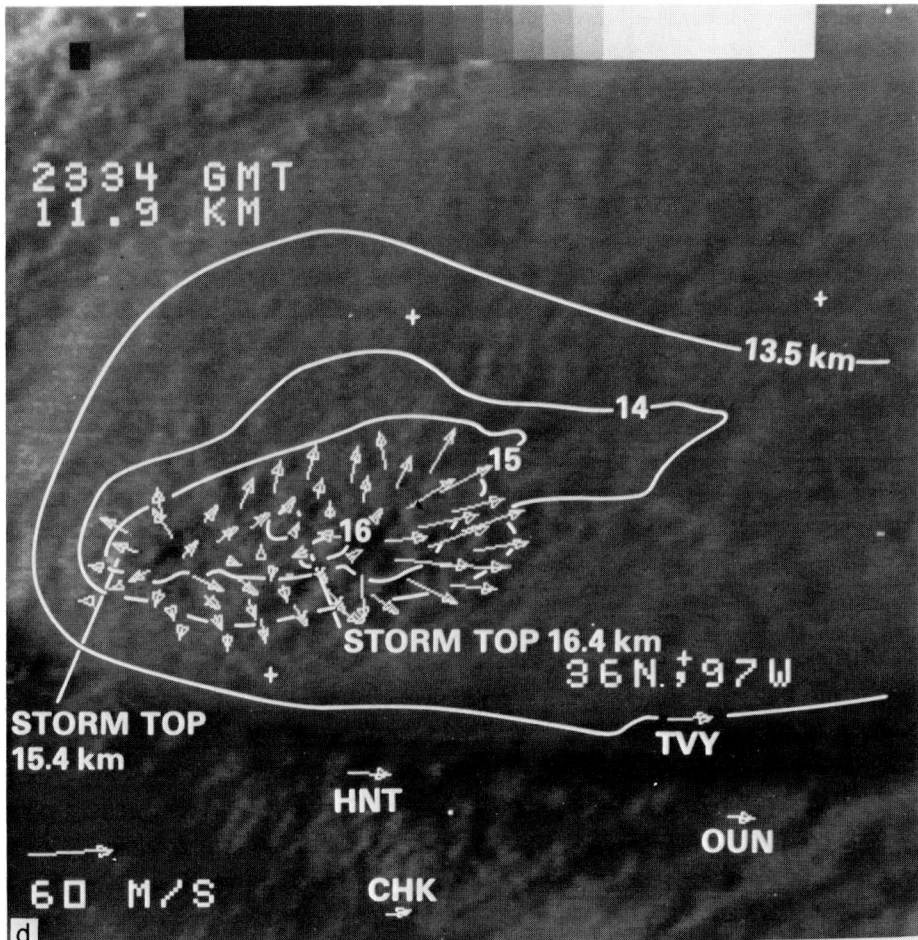

Fig. 9.4d. Dual Doppler synthesized winds at 11.9 km superimposed on the anvil cloud photographed from a satellite at 17:34 C.S.T. The height contours of the cloud tops are obtained from a pair of stereo satellite images. The vector at 60 m s^{-1} scales the wind vectors. Winds at 11.9 km but measured from rawinsonde observations from four stations [at Norman (OUN), Hinton (HNT), Chickasha (CHK), and Oklahoma City (TVY)] are also displayed. The diverging outflow is near the bubbles of cloudy air overshooting the tops of the anvil cloud. (Courtesy of Dr. G. Heymsfield, GLAS, NASA.)

Fig. 9.4e. Tornado embedded in the larger-scale mesocyclone. The mesocyclone cloud base was seen to be rotating counterclockwise. (Courtesy of Don Burgess, NSSL.)

section on Fig. 9.4b. Large gradients of reflectivity mark the region of inbound air, and strong outflow occurs in the anvil (Fig. 9.4d). These features are similar to those shown in Fig. 9.1. Although the cross section in Fig. 9.4b shows mostly updrafts, other regions of the storm have downdrafts.

The second example of Doppler-synthesized wind fields is from a storm that occurred 20 May 1977. Later, in Section 9.5.3, we shall give the Doppler spectra of winds around the tornado that developed in this storm. On the morning of this day a pool of cold air, resulting from the outflow of nocturnal thunderstorms in northwest Oklahoma, formed a boundary that extended southwestward from northeast Oklahoma. A morning rawinsonde released from Oklahoma City revealed a potentially very unstable air mass with strong vertical shear. Southeasterly boundary layer flow brought warm moist air over the cold air. A series of thunderstorms was initiated along the cold air boundary, and by 14:00 CST the storms were intense. Individual cells propagated northeastward, whereas the line of storms progressed slowly eastward. At 18:47 CST the Del City tornadic storm was located about 30 km north–northeast of Norman, Oklahoma, and in an excellent position (Fig. 9.5) to be sampled by NSSL's Norman and Cimarron 10-cm Doppler radars. Two other Doppler radars farther to the southwest also sampled the Del City storm.

9.2 Observations with Two Doppler Radars

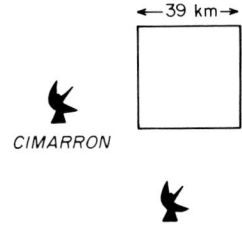

Fig. 9.5. Location of NSSL's Cimarron and Norman Doppler radars. The box outlines the area within which storm winds (Fig. 9.6) were synthesized from Doppler radar observations.

Data from all four radars entered into the synthesis of the storm's vector wind fields portrayed in Fig. 9.6, but most weight was given to the radial velocities measured by the Norman and Cimarron radars, which had the best aspect angles and resolution. The method of synthesis is given by Ray et al. (1981), Although a dual Doppler analysis was not used to arrive at the winds depicted in Fig. 9.6, a dual Doppler analysis, as outlined in Section 9.1, should produce similar wind fields because of the small weight given to data from the more distant radars. For example, one can compare the storm wind fields produced by dual Doppler analysis (Ray et al., 1980, Fig. 11, method B′) and that by four radars (Ray et al., 1980, Fig. 11, method E). Upper boundary conditions of $w = 0$ and $\bar{w}_t = 2 \text{ m s}^{-1}$ were assumed for altitudes 1 km above the highest data level, and the divergence at the data level is assumed to be representative of the 1-km layer above.

At 18:14 the Del City storm passed nearly over Norman, and although there was some indication of descending air south of the updraft, ground clutter prevented reliable measurements. However, strong downdrafts remained to the north of the updraft, suggesting the existence of the forward flank downdraft (Fig. 9.1a). There is an indication (Fig. 9.6a) of descending air being drawn around the western portion of the updraft. Earlier the updraft was confined to a single cyclonically rotating region, whereas at 18:47 two major updraft branches have formed. The western branch (A; coordinates 3, 31) contained a vorticity maximum of $2.5 \times 10^{-2} \text{ s}^{-1}$ at an altitude of 2 km, while the eastern branch (B; 8, 29) had a $1.5 \times 10^{-2} \text{ s}^{-1}$ vorticity maximum.

Another updraft (coordinates 12, 34) formed northeast of the updraft associated with the Del City storm. This updraft was caused by the strong convergence along the line indicated in the 0-km height wind fields (Fig. 9.6b). Updraft B in Fig. 9.6a appears to have been caused by the convergence along the outward propagating gust front (delineated in Fig. 9.6b), where feeder cells form (Fig. 9.2a). The tornado's location is between the updraft air of branch A and the downdraft forming the gust front in the region of strong horizontal gradient of vertical wind. The strongest inflow (Fig. 9.6b)

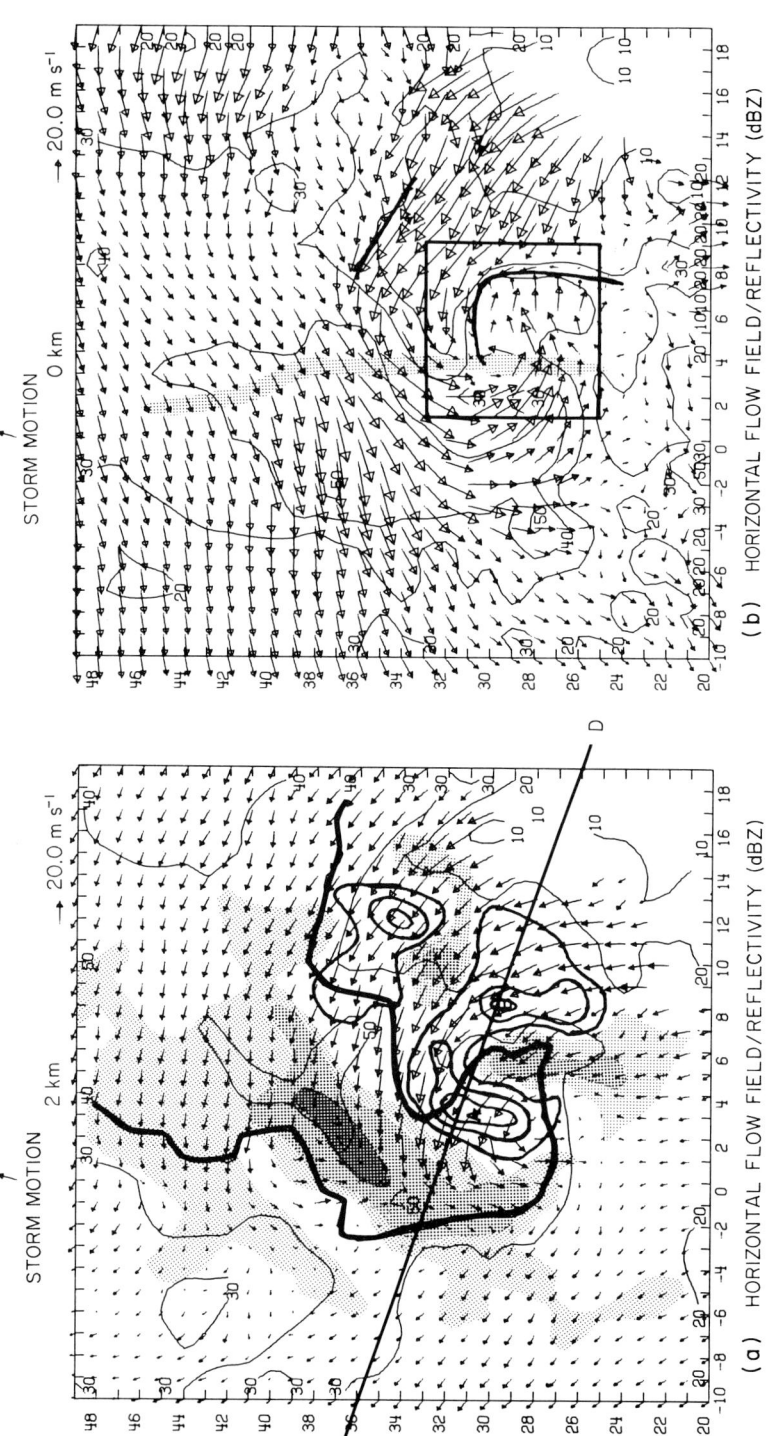

Fig. 9.6a,b. Horizontal cross sections of storm relative winds, synthesized from Doppler radar observations at 18:47 C.S.T. on 20 May 1977. The origin is at the Norman Doppler radar. (a) Height above ground $h = 2$ km; (b) $h = 0$ km. The storm velocity is indicated by the arrow at the top of each part of the figure, and velocity scale is the arrow labeled 20 m s^{-1} in the upper-right corner. Reflectivity (in dBZ) is contoured in thin lines although the 40-dBZ contour in (a) is a bold line showing a hooklike feature. The lines of medium thickness in (a) are contours of updrafts in steps of 10 m s^{-1} from the first contour at 5 m s^{-1}. Downdrafts more than 1 m s^{-1} are stippled, and larger ones have an increased density of stippling for speeds higher than 5 and 15 m s^{-1}. The elongated north–south lightly stippled area in (b) is the tornado damage path. Tornado is at 4 km, 29.5 km at this time and the gust front is the curved line. Line CD in (a) is the location of the cross section shown in Fig. 9.6c,d. (From Ray *et al.*, 1981.)

9.2 Observations with Two Doppler Radars

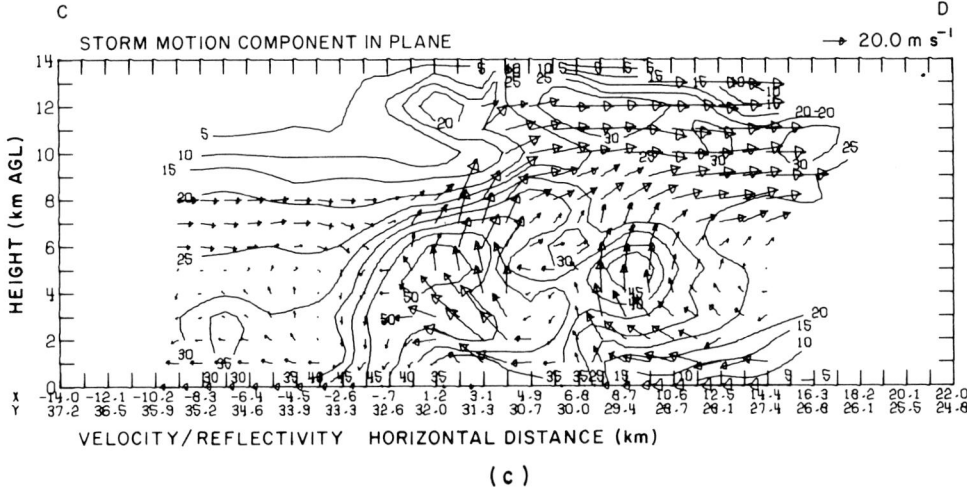

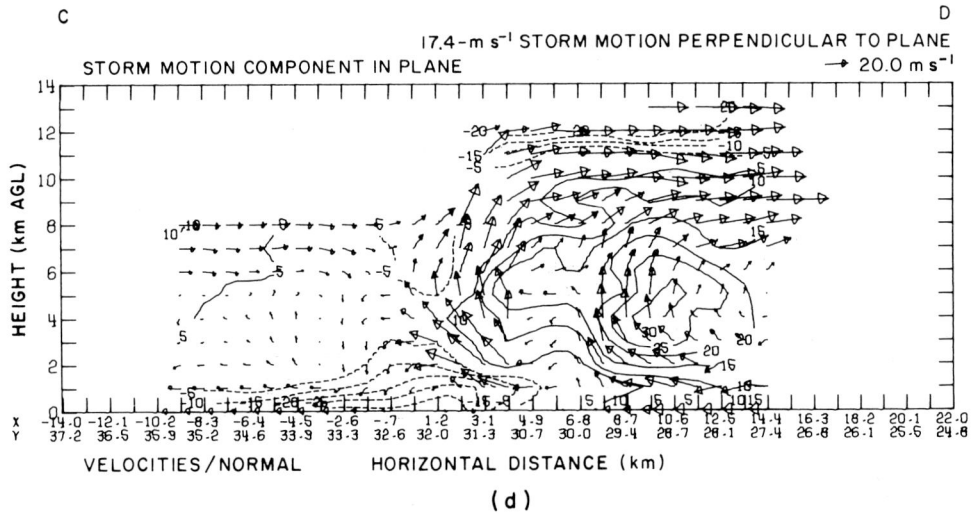

Fig. 9.6c,d. Vertical cross section of storm winds relative to storm motion in the plane CD. Contours in (c) are dBZ levels, whereas those in (d) are wind speeds into (solid lines) and out of (dashed lines) the cross section CD. Speed contours are in steps of 5 m s^{-1} starting from ± 5 m s^{-1}. (From Ray et al., 1981.)

approaches from the southeast, although environmental air from the east and northeast enters into this storm.

The wind in the vertical cross section CD in Fig. 9.6a is shown in Figs. 9.6c and 9.6d. Clearly in Fig. 9.6c there are two branches of updraft, with high-reflectivity regions within each. The contours in Fig. 9.6d show the airflow perpendicular to the plot, whereas the in-plane wind is the same as in Fig. 9.6c. It is evident that the updraft stream (B) is twisted and inclined from the vertical and veers anticyclonically (as also seen on Fig. 9.1a), but the vorticity of air within it is cyclonic. The strong surface wind from the north is quite evident in the lowest levels of Fig. 9.6d.

9.2.3 Errors in Synthesized Wind Fields

The wind fields derived from dual Doppler radar measurement contain errors from several sources. Some of these are (1) variance in the mean Doppler velocity and Z estimates caused by the statistical nature of the weather echo, (2) a nonuniform reflectivity factor Z within a resolution volume, (3) the use of an incorrect w_t, Z relationship, (4) inaccuracies in the location of the resolution volume, (5) increases in the vertical velocity variance with each integration step owing to error in the divergence estimates used in the continuity equation, (6) nonstationarity of the storm during a data collection scan, and (7) echoes received through side lobes that contaminate signals associated with the resolution volume. How these errors affect the estimates of horizontal and vertical wind is discussed by Doviak *et al.* (1976) and Doviak and Strauch (1980).

9.3 LINEAR WIND MEASUREMENTS WITH A SINGLE DOPPLER RADAR

Lhermitte and Atlas (1961) described how a single Doppler radar could be used to determine the vertical profile of horizontally uniform wind. They proposed a data collection mode in which the radar beam is directed at a constant elevation angle and the radial velocity is continuously recorded at several ranges (to obtain soundings at different heights) as the radar beam sweeps through a full circle; this produces on an oscilloscope a series of traces called a *velocity azimuth display* (VAD). Lhermitte and Atlas showed how trace properties (e.g., the radial velocity offset, peak velocity, and zero crossings) can be used to estimate the horizontal wind and precipitation terminal velocity. Although analyses of data collected in this way have increased in sophistication (Caton, 1963; Browning and Wexler, 1968), the data collection method is still often referred to as VAD. However, the radar

9.3 Linear Wind Measurements with a Single Doppler Radar

measures radial velocities over a volume of space so we are led to consider the analysis of linear wind fields over a three-dimensional volume. This analysis technique was first studied by Waldteufel and Corbin (1979), who introduced the term *volume velocity processing* (VVP).

In this section we examine the linear wind model in three dimensions. The theory of multivariate regression analysis will be used to determine the accuracy (bias and variance) of our wind estimates. We shall show that the accuracy of estimates of the kinematic properties (divergence, deformation, etc.) of the linear wind depends on the size and shape of the analysis volume, the measurement error in the radial velocity, the number of parameters in the linear wind model, and the linearity of the actual wind.

We begin by deriving an expression relating the radial velocity measured by the radar to the wind at the measurement point (r, θ_e, ϕ). A spherical earth coordinate system is used with x and y as arc distances from the radar along two orthogonal great circle paths. We chose y to be along a longitude, with north the positive direction, as shown in Fig. 9.7. The earth's radius vector, drawn through the measurement location defines the vertical axis, along which z measures the height above the surface. The earth is assumed to have an effective radius a_e (Chapter 2) due to the mean vertical gradients of the refractive index. The horizontal components u and v of the vector wind $\mathbf{v}$ are

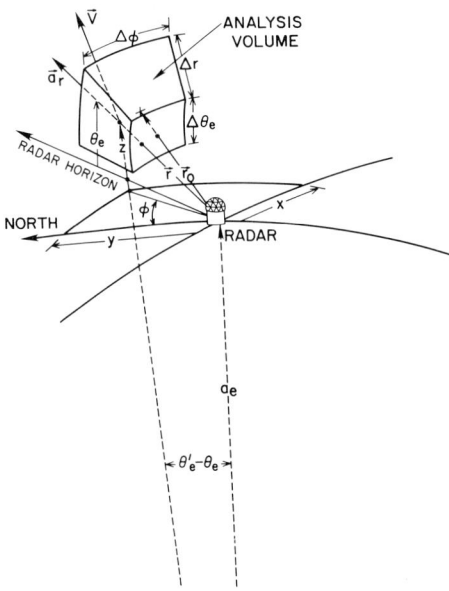

Fig. 9.7. The spherical coordinate system chosen for linear wind analysis. $\Delta r \simeq 20\text{--}30$ km; $\Delta\theta'_e \simeq 1\text{--}2°$; $\Delta\phi \simeq 30\text{--}40°$.

tangent to the great circle arcs at x and y and are directed eastward and northward, respectively. The vertical component w of $\mathbf{v}$ is along z.

The vector wind $\mathbf{v}$ at r, θ_e, ϕ is assumed to be well represented by a first-order (linear) Taylor series in a region (i.e., analysis volume) about some point (x_0, y_0, z_0), usually the center of the analysis volume. Therefore,

$$\mathbf{v}(x, y, z) = \mathbf{v}(x_0, y_0, z_0) + \frac{\partial \mathbf{v}}{\partial x}(x - x_0) + \frac{\partial \mathbf{v}}{\partial y}(y - y_0) + \frac{\partial \mathbf{v}}{\partial z}(z - z_0). \quad (9.7)$$

If $r \ll a_e$, the coordinates $x, y,$ and z are related to the radar coordinates $r, \theta_e,$ and ϕ by

$$x \approx r \cos \theta_e' \sin \phi,$$
$$y \approx r \cos \theta_e' \cos \phi, \quad (9.8)$$
$$z = (a_e^2 + r^2 + 2ra_e \sin \theta_e)^{1/2} - a_e.$$

The angle θ_e' is the elevation angle θ_e corrected for the angle subtended by the verticals at the radar and at the measurement (i.e., data) point:

$$\theta_e' = \theta_e + \tan^{-1}[r \cos \theta_e/(a_e + r \sin \theta_e)]. \quad (9.9)$$

The true radial velocity v_r is the projection of $\mathbf{v}$ onto $\mathbf{r}$, the vector from the radar to the data point (r, θ_e, ϕ). Again, if $r \ll a_e$,

$$v_r = \mathbf{v} \cdot (\hat{\mathbf{i}} \cos \theta_e' \sin \phi + \hat{\mathbf{j}} \cos \theta_e' \cos \phi + \hat{\mathbf{k}} \sin \theta_e'), \quad (9.10)$$

where $\hat{\mathbf{i}}, \hat{\mathbf{j}}, \hat{\mathbf{k}}$ are unit vectors at $\mathbf{r}$ in the x, y, z directions, respectively. Substitution of the first-order expansion (9.7) and the relations for x and y, from (9.8), into (9.10) and rearranging terms as recommended by Easterbrook (1974), we obtain

$$\begin{aligned} v_r = &\ u_0' \cos \theta_e' \sin \phi + u_x \cos \theta_e' \sin \phi \, (r \cos \theta_e' \sin \phi - x_0) \\ &+ u_z \cos \theta_e' \sin \phi \, (z - z_0) + v_0' \cos \theta_e' \cos \phi \\ &+ v_y \cos \theta_e' \cos \phi \, (r \cos \theta_e' \cos \phi - y_0) + v_z \cos \theta_e' \cos \phi \, (z - z_0) \\ &+ (u_y + v_x) \cos \theta_e' \left[r \cos \theta_e' \sin \phi \cos \phi - \tfrac{1}{2}(x_0 \cos \phi + y_0 \sin \phi)\right] \\ &+ w_0 \sin \theta_e' + w_x \sin \theta_e' \, (r \cos \theta_e' \sin \phi - x_0) \\ &+ w_y \sin \theta_e' \, (r \cos \theta_e' \cos \phi - y_0) + w_z \sin \theta_e' \, (z - z_0), \end{aligned} \quad (9.11)$$

where $u_0' \equiv [u_0 + \tfrac{1}{2}y_0(v_x - u_y)]$, $v_0' \equiv [v_0 - \tfrac{1}{2}x_0(v_x - u_y)]$ are termed the *modified wind components* and the subscripts x, y, z denote partial derivatives. Because each of the 11 unknowns $u_0', v_0', u_y + v_x, u_x, v_y, u_z, v_z, w_0, w_x, w_y, w_z$ is multiplied by a unique function of r, θ_e, ϕ, in principle we can discriminate some of these kinematic properties of the linear wind through the different

9.3 Linear Wind Measurements with a Single Doppler Radar

dependencies they predict on the measured radial component $\hat{v}_r$ as **r** scans the analysis volume. Because the linear wind field is described by 12 parameters $(u_0, u_x, u_y, u_z, v_0, v_x, v_y, v_z, w_0, w_x, w_y, w_z)$, four of which appear in combined forms multiplied by unique trigonometric functions, additional assumptions must be made in order to determine all 12 components. For example, one difficulty is that vorticity $v_x - u_y$ of horizontal wind cannot be discriminated from u_0, v_0 because they share a common trigonometric dependency. If x_0, y_0 were zero, then u_0 and v_0 would be uncoupled from vorticity and hence can be retrieved, but this would restrict the analysis to linear wind in an analysis volume with its center at the radar site. For many situations it is preferable not to choose such an x_0, y_0, because the linearity approximation may be better satisfied at points away from the radar, and furthermore other important kinematic properties (e.g., the horizontal divergence, $u_x + v_y$) can be mapped if the location of the analysis volume is not restricted. Unless u_0 or v_0 are known, the vorticity cannot be estimated.

To facilitate manipulation of (9.11), we introduce matrix notation. Define the vector **K** by

$$\mathbf{K}^T = (u_0', u_x, u_z, v_0', v_y, v_z, u_y + v_x, w_0, w_x, w_y, w_z), \qquad (9.12)$$

where the T indicates the transpose. For later derivations we shall find it useful to define the vector $\mathbf{K}_m$ containing a subset of m elements of **K**. For example, in the VAD analysis scheme only the elements $u_0, u_x, v_0, v_y, u_y + v_x, w_0, w_x$ are required (i.e., $m = 8$) to specify $\hat{v}_r$ on the measurement circle if the wind is linear. Let the predictor vector $\mathbf{P}_i$ contain the following functions of $\mathbf{r}_i$, associated with the ith observation:

$$\mathbf{P}_i^T = \begin{bmatrix} p_{i,1} \\ p_{i,2} \\ p_{i,3} \\ p_{i,4} \\ p_{i,5} \\ p_{i,6} \\ p_{i,7} \\ p_{i,8} \\ p_{i,9} \\ p_{i,10} \\ p_{i,11} \end{bmatrix} = \begin{bmatrix} \cos\theta_e' \sin\phi \\ \cos\theta_e' \sin\phi \, (r\cos\theta_e' \sin\phi - x_0) \\ \cos\theta_e' \sin\phi \, (z - z_0) \\ \cos\theta_e' \cos\phi \\ \cos\theta_e' \cos\phi \, [r\cos\theta_e' \cos\phi - y_0] \\ \cos\theta_e' \cos\phi \, (z - z_0) \\ \cos\theta_e' [r\cos\theta_e' \sin\phi \cos\phi - \frac{1}{2}(x_0 \cos\phi + y_0 \sin\phi)] \\ \sin\theta_e' \\ \sin\theta_e' (r\cos\theta_e' \sin\phi - x_0) \\ \sin\theta_e' (r\cos\theta_e' \cos\phi - y_0) \\ \sin\theta_e' (z - z_0) \end{bmatrix}. \qquad (9.13)$$

With this notation we can express the true radial velocity of air as $v_r = \mathbf{P} \cdot \mathbf{K}$. $\mathbf{P}_{im}$ will denote a subset of the predictor functions of $\mathbf{r}_i$ corresponding to $\mathbf{K}_m$.

The estimates $\hat{v}_r$ of the reflectivity-weighted mean radial velocity will contain some error $\varepsilon(\hat{v}_r)$. In addition, if the measurement is made in precipitation, it will contain a component $\bar{w}_t \sin \theta'_e$ due to the terminal fall speeds whose estimate will also contain error, or $\varepsilon(\bar{w}_t)$. If (9.2) is used to estimate $\bar{w}_t$, then $\bar{Z}$ estimate errors will contribute to $\varepsilon(\bar{w}_t)$. Doviak et al. (1976) show that the $\bar{Z}$ estimate variance produces much smaller error in $\hat{w}_t$ than that caused by uncertainties in (9.2). Unless an accurate measurement of the drop size distribution can be made, we should assume that $\varepsilon(\bar{w}_t) = 1 \text{ m s}^{-1}$. Finally, the wind field is very likely to deviate from linearity; let this error be denoted $\varepsilon(\delta \bar{v})$. Thus, the relationship between the estimated radial velocity and the true radial component of the wind at the ith measurement point becomes

$$\hat{v}_{ri} + \hat{w}_{ti} \sin \theta'_{ei} = \mathbf{P}_i \mathbf{K} + \varepsilon_i, \tag{9.14}$$

where $\varepsilon_i = \varepsilon(\hat{v}_r) - \varepsilon(\bar{w}_t) \sin \theta'_e + \varepsilon(\delta v)$ is the combined error.

Equation (9.14) can be used to find linear wind parameters from an estimate of $\mathbf{K}$. There are as many as 11 quantities in $\mathbf{K}$ to be estimated. The number of radial velocity measurements n will generally be at least several hundred. The most common technique for fitting such an overdetermined set of equations is least squares. In Section 9.3.1 we shall present a matrix equation for least squares estimates of $\mathbf{K}$ and study the bias and variance of these estimates.

9.3.1 Least Squares Fitting of the Wind Field

The fitting equations are concisely represented with the matrix notation. The n observed radial velocities are collected into a vector $\hat{\mathbf{V}}_n^T = (\hat{v}_{r1}, \hat{v}_{r2}, \ldots, \hat{v}_{rn})$; the associated predictor functions are collected into an $n \times m$ matrix $\mathbf{P}_{nm}^T = (\mathbf{P}_{1m}^T, \mathbf{P}_{2m}^T, \ldots, \mathbf{P}_{nm}^T, \ldots)$, where $\mathbf{P}_{im}$ represents (9.13) for m of the predictors. As shown in Draper and Smith (1966, Ch. 2), least squares estimates of $\mathbf{K}_m$, denoted $\hat{\mathbf{K}}_m$, are computed from

$$\hat{\mathbf{K}}_m = (\mathbf{P}_{nm}^T \mathbf{P}_{nm})^{-1} (\mathbf{P}_{nm}^T \mathbf{V}_n). \tag{9.15}$$

We would like expressions for the bias and variance of the estimates $\hat{\mathbf{K}}_m$. $\hat{\mathbf{K}}_m$ is an unbiased estimate, i.e., $\mathrm{E}(\hat{\mathbf{K}}_m) = \mathbf{K}_m$, if (9.14) is an adequate model [i.e., it contains all the data attributes of the radial velocities; (Draper and Smith, 1966, p. 59]. Two reasons why (9.14) would not be an adequate model are now considered. First, if the wind field is nonlinear, the radial velocities will show variations associated with terms of higher order than in (9.7). If examination of the residuals, that is,

$$\hat{\varepsilon}_i = \hat{v}_{ri} + \hat{w}_{ti} \sin \theta'_{ei} - \mathbf{P}_{im} \cdot \hat{\mathbf{K}}_m, \tag{9.16}$$

9.3 Linear Wind Measurements with a Single Doppler Radar

shows large scale features within the analysis volume, then the linearity assumption is not valid and the estimates $\hat{\mathbf{K}}_m$ may be biased.

The second reason is relevant when the wind field is linear but m is chosen to be less than 11 [i.e., certain terms of (9.11) are ignored], so that all of the variations of v_r vs $\mathbf{r}$ are not modeled. This situation is considered because, as discussed later, for some practical cases $m < 11$ reduces the variance of the estimates. With this in mind, the bias is

$$E(\hat{\mathbf{K}}_m - \mathbf{K}_m) = (\mathbf{P}_{nm}^T \mathbf{P}_{nm})^{-1}(\mathbf{P}_{nm}^T \mathbf{P}_{nl})\mathbf{K}_l^T, \qquad (9.17)$$

where $\mathbf{K}_l$ contains the remaining $l = 11 - m$ parameters and $\mathbf{P}_{nl}$ the associated predictors. Evaluation of (9.17) is difficult in practice because it requires an estimate $\mathbf{K}_l$. However, one can assign an upper value to the elements of $\mathbf{K}_l$ in order to estimate the maximum bias.

Even when $\hat{\mathbf{K}}_m$ is unbiased, it will contain random errors, termed the variance errors. Assuming an adequate model, the covariance matrix of $\hat{\mathbf{K}}_m$ is

$$\mathbf{C}_{mm} = E[(\hat{\mathbf{K}}_m - \mathbf{K}_m)(\hat{\mathbf{K}}_m - \mathbf{K}_m)^T] = (\mathbf{P}_{nm}^T \mathbf{P}_{nm})^{-1}\sigma_\varepsilon^2, \qquad (9.18)$$

where σ_ε^2 is the variance of the measurement error. In practice, σ_ε^2 is estimated by the residual variance $s^2 = \hat{\mathbf{E}}_n^T \hat{\mathbf{E}}_n/(n-m)$, where $\hat{\mathbf{E}}_n^T = (\hat{\varepsilon}_1, \hat{\varepsilon}_2, \ldots, \hat{\varepsilon}_n)$.

The $\mathbf{P}_{nm}^T \mathbf{P}_{nm}$ matrix appearing in these equations is the covariance matrix of the predictors. Since the predictors contain sine–cosine products, the off-diagonal terms can be large compared to the diagonal terms, especially for small analysis volumes. Because the variance of the estimates $\hat{\mathbf{K}}_m$ are proportional to the diagonal elements of the inverse matrix, the accuracy of $\hat{\mathbf{K}}_m$ improves with the size of the analysis volumes, as is intuitively expected. Equation (9.18) allows a quantitative measure of the variances. Thus $(\mathbf{P}_{nm}^T \mathbf{P}_{nm})^{-1}$, which depends on the size and shape of the analysis volume and distribution of data therein, determines how well the chosen analysis predicts the components of a linear wind field.

We now have sufficient theoretical development to allow quantitative evaluation of the errors in the estimates of the linear wind components.

9.3.2 Analysis on a Circular Arc

Because of its simplicity, we shall consider a uniform horizontal wind in some detail. For this case, $m = 2$, $\mathbf{K}_m^T = (u_0, v_0)$, and we assume the remaining parameters are negligible. The n radial velocity estimates are made at the same elevation angle along an arc of length $\Delta\phi$ centered at the azimuth ϕ_0, so the covariance matrix becomes

$$\mathbf{P}_{n2}^T \mathbf{P}_{n2} = \cos^2\theta_e' \begin{bmatrix} a & b \\ b & c \end{bmatrix},$$

where

$$a = \sum_{i=1}^{n} \sin^2 \phi_i \approx \frac{n}{\Delta\phi} \int_{\phi_0-\Delta\phi/2}^{\phi_0+\Delta\phi/2} \sin^2 \phi \, d\phi$$

$$= (n/2\,\Delta\phi)(\Delta\phi - \cos 2\phi_0 \sin \Delta\phi),$$

$$c \approx (n/2\,\Delta\phi)(\Delta\phi + \cos 2\phi_0 \sin \Delta\phi),$$

$$b \approx (n/2\,\Delta\phi) \sin 2\phi_0 \sin \Delta\phi.$$

The determinant

$$\det[(\mathbf{P}_{n2}^T \mathbf{P}_{n2})] = (\cos^4 \theta_e' \, n^2/4\,\Delta\phi^2)(\Delta\phi^2 - \sin^2 \Delta\phi)$$

is independent of the center azimuth ϕ_0. The trace of the covariance matrix (9.18),

$$\text{tr}(\mathbf{C}_{22}) = (a + c)\sigma_\varepsilon^2 \cos^2 \theta_e'/\det(\mathbf{C}_{22}) = n\sigma_\varepsilon^2 \cos^2 \theta_e'/\det(\mathbf{C}_{22})$$

is proportional to the sum of the variances of u_0 and v_0 and provides a measure of the accuracy of the uniform wind analysis. Simplifying somewhat,

$$\text{tr}(\mathbf{C}_{22}) = \frac{n}{(\cos^2 \theta_e' \, n^2/4\,\Delta\phi^2)(\Delta\phi^2 - \sin^2 \Delta\phi)} \approx \frac{12 \sec^2 \theta_e'}{n\,\Delta\phi^2},$$

where $\sin \Delta\phi$ has been approximated by $\Delta\phi - \frac{1}{6}\Delta\phi^3$.

It can be seen from this result that the most accurate uniform horizontal wind analysis is for low elevation angles (where $\sec^2 \theta_e' \approx 1$). The accuracy depends inversely on the number of data points n and on the square of the arc length (i.e., azimuthal width). Thus the variance of the estimates of the horizontal wind increases rapidly as $\Delta\phi$ decreases (below 1 rad).

The more complicated and complete models (i.e., $m > 2$) have similar (square) increases in the variance of estimates of linear wind components with decreasing analysis volume. However, the algebra becomes more complicated, and a numerical evaluation of the inverse matrix by a computer is necessary.

9.3.3 Analysis on a Complete Circle

The radial component of wind along a circle of measurement (Fig. 9.8) has the ϕ dependence

$$v_r = w \sin \theta_e + v_h \cos \theta_e \cos(\delta - \phi), \tag{9.19}$$

where $D = \delta + \pi$ is the direction from which the wind blows, and w and v_h respectively are the wind's vertical and horizontal components. On expanding the second term and using the relation between Cartesian and polar wind

9.3 Linear Wind Measurements with a Single Doppler Radar

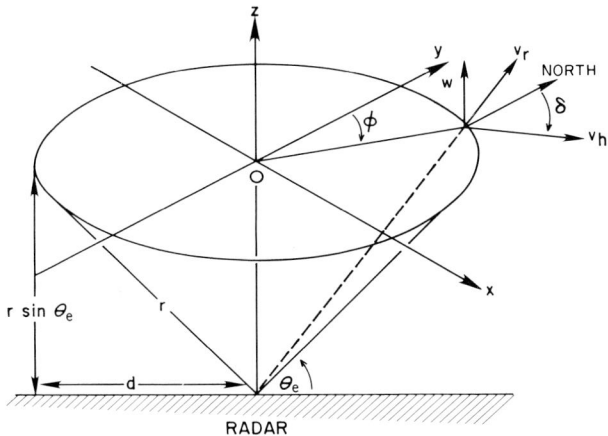

Fig. 9.8. Geometry for the scan of velocities on a circle.

components, (9.19) is reformulated as

$$v_r = w \sin \theta_e + u \sin \phi \cos \theta_e + v \cos \phi \cos \theta_e. \tag{9.20}$$

Now let us assume the wind field is nearly linear within the measurement circle so that the components u, v, and w are well approximated by the zeroth- and first-order terms of a Taylor series about the radar location, as in (9.7). Because the circle of measurement lies on a horizontal plane, we need only consider the x and y expansions.

Combining (9.7) and (9.20) and expressing x, y in terms of the radar coordinates, one obtains

$$v_r = w_0 \sin \theta_e + \frac{1}{2}\left(\frac{\partial u}{\partial x} + \frac{\partial v}{\partial y}\right) r \cos^2 \theta_e$$

$$+ \left(\frac{\partial w}{\partial x} r \sin \theta_e + u_0\right) \cos \theta_e \sin \phi + \left(\frac{\partial w}{\partial y} r \sin \theta_e + v_0\right) \cos \theta_e \cos \phi$$

$$+ \frac{1}{2}\left(\frac{\partial v}{\partial x} + \frac{\partial u}{\partial y}\right) r \cos^2 \theta_e \sin 2\phi + \frac{1}{2}\left(\frac{\partial v}{\partial y} - \frac{\partial u}{\partial x}\right) r \cos^2 \theta_e \cos 2\phi. \tag{9.21}$$

Examination of (9.21) reveals that a linear wind field contributes only to the first three components of a Fourier expansion of v_r over the fundamental period $0 \le \phi \le 2\pi$. The measured Doppler radar mean velocity $\hat{v}_r$ contains errors and a bias due to the particle terminal velocity. From (9.14), the unbiased estimate $\hat{v}_r$ (for $m = 8$) is

$$\hat{v}_r = v_r - [(\hat{w}_t + \varepsilon(\overline{w}_t)] \sin \theta_e + \varepsilon(\hat{v}_r) + \varepsilon(\delta \overline{v}). \tag{9.22}$$

A series of n uniformly spaced $\hat{v}_r$ and $\hat{w}_t$ data along the circle of measurement is, in general, expressible in terms of the discrete Fourier transform (Chapter 5),

$$\hat{v}_r(\phi) + \hat{w}_t(\phi)\sin\theta_e = \sum_{k=0}^{n-1} C_k e^{jk\phi}, \qquad (9.23)$$

where the C_k are complex Fourier coefficients. Note that spectra coefficients symetrically placed about $n/2$ are complex conjugates (Section 5.1), so their sums result in real harmonics at wavenumbers less than $n/2$. For noiseless linear wind, only the first three coefficients differ from zero. If we use the properties, introduced in Chapter 5, that relate the complex amplitude of the various spectral coefficients, and assume the terms $\partial w/\partial x$, $\partial w/\partial y$ are small (their contribution is 0 at $\theta_e = 0$) compared to others in (9.21) and that the errors are negligible, comparison of (9.23) and (9.21) reveals the following:

horizontal divergence:

$$\operatorname{div} v_h \equiv \left(\frac{\partial u}{\partial x} + \frac{\partial v}{\partial y}\right) = \frac{2}{r\cos^2\theta_e}(C_0 - w_0\sin\theta_e); \qquad (9.24a)$$

horizontal wind speed:

$$v_h = 2|C_1|/\cos\theta_e; \qquad (9.24b)$$

horizontal wind direction:

$$D = \arg C_1 - \pi; \qquad (9.24c)$$

deformations:

$$\frac{\partial u}{\partial x} - \frac{\partial v}{\partial y} = -\frac{4}{r\cos^2\theta_e}\operatorname{Re}(C_2), \qquad (9.24d)$$

$$\frac{\partial u}{\partial y} + \frac{\partial v}{\partial x} = \frac{4}{r\cos^2\theta_e}\operatorname{Im}(C_2). \qquad (9.24e)$$

The divergence, the wind speed, and the sum of the squares of each deformation are invariant with respect to rotation of the x, y axes about z and hence are fundamental properties of the linear wind. Fourier analysis of $\hat{v}_r + \hat{w}_t\sin\theta_e$ directly determines the wind speed and direction and the deformation. Other important properties of motion, such as the vorticity, divergence, and vertical velocity, cannot be directly determined using the VAD method. However, by inserting the mass continuity equation into (9.24a) and assuming stationarity of the mass density (i.e., $\partial\rho/\partial t = 0$),

$$\frac{\partial}{\partial z}(\rho w_0) = -\frac{2\rho}{r\cos^2\theta_e}(C_0 - w_0\sin\theta_e), \qquad (9.25)$$

9.3 Linear Wind Measurements with a Single Doppler Radar

we can solve for vertical wind if we have an estimate of w_0 at some height $z = z_1$.

Equation (9.25) can be solved analytically:

$$\rho w_0 = \exp\left(-\int_{z_1}^{z} P\, dz\right) \int_{z_1}^{z} Q(z) \exp\left(\int_{z_1}^{z} P\, dz\right) dz$$

$$+ \rho w_0(z_1) \exp\left(-\int_{z_1}^{z} P\, dz\right), \tag{9.26}$$

where

$$P(z) = -2 \sin \theta_e / r \cos^2 \theta_e \quad \text{and} \quad Q(z) = 2\rho C_0 / r \cos^2 \theta_e.$$

Usually one has C_0 values from near the surface to levels through the convective boundary layer (1–2 km) in clear air (Section 11.6) and much higher in showery weather, so $w_0(z_1) = 0$, where z_1 is the height above sea level of the ground. Thus a single Doppler radar can measure the vertical profiles of the three components of wind when it is linear over a sampling circle of radius d (Fig. 9.8).

As an example, the velocity trace obtained from the azimuthally scanning

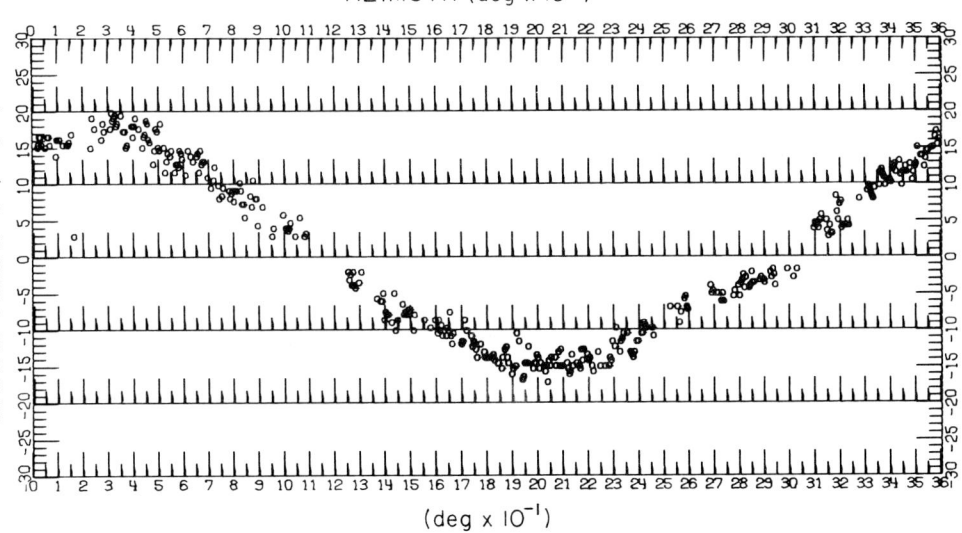

Fig. 9.9. Radial velocity versus azimuth of Cimarron radar, 14:11–14:14 C.S.T., 27 April 1977, for a range of 40 km. The elevation is 0.5°.

Cimarron radar (located 40 km northwest of Norman, Oklahoma) is presented in Fig. 9.9. The mean velocities were calculated from the pulse pair algorithm (Chapter 6) acting on samples of echoes from clear air refractivity fluctuations. (For relations between reflectivity and the refractive index fluctuations, see Chapter 11.) It is visually evident that there is no significant power other than in the zeroth and first harmonic. The large data gaps are the results of editing uncertain radial velocities because of low signal-to-noise ratios. Substantial scatter of the accepted velocities exists even though velocities from three consecutive range gates were averaged to give an effective range interval of 450 m. Some of this scatter is attributed to weak echoes. Gaps occur, for example, at 20° and 110°, where tall trees interfere with the low-elevation radar beam. Because the data are spaced nonuniformly, the Fourier coefficients were obtained by least squares fitting the zeroth harmonic and the fundamental to the data (Rabin and Zrnić, 1980).

Because the method is most useful for determining the magnitude and direction of wind at scales large compared to the diameter of the scan, it is interesting to compare the results obtained from two closely spaced radars. In this comparison, the radius of the circle used for the velocity data was changed in steps of 0.3 km, and the elevation angle θ was fixed at 0.5°. Hence the following discussion will refer to curves of the wind speed and direction (Fig. 9.10) and divergence (Fig. 9.11) plotted against the radius and area of the azimuthally scanned circle, for the Cimarron and Norman radars spaced 40 km part.

The apparent decrease in wind speed at short range is due to ground clutter, which biases the pulse pair velocity estimate toward zero. The slight difference in observed wind speeds between the Norman and Cimarron radar sites (Fig. 9.10) indicates a wind speed gradient of about $1 \text{ m s}^{-1}/40 \text{ km}$ to the northwest. (The difference in wind speeds was even smaller at an earlier time.) This is of the same order as the synoptic-scale gradient ($0.6 \text{ m s}^{-1}/40 \text{ km}$) observed at that time from a surface network of anemometers. The difference in wind direction between Norman and Cimarron is about 5° (Fig. 9.10) owing to a curved flow.

The horizontal divergence as a function of range for both radars (Fig. 9.11) shows excellent consistency between 40 and 65 km. At shorter ranges ground clutter and the noncommon area are responsible for the differences.

The gradual decrease of the divergence for areas $\geq 5 \times 10^3 \text{ km}^2$ may be due to an increase in height of the sampling circle rather than an increase in sampling area. (This conclusion is based on the divergence profile obtained from the detailed dual radar analysis for a smaller area contained within the circle of measurement.) The positive divergence measured on 27 April 1977 is consistent with the anticyclonic geostrophic wind flow and subsidence over central Oklahoma.

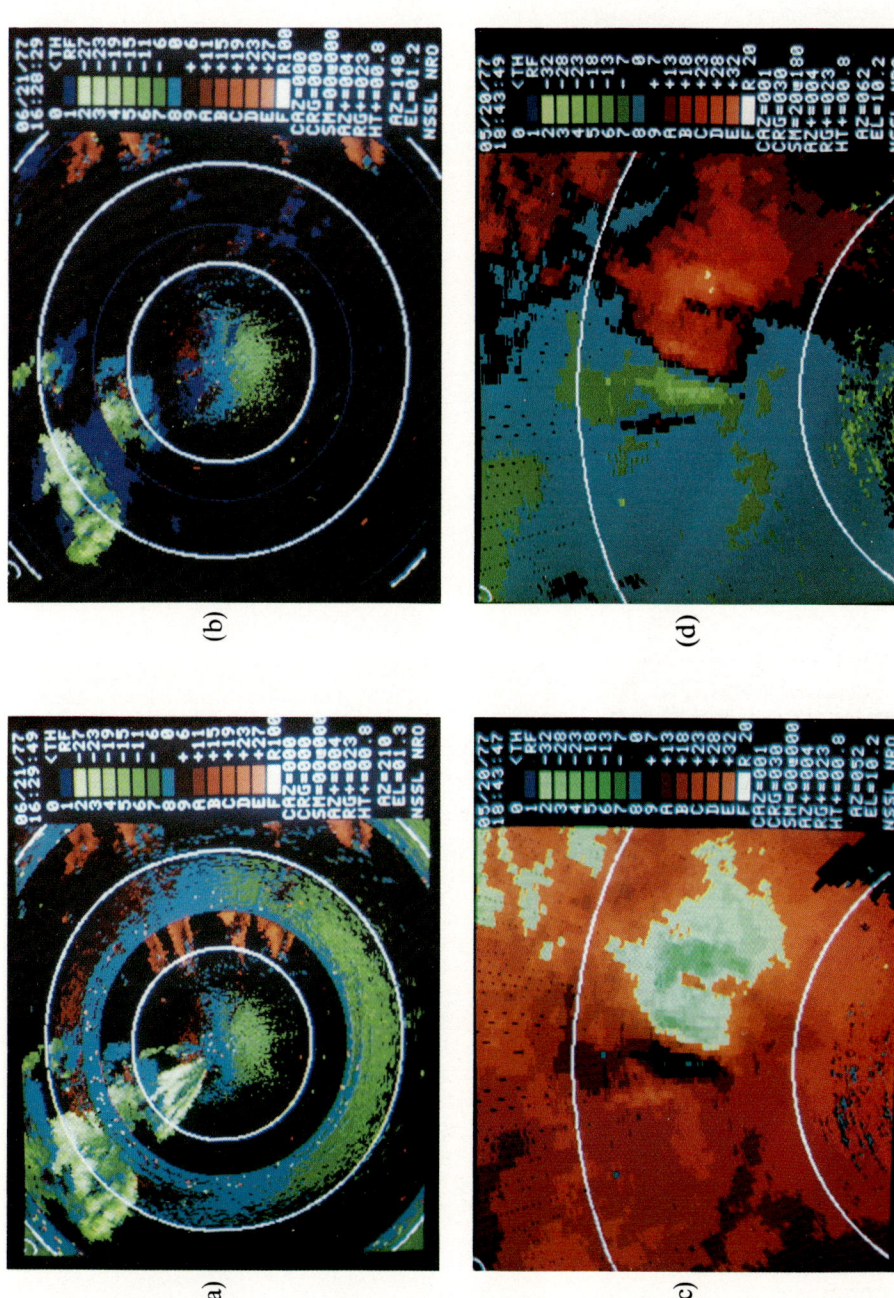

Color Plate 1. (a) Storm echoes without the range dealiasing algorithms and (b) with range dealiasing adjustment. (c) The Doppler velocity field without storm motion removed and (d) with it removed using algorithm (7.9). Notice in (c) the large region of velocity aliasing, whereas in (d) the velocity aliasing is practically eliminated.

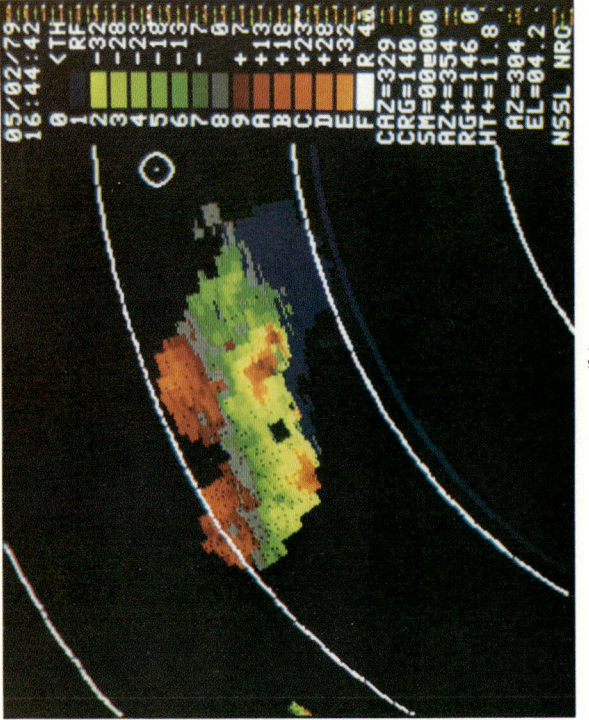

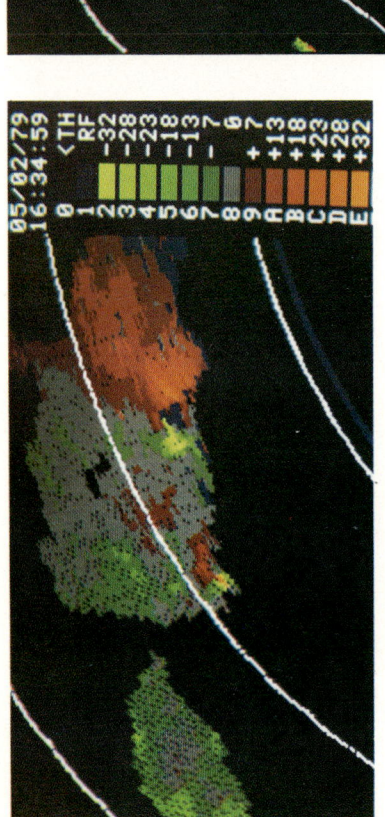

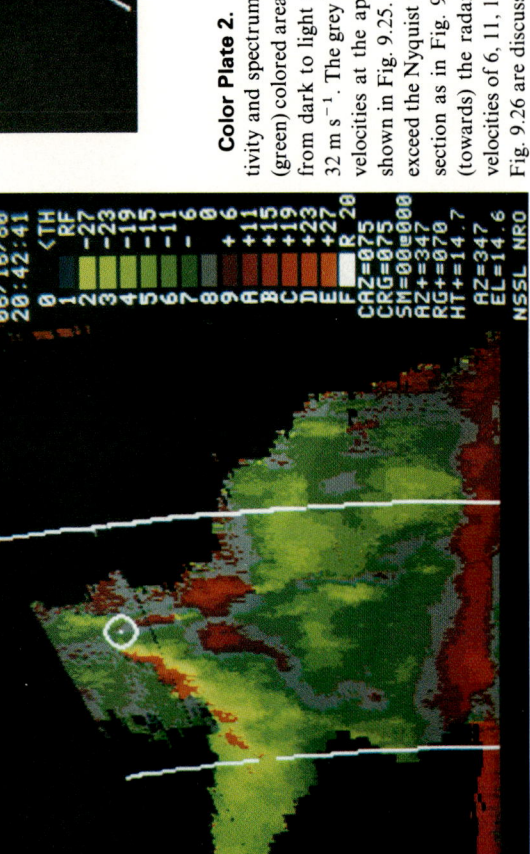

Color Plate 2. (a) PPI display of Doppler velocities for the storm whose reflectivity and spectrum width fields are shown in Fig. 9.24. The elevation is 0.8°. Red (green) colored areas are radial velocities away from (towards) the radar. The shades from dark to light correspond to median radial velocities of 7, 13, 18, 23, 28, and 32 m s^{-1}. The grey hue is for a median value of 0 m s^{-1}. (b) PPI display of Doppler velocities at the approximate height at which reflectivity and spectrum width are shown in Fig. 9.25. Strong divergence at the storm top is evident as radial velocities exceed the Nyquist values of ± 34 m s^{-1}. (c) Radial velocity field in the same vertical section as in Fig. 9.26. Red (green) colored areas are radial velocities away from (towards) the radar. The shades from dark to light correspond to median radial velocities of 6, 11, 15, 19, 23, and 27 m s^{-1}. Differences in storm outline from that in Fig. 9.26 are discussed in the text.

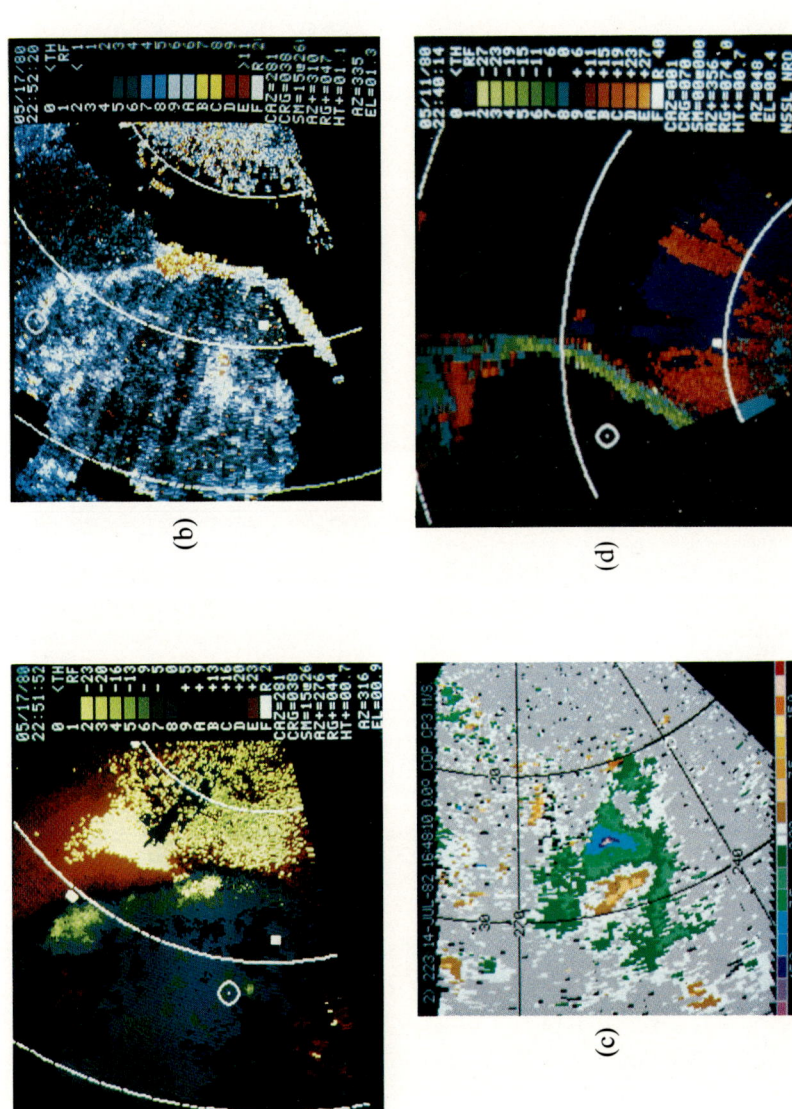

Color Plate 3. (a) Doppler velocity field of a gust front with several downdrafts behind it. Velocity scale is in m s^{-1} and a 20 m s^{-1} contribution of mean motion from 260° has been subtracted from each radial. Range marks are 20 km apart. (b) Doppler spectrum width field of the gust front. Width scale is in m s^{-1} and data with signal-to-noise ratios above 20 dB are displayed. (c) Doppler velocity field of a downburst. Clearly visible is the symmetrical signature of the divergent outflow. The elevation angle is 0°, range marks are at 20 and 30 km, and the velocity scale on the bottom is in m s^{-1}. (Courtesy of J. Wilson, National Center for Atmospheric Research, Boulder, Colorado.) (d) Doppler velocity field of a solitary gust at the same time as the reflectivity field in Fig. 9.34. Arcs are at constant range of 40 and 80 km. The dark blue color depicts areas where velocities are contaminated by multiple-trip echoes (see Chapter 7). The elevation is 0.4°.

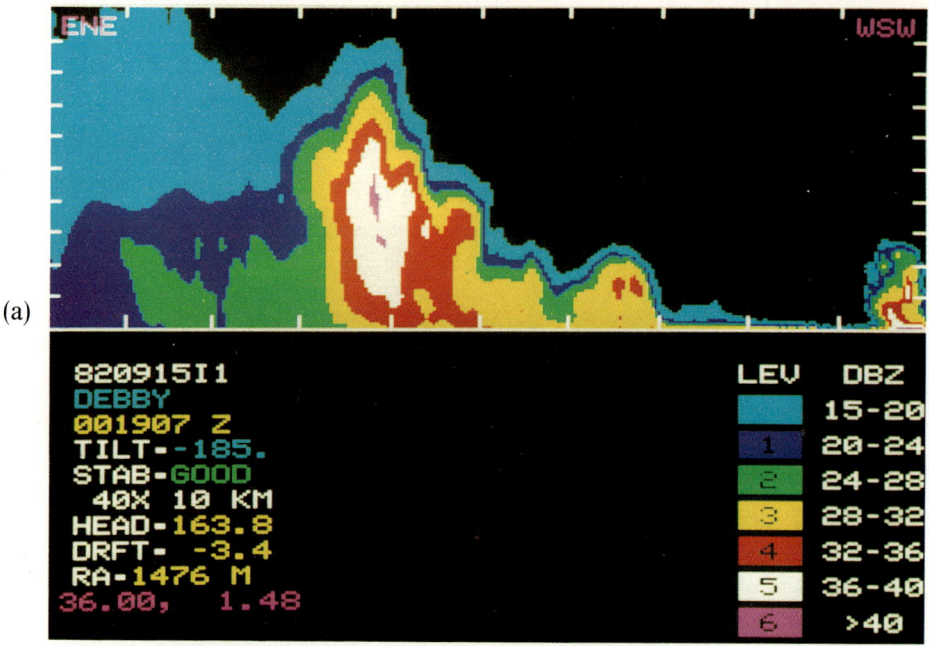

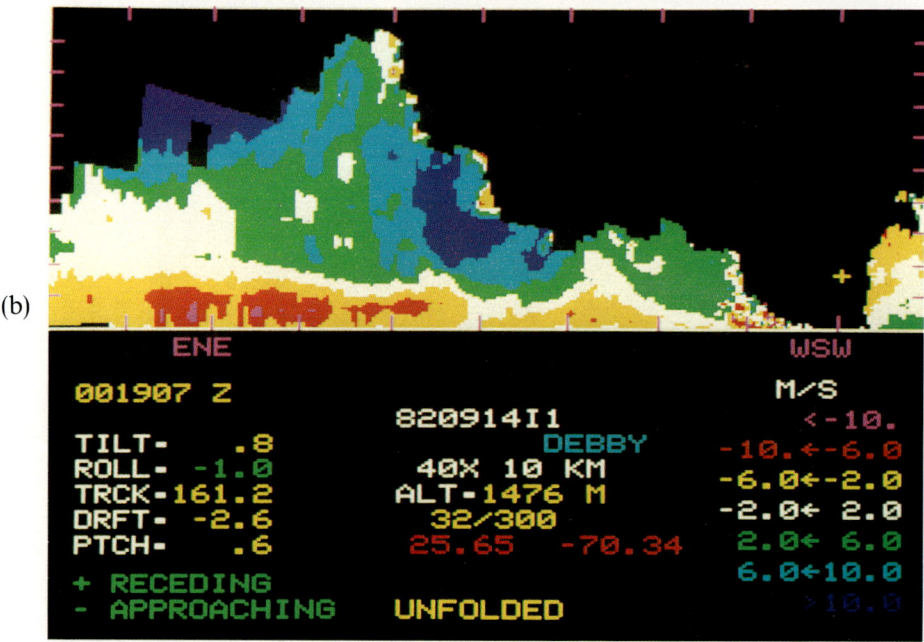

Color Plate 4. (a) Reflectivity in the cross section A–A indicated in Fig. 9.41. These data were obtained with a radar whose wavelength was 3.2 cm and which was located in the tail section of the aircraft. The antenna beamwidths are 1.9° vertical and 1.4° horizontal. Tick marks are 4 km apart along the horizontal and 1 km apart along the vertical. In this presentation the left side is east–northeast from the aircraft. (b) Doppler velocity field in cross section A–A. Yellow-red colors are winds toward the aircraft located at the cross (+). (Courtesy of D. Jorgensen, National Hurricane Center, Coral Gables, Florida.)

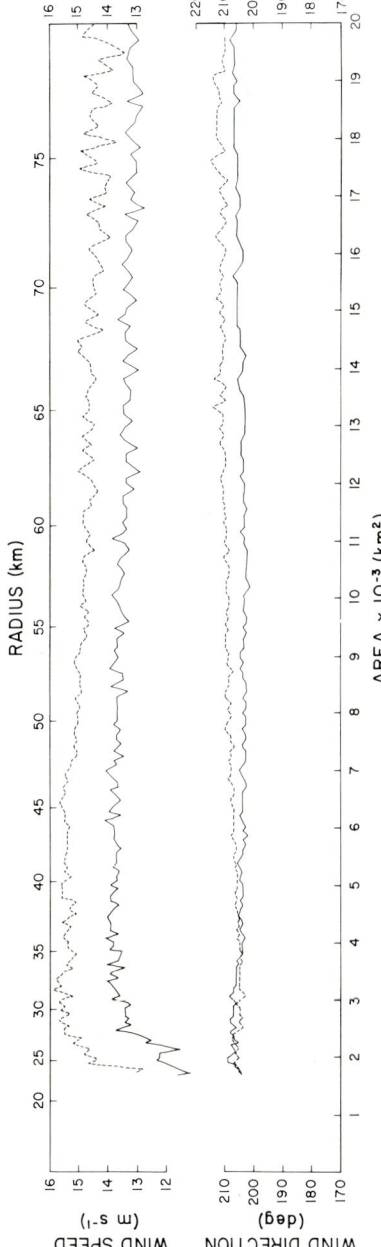

Fig. 9.10. Wind direction and speed obtained from the analysis of the azimuthally scanned velocities for data from two radar separated 40 km (27 April 1977); ———, Norman; ---, Cimarron.

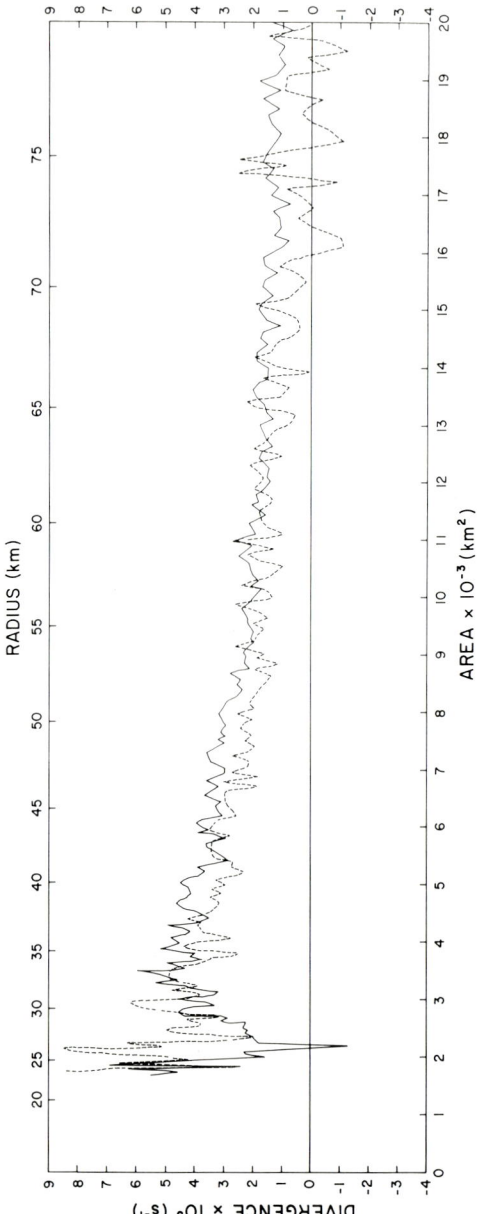

Fig. 9.11. Divergence over the circular areas above the Norman (———) and Cimarron (---) radars at 15:25–15:29 C.S.T. on 27 April 1977.

9.3.4 Analysis on Sections of a Conical Surface

If r and ϕ are varied but the elevation angle θ_e is held constant, then all 11 predictors are unique. In principle, then, the 11 kinematic parameters can be estimated. However, because of the linearity approximation, we usually want an analysis volume less than 360° in azimuth. For small analysis volumes, predictors containing such terms as $r \sin \phi$ or $r \sin^2 \phi$ are highly correlated. As discussed in Section 9.3.1, this leads to large variances for the estimates. To reduce the estimator variance, some of highly correlated predictors must be excluded. However, the parameters associated with the excluded predictors will bias the estimated ones, as described by (9.17), so one should retain those predictors of kinematic parameters having large magnitude. Since vertical gradients are generally much larger than horizontal ones, a more suitable set of six predictors is obtained by excluding P_2, P_5, P_7, P_9, and P_{10}:

$$\mathbf{K}_6^T = (u_0', u_z, v_0', v_z, w_0, w_z).$$

Easterbrook (1975) has analyzed data on sectors of a conical surface to extract, among other components, the horizontal gradients of u and v. However, Koscielny et al. (1982) find that if these gradients are to be used to estimate divergence over narrow sectors ($<45° \times 20$ km) in prestorm boundary layers, the analysis suffers either from large bias errors because vertical shear has been neglected or from large variance because of the high correlation between predictors. There is keen interest in mapping horizontal divergence so we need to expand $\mathbf{K}^T$ to include the horizontal gradients of u and v. In Section 9.3.5 we make such an expansion, but this requires an analysis within a volume in order not to ignore important vertical shear components.

9.3.5 Analysis within a Volume: The VVP Method

With a volume of data, all 11 parameters can be discriminated. Because of the linearity approximation, we want to keep the volume as small as possible. We are free to select both m and the size and shape of the analysis volume to control the bias and variance errors.

One interesting application of linear wind field analysis is estimation of the low-altitude divergence. If only the low altitudes are of interest, we can neglect the predictors for the vertical velocity terms (which contain $\sin \theta_e$), but not the vertical shear which for many meteorological conditions will be much larger than the horizontal gradients. If the analysis volume includes at least two elevation angles, then the vertical and horizontal gradients can be discriminated.

The bias caused by the excluded vertical velocity term depends on the analysis volume. Before (9.17) can be applied, we must specify the analysis volume; for convenience we shall choose one 30° wide in azimuth, 30 km in range, and containing the elevation angles 0.4° and 0.8°. In addition, we assume that the data are spaced 1° in azimuth and 450 m in range. Evaluation of (9.17) gives

$$E \begin{bmatrix} \hat{u}'_0 - u'_0 \\ \hat{u}_x - u_x \\ \hat{u}_z - u_z \\ \hat{v}'_0 - v'_0 \\ \hat{v}_y - v_y \\ \hat{v}_z - v_z \\ (\hat{u}_y + \hat{v}_x) - (u_y + v_x) \end{bmatrix} = \begin{bmatrix} 0.004 & 532.0 & 141.0 & 0.162 \\ 4 \times 10^{-8} & 0.0015 & 0.0005 & 1 \times 10^{-5} \\ 6 \times 10^{-6} & 0.935 & -0.246 & 0.0043 \\ 0.015 & -141.0 & 44.30 & 0.603 \\ -2 \times 10^{-7} & 0.0025 & 0.0142 & 2 \times 10^{-7} \\ 2 \times 10^{-5} & -0.2460 & 0.0841 & 0.0160 \\ -2 \times 10^{-7} & 0.0064 & 0.0064 & -6 \times 10^{-6} \end{bmatrix} \times \begin{bmatrix} w_0 \\ w_x \\ w_y \\ w_z \end{bmatrix}.$$

Although there is substantial bias in some of the parameters, u_x and v_y are significantly less biased ($< 10^{-5}$ s^{-1}) if $w_0 < 10^2$ m s^{-1}, $w_x, w_y < 10^{-3}$ s^{-1}, and $w_z < 1$ s^{-1}. Evaluating (9.18) for this analysis volume gives the variance of the estimates. For the divergence, $u_x + v_y$, the variance is given by $c_{22} + c_{55} + 2c_{25}$, which, with about 4000 data points for $\sigma_\varepsilon = 1$ m s^{-1}, is $(1.6 \times 10^{-5}$ s$^{-1})^2$. The prestorm divergence has a magnitude of 10^{-4} (Ogura and Chen, 1977). Thus if the wind is linear, a VVP analysis over the sector (30 km, 30°, 1°) will allow the estimation of low-level mesoscale divergence with an accuracy on the order of 10^{-5} s^{-1}.

9.3.6 Prestorm Observations

The VVP technique described in Section 9.3.5 was used to analyze the kinematic wind structure in the clear-air convective boundary layer before the onset of thunderstorms. On 19 June 1980 a nearly stationary front was located just southwest of the Norman radar. With the possibility of thunderstorm development near the front, radar data collection began at 15:30 CST.

Figure 9.12 shows an analysis of the wind field at 15:30 CST, as determined from a single Doppler radar at Norman, Oklahoma. This wind analysis requires the assumption that the wind is uniform over each analysis volume, so only two parameters (wind speed and direction) are estimated to produce a least squares fit to the observed radial velocity data. The sky was generally free of clouds; however, there was sufficient backscattering to obtain wind data to a range of 120 km and a height of about 2 km. The arrows indicate the direction and speed of the horizontal wind on a conical surface elevated 0.6° from the radar horizon. The position of the front is clearly evident; winds are easterly to its north, and southerly diffluent flow appears to its south. The three rawinsondes located to the north of the front showed easterly winds

9.3 Linear Wind Measurements with a Single Doppler Radar 275

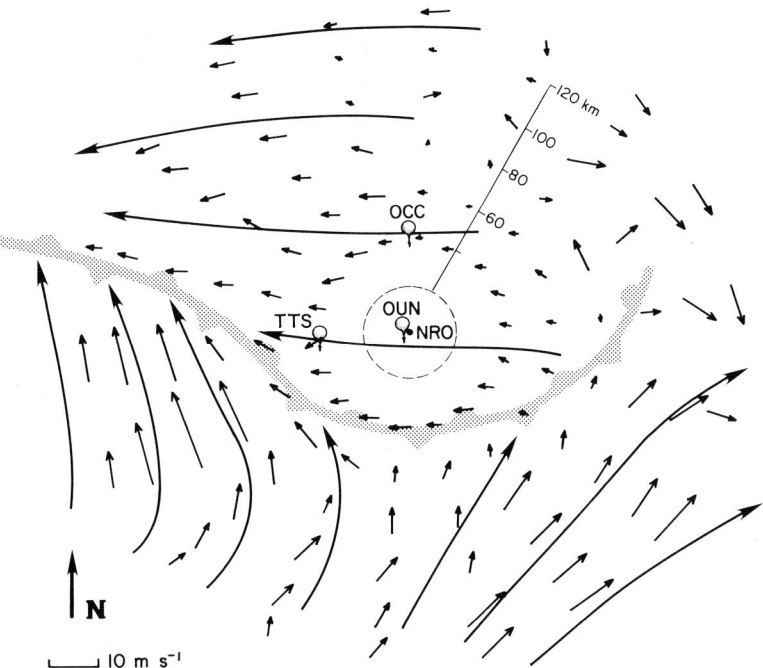

Fig. 9.12. Wind field obtained from a VVP analysis of radar data in a prestorm boundary layer, 19 June 1980, 15:30 C.S.T. (Single Doppler winds; 0.4–0.8°.)

below the frontal surface (0.6–0.9 km AGL) above which the wind had a southerly component, in agreement with the VVP results.

The National Weather Service's surface and 850 mbar data support the wind pattern shown here. However, due to the slope of the front, its position at the earth's surface appears south of where it is intersected by the conical surface scanned by the radar beam (~ 500 m AGL).

Results of the VVP technique with seven parameters from data at 15:30 CST are shown in Fig. 9.13. There is a band of maximum convergence just southwest of the frontal boundary in Fig. 9.12. This convergence zone, as well as the one northeast of the radar, are superimposed on a satellite photograph of clouds 2 hr later (17:30 CST) in Fig. 9.14. The band of cumulus to the southwest of NRO persisted during the 2 hr without any perceptible movement. Satellite photographs showed that the cloud to the northeast developed between 17:00 and 17:30 in the vicinity of the convergent area. The displacement of the cloud line to the northeast of maximum convergence in Fig. 9.14 can be accounted for by the trajectory of the air as it is lifted to the height of the condensation level. An area of divergence appears in Fig. 9.13

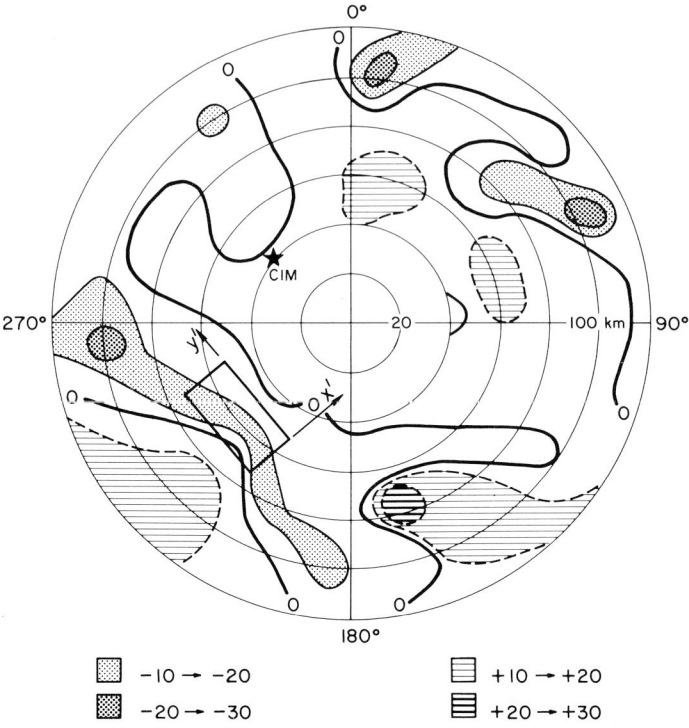

Fig. 9.13. Horizontal divergence $\times 10^5$, 19 June 1980, at 15:30 C.S.T. on a conical surface with $\theta_e = 0.6°$. (From Koscielny *et al.*, 1982.)

extending from northwest to southeast over the radar. This is consistent with the lack of clouds in central Oklahoma as seen from satellite.

An area of deeper convection (ADC; Harrold and Browning, 1971), a region of increased depth of the convective boundary layer, formed from the convergence maximum (Fig. 9.13) to the west of the radar. At 18:00 CST this ADC did not extend above 3 km, reflectivities were less than -3 dBZ, and no outstanding cloud appeared on satellite photographs at this time; 100 min later 19:40 CST, clouds had developed from this ADC to a height of 10 km and attained reflectivities in excess of 25 dBZ.

In order to evaluate the performance of the VVP method with actual data, the estimated divergence field is compared with one obtained 3 min later. Figure 9.15 is a comparison of the divergence for the two times for analysis volumes between 40 and 135 km from the radar. Let us assume that the actual divergence changes less than 10^{-5} s^{-1} in 3 min over each analysis volume. Also, no system bias is expected between the two sample sets, because they

9.3 Linear Wind Measurements with a Single Doppler Radar

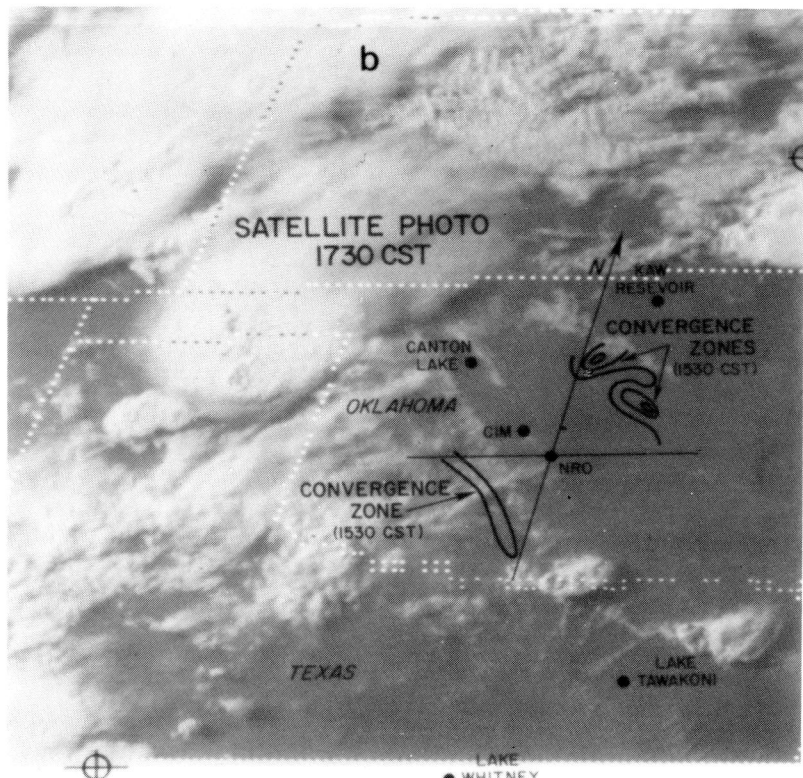

Fig. 9.14. Satellite photograph (17:30 C.S.T.) of the cloud cover superimposed on the convergence patterns measured by Doppler radar.

are obtained with the same radar. With these assumptions the data can be compared with a regression line with slope of 1 passing through the origin to obtain the noise N_D and signal S_D in the divergence estimates:

$$N_D = \frac{1}{2k}\sum_{i=1}^{k}(y_i - x_i)^2,$$

$$S_D = \tfrac{1}{2}(\sigma_x^2 + \sigma_y^2 + \bar{x}^2 + \bar{y}^2) - N_D,$$

where x_i and y_i are estimates at times 1 and 2, respectively, σ_x and σ_y are their standard deviations, $\bar{x}, \bar{y}$ are their means, and k is the number of estimates.
  Using the data of Fig. 9.15 we obtain

$$\sqrt{N_D} = 4.0 \times 10^{-5}\,\text{s}^{-1}, \qquad \sqrt{S_D} = 9.8 \times 10^{-5}\,\text{s}^{-1}.$$

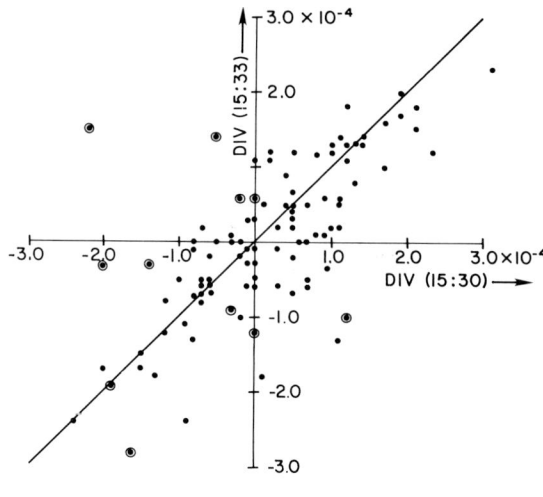

Fig. 9.15. Comparison of the divergence at 15:30 and at 15:33 C.S.T., 19 June 1980. Circled points are for data at ranges larger than 130 km.

The noise in the divergence estimate is about four times larger than that expected from the model as discussed in the preceding section. Some of this increased uncertainty of the divergence estimate may be attributed to the inadequacy of the linear wind assumption, some to unexplained variance in the Doppler velocity estimates. However, the magnitudes of divergence in prestorm environments are sufficiently large to be accurately measured.

9.4 NONLINEAR WIND

We now examine two special cases in which wind need not be linear: average vertical wind estimates and linear wind with superimposed waves.

9.4.1 Vertical Wind

Estimates of the vertical velocity averaged over the measurement circle do not necessarily require the assumption of linear wind. We apply Gauss's theorem,

$$\oint \mathbf{v} \cdot \mathbf{n} \, dS = \int_V \mathbf{\nabla} \cdot \mathbf{v} \, dV, \tag{9.27}$$

9.4 Nonlinear Wind

to the volume V (see Fig. 9.16) enclosed by the area S_1 sampled by the radar and an area S_2 at constant height, over which the vertical velocity is to be averaged. Now assuming no time change in the mass density and further assuming that it is height dependent only, we obtain from the continuity equation

$$\mathbf{V} \cdot \mathbf{v} = -\frac{1}{\rho}\frac{\partial \rho}{\partial z} w,$$

so that (9.27) can be written

$$\iint_{S_1} v_r \, dS_1 + \iint_{S_2} w \, dS_2 = -\int_V \frac{1}{\rho}\frac{\partial \rho}{\partial z} w \, dV. \tag{9.28}$$

The average vertical velocity is defined as

$$\bar{w} = \frac{1}{S_2} \iint_{S_2} w \, dS_2, \tag{9.29}$$

and differentiating (9.28) with respect to z we obtain

$$\frac{dS_2 \bar{w}}{dz} + \frac{1}{\rho}\frac{\partial \rho}{\partial z} S_2 \bar{w} = -\int_0^{2\pi} v_r r \, d\phi. \tag{9.30}$$

Expressing S_2 in terms of r and z and using C_0 for the average of v_r around the circle of measurement we finally obtain

$$\frac{d\bar{w}}{dz} + \left(\frac{1}{\rho}\frac{\partial \rho}{\partial z} - \frac{2 \sin \theta_e}{r \cos^2 \theta_e}\right)\bar{w} = \frac{-2C_0(z)}{r \cos^2 \theta_e}. \tag{9.31}$$

It is easy to show that this differential equation has exactly the same form as (9.25) except that in place of w_0 we now have $\bar{w}$. Thus the solution for $\bar{w}$ is also given by (9.26). Errors in estimating C_0 lead to errors in the estimates of $\bar{w}$. Because ρ decreases exponentially with height, it can be shown that these errors increase exponentially with the height to which $\bar{w}(z)$ is estimated (Section 9.2.1). The average wind speed and direction are not, in general,

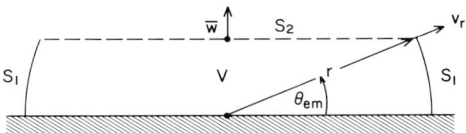

Fig. 9.16. Geometry for estimating the vertical velocity averaged over the circular area S_2.

obtained from analysis of azimuthally scanned velocities when the wind is not linear.

9.4.2 Waves

Another meteorological feature that can be observed from azimuthally scanned velocity data when the wind is not linear is a horizontal wave. Testud *et al.* (1980) have analyzed the residuals of the least squares fitted data obtained by subtracting the first three Fourier components, i.e., subtracting the linear wind. They assume a wave perturbation $\delta \mathbf{v}(h)$ of the horizontal wind of the form

$$\delta \mathbf{v}(h) = A(h)(\mathbf{K}_H/K_H) \exp[j\mathbf{K}_H \cdot \mathbf{d} + \psi(h) - \omega t], \qquad (9.32)$$

where $A(h)$ is the amplitude, which is a function of height h; $\mathbf{K}_H$ the horizontal wave vector; $\psi(h) - \omega t$ a phase term; and $\mathbf{d}$ the horizontal distance vector whose origin is the center of the measurement circle. The factor $\mathbf{K}_H/K_H$ expresses the fact that $\delta \mathbf{v}$ is colinear with $\mathbf{K}_H$, which is to be expected if the wave is a compressional type.

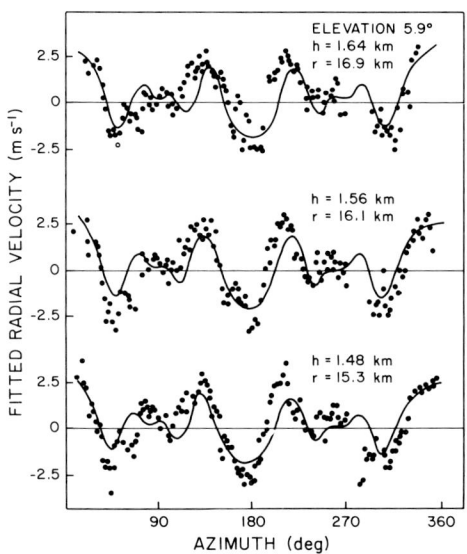

Fig. 9.17. Examples of the δv_r fluctuations observed at three range locations while the beam was scanned in azimuth. Dots are experimental points; continuous lines are theoretical curves realizing the best fit. (From Testud *et al.*, 1980.)

Under the assumption that the period of the wave is large compared to the acquisition time, the corresponding perturbation in the radial velocity becomes

$$\delta v_r(h, \phi) = A(h)\cos[K_H r \cos\theta_e \cos(\phi - \phi_0) + \psi(h) - \omega t]\cos(\phi - \phi_0)\cos\theta_e,$$

(9.33)

where ϕ_0 is the azimuth in the **K** direction. This model does not exclude the existence of vertical motions, but their contributions can be neglected for small elevation angles θ_e.

The wind perturbation δv_r depends on the four parameters A, K, ϕ_0, and ψ. By using a least squares fit analysis, Testud *et al.* (1980) have obtained the four parameters as a function of height. An example of their result for three ranges (Fig. 9.17) shows a remarkably good resemblance between the fitted curve and data. For the 83 scans they analyzed, the standard deviation of the experimental points never exceeds 1 m s^{-1}, which is satisfactory considering that the error in the radial velocity was about 0.5 m s^{-1}. The authors conclude that the three parameters A, K, ϕ_0 can be precisely determined but that the error in the phase term ψ is very large ($\pm 100°$) so that the direction of propagation along the azimuth ϕ_0 cannot be determined (has an 180° uncertainty).

9.5 WEATHER PHENOMENA OBSERVED WITH A SINGLE DOPPLER RADAR

Although the Doppler radar measures only the radial wind component, its spatial distribution can signify important meteorological events such as mesocyclones. Moreover, high straight winds can often be measured, as can turbulent regions. Thus a single Doppler radar offers good promise for severe weather warning and will most likely become a useful operational tool of the National Weather Service.

9.5.1 Vortices

Because the radar maps the distribution of Doppler velocity inside a storm, significant meteorological events (unseen by the eye) such as tornado cyclones (see Fig. 9.4a) can produce tell-tale signatures. Donaldson (1970) stipulated criteria whereby a vortex signature can be identified from single-radar observations. Briefly, there must be a localized region of persistently high, 5×10^{-3} s^{-1}, azimuthal shear (i.e., the Doppler velocity gradient along an arc at constant range) that has a vertical extent equal to or larger than its diameter.

It can be shown that nontranslating cyclones have isodops encircling positive and negative velocity maxima (Fig. 9.18). If the inner portion of the vortex is a solidly rotating core, its tangential velocity increases linearly with radius to a maximum. Outside this maximum the velocity decreases (roughly) inversely with the radius. The isodop contours of such a combined Rankine vortex are circular arcs connected with straight lines. If there is inflow into the cyclone, and if the functional dependence of inflow velocity on distance from the vortex center is the same as for the tangential component, the pattern of isodops remains unchanged except for an angular displacement equal to α (Fig. 9.18) and proportional to the strength of convergence. If there is no rotation but axially symmetric divergence, then $\alpha = -90°$. This pattern has been observed many times. The pattern of radial velocities for an actual mesocyclone appears in Fig. (9.19). Although a tornado was embedded in this cyclone, its velocity pattern is not resolved, because its size is small compared to the radar's resolution volume. However, positive evidence of tornado wind speeds has been obtained and will be given in Section 9.5.3. The larger-scale mesocyclone (this can also be called a tornado cyclone because a tornado occurs within it) produces radial velocity contours that bear a strong

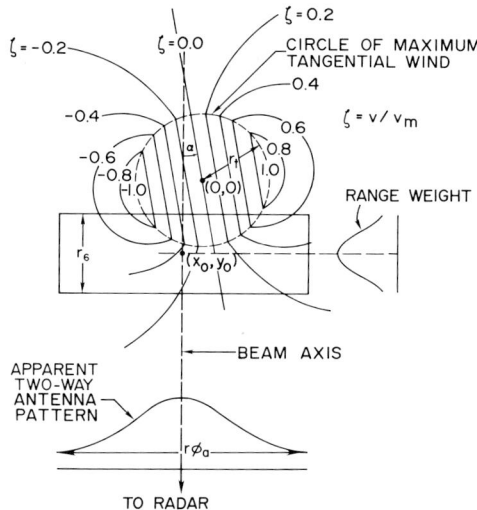

Fig. 9.18. Plan view of an idealized isodop pattern (i.e., one that would be obtained with a radar of infinitely fine resolution) for a stationary modified Rankine vortex located at range large compared to vortex diameter. ζ is the Doppler velocity normalized to its peak value. The radar is located beyond the bottom of the figure. The resolution volume and antenna and range weighting functions are depicted. The angle α determines the radial inflow ($\alpha < 0$) or outflow ($\alpha > 0$).

9.5 Weather Phenomena Observed with a Single Doppler Radar

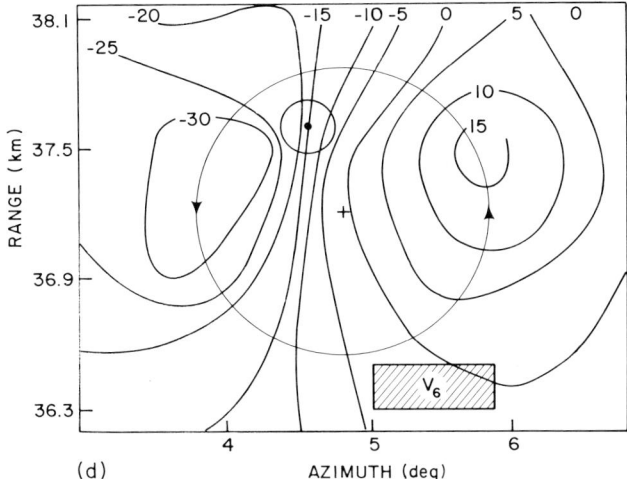

Fig. 9.19. Position of the Del City tornado (small circle drawn to scale) with respect to the mesocyclone signature. Radial velocity contours are drawn from data spaced 0.6 km in range and 0.2° in azimuth. The mean radial motion of the mesocyclone is removed. The hatched area gives the size of the resolution volume V_6. 18:57:48–18:57:52 C.S.T.; elevation: 1.0°; height: 0.7 km.

resemblance to the idealized isodop pattern depicted in Fig. 9.18. However, notice the large asymmetry in the magnitude of the velocity even though the translation speed has been subtracted. This asymmetry is probably due to the additive effects of cyclonic shear associated with the forward flank downdraft air that is west of the tornado cyclone and is flowing southward (see Fig. 9.6b).

9.5.1.1 Smoothing of the Vortex Signature by the Resolution Volume

Recognition of atmospheric vortices from the mean Doppler velocities is not feasible if the resolution volume size is larger than the vortex radius r_t (Fig. 9.18). To illustrate this point, we present a simplified example of a resolution volume. We assume the beam axis is horizontal, the reflectivity η is constant, and the two-way gain pattern is Gaussian and one dimensional (i.e., independent of z) with the following normalized azimuthal dependence:

$$f^4(\phi - \phi_0) = \exp[(4 \ln 4/\phi_a^2)(\phi - \phi_0)^2], \qquad (9.34)$$

where ϕ_a is the apparent beamwidth (Fig. 7.20) that takes into account antenna rotation and postdetection integration. The range weighting function is assumed to be narrow (i.e., $r_6 \ll r_t$) and to cut through the vortex center so that the radar senses only the true tangential velocities. The tangential velocities v_t follow the Rankine profile:

$$v_t = \begin{cases} v_m x/r_t, & x < r_t, \\ v_m r_t / x, & x > r_t, \end{cases} \quad (9.35)$$

where v_m is the peak tangential wind, and $x = r\phi$ is the distance along an azimuthal arc. Using Eq. (5.52), the normalized velocity spectrum for an antenna pointing at ϕ_0 and vortex center at ϕ_v is calculated to be

$$S(v) = \frac{4\sqrt{\ln 2}}{\sqrt{2\pi}} \frac{r_t}{r\phi_a} \left\{ \frac{1}{v_m} \exp\left[-\frac{8\ln 2 \, r_t^2}{r^2 \phi_a^2} \left(\frac{v}{v_m} - \frac{x_0}{r_t} \right)^2 \right] \right.$$
$$\left. + \frac{v_m}{v^2} \exp\left[-\frac{8\ln 2 \, r_t^2}{r^2 \phi_a^2} \left(\frac{v_m}{v} - \frac{x_0}{r_t} \right)^2 \right] \right\} \quad v \leq v_m, \quad (9.36)$$

where $x_0 = (\phi_0 - \phi_v)r$ is the distance between the resolution volume center and the vortex center. The first term in (9.36) is the spectral power contribution from the region inside r_t, whereas the second term is from outside. When the beam is centered on the tornado, (9.36) also describes the tornado spectral signature (see Section 3), which is a double-peak spectrum having a total velocity span equal to the maximum Doppler velocity difference across the vortex.

Integration between $-v_m$ and v_m of $vS_n(v)$ produces the mean Doppler velocity. In this simplified model the mean velocity depends on the radius of maximum winds r_t, the distance r to the vortex, the apparent beamwidth ϕ_a, the maximum wind speed v_m, and the relative (azimuthal) distance between the centers of the resolution volume and the vortex. With proper normalization it is possible to reduce the number of parameters to three; these *normalized parameters* are plotted in Fig. 9.20. Accepting (somewhat arbitrarily) that the vortex will be recognized if the peak in radial velocity is pronounced, we note that the resolution $r\phi_a$ should be less than or equal to the vortex diameter. Figure 9.20 represents the mean velocities as if they were densely sampled by the radar. In reality, discrete azimuthal measurements are taken, and therefore the peak of the velocity is often missed. Also, velocity aliasing may be present, further complicating the pattern.

As an example, consider a vortex at a range of 240 km with a radius of 2 km and a radar having an apparent beamwidth of 1°. Then the normalized parameter $r_t/r\phi_a$ would be about 0.5, showing significant loss in detecting the peak wind. Furthermore, the patterns of velocity in Fig. 9.20 for $r_t/r\phi_a < 0.2$ could easily be produced by shear without rotation. Worse yet, if samples were collected every beamwidth, the spacings along the axis x_0/r_t (i.e., in the azimuthal direction) would be 2 km, making the probability even greater that significant wind speeds would be missed! All these effects somewhat

9.5 Weather Phenomena Observed with a Single Doppler Radar

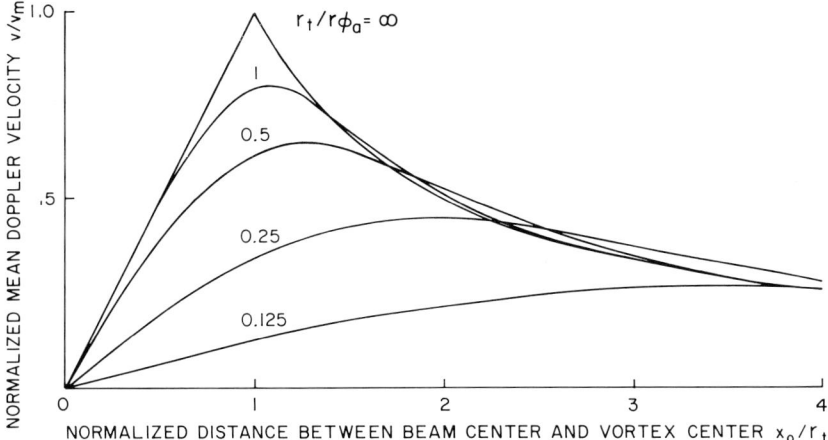

Fig. 9.20. Smoothing of the vortex signature by the resolution volume. (From Brown and Lemon, 1976.)

reduce one's ability to recognize vortices from mean velocity signatures. In order to reduce significantly these sampling errors, the velocity should be estimated at intervals much smaller than the apparent beamwidth (e.g., $\phi_a/2$). However, mean Doppler velocities of a vortex sampled in both azimuth and range create significantly more data and do produce a characteristic pattern similar to the idealized one of Fig. 9.18. Thus pattern recognition techniques should enable one to identify important atmospheric vortices among the Doppler data.

9.5.2 Severe Storms

The reflectivity factor is now routinely displayed by National Weather Service radars on the PPI scope and by some television stations on a color PPI display. Although the reflectivity cannot be reliably used for tornado detection, it has proved valuable for hydrological studies and severe weather warnings. Those warnings are primarily based on reflectivity values, storm top heights, and sometimes on circulatory features or hook echoes (Fig. 9.21).

The reflectivity indentations or weak echo regions penetrating the southern edge of the cells sometimes indicate the presence of a mesocyclone and hence tornado potential. However, at best this signature is a poor indicator of circulation because of the large number of false alarms (as we shall see in Table 9.1).

Large changes in the first Doppler moment from resolution volume to resolution volume, and large second-moment magnitudes have been judged

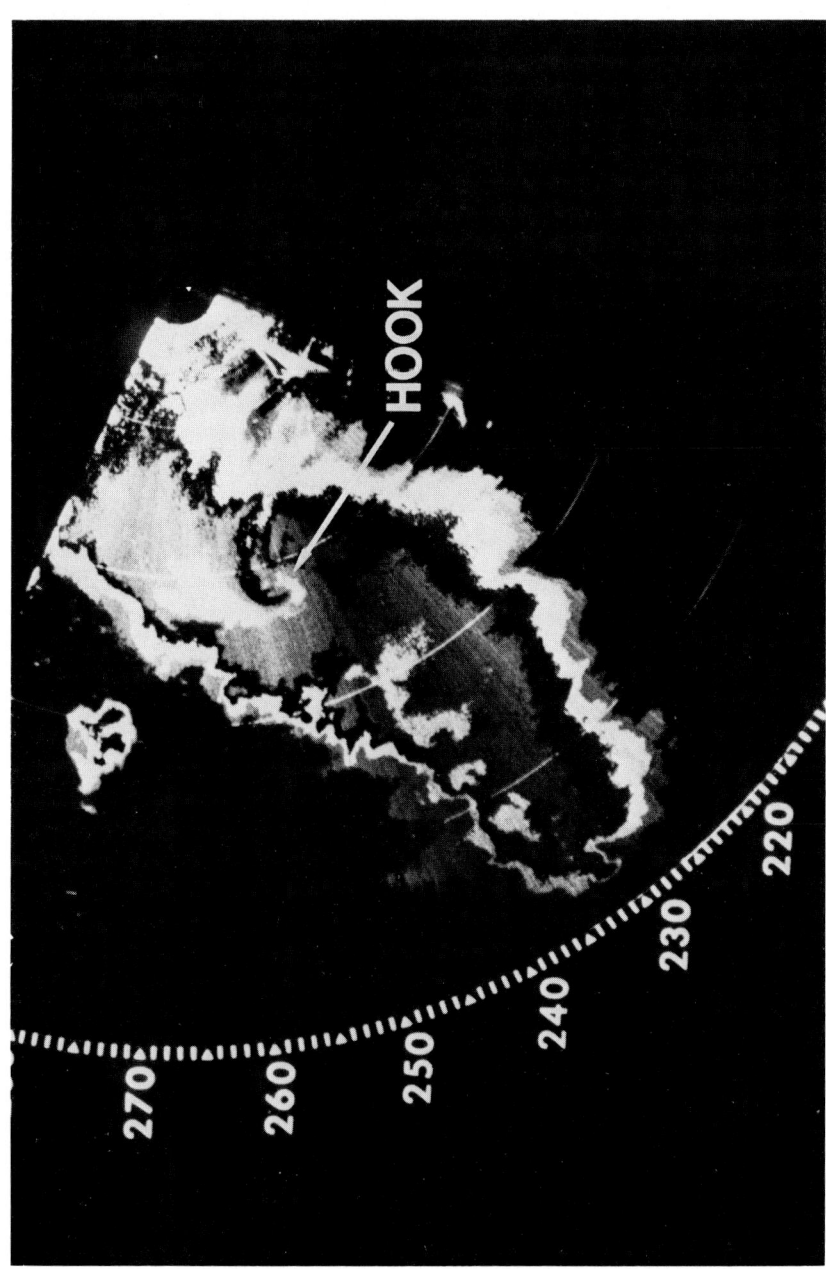

Fig. 9.21. Plan position indicator display of contoured echo power for the Tabler storm (16:24 C.S.T. at 6 June 1974), within which a hook is seen. Arcs at constant range are spaced 20 km apart. 10 log brightness catagories (dim, bright, black, dim, etc.) start at 10 dBZ and are incremented in 10-dBZ steps. (From Brandes, 1977.)

9.5 Weather Phenomena Observed with a Single Doppler Radar

potentially important as tornado signatures. Storms having such signatures have either a tornado or the potential to produce one and are considered dangerous. Sufficiently intense azimuthal shear is the feature used to locate the tornado vortex signature (TVS); this has been correlated with many tornados (Brown et al., 1978). Fields of large spectrum width are a distinguishing feature of a severe storm. The various signatures of mesocyclones, tornados, and severe storms in general may be revealed simultaneously by using three separate displays or, as we shall now describe, with a single *multimoment Doppler display* (Fig. 9.22).

To present simultaneously the three principal Doppler moments for each resolution volume, a field of arrows is displayed with arrow length proportional to the logarithm of the echo power, arrow direction to radial velocity, and arrowhead size to spectrum width (see the inset of Fig. 9.22; Burgess et al., 1976). Zero radial velocity is a horizontal arrow pointing to the right, and nonzero velocities are proportional to the angle between the arrow direction and its zero position. A horizontal arrow pointing to the left corresponds to the Nyquist velocity (± 34 m s^{-1}). As the velocity increases beyond ± 34 m s^{-1}, the arrow rotates smoothly through the Nyquist limits and appears as a lower velocity of opposite sign (e.g., 38 m s^{-1} appears as -30 m s^{-1}). In these displays the radar is always toward the bottom of the figure so that arrows in the upper half of the circle denote (if velocity aliasing has not occurred) flow away from the observer, whereas arrows in the lower half denote flow toward the radar. Thus the field of arrows in Fig. 9.22 illustrates quite nicely the signature of the circulation (centered at 187° azimuth, 70-km range) or convergence (188° azimuth and 75-km range). The radial component of storm motion (10 m s^{-1} from 225°) has been subtracted from all velocities.

A tremendous advantage is obtained with Doppler radar because it can sort out among many storms those that have intense circulation and hence the potential for tornado development. Figures 7.1 and 7.2 show a large storm system composed of many individual convective cells. The multimoment display depicts the principal moments in a sector of space that can be placed over any storm so that the principal moments can be simultaneously examined in detail for evidence of significant meteorological phenomena. Each storm can be systematically interrogated, and for the example shown in Fig. 7.1 only the storm outlined by the box produced a tornado. This tornado's mesocyclone signature was tracked for almost an hour, and Fig. 9.23 shows the position of the signature relative to the damage path. Even though the storm was in the Doppler radar's second trip, the beamwidth was sufficiently small (~ 2 km) to track its circulation from 170 km, where the signature was first noticed. The Doppler radar had an unambiguous range of 115 km on this day, but the distribution of storms was such that no storm was range overlaid onto this mesocyclone, thus

288 9 Observations of Winds, Storms, and Related Phenomena

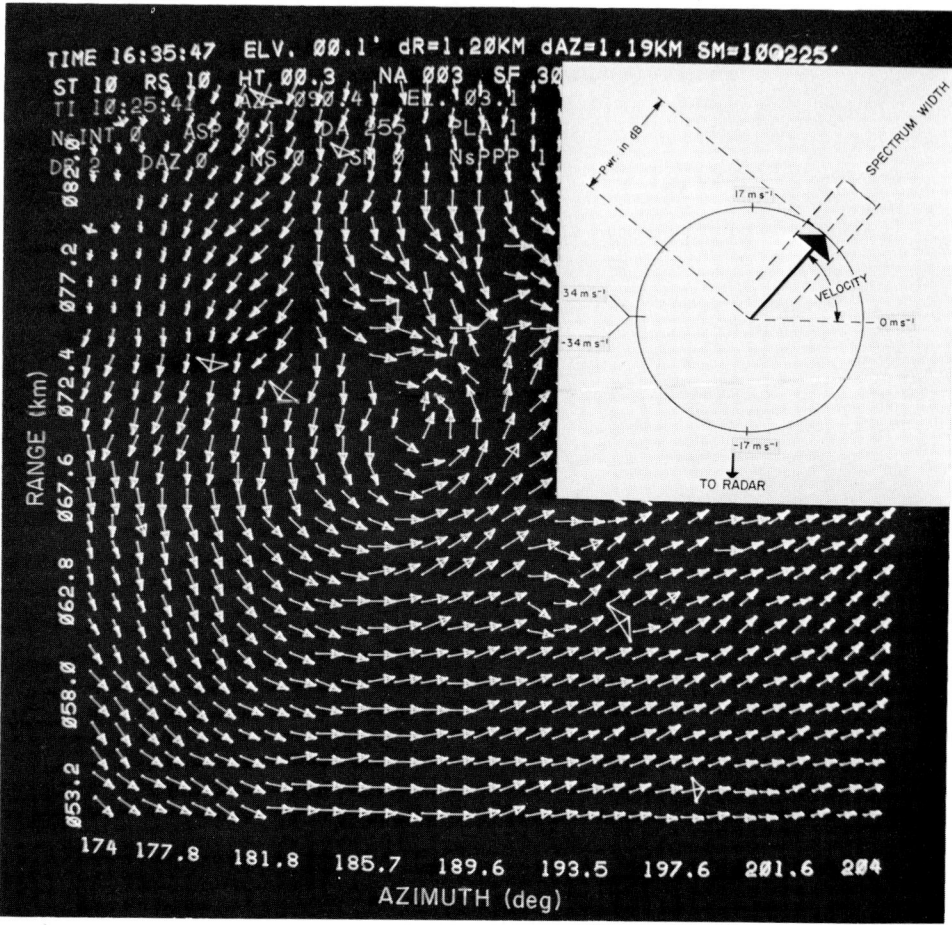

Fig. 9.22. The multimoment Doppler display of a mesocyclone. Each arrow contains information about the three principal Doppler spectrum moments for a resolution volume. The abscissa is the azimuth and the ordinate scale denotes the range (km) from the radar. At the top of the screen is time 16:35:47; elevation angle, ELV = 00.1°; spacing of data along range $dR = 1.2$ km and azimuth $dAZ = 1.19$ km. SM is the storm velocity at 10 m s^{-1} from 225°, whose component along each radial was subtracted from each measured data. ST is a SNR threshold set at 10 dB to eliminate data associated with weak signals. RS is the relative magnitude for suppressing overlaid data due to multiple-trip echos and is set at 10 dB. Any overlaid echos within 10 dB of one another are automatically deleted from the display and replaced with a dot (none appear in this data field). HT = 00.3 is the height above the ground in kilometers. The inset gives the key to interpreting the data display (see text).

9.5 Weather Phenomena Observed with a Single Doppler Radar

allowing an unobscured measurement of its velocity signature. The overlaying of storms due to the small unambiguous range (see Chapter 7) associated with Doppler radars can result in obscuration of signatures.

Figure 9.24 and Color Plate 2a display Doppler radar data on a tornadic thunderstorm in central Oklahoma during the afternoon of 2 May 1979. These depict the fields of the three principal moments of the Doppler spectra: reflectivity (Fig. 9.24a), spectrum width (Fig. 9.24b), and Doppler velocity (Color Plate 2a). The gray scale in Fig. 9.24a is the reflectivity factor, in dBZ, as indicated to the right of the scale. Three cells are evident, two of which are closely spaced and appear as a merged reflectivity and all of which have reflectivities of 46 dBZ or more.

Although the reflectivity display shows no evidence of a mesocyclone, the mean Doppler velocity display (Color Plate 2a) depicts it remarkably well.

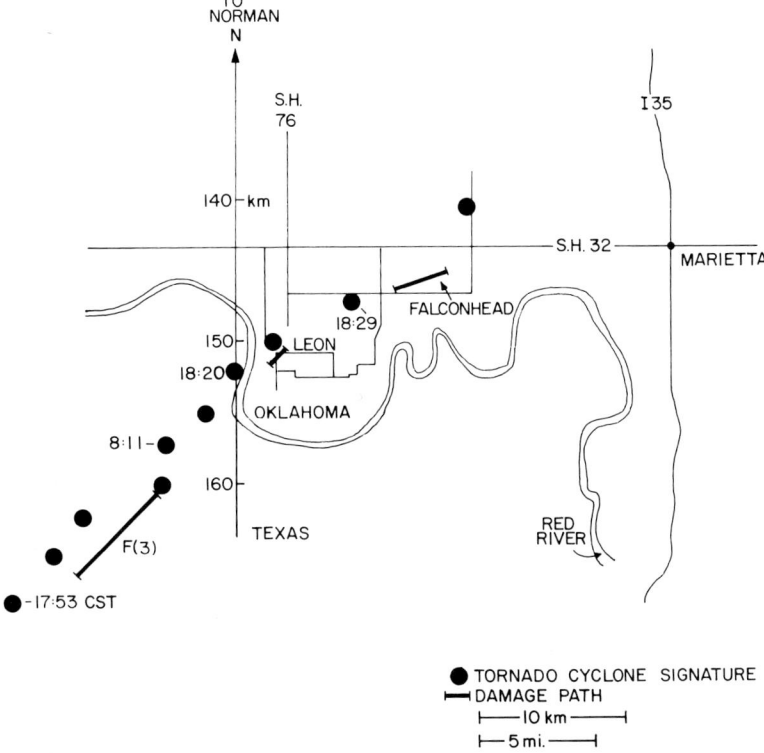

Fig. 9.23. Mesocyclone signature track for the Falconhead tornado showing the damage path location relative to the cyclone center (19 April 1976). The reflectivity field of the storms at 18:16 is shown in Figs. 7.1 and 7.2, where the boxed area contains the storm that produced the only tornado on this day.

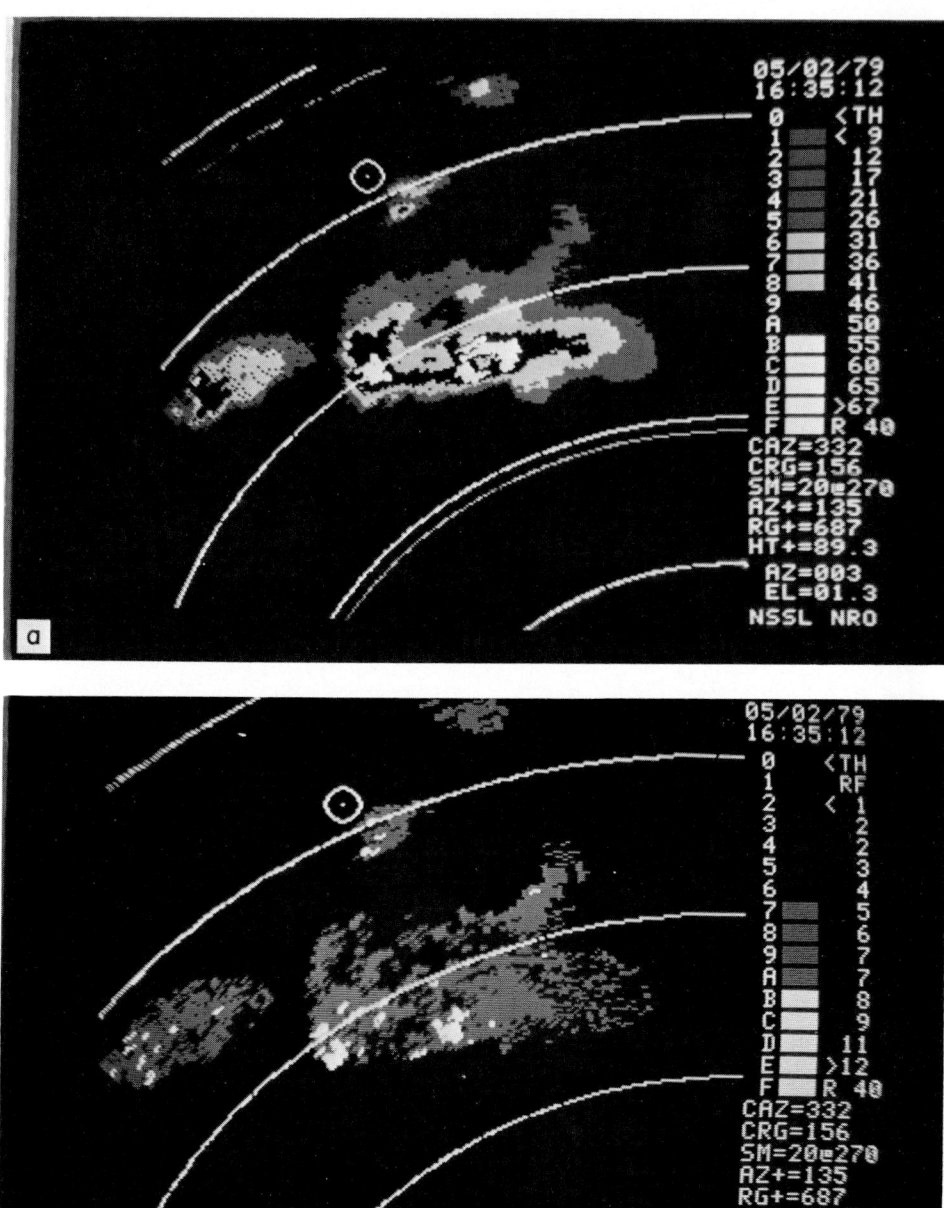

Fig. 9.24. Plan position indicator display is of (a) the reflectivity and (b) the spectrum width for an Oklahoma tornadic storm on 2 May 1979, at 16:35 C.S.T. Range marks are 40 km apart, and the two closely spaced cells are 140 km from the radar. The height above ground is about 3 km. The elevation angle $\theta = 1.3°$. Data having SNR < 15 dB have been edited. The display of velocity fields corresponding to these data is presented in Color Plate 2a.

9.5 Weather Phenomena Observed with a Single Doppler Radar

The velocity is scaled in meters per second, with positive values (red) indicating motion away from the radar. The signature of a couplet of closed contours of opposing Doppler velocities (bright green next to bright red) suggest the presence of a strong circulation. Both eastern cells have strong circulation, and the velocity in the center cell exceeds the Nyquist limit, where green areas surround an abrupt change to red. The western cell also shows evidence of weak circulation. However, a circulation signature may not always indicate storm rotation. Only if the signatures have continuity in height is there positive evidence of storm rotation. The two eastern cells did show this continuity with height, and both produced tornados that remained on the ground for over an hour (approximately 16:20–17:20 CST). The easternmost tornado produced considerable damage. The display (Fig. 9.24b) of the spectrum width (the square root of the second moment about the mean Doppler) is an indicator of turbulence or large shear within the radar's resolution volume. We note high spectrum width values in the vicinity of the circulation. The spectrum width is in meters per second, with its mean value indicated by the numbers to the right of the stepped brightness categories.

A signature pattern for circularly symmetric divergence (or convergence) is similar to the cyclonic vortex pattern but rotated $+90°$ (or $-90°$). As we move to higher altitudes in the storm, the circulation signature gives way to an almost perfect divergence signature (Color Plate 2b). At an altitude of about 11 km, the reflectivity (Fig. 9.25a) of each of the two eastern cells is greater than 50 dBZ, while the western cell has tops less than 12 km. The divergence velocity signature (Color Plate 2b) has a center almost directly above the vortex signature center of Color Plate 2a. The high speed outflow toward the radar exceeds the negative Nyquist velocity limit of -34 m s^{-1} and aliases into the positive velocities, shown by the red areas surrounded by green. Thus the Doppler velocities in the eastern storm range from about -60 to $+40$ m s^{-1}. The velocities toward the radar are higher than those away because the storm's outflow is embedded in an environmental wind that is out of the northwest. Most spectrum widths on the display (Fig. 9.25b) exceed 5 m s^{-1}, indicating strong shear or turbulence (Section 10.4).

The storm structure is well illustrated when data from a single radar are displayed on a range height indicator (RHI) that gives vertical cross sections of Z, v, and σ_v. Figure 9.26 and Color Plate 2c show, respectively, Z (scaled in decibel units) and v, in cross sections along a $347°$ azimuth, for a severe thunderstorm. Visible in the Z field is an anvil of cloud particles extending to the left (south–southeast) at ~ 10 km AGL. More of the anvil is evident in the velocity display, for which the SNR threshold has been reduced to show the motion of the weakly reflective particles, which are probably ice crystals. Storm overhang and strong reflectivity gradients below 2 km AGL on the

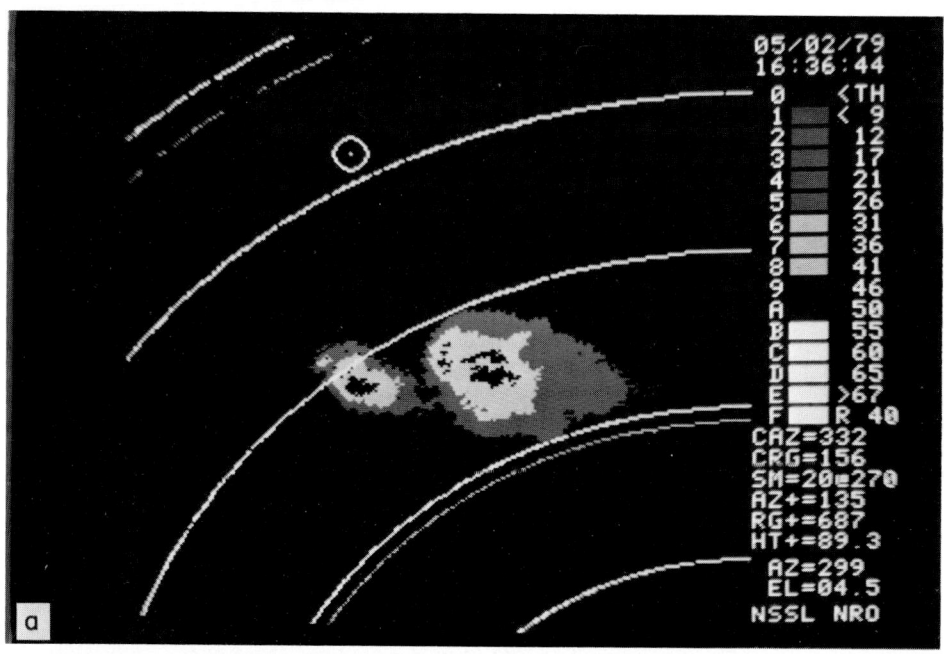

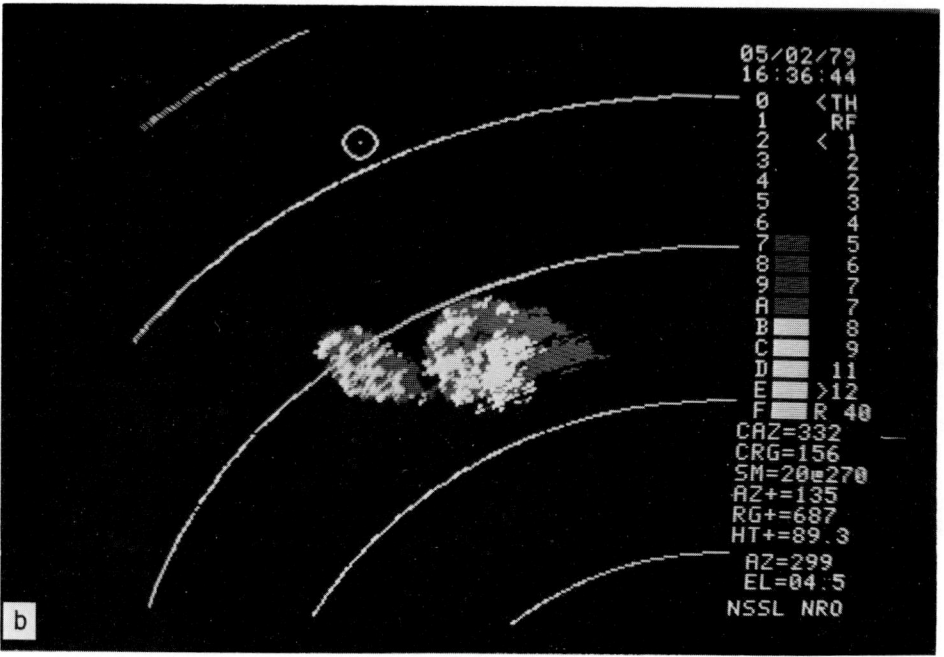

Fig. 9.25. Same as in Fig. 9.24 but at an altitude of about 12 km and at time 16:36 C.S.T. Elevation angle, 4.5°. Strong outflow is indicated by the divergence signature in Color Plate 2b.

9.5 Weather Phenomena Observed with a Single Doppler Radar

storm's south side suggest strong inflow into the storm. If we assume that the hydrometeors are liquid and have a Marshall–Palmer exponential drop size distribution, then the 20-dBZ contour near the ground corresponds to a rainfall rate of <1 mm hr^{-1} and 30, 40, and 50 dBZ to ~ 3, 10, and 50 mm hr^{-1}, respectively. Often the high-reflectivity regions (>50 dBZ) extending to the ground contain wet hail, causing errors in the estimate of rainfall rate.

The Doppler velocity field in Color Plate 2c depicts motion to the right and away from the radar (positive velocities) as well as motion toward the radar (negative velocities). The storm outflow, assisted by northwesterly winds outside the storm, carried cloud particles in the anvil, causing it to grow southeastward, toward the radar. In the anvil, air velocities were more negative than -27 m s^{-1}, the Nyquist limit of the radar, so they appear as high positive values (but less than 27 m s^{-1}) and are seen as patches of lightest red embedded in regions of lightest green. The momentum of updraft air

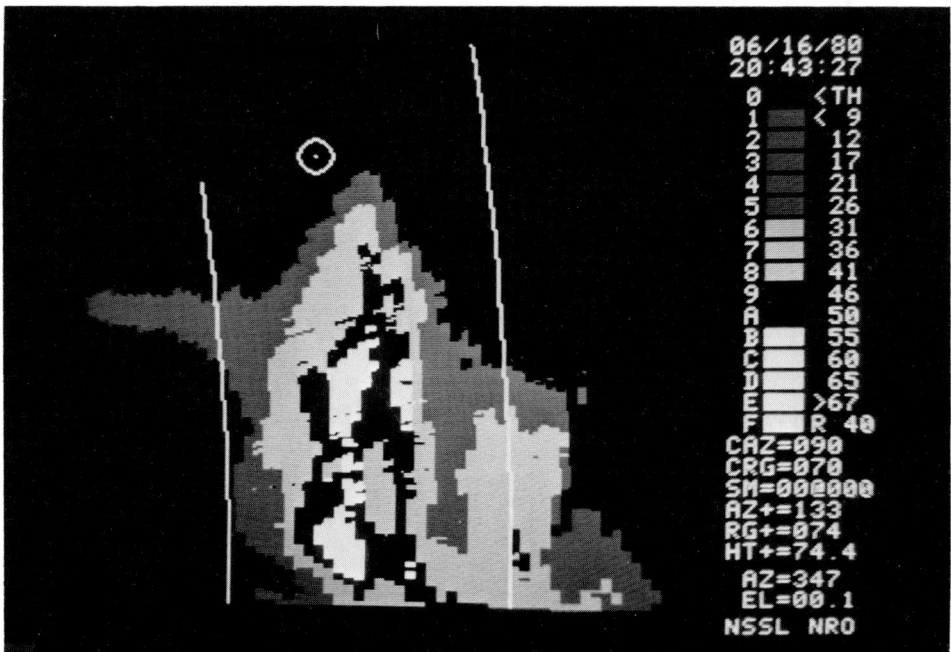

Fig. 9.26. Vertical cross section of the reflectivity factor Z through a severe storm that occurred in central Oklahoma on 16 June 1980. The storm top (near the dot in the white circle) is at 14.7 km above ground level. The grey shades of reflectivities are in dBZ, as indicated at the right. Reflectivities above 20 dBZ are displayed. The radar location is to the left, and the vertical arcs are at 60 and 80 km from the radar. The display of Doppler velocities corresponding to this reflectivity field is presented in Color Plate 2c.

carried condensate into the lower stratosphere, where negative buoyancy forced air downward and outward around the updraft. This latter motion is evident in Color Plate 2c above the anvil on the storm's southern side, where we see the highest radial velocities.

In Color Plate 2c the velocity data above the highest altitudes of the reflectivity field shown in Fig. 26 are due to antenna side lobes that receive echoes from the high-reflectivity regions when the main beam lobe is pointed above the storm. These falsely mapped velocities do not portray the true wind there. However, if there are no strong echoes at a given range in the side lobes, then the Doppler radar can accurately sense the velocities in weak-echo regions in the main beam at that range. For example, weakly reflecting targets drifting with the flow in the clear-air boundary layer (at low altitude—the red region in Color Plate 2c between ranges of 50 and 60 km) trace the air moving into the storm, consistent with Fig. 9.1.

9.5.3 Doppler Spectra of Tornados

Radar usually senses that portion of circulation that lies within the resolution volume, but all tracers moving with the same velocity contribute to a spectral coefficient according to their reflectivity, their number (i.e., in proportion to isodop spacing), and the antenna pattern illumination [Eq. (5.52)]. Figure 9.27 illustrates the locations of 16 resolution volumes along the beam intersecting a tornado vortex. Echoes from each resolution volume are Fourier analyzed to obtain the spectrum of Doppler velocities. Only those targets whose spectral power is above the receiver noise level may be positively identified. However, it has been our experience that tornados in a resolution volume offer enough reflectivity, due to debris and hydrometeors, that a large velocity span can be observed. Currently it is not known whether tracers moving at the peak tornado wind speed can be reliably resolved.

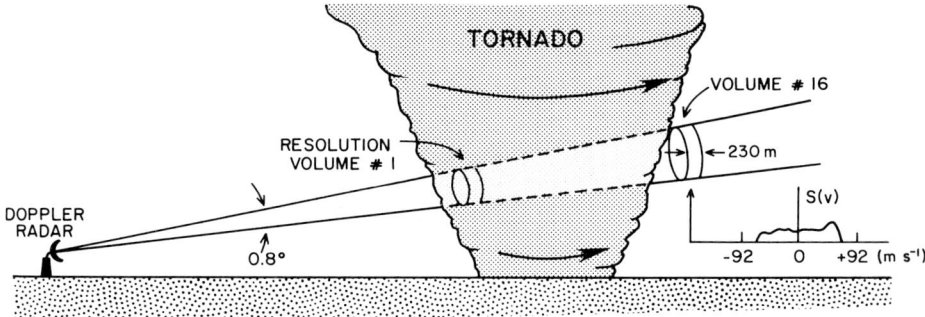

Fig. 9.27. Schematic of radar beam and resolution volumes intersecting a tornado vortex.

9.5 Weather Phenomena Observed with a Single Doppler Radar

In order to relate the expected Doppler spectra of tornados to radar measurements, we shall show spectra for a model tornado circulation (Zrnić and Doviak, 1975). The vortex is approximated by a combined Rankine model, with provision made for particle inflow or ejection. Positions of maximum radial and tangential velocities are assumed to be coincident, and the radius of their location is referred to as maximum wind radius r_t. Given the model reflectivity and Doppler velocity fields (Fig. 9.18), one can calculate using (5.52) the mean Doppler spectra for specified resolution volumes. Although the tornado may have nonuniform reflectivity, we assume (for simplicity) the reflectivity to be either uniformly distributed throughout the resolution volume or to have a Gaussian-shaped profile (as a function of distance from the vortex center with control over the peak position and width). This allows us to account for nonuniformities, at least from resolution volume to resolution volume, whereas with the Gaussian profile, radial nonuniformities can be accommodated. Other parameters, such as tornado radius and position within the resolution volume, are variable.

When centered on the beam axis, the Rankine vortex model predicts a bimodal spectrum, called a *tornado spectral signature* (TSS), which has been verified experimentally several times (see the dashed line in Fig. 9.28b). The TSS is also predicted by the analytic formula (9.36) when the beam is centered on the vortex and the beamwidth is large compared to the vortex diameter. The spectral peaks are caused by targets outside the tornado radius r_t, where the velocity gradient is less and hence the scattering volume contribution per unit velocity interval is larger, as predicted by (5.52). However, this increase in spectral power is attenuated by the antenna gain, which leads to greater decreases in the contribution from those targets that are farther away from beam centered on the vortex. The velocity at the peak is related to the parameters of the tornado and beamwidth and can be found by locating the maxima of (9.36).

Shown in Fig. 9.28 are Rankine model vortex spectra (dashed lines), matched to the data by a least squares fit. A von Hann weight (raised cosine) was applied to the data prior to a discrete Fourier transform. The von Hann weight offers a good compromise between the width of the main spectral window and the size of the window side lobes (see Table 5.2). Specifically, the rapid side lobe decay reduces contamination of high-velocity spectral coefficients by strongly reflecting low-velocity ones. The mean square difference between the data and the simulated spectra is simultaneously minimized for spectra closest to the tornado. The resolution volume depth r_6 is 230 m, the range gate spacing 600 m, and the antenna beamwidth 0.8°. Although a 0.3-μs pulse was transmitted, the receiver filter bandwidth is 0.6 MHz, and from (4.26) a range resolution of 230 m is obtained. The tornado is at azimuth 4.6°. The beam height above ground level is 640 m.

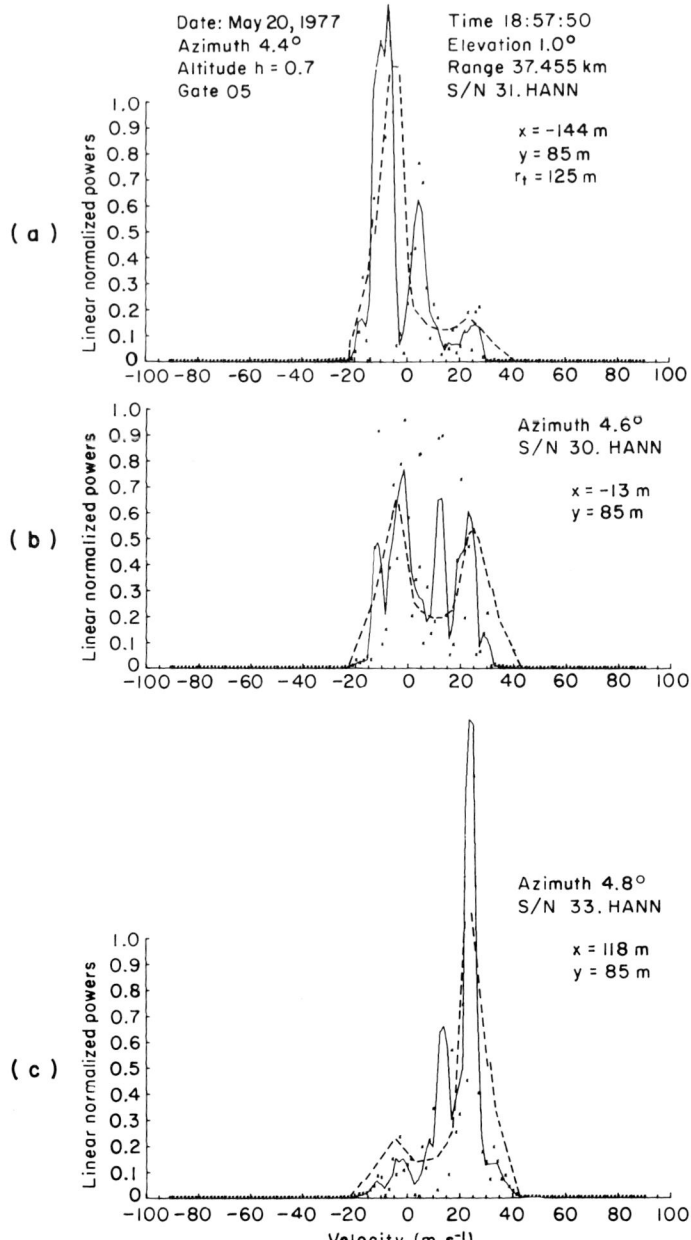

Fig. 9.28. Spectra from three consecutive azimuthal locations for the Del City tornado. The dots are squared magnitudes of Fourier coefficients for recorded time series data weighted with a von Hann window. The solid lines are three-point running averages. Dashed lines are simulated spectra. The SNR is in decibels. x (azimuthal distance) and y (range distance) are the coordinates of the tornado center with respect to the resolution volume center; the altitude h (km) is to the beam center from the ground level.

9.5 Weather Phenomena Observed with a Single Doppler Radar

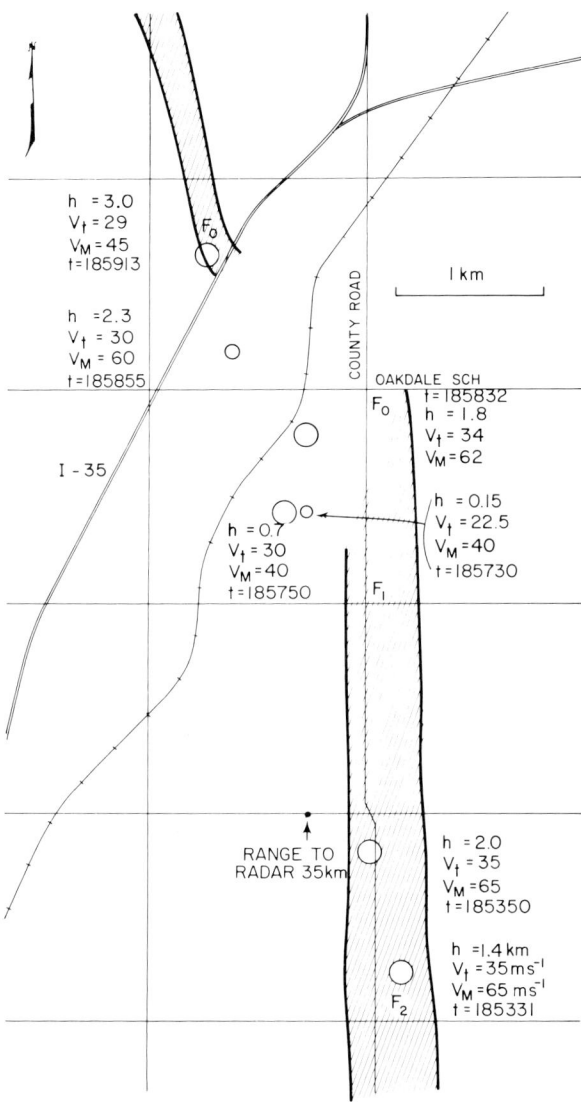

Fig. 9.29. The Del City tornado positions and size (circles drawn to scale), as deduced from the Doppler spectra, are superimposed onto the damage path. The height of the beam center with respect to the ground is h (km), V_t (m s^{-1}) is the speed of rotation, and V_M is the absolute maximum speed. The damage scale ($F_0 \to F_2$) is according to Fujita (1981). County roads (the square grid) are 1 mile apart.

The data from the Del City tornado were collected with an unambiguous velocity of ± 91 m s^{-1}. Uniform and Gaussian-type reflectivity profiles were used in the spectra models. More often the toroidal profile resulted in a somewhat better fit because the two degrees of freedom (diameter and thickness) allow easier adaptation of the reflectivity to nonuniformities caused by debris. The model fitted in Fig. 9.28 had a toroidal reflectivity. The tornado parameters deduced from the fit of the Del City tornado (20 May 1977) are a diameter of between 130 and 250 m and rotation speeds between 22 and 35 m s^{-1}.

The tornado location determined from the spectral fit is superimposed in Fig. 9.29 on the damage path obtained by survey teams. Also in Fig. 9.29 are maximum measured Doppler velocities V_M, maximum rotational (tangential)

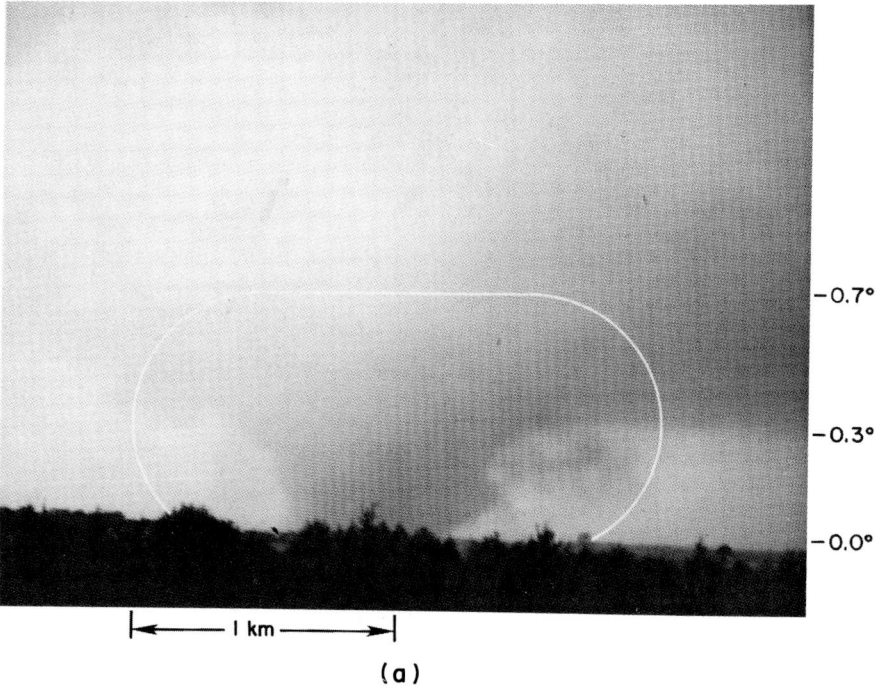

Fig. 9.30a. The tornado near Binger, Oklahoma, at $\sim 16{:}07$ C.S.T. on 22 May 1981. View is to the northwest and the range from the photographer is 5.2 km. (From R. Davies-Jones and D. Burgess, personal communication.) The white contour is the approximate size of the radar measurement volume for the data shown in (b). The radar beam is blocked by nearby terrain up to elevation angle 0.1°.

9.5 Weather Phenomena Observed with a Single Doppler Radar

velocities V_t, and heights of the scans above ground. Estimated radii of wind maxima range from 65 to 125 m; they are drawn to scale in Fig. 9.29. The relative position of the tornado (whose spectra are shown in 9.28b) with respect to the isodops of the mesocyclone is drawn in Fig. 9.19.

We emphasize that the observed Del City tornado produced moderate damage although it passed through a populated area. Corresponding maximum rotational winds deduced from Doppler spectra were 35 m s^{-1}. With added translational motion the peak winds on the tornado's east side (positive radial velocities) are no more than 65 m s^{-1}. These peak winds are deduced from plots of spectra such as those of Fig. 9.28 (Zrnić and Istok, 1980). Although higher wind speeds cannot be ruled out on the basis of spectral measurements, the damage survey indicates that radar-estimated values are quite realistic. Nonuniformities in reflectivity, tilting of the vortex with height, and targets in sidelobes are just a few effects that do occur but are not accounted for in the model used here.

The first scanning of an intense tornado (Fig. 9.30a) with a pulsed-Doppler radar having a Nyquist interval sufficiently large (180 m s^{-1}) to measure unambiguously the high tornadic wind speeds was in the Spring of 1981 at the National Severe Storms Laboratory in Norman, Oklahoma. The contour in Fig. 9.30a shows the cross section for each of 16 consecutive (in range) measurement volumes (see Fig. 9.27) within which targets produced the 16

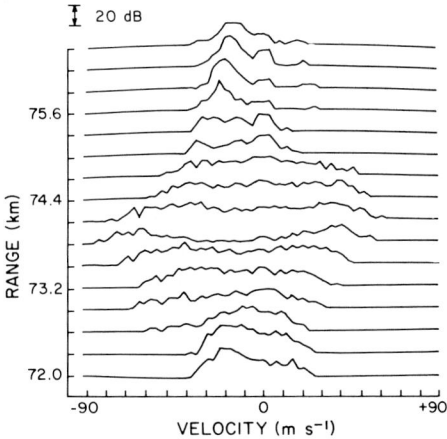

Fig. 9.30b. Reflectivity spectral densities (in dB) versus the Doppler (radial) velocity at 16 range locations in the tornado's circulation. The beam elevation angle is 0.3°. The nearly horizontal lines at each end of the spectra are thresholds at 10 dB above the receiver noise level and correspond to a reflectivity spectral density of ~ 3.4 mm^6 m^{-3} m^{-1} s. The vertical scale is 20 dB per division.

Doppler spectra in Fig. 9.30b. These spectra were recorded within a minute of the time of the tornado photograph in Fig. 9.30a. Because the antenna was scanning in azimuth while data were collected to form these spectra, the measurement volume has an azimuthal dimension of 1.5°, although the beamwidth is 0.8° (see Section 7.6); the range resolution $r_6 = 230$ m. The 16 volumes are aligned along the beam, but only two or three of these cross the tornado's circle of maximum wind. In Fig. 9.30b the widest spectra denote the location of the tornado. Maximum tornado wind speeds have been estimated from these data. At this time the radar-measured maximum winds were ~ 90 m s^{-1}. The asymmetry in the spectra (e.g., the peak positive velocity occurs in a resolution volume further in range than the peak negative velocity) is caused by targets centrifuging out of the vortex.

Acquiring in situ data in and around the tornado is very difficult, but Doppler radar provides a remote and relatively safe method of surveying storms for tornados. In view of the variety of displays and signatures associated with tornados, it is natural to ask which technique is most promising for their detection. It became apparent during several years of experimental work that the mesocyclone circulations (the parent circulations of the tornado) showed very nicely on the color display of radial velocities. However, the multimoment display proved more suitable for quick vortex recognition. Although a display of the mean Doppler velocity fields depicts rather nicely the circulation of the tornado's parent cyclone, it does not give a positive indication that a tornado has formed. This is because most tornado diameters are much smaller than the resolution volume, so that the tornado's wind profile cannot be measured accurately. However, the display of the spectrum of all detected radial velocities gives a more definite indication of the tornado's presence because the peak circulating winds of the tornado are directly observed. Nevertheless, all large tornados are preceded by a larger-scale circulation that starts at middle levels (6–8 km) and works its way toward the ground; this incipient circulation within which the tornado forms is usually much larger, so its rotation can be resolved. Using criteria discussed by Brown *et al.* (1978), scientists at NSSL were able to detect a number of tornado cyclones. In fact, all tornados that occurred within a range less than 115 km were detected. At further ranges the signatures of small mesocyclones are lost due to poor resolution; however, circulations associated with large destructive tornadoes were detected up to 240 km with the 0.8° beam of NSSL's Doppler radar. Spectrum width has not been a reliable indicator of tornados, because turbulent areas in storms exhibit large widths and are easily mistaken for tornados.

A test was conducted in Oklahoma to determine how tornado warnings are improved over the present non-Doppler warning system when Doppler information alone is used to forecast tornados (Burgess *et al.*, 1979). Table 9.1

9.5 Weather Phenomena Observed with a Single Doppler Radar

TABLE 9.1 Comparison of Tornado Warning Performance, 1977–1978[a]

	Present system	Doppler radar alone
Probability of detection[b]	0.64	0.69
False alarm rate[c]	0.63	0.25
Critical success index[d]	0.30	0.56
Lead time (min)[e]	−0.8	21.4

[a] Approximately 200 cases were examined.
[b] $X/(X + Z)$, where X is the number of predicted tornados that occurred, Y the number of predicted tornados that did not occur, and Z the number of tornados that occurred without a prediction.
[c] $Y/(X + Y)$.
[d] $X/(X + Y + Z)$.
[e] Between issuing of warning and tornado occurrence.

compares several measures of effectiveness of tornado warnings using Doppler radar alone with present techniques that rely heavily on visual sightings. The critical success index (CSI) is higher the higher the detection probability and the lower the false alarm rate, CSI = 1.0 being perfect.

Doppler radar can increase the average lead time to tornado occurrence by 20 min, thus offering the population a longer time to take cover if warnings are communicated properly. Although the probability of detection at the instant of issuance of warning is not significantly improved, the false alarm rate is significantly decreased with Doppler radar. Therefore, Doppler radar can reduce the overwarning inherent in the present system while giving longer lead times, and warning areas may be reduced substantially by knowledge of the precise location of the vortex.

9.5.4 Downdrafts

Because most storms are observed with the radar's beam axis almost perpendicular to the vertical, direct observation and detection of vertical motion are quite unlikely. However, because the stratosphere impedes ascending air and the ground stops descending air, diverging outflows are created and can often be seen as characteristic divergence signature (i.e., Fig. 9.18 with $\alpha = 90°$). Whereas strong updrafts can be located reasonably well by association with divergence signatures at the storm top, downdrafts are not so easily identified, because diverging outflow, confined to a shallow layer near the earth's surface, may not be sampled (ground clutter and earth's curvature obscure observation) and also because downdraft air is sometimes found in regions of weak reflectivity (note the RFD location in Fig. 9.1 and precipitation-free volume in Fig. 9.2c) and may not be detected by the radar. The outflow air associated with down drafts is most easily detected when precipitation is entrained within it, but when evaporatively cooled outflow moves far from the storm, weak reflectivity usually accompanies it, making detection more difficult. However, if the outflow is strong, winds along the front will be gusty, and hence debris can be made airborne, becoming targets visible to the radar. Gust fronts may also be visible because turbulence may produce irregularities in the refractive index that scatters microwaves (see Chapter 11).

Color Plates 3a and 3b show the Doppler velocity and spectrum width fields in a thunderstorm and in its gust front at the lower elevations observed by radar. This front was produced by a strong squall line with reflectivities of 60 dBZ. Peak radial velocities of about 38 m s^{-1} relative to the ground were measured in the yellow-green patch in the midst of blue at 30 km and 285°. Note that the spectrum width field (Color Plate 3b) depicts the frontal discontinuity, even in a region where the edge of the front is aligned along the radial (i.e., at the cursor's location). Thus even in cases when the front moves perpendicular to the radial, spectrum width data may offer a good signature to the gust front location. In this one and most other gusts associated with intense Oklahoma storms, the presence of downdrafts is also accompanied by very turbulent eddies, which are evident on the spectrum width display. Several signatures of divergent flow are apparent in the velocity field. Sizes of these range from 2 km just south of the cursor location to 10 km (just west of the top white square; position of the Cimarron radar). The large divergence signature also exhibits some anticylonic rotation. The strong shears along and behind the front are produced by the constantly evolving and interacting cells that generate shortlived up/down drafts. Maximum radial shear of 2×10^{-2} s^{-1} was produced by the small downdraft (near the cursor) and was measured with the beam center at 800 m above ground. Maximum azimuthal shear of 4.7×10^{-2} s^{-1} was detected in the gust front (30 km, 290°). At several locations behind the front we have measured radial shears of 1×10^{-2} s^{-1}. Although the maximum difference of outflow velocities produced by these downdrafts depends little on the downdraft size, the maximum measured shear is associated with smallest sizes!

An example of nearly vertical descent of air is seen in Fig. 9.31, showing rain embedded in the downdraft. The outrush of air as it deflects from the ground carries precipitation along with the air, producing relative symmetry on both sides of the rain shaft. Intense downdraft air impinging on the ground does not necessarily have precipitation embedded within it. Cloud particles and smaller drops may have completely evaporated when the air reaches the ground; in Fig. 9.32 the presence of downdraft air is delineated by the streaks of dust made airborne by the outward-moving air.

9.5 Weather Phenomena Observed with a Single Doppler Radar

Fig. 9.31. Downdraft in an Oklahoma thunderstorm with embedded precipitation. (Courtesy of H. Bluestein, University of Oklahoma.)

Fig. 9.32. Downdraft without precipitation near Denver, Colorado. (Courtesy of T. Fujita, University of Chicago.)

When the horizontal dimensions of a downdraft are small but vertical velocities intense, the downdraft is called a *microburst* (Fujita, 1981). More specifically, a microburst can be defined as a downdraft when the associated radial outflow along the ground has horizontal dimensions smaller than several kilometers and winds are intense enough (e.g., ≥ 20 m s^{-1}) to cause damage. The Doppler velocity field of a microburst is shown in Color Plate 3c at a beam height of 50 m. Strong gradients of horizontal flow can cause aircraft to markedly depart from their intended flight altitude, which can be diastrous for low-flying aircraft on their approach or departure from airports, even if winds are not intense enough to cause appreciable damage to ground-based equipment and facilities (see Section 9.5.5).

9.5.5 Gust Fronts

A *gust front* is the leading edge of air produced when evaporatively cooled downdraft turns into an outward-moving ground-based flow. The warm moist boundary layer air, which in the central United States usually flows from the southeast, is lifted as it flows over the top of the pool of cooled denser air and often forms a conspicuous arcus cloud (Fig. 9.2c) that appears near the front. The front is marked by shifts (shear) in the wind, both in the vertical and horizontal directions. A gust front can propagate in the clear air many tens of kilometers away from the thunderstorm that caused it and yet harbor shear forces that can be destructive to aircraft, especially when a flight crew is unaware of its presence. The wind behind the front is usually strong and turbulent. Gust fronts often appear as thin lines on radar displays (Fig. 9.33a), and reflectivity can be so weak (≤ 10 dBZ; see Table 9.2) that some radars fail to detect them. However, moderately sensitive Doppler weather radars can sense reflectivities as low as -10 dBZ at ranges ≤ 60 km.

The thin line seen in Fig. 9.33a usually marks the leading edge of the front and is thought to be generated by debris made airborne by the strong gust winds immediately behind the front. Further back from the front the flow is often less turbulent and debris may settle out so that the reflectivity becomes weak. A vertical cross section of the reflectivity field is shown in Fig. 9.33b. These reflectivities are averages over the azimuth sector $305°-310°$.

Even though the single Doppler radar measures only the radial component of the vector wind, we can, by assuming the gust winds are directed perpendicular to the front, obtain the horizontal and vertical wind components from the continuity equation. Radial velocities averaged over the azimuthal interval $305°-310°$ were used to obtain the vector winds plotted on the vertical cross section in Fig. 9.33b. This gust front, observed shortly after midnight, was extremely strong, with vertical winds in excess of 20 m s^{-1}. The turbulent winds associated with this gust extended to altitudes of at least

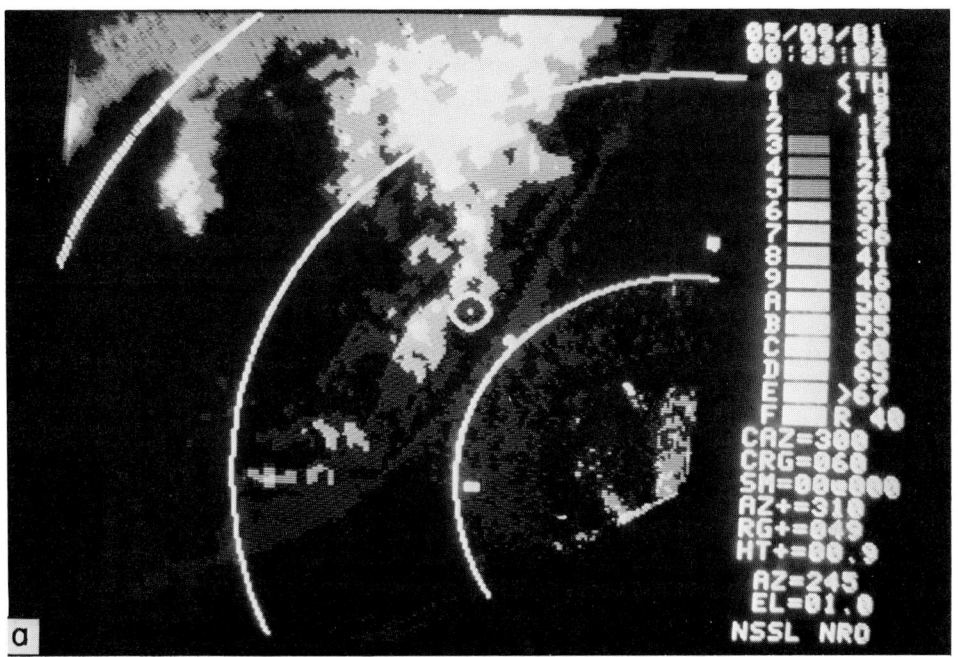

Fig. 9.33a. Plan position indicator display of Z for the gust front of 9 May 1981. Range-marking arcs are at 40-km intervals. Elevation, 1.0°.

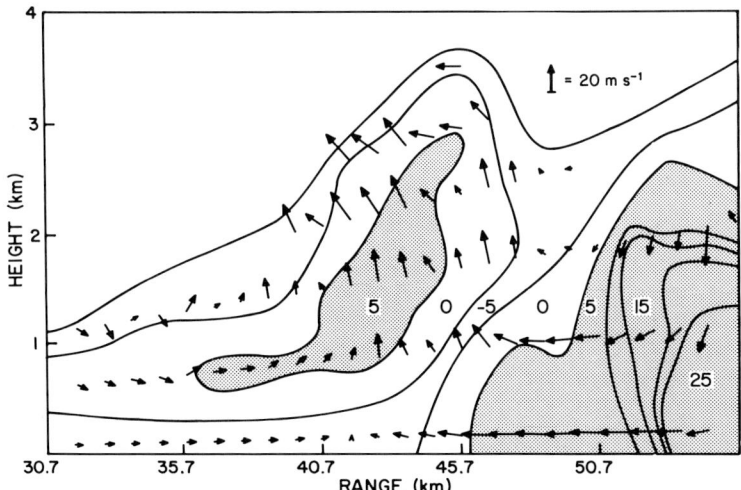

Fig. 9.33b. A vertical cross section of the gust front (9 May 1981 at 00:31:32 C.S.T.). Arrows are the horizontal and vertical wind components in the plane of observation, as scaled by the 20 m s^{-1} arrow at the upper right. The reflectivity factor contours are in steps of 5 dBZ, and the stippled areas start at 5 dBZ. Azimuth, 305°–310°. (Courtesy of Robin King, Finnish Meteorology Institute, Helsinki, Finland.)

TABLE 9.2 Clear-Air Gust Fronts

Date (1981)	Maximum height (km)	Peak reflectivity[a] (dBZ)	Radial speed difference[b] (m s^{-1})	Width of front[c] (km)	Length (km)
10–11 April	1.2	8	40	1–4	50–80
13 April	1.3	9	35	1–5	25–110
30 April	1.6	11	36	2–7	120
9 May	3	7	30	7	120
13 May	0.6	2	21	0.4–1.0	80–100

[a] The largest Z_e over a patch of several resolution volumes. Because wind tracers are not known, the reflectivity factor is the equivalent one (see Chapter 4).

[b] The difference in the peak radial velocity in the gust and the radial velocity of the environmental flow immediately in advance of the front.

[c] Measured from the peak gust velocity location to the leading edge of the front.

3 km, above which weak reflectivity precluded measurements. In this case the strong shear regions were 10 km away from the higher-reflectivity regions associated with precipitation.

The farther away a gust front is from the storm, the more likely it is to pass undetected. Although the gust front is usually connected to the storm through the stream of cool outflow air, it can apparently detach itself from the storm and propagate away. In Fig. 9.34 the reflectivity field depicting a thin line marks the position of a solitary gust that propagated more than 60 km from a storm (seen at the top of the figure) that generated it.

The Doppler velocity field associated with the reflectivity field of Fig. 9.34 is seen in Color Plate 3d. The radar beam elevation was fixed at 0.4° while the radar scanned in azimuth from 330° to 45°. Note the abrupt change in radial velocity across the gust line. The wind on the radar side (southeast) of the line has a component away from the radar and just beyond the line, toward the radar. The region of approaching velocities along the gust is rather narrow. The air flow immediately behind the 4-km-wide band of approaching air is again southerly, although it is not evident in Color Plate 3d because the SNR threshold eliminated from the display the weaker targets behind the gust. The dark blue area near the 40-km-range arc indicates regions where the velocity data are contaminated with range-overlaid echoes from distant (≥ 150 km) storms along the same bearing.

The time sequence of the thunderstorm that generated the gust in Fig. 9.34 and Color Plate 3d is shown in Fig. 9.35. By extrapolating the positions of the gust and thunderstorm backward in time, we deduced that the portion of the gust, 40 km northwest of the Norman Doppler radar, coincided with

9.5 Weather Phenomena Observed with a Single Doppler Radar

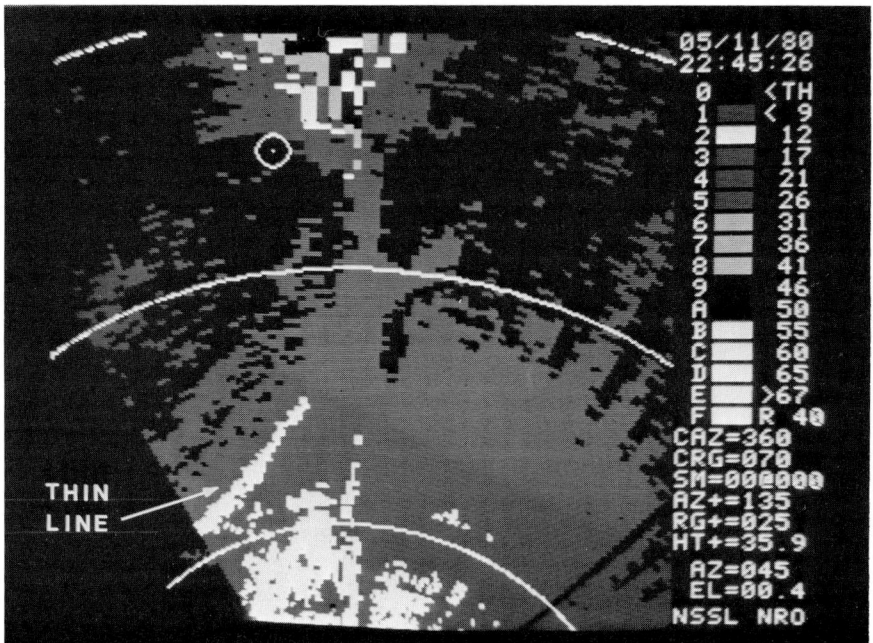

Fig. 9.34. Photograph of a radar reflectivity PPI display showing a thin line at 22:45 C.S.T., 11 May 1980. The center of the beam is at an altitude of 0.8 km at the range indicated by the small white circle. The arcs are at the ranges of 40 and 80 km. The gray code gives the reflectivity factor (dBZ). Elevation, 0.4°.

the storm circled in Fig. 9.35. Because storm outflows are usually confined to the first kilometer of the atmosphere, they cannot be seen at large distances owing to the earth's curvature. The gust in this case was first detected by Doppler radar when it was about 65 km away and after it had propagated over 40 km from the storm.

The first detection of gust fronts is conditioned by several factors the gust front reflectivity, height, and range and the radar characteristics of transmitted power, receiver sensitivity, and beam elevation angle. Usually, beam elevation is one half beamwidth above the horizon in order to clear obstacles blocking the beam and to minimize beam pattern deformation caused by surface reflections and diffraction. In the case under discussion, the beam center was at 0.4° elevation angle and, for a target at 65 km, places the location of maximum detection sensitivity at about 700 m AGL. Because, as we show later, the gust was confined to altitudes below 700 m, it is no surprise that the gust was not evident earlier.

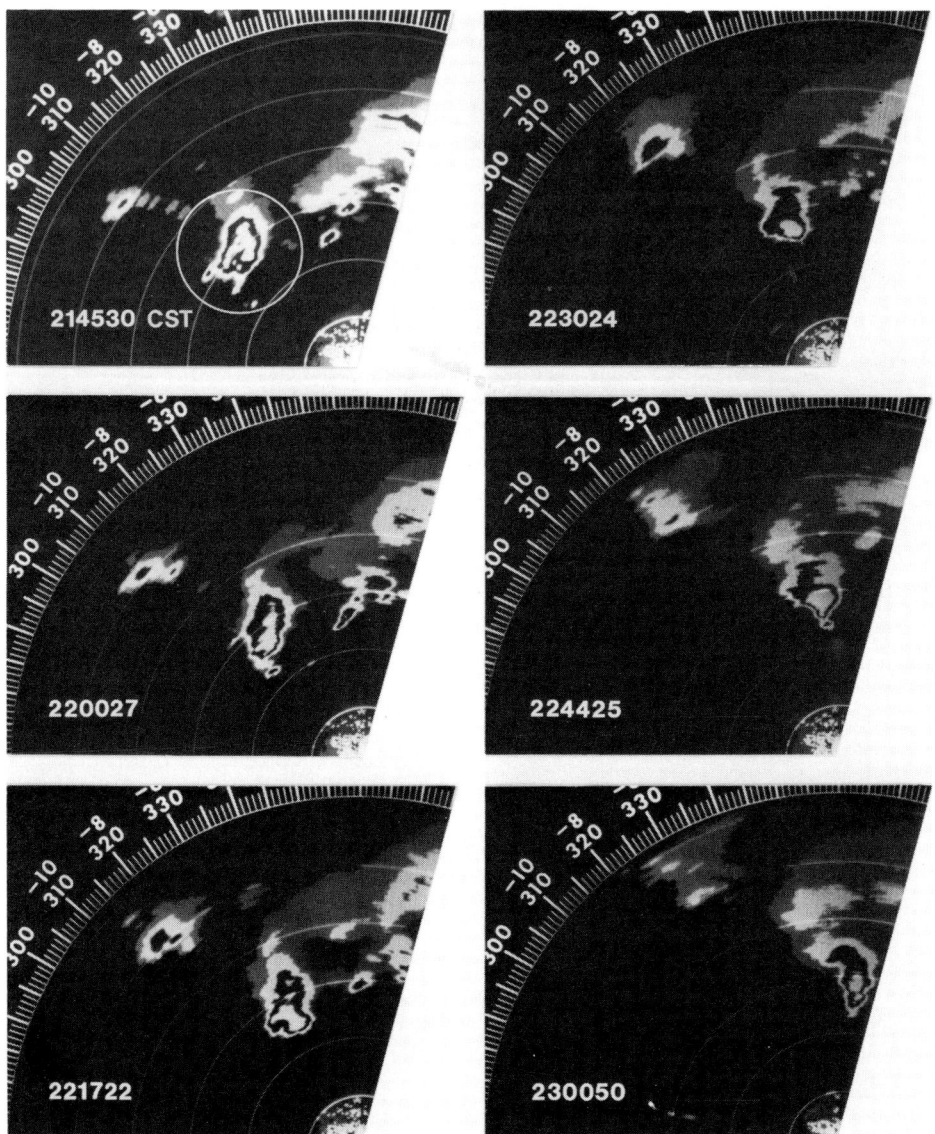

Fig. 9.35. Time sequence of NSSL's WSR-57 radar scope photographs for the thunderstorms on 11 May 1980 that generated the gust front in Fig. 9.34. The range marks are at 40-km intervals. The circle locates the storm at the time the gust front was generated.

9.5 Weather Phenomena Observed with a Single Doppler Radar

Figure 9.36 shows isochrones (lines of equal time) for the gust as determined by Doppler radar observations of the leading edge of the radial velocity shift. Superimposed on this plot are locations of surface sites where the time of gust passage was determined when surface wind had a sudden rise. The time of arrival of the front is indicated at each site, and good agreement is found between the positions of the isochrones and time of arrival of the gust at the surface sites. The upper left-hand photograph shows the gust position vs time and implies that the gust was propagating with a relatively uniform velocity of 13 m s^{-1}.

Although NSSL's Doppler radar has a relatively small beamwidth (0.8°), its spatial resolution at 30 or 40 km is about 500 m, too coarse to determine the fine structure of the gust. Fortunately, the gust passed the 500-m-tall KTVY TV tower instrumented by NSSL (the star in Fig. 9.36), so that in situ measurement of the vertical structure of the wind as well as the temperature was possible. Both these parameters affect the aircraft lift in the critical

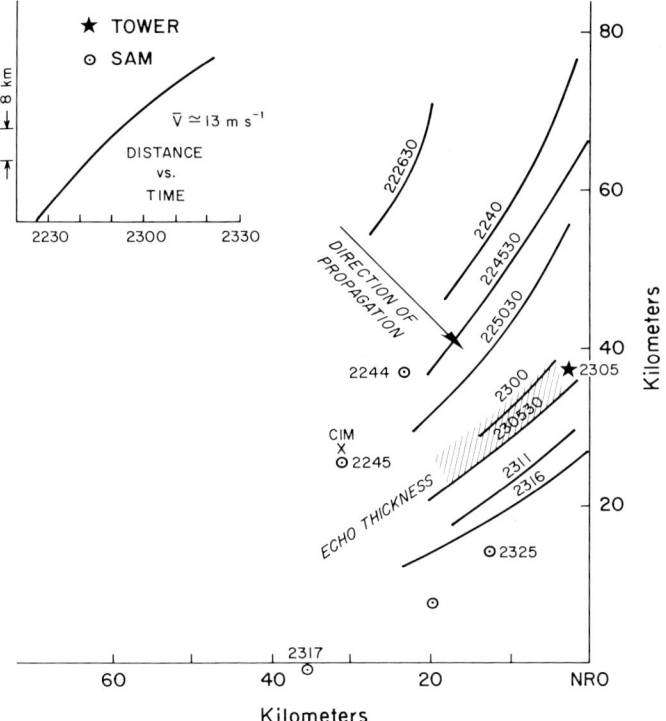

Fig. 9.36. Position of the leading edge of the wind gust on 11 May 1980, showing isochrones of gust front movement as determined by the Norman Doppler radar.

approach phase of landing. Table 9.3 includes the vertical winds and shear of horizontal wind, as well as the associated intensity categories, and suggests what the corresponding effect on aircraft control might be. It is quite evident (Fig. 9.37) that at the uppermost level of the tower (444 m) the wind was from the south and abruptly switched to a north–northwest direction at about 23:08 CST. The wind speed also changes abruptly from 10 m s^{-1} to nearly zero in about 2 min. Assuming stationarity of gust structure and a propagation speed of 13 m s^{-1}, there is an equivalent 10 m s^{-1} change in velocity in a horizontal distance of about 1.5 km or a horizontal shear of 7×10^{-3} s^{-1}. Horizontal shears of horizontal wind as large as 2.2×10^{-2} s^{-1} in gust fronts have also been measured with instrumented towers (Goff et al., 1977). McCarthy and Blick (1979) show, using computer simulation, that shears of this magnitude can cause 100-m vertical deflections of a jet transport from its flight path on its descent to touchdown. Horizontal gradients of air speed can be experienced when the aircraft descends (or ascends) through a vertically sheared horizontal flow. A vertical shear of 2×10^{-1} s^{-1} in horizontal wind will cause the equivalent horizontal shear of head- or tailwind equal to 10^{-2} s^{-1} along a 3° glide slope. A vertical shear of horizontal wind in excess of 7×10^{-2} s^{-1} may perturb aircraft control (Table 9.3). Vertical shears of this magnitude are found in this gust. Horizontal gradients of horizontal wind are accompanied by vertical wind, and both act to perturb aircraft flight. In an incident involving a wide body jet (B-747), a 5 m s^{-1} updraft followed by a 20 m s^{-1} downdraft caused the aircraft to descend 180 m below its intended path along a 3° glide slope (Membery, 1982). Vertical velocity

TABLE 9.3 Proposed Classifications for Wind Shear along a 3° Glide Slope

Intensity of wind shear	Effect on aircraft control	Vertical shear of horizontal wind[a] (s^{-1})	Updraft or downdraft velocity (m s^{-1})
light	little	$(0-7) \times 10^{-2}$	0–2
moderate	significant[b]	$(7-13) \times 10^{-2}$	2–4
strong	considerable difficulty	$(1.3-2) \times 10^{-1}$	4–6
severe	hazardous	$>2 \times 10^{-1}$	>6

[a] Along a glide slope. This is based on criteria given by Brown (1982).

[b] Greene et al. choose a wind shear of 8.4×10^{-2} as significant to aircraft operation.

9.5 Weather Phenomena Observed with a Single Doppler Radar

perturbations larger than ± 3 m s^{-1} at 444-m altitudes (Fig. 9.37) may also affect aircraft control. Thus, the shears and vertical wind in this gust are significant, even though the gust is some 80 km from its source!

The surface wind increased from zero and peaked at about 8 m s^{-1}. The temperature traces at 444 and 266 m have a noticeable drop of 5°C and a return to nearly the ambient temperature as the gust passed the tower.

The structure of the wind and temperature in a southeast-to-northwest cross section obtained from a composite of radar and tower data is in Fig. 9.38. In this cross section the gust appears to be a solitary tube of rolling air, but if an observer remains in a frame moving at the propagation speed of the gust (13 m s^{-1}), the streamlines show the gust to be a cylindrical pool of cool air that slips southeastward with very little, if any, rolling. The gust has many of the characteristics of a solitary wave that harbors wind shears and may present a significant problem for aircraft during the landing and takeoff phase (Christie and Muirhead, 1982).

Because the gust appears to be cut off from its source, the long lifetime of this solitary pool of air should indicate that the magnitude of dissipating

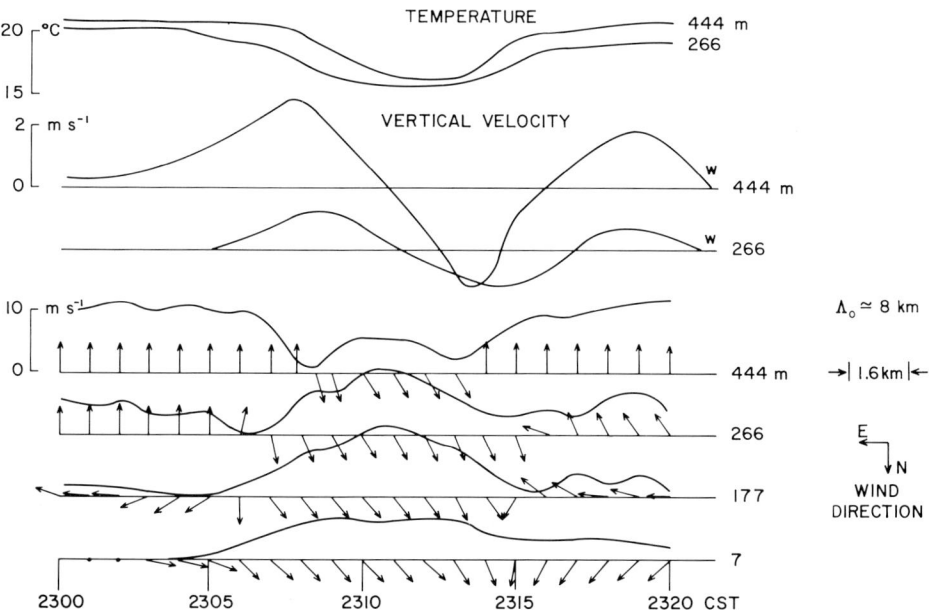

Fig. 9.37. Samples of the time–space cross section of the wind velocity, temperature, and vertical velocity recorded at several levels on the KTVY TV tower during one gust front passage (thin line) on 11 May 1980.

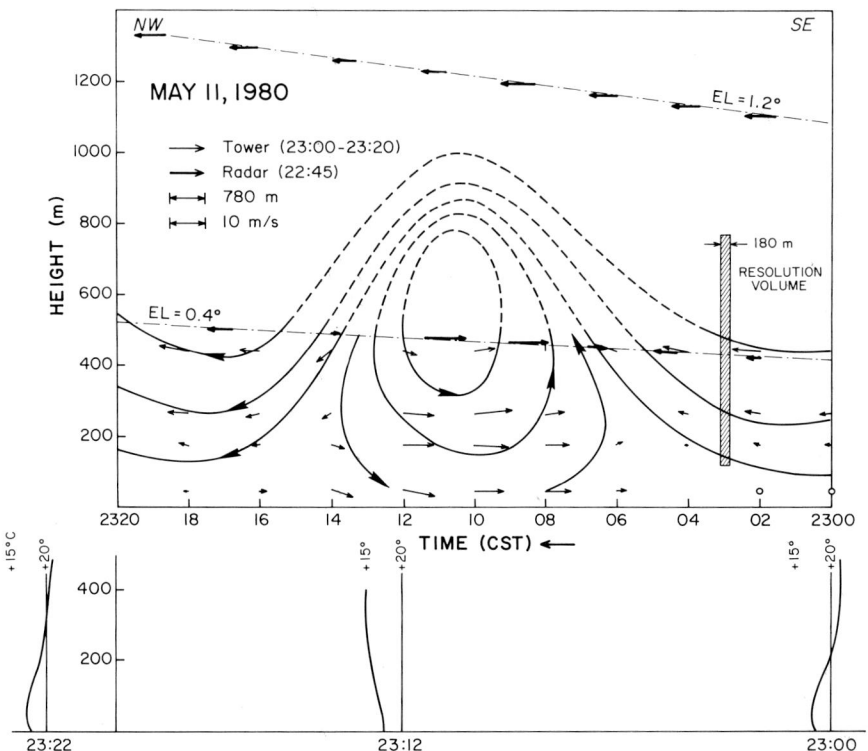

Fig. 9.38. Northwest to southeast cross section of a gust on 11 May 1980. Lightly traced arrows are the tower measured winds projected onto this plane; bold arrows the Doppler velocity. The lower figure gives the temperature profile at three selected times. The hatched area shows the radar's resolution volume cross section in the plane of the beam axis. Stream lines have been drawn subjectively.

forces are weak. Most probably, the stable boundary layer without surface-based convection in the evening has allowed the gust a long life.

9.5.6 Lightning

Lightning flashes are either from cloud to ground (CG) or intracloud (IC). A CG flash consists of one or more return strokes that have large electrical current in the channel between the cloud and ground, whereas all flashes not reaching the ground are termed IC. [See Uman (1969) for a general description of lightning and its measurement.] Both IC and CG flashes create ionized channels that carry electrical currents. Backscattering from these ionized channels is sufficient to produce radar echoes.

9.5 Weather Phenomena Observed with a Single Doppler Radar

Radar detection of lightning has been known for many years (for instance, Ligda, 1956; Miles, 1953). Although some early studies included estimates of values for lightning parameters such as electron density, it was not until 1972 that a theory of lightning echoes from return-stroke channels was proposed (Dawson, 1972). Since then, new measurement techniques and theories have been developed to help unravel the complexities of lightning echoes.

9.5.6.1 Characteristics of Lightning Echoes

In a few early studies of lightning echoes, attempts were made to relate the echoes to known lightning processes by comparing the sferics from lightning (the transient electromagnetic field radiated by a flash) with the radar echoes of the same flashes (Hewitt, 1957). More recently, coincident measurements of lightning echoes (Fig. 9.39) and changes in the electrostatic field, which allow reasonably conclusive identification of the flash type and of interflash events, showed that lightning echoes from a CG flash often have sudden increases in amplitude associated with return strokes, followed by a decay to preflash levels (Szymanski and Rust, 1979). This feature has also been noted during some interstroke intervals. The radar echoes from an IC flash usually have a relatively broad peak with variations superimposed, which may be seen in Fig. 9.39 during the first part (0.6–1.5 s) of the flash. The abrupt changes in the electric field (R in Fig. 9.39b) are typical of return strokes. This flash was independently verified by a lightning strike location system as a CG with at least eight return strokes. Note that abrupt increases in the lightning echoes in a particular resolution volume may not always coincide with the change in the electric field produced by the return stroke.

Comparison of lightning echoes with electric field changes suggests that a continuing discharge process often keeps the channels ionized (Holmes *et al.*, 1980; Zrnić *et al.*, 1982). We define the duration of lightning echoes as the time interval from the instant the echoes are first discernible until they decay back to the predischarge amplitude. This duration forms a lightning echo event. In Fig. 9.39 there are at least nine echo events. Typical lightning echo events have durations of ~ 200–300 ms, but extreme values of 10 ms and 3 s have been observed.

The rise time (the time for the radar received signal to reach peak amplitude) of a lightning echo event is generally a few tens of milliseconds. Although it was first thought that the rise time indicates increasing current flow in the channel (Hewitt, 1957), more recent analysis (Mazur, 1981) shows that observed rise times agree well with ones determined theoretically when the lightning is modeled as an ionized channel or streamer propagating across the antenna beam at a rate of ~ 100 km s^{-1}. Propagation speeds of this order of magnitude for streamers within clouds have been inferred by several investigators, using different measurement techniques (Brook and Ogawa,

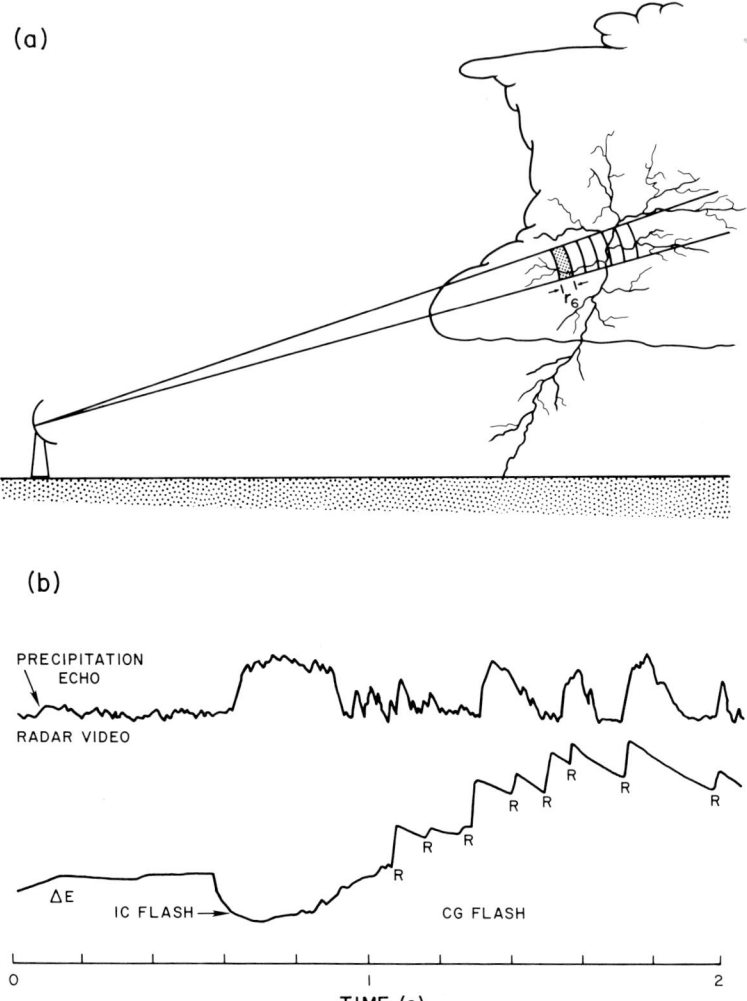

Fig. 9.39. (a) Depiction of a technique for observing lightning and (b) an example of lightning echoes for a long flash that began as an IC and then became a CG. The stippled region in (a), r_6, represents one of the radar's resolution volumes in which the time variation of echo intensity may be studied. The top trace of (b) is the time history of the lightning echo intensity in a resolution volume at a range of 86 km and at a 4-km altitude. Although depicted as a continuous line, the trace is actually a series of points separated in time by the period T_s ($\simeq 1$ ms) between transmitted pulses. Precipitation echo denotes a preflash radar echo intensities due to precipitation. The bottom trace is the electrostatic field change ΔE at the radar site due to the same flash. (Reprinted from Rust *et al.*, 1981.)

1977; Proctor, 1981). When streamers propagate radially along the beam, their echoes can be seen developing in contiguous resolution volumes. Radar observations of several thousand lightning flashes in Oklahoma storms have yielded a maximum propagation speed of ~ 250 km s^{-1}, and streamer propagation speeds of >100 km s^{-1} have been measured within small storms (Holmes et al., 1980).

The radar cross section (RCS) of lightning has been estimated both theoretically and experimentally. The lightning RCS depends on two factors: (1) the size and number of reflecting lightning channel elements within the radar resolution volume and their orientations relative to the incident wave, and (2) the channel plasma characteristics. The initial theoretical work was principally done by Dawson (1972), who assumed that only those ionized channels perpendicular to the beam axis contributed significantly to lightning echoes, and Divinsky (1976), who considered the return echoes irrespective of channel orientation. Both treated the limiting cases of low density and overdense plasma (the plasma is considered overdense when the radar signals are reflected from it as if it were a metallic conductor). More recently Mazur (1981) estimated the RCS as a function of channel orientation relative to the incident wave and any level of electron density and temperature by treating lightning elements as dielectric cylinders of finite length. Experimentally determined estimates of the lightning RCS reported by Mazur (1981), Holmes et al. (1980), and Zrnić et al. (1982) are reasonably consistent. Holmes et al. (1980) conclude that the in-cloud portion of their flashes, which were at altitudes above 6 km, had channels with diameters <2 cm and electron densities of $\sim 10^{16}$ m^{-3}.

9.5.6.2 Lightning and Storm Structure

The fact that radars of wavelength >10 cm can easily locate lightning, even in intense precipitation, out to ~ 300 km makes them useful in the study of lightning relative to storm structure. The use of Doppler radar to observe lightning is a promising new area of research.

A phenomenon of long-standing interest and controversy is the *rain gush*, a dramatic increase in rainfall soon after lightning. Controversy has arisen as a result of different interpretations of the gush. One is that lightning causes the rapid growth of precipitation (Vonnegut and Moore, 1960); another is that the precipitation is in electrostatic levitation before the flash and is merely released by the change in the electric field caused by the flash. Other interpretations have also been suggested.

Recently two radars have been used by Szymanski et al. (1980) to search for a relationship between rain gushes and lightning. A 3-cm-wavelength radar was used to observe precipitation reflectivity while an 11-cm-wavelength radar was used to record the location of lightning echoes. They report that of more than 150 flashes observed by radar, only one flash, which occurred in very low precipitation reflectivities, seemed to cause precipitation growth in the volume around the lightning.

Zrnić et al. (1982) found, as had Szymanski et al. (1980), that lightning in heavy precipitation usually causes no discernible change in precipitation reflectivity. However, Zrnić et al. (1982) did find that lightning altered the reflectivity of precipitation in regions of low Z and suggested that this was a result of changes in hydrometer orientation brought about by changes in the electrostatic field. These changes occurred within 0.1–0.5 s after the flash began. McCormick and Hendry (1979) give convincing evidence that both aerodynamic forces and the storm's electric field control the orientation of ice crystals.

At the NSSL, researchers have combined measurements of precipitation, using a 10-cm wavelength Doppler radar, and lightning echoes, using a 23-cm wavelength radar, to study the coevolving lightning and precipitation structure of large storms (Mazur, 1981). In a squall line (that is, storm cells along a line) on 21 June 1980, over 1,000 lightning flashes were observed during a 44-min period. The storm cells were aligned nearly radially from the colocated radars, and lightning flashes from several cells were recorded. For this one squall line the following were noted: Although lightning is observed throughout the storm, most of the lightning occurs close to the leading edge (relative to storm movement) of the heavy precipitation; the incidence of long lightning flashes that propagate between growing and dissipating cells increases as the precipitation echo intensity decreases in the dissipating cell.

The length of lightning flashes determined from the radial extent of lightning echoes varies greatly. Holmes et al. (1980) report lengths of up to 2 km in small isolated mountain thunderstorms. Ligda's (1956) photographs show lightning with lengths of up to ~ 160 km in large storms. Most lightning observed in Oklahoma is several tens of kilometers long, and some lengths exceed 100 km. These very long flashes have occurred in severe storms or along a line of storms.

Zrnić et al. (1982) have used a vertically pointing 10-cm wavelength Doppler radar to study interactions between lightning and precipitation. Figure 9.40 is an example of the Doppler spectrum of lightning and precipitation observed simultaneously by Doppler radar.

Doppler spectra obtained with vertically pointed radars have been used to determine the drop size distributions (Hauser and Amayenc, 1981). Unless the vertical air velocity and turbulence are estimated to better than 1 m s^{-1}, large errors in the drop size distribution will result. Although several methods have been proposed, none provide correct values of vertical air motion (Joss and Dyer, 1972). If lightning produces ionized channels that permeate the measurement volume, these channels should be perfect tracers of the updraft and turbulence because they have zero terminal velocity, and they should

9.5 Weather Phenomena Observed with a Single Doppler Radar

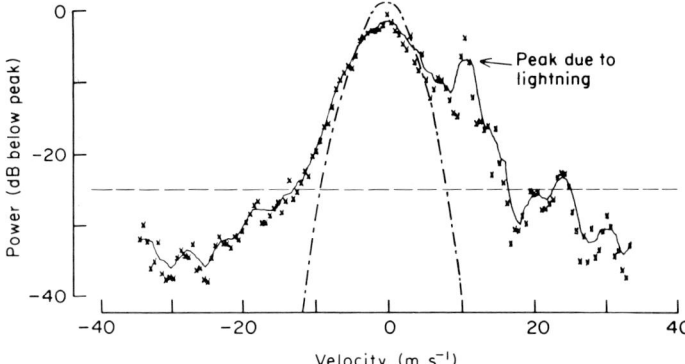

Fig. 9.40. Reflectivity spectral density for precipitation and lightning at 7.1 km versus vertical velocity. The dash-dotted curve is the estimated true spectrum. The solid curve is a five-point running average of the data ($\times$), which were obtained by averaging 12 spectra. Powers below the dashed line and the peak at 24 m s^{-1} are considered to be unreliable (i.e., because of window effects and noise sources; see Chapter 3), so the intersection of this line and the estimated true spectrum yields the assumed reliable maximum fall velocity of about 14 m s^{-1}.

enable one to obtain the distribution of hydrometeors (Fig. 8.27). The resolution of a Doppler spectrum of echos reflected from an ionized channel requires the channel to be reflecting for at least the dwell time (e.g., 50 ms for at 1 m s^{-1} resolution with a 10-cm-wavelength radar). Lhermite (1982) has shown that a channel remains reflecting when the ionization is sustained by current continuing in the channel. Uman (1969, p. 230) calculates that ionization can also be sustained by molecules that are thermally agitated by high temperature, which have decay times of tens of milliseconds.

A subjectively estimated true spectrum for the hydrometeors is also plotted in Fig. 9.40 to show how the spectral density depends on velocity. The true spectrum has an upper speed limit equal to the fall speed of the smallest detectable drops. Departures of the observed spectrum from the true spectrum is caused by turbulence, window broadening, possibly returns received via sidelobes, or other nonmeteorological effects. Assuming that the ionized channels drift with the wind, Zrnić et al. (1982) estimated the updraft velocity to be 11 m s^{-1}, and so the terminal velocity estimate for the largest hydrometeor is 21 m s^{-1}. Such a high velocity was attributed to hail, which indeed fell in the vicinity of the radar.

9.5.7 Hurricanes

Hurricanes are large rotating storms (diameters of 100–300 km) that form over warm tropical seas (Gray, 1978). They have a warm core of low pressure associated with the rotation. In the mature stage, surface friction acts on the low-level air causing it to flow toward the center. As this low-level air spirals inward toward lower pressure, sensible and latent heat energy is picked up from the sea. This increase in energy is sufficient to maintain a near-isothermal condition against the adiabatic cooling (3°–5°C) induced by the lowering pressure as the air approaches the center (Byers, 1944). The low-level inflow does not penetrate to the center of the mature hurricane, but only to the outer edge of the eye. The convergence of low-level air caused by the deceleration of the inflow supports vigorous convection in a band surrounding the eye (the *eyewall*). Maximum winds are found within this band.

The eye is a region of low wind velocity generally free of clouds and precipitation. The eye is maintained by the subsidence of upper tropospheric air in response to the vertical motion (convective heating) in the eyewall (Shapiro and Willoughby, 1982). The eye may be from 5 to 50 km wide, and its translational speed is usually slow, 5–10 m s^{-1}. Most of the energy supplied by the sea is released as latent heat of condensation in the main updrafts of the eyewall. Some rainfall also occurs within spiral bands surrounding the eyewall (Fig. 9.41). The air usually flows outward near the tropopause after rising in the inner region.

As the hurricane passes over land, it diminishes in intensity, since its source of moisture has been shut off and surface friction increases. Severe weather such as tornados, gust fronts, or hail is often a by-product of a hurricane landfall (Novlan and Gray, 1974). The immediate coastal areas are most affected because of the strong winds (from 50 to over 70 m s^{-1}), storm surge, and flooding. Because hurricanes move relatively slowly, there is usually enough time to issue warnings, but this slow motion causes a large accumulation of rainfall (typically 10 in.) and inland flooding. Typical rain rates within the largest convective cells are about 75 mm hr^{-1} (50 dBZ; Jorgensen and Willis, 1982).

As yet there have been no observations of hurricanes with a ground-based Doppler radar, although such a scenario has been described by Baynton (1979). Recently researchers from the National Hurricane Research Laboratory have made Doppler measurements of hurricane Debby on 14 September 1982 using a radar carried on one of the NOAA WP-3D research aircraft. This airborne Doppler radar operates at 3.2-cm wavelength and rotates in a vertical plane perpendicular to the aircraft's ground track.

Hurricane Debby's reflectivity in a horizontal plane 1.5 km above the sea surface (Fig. 9.41) was obtained from the aircraft's horizontally scanning, incoherent 5-cm radar located below the fuselage. The aircraft was flying at an altitude of 1.5 km, and the radar collected data over a full conical scan at 1° elevation. The antenna is stabilized to a horizontal plane against aircraft pitch and roll variations. At this stage of its development, hurricane Debby

9.5 Weather Phenomena Observed with a Single Doppler Radar

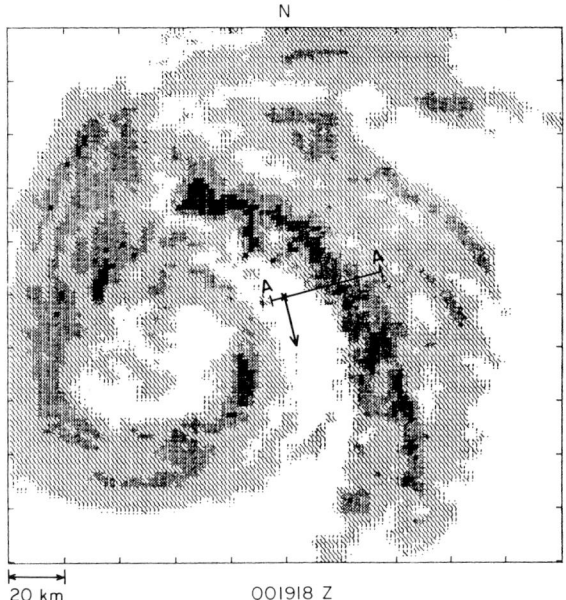

Fig. 9.41. Reflectivity field of hurricane Debby mapped by an airborne 5 cm radar located in the center of the figure (×). Antenna beamwidths are 4.1° (vertical) and 1° (horizontal). The aircraft path at an altitude of 1.5 km is to the south–southeast, as indicated by the arrow. The reflectivity contours are from 20 to 40 dBZ in 5-dB steps. The eye (center at 40 km southwest of the aircraft) has a diameter of about 20 km. The location AA of the vertical cross section plane presented in Color Plate 4 is indicated. The domain is 200 × 200 km. (Courtesy of D. Jorgensen, National Hurricane Center, Coral Gables, Florida.)

had maximum winds about 35 m s^{-1}, barely hurricane force. The peak reflectivity in the rain band 40 km to the east of the center was only 40 dBZ. Incomplete beam filling owing to the large vertical beamwidth of 4.1°, coupled with the 1° tilt to avoid sea clutter, and intervening rainfall attenuation leads to underestimates of the reflectivity factor on the far west side of the storm. The Doppler radar collected vertical cross sections of reflectivity and radial velocity (Color Plate 4). The cross sections are through the outer strong rainband and the eyewall. Convective tops are only 6–8 km in height. The velocity display in this cross section (Color Plate 4) portrays a low-level inflow (below 1.5 km, on the left side of the display) toward the hurricane center (the right side of the display). This inflow is disrupted by the band with pronounced convergence indicated on the inside edge of the band. Outflow is seen near the top of the band. In spite of this relatively weak convection, Hurricane Debby intensified rapidly, becoming a moderately intense hurricane (maximum winds of about 50 m s^{-1}) the next day.

REFERENCES

Alberty, R. L., Burgess, D. W., Hane, C. E., and Weaver, J. F. (1979). "SESAME 1979 Operations Summary." NOAA, U.S. Dept. of Commerce, U.S. Govt. Printing Office, Washington, D.C.

Anthes, R. A., Orville, H. D., and Raymond, D. J. (1982). Mathematical modeling of convection. In "A Social, Scientific, and Technological Documentary" (E. Kessler, ed.), Vol. 2, pp. 495–579. NOAA, U.S. Dept. of Commerce, U.S. Govt. Printing Office, Washington, D.C.

Armijo, L. (1969). A theory for the determination of wind and precipitation velocities with Doppler radars. *J. Atmos. Sci.* **26,** 570–573.

Atlas, D., Srivastava, R. C., and Sekhon, R. S. (1973). Doppler radar characteristics of precipitation at vertical incidence. *Rev. Geophys. Space Phys.* **2,** 1–35.

Barnes, S. L. (1978). Oklahoma thunderstorms on 29–30 April 1970. Pt. I. Morphology of a tornadic storm. *Mon. Weather Rev.* **106,** 673–684.

Baynton, H. W. (1979). The case for Doppler radars along our hurricane affected coasts. *Bull. Am. Meteorol. Soc.* **60,** 1014–1023.

Bohne, A., and Srivastava, R. C. (1975). "Random Errors in Wind and Precipitation Fall Speed Measurement by a Triple Doppler Radar System," Rep. No. 37. Lab. Atmos. Probing, University of Chicago, Chicago, Illinois.

Brandes, E. A. (1977). Severe thunderstorms observed by dual-Doppler radar. *Mon. Weather Rev.* **105,** 113–120.

Brook, M., and Ogawa, T. (1977). The cloud discharge. In "Lightning" (R. H. Golde, ed), Vol. 1, pp. 191–230. Academic Press, New York.

Brown, H. A. (1982). "Analysis and Specification of Slant Wind Shear," AFGL-TR-82-0366. Air Force Geophys. Lab., Hanscom AFB, Massachusetts.

Brown, R. A., and Lemon, L. R. (1976). Single Doppler radar vortex recognition. Part II. Tornadic vortex signatures. *Prepr., Radar Meteorol. Conf., 17th, 1976* pp. 104–109.

Brown, R. A., Lemon, L. R., and Burgess, D. W. (1978). Tornado detection by pulsed Doppler radar. *Mon. Weather Rev.* **106**(1), 29–38.

Browning, K. A. (1982). General circulation of middle-latitude thunderstorms. In "A Social, Scientific, and Technological Documentary" (E. Kessler, ed.), Vol. 2, pp. 211–247. NOAA, U.S. Dept. of Commerce, U.S. Govt. Printing Office, Washington, D.C.

Browning, K. A., and Wexler, R. (1968). A determination of kinematic properties of a wind field using Doppler-radar. *J. Appl. Meteorol.* **7,** 105–113.

Burgess, D. W., Hennington, L. D., Doviak, R. J., and Ray, P.S. (1976). Multimoment Doppler display for severe storm identification. *J. Appl. Meteorol.* **15,** 1302–1306.

Burgess, D. *et al.* (1979). Final report on the Joint Doppler Operational Project (JDOP), 1976–1978. NOAA Tech. Memo ERL NSSL-86.

Byers, H. R. (1944). "General Meteorology." McGraw-Hill, New York.

Caton, P. A. F. (1963). Wind measurement by Doppler radar. *Meteorol. Mag.* **92,** 213–222.

Christie, D. R., and Muirhead, K. J. (1982). "Solitary Waves: A Hazard to Aircraft Operating at Low Altitudes," Rep. ACT 2600. Research School of Earth Sciences, Australian National University, Canberra.

Crane, R. K. (1979). Automatic cell detection and tracking. *IEEE Trans. Geosci. Electron.* **GE-17** (No. 4), 250–261. Special Issue *Radio Meteorol.*

Cressman, G. P. (1959). An operational objective analysis system. *Mon. Weather Rev.* **87,** 367–374.

Dawson, G. A. (1972). Radar as a diagnostic tool for lightning. *J. Geophys. Res.* **7,** 4518–4528.

References

Divinsky, L. J. (1976). An effective radar cross section of a lightning channel. *J. Atmos. Electric.*, Vol. 1, November, 177–185.

Donaldson, R. J., Jr. (1970). Vortex signature recognition by a Doppler radar. *J. Appl. Meteorol.* **9,** 661–670.

Doviak, R. J., and Strauch, R. G. (1980). Pt. III. Single radar data acquisition. The multiple Doppler radar workshop. *Bull. Am. Meteorol. Soc.* **10,** 1178–1183.

Doviak, R. J., Ray, P. S., Strauch, R. G., and Miller, L. J. (1976). Error estimation in wind fields derived from dual-Doppler radar measurement. *J. Appl. Meteorol.* **15,** 868–878.

Draper, W. R., and Smith, H. (1966). "Applied Regression Analysis." Wiley, New York.

Easterbrook, C. C. (1974). Estimating horizontal wind fields by two-dimensional curve fitting of single Doppler radar measurements. *Prepr., Radar Meteorol. Conf., 16th, 1974* pp. 214–219.

Fritsch, J. M. (1975). Cumulus dynamics: Local compensating subsidence and its implications for cumulus parameterization. *Pure Appl. Geophys.* **113,** 851–867.

Fujita, T. T. (1981). Tornadoes and downbursts in the context of generalized planetary scales. *J. Atmos. Sci.* **38,** 1511–1534.

Goff, R. C., Lee, J. T., and Brandes, E. A. (1977). "Gust Front Analytical Study," Final Rep. No. FAA-RD-77-119. FAA Syst. Res. Dev. Serv., Washington, D.C.

Gray, W. M. (1978). Hurricanes: Their formation, structure and likely role in the tropical circulation. *Meteorol. Trop. Oceans, R. Meteorol. Soc.* pp. 155–218.

Greene, G. E., Frank, H. W., Bedard, A. J., Jr., Korell, J. A., Cairns, M. M., and Mandics, P. A. (1977). "Wind Shear Characterization," FAA Rep. No. FAA-RD-77-33. FAA Syst. Res. Dev. Serv., Washington, D.C.

Harrold, T. W., and Browning, K. A. (1971). Identification of preferred areas of shower development by means of high power radar. *Q. J. R. Meteorol. Soc.* **97,** 330–339.

Hauser, D., and Amayenc, P. (1981). A new method for deducing hydrometeor-size distributions and vertical air motions from Doppler radar measurements at vertical incidence. *J. Appl. Meteorol.* **20,** 547–555.

Hewitt, F. J. (1957). Radar echoes from interstroke processes in lightning. *Proc. Phys. Soc. London, Sect. B* **70,** 961.

Holmes, C. R., Szymanski, E. W., Szymanski, S. J., and Moore, C. B. (1980). Radar and acoustic study of lightning. *J. Geophys. Res.* **85,** 7517–7532.

Jorgensen, D. P., and Willis, P. T. (1982). A Z-R relationship for hurricanes. *J. Appl. Meteorol.* **21,** 356–366.

Joss, J., and Dyer, R. (1972). Large errors involved in deducing drop-size distributions from Doppler radar data due to vertical air motion. *Prepr., Radar Meteorol. Conf., 15th, 1972* pp. 179–180.

Joss, J., and Waldvogel, A. (1970). Raindrop size distributions and Doppler velocities. *Prepr., Radar Meteorol. Conf., 14th, 1970* pp. 153–156.

Kessler, E., ed. (1982). "Thunderstorms: A Social, Scientific, and Technological Documentary," Vols. I, II, and III. NOAA, U.S. Dept. of Commerce, U.S. Govt. Printing Office, Washington, D.C.

Klemp, J. B., Wilhelmson, R. B., and Ray, P. S. (1981). Observed and numerically simulated structure of a mature supercell thunderstorm. *J. Atmos. Sci.* **38,** 1558–1580.

Koscielny, A. J., Doviak, R. J., and Rabin, R. (1981). Statistical considerations in the estimation of divergence from single-Doppler radar and application to prestorm boundary-layer observations. *J. Appl. Meteorol.* **21,** 197–210.

Lemon, L. R., and Doswell, C. A., III (1979). Severe thunderstorm evolution and mesocyclone structure as related to tornadogenesis. *Mon. Weather Rev.* **107,** 1184–1197.

Lhermitte, R. M. (1970). Dual-Doppler radar observations of convective storm circulation. *Prepr., Radar Meteorol. Conf., 14th, 1970* pp. 153–156.

Lhermitte, R. M. (1982). Doppler radar observations of triggered lightning. *Geophys. Res. Lett.* **9** (No. 6), 712–715.
Lhermitte, R. M., and Atlas, D. (1961). Precipitation motion by pulse Doppler radar. *Proc. Weather Radar Conf., 9th, 1961* pp. 218–223.
Ligda, M. H. (1956). The radar observation of lightning. *J. Atmos. Terr. Phys.* **9**, 329–346.
McCarthy, J., and Blick, E. F. (1979). An airport wind shear detection and warning system. *Prepr., WMO Tech. Conf. Aviat. Meteorol. (TECAM), 1979* pp. 1–35.
McCormick, G. C., and Hendry, A. (1979). Radar measurement of precipitation-related depolarization in thunderstorms. *IEEE Trans. Geosci. Electron., Spec. Issue* **GE-17**, 142–150.
Mazur, V. (1981). Investigation of radar returns from lightning discharges. Ph.D. Dissertation, University of Oklahoma, Norman.
Membery, D. A. (1982). A documented example of strong wind-shear. *Weather* **37**, 19–22.
Miles, V. G. (1953). Radar echoes associated with lightning. *J. Atmos. Terr. Phys.* **3**, 258–262.
Novlan, D. J., and Gray, W. M. (1974). Hurricane spawned tornadoes. *Mon. Weather Rev.* **102**, 476–488.
O'Brien, J. J. (1970). Alternative solutions to the classical vertical velocity problem. *J. Appl. Meteorol.* **9**, 197–203.
Ogura, Y., and Chen, Y. L. (1977). A life history of an intense mesoscale convective storm in Oklahoma. *J. Atmos. Sci.* **34**, 1458–1476.
Proctor, D. E. (1981). Radar observations of lightning. *J. Geophys. Res.* **86**, 4041–4071.
Rabin, R., and Zrnić, D. S. (1980). Subsynoptic-scale vertial wind revealed by dual-Doppler radar and VAD analysis. *J. Atmos. Sci.* **37**, 644–654.
Ray, P. S., Ziegler, C. L., Serafin, R. J., and Bumgarner, W. (1980). Single- and multiple-Doppler radar observations of tornadic storms. *Mon Weather Rev.* **108**, 1607–1625.
Ray, P. S., Johnson, B. C., Johnson, K. W., Bradberry, J. S., Stephens, J. J., Wagner, K. K., Wilhelmson, R. B., and Klemp, J. B. (1981). The morphology of several tornadic storms on 20 May 1977. *J. Atmos. Sci.* **38**, 1643–1663.
Reinhart, R. E. (1979). Internal storm motion from a single non-Doppler weather radar. Ph.D. Dissertation, Colorado State University, Ft. Collins.
Rust, W. D., Taylor, W. L., MacGorman, D. R., and Arnold, R. T. (1981). Research on electrical properties of severe thunderstorms in the great plains. *Bull. Am. Meteorol. Soc.* **62**, 1286–1293.
Shapiro, L. J., and Willoughby, H. E. (1982). The response of balanced hurricanes to local sources of heat and momentum. *J. Atmos. Sci.* **39**, 378–394.
Sinclair, P. C. (1973). Severe storm air velocity and temperature structure from penetrating aircraft. *Prepr., Conf. Severe Local Storms, 8th, 1973* pp. 25–32.
Stackpole, J. D. (1961). The effectiveness of raindrops as turbulence sensors. *Proc., Weather Radar Conf., 9th, 1961* pp. 212–217.
Szymanski, E. W., and Rust, W. D. (1979). Preliminary observations of lightning radar echoes and simultaneous electric field changes. *Geophys. Res. Lett.* **6**, 527–530.
Szymanski, E. W., Szymanski, S. J., Holmes, C. R., and Moore, C. B. (1980). An observation of a precipitation echo intensification associated with lightning. *J. Geophys. Res.* **85**, 1951–1953.
Testud, J., Breger, G., Amayenc, P., Chong, M., Nutten, B., and Sauvageot, A. (1980). A Doppler radar observation of a cold front: Three-dimensional air circulation, related precipitation system and associated wavelike motions. *J. Atmos. Sci.* **37**, 78–98.
Uman, M. A. (1969). "Lightning." McGraw-Hill, New York.
Vonnegut, B., and Moore, C. B. (1960). A possible effect of lightning discharge on precipitation formation process. *In* "Physics of Precipitation," Monogr. No. 5, pp. 287–290. Am. Geophys. Union, Washington, D.C.

References

Waldteufel, P., and Corbin, H. (1979). On the analysis of single Doppler data. *J. Appl. Meteorol.* **18,** 532–542.

Ziegler, C. L. (1978). "A Dual-Doppler Variational Objective Analysis as Applied to Studies of Convective Storms," NOAA Tech. Memo. ERL NSSL-85. Natl. Severe Storms Lab., Norman, Oklahoma. Available Natl. Tech. Inf. Serv., Operations Div., Springfield, Virginia.

Zrnić, D. S., and Doviak, R. J. (1975). Velocity spectra of vortices scanned with a pulse-Doppler radar. *J. Appl. Meteorol.* **14,** 1531–1539.

Zrnić, D. S., and Istok, M. J. (1980). Wind speeds in two tornadic storms and a tornado, deduced from Doppler spectra. *J. Appl. Meteorol.* **19,** 0065–0075.

Zrnić, D. S., Rust, W. D., and Taylor, W. L. (1982). Doppler radar echoes of lightning and precipitation at vertical incidence. *J. Geophys. Res.* **87,** 7179–7191.

10
Measurement of Turbulence

It was pointed out in Chapter 5 that the mean velocity and spectrum width measured by a Doppler radar are weighted averages of point velocities in the resolution volume. Mean velocity measurements are quite adequate to depict motion on scales larger than the resolution volume V_6, and the spectrum width gives a rough estimate of the variance of motion on scales smaller than V_6, but we cannot infer the details of the flow inside the volume. However, in some special instances, such as when a vortex is within the resolution volume, certain attributes of motion can be deduced from the Doppler spectra (see Chapter 9). Doppler radar offers intriguing possibilities for the measurement and study of turbulence on scales larger or smaller than the resolution volume. The purpose of this chapter is to introduce the basic concepts of turbulence and to establish a firm connection between the physical (statistical) properties of the atmosphere and Doppler-derived measurements. We present relationships between turbulence and the mean Doppler velocity and spectrum width.

10.1 STATISTICAL THEORY OF TURBULENCE

Thus far in our analysis of weather echo signals it has been assumed that statistical fluctuations are a result of the random motion of scatterers in the atmosphere. However, no characteristics of such motion were given. In order to relate measurements of the Doppler velocity and spectrum width to turbulent flow that may affect aircraft, it is necessary to introduce elements of the statistical theory of turbulence. Interpretating Doppler radar observations of thunderstorms will be easier if theories of turbulence apply. After all, most thunderstorm velocity fields are turbulent. Not only will this theory help establish a quantitative relationship between the Doppler spectrum width and turbulence, but it will also prove to be very useful in determining the intensity of echoes from fluctuations in the refractive index. Such echoes are treated in Chapter 11.

10.1 Statistical Theory of Turbulence

10.1.1 Turbulence Spectra and the Correlation Function

Some measurable properties of turbulence can best be described in statistical terms. A random variable that is a function of position **r** and time defines a *random field*. If the random variable is a scalar quantity such as the refractive index $n(\mathbf{r}, t)$, the field is called a *scalar field*. When the random variable is a vector, it forms a *vector field*. Thus, the velocities in a turbulent flow form a random vector field such that at any point **r** (and time t) the velocity vector is a random variable $\mathbf{v}(\mathbf{r}, t)$. In this chapter, various formulas are obtained for random vector and scalar fields.

A great deal of information about the vector field can be obtained if the first and second statistical moments are known. A random field is statistically homogeneous if its mean value is constant and if its correlation function does not change when the pair of points **r**, **r**′ are both displaced by the same amount in the same direction. Thus, for a homogeneous random field in which the mean value has been removed,

$$\langle v_i(\mathbf{r}) \rangle = 0. \tag{10.1}$$

In this and the next chapter brackets denote ensemble averages, whereas the overbar will signify a spatial average weighted by the radar resolution volume. The correlation function of two zero-mean components v_i, v_j depends only on the distance $\boldsymbol{\rho} = \mathbf{r}' - \mathbf{r}$ between the two field points:

$$R_{ij}(\mathbf{r}, \mathbf{r}') \equiv \langle v_i(\mathbf{r}) v_j(\mathbf{r}') \rangle = R_{ij}(\boldsymbol{\rho}). \tag{10.2}$$

The time dependence is implicit in the vectors **r** and $\boldsymbol{\rho}$; i.e., these are four-dimensional vectors, and $i = 1, 2$, or 3 identifies the velocity component along one of three orthogonal directions in a frame of reference moving with the mean flow.

If the correlation does not depend on the direction of the vector $\boldsymbol{\rho}$, the field is *isotropic*. The concept of isotropy is significantly different for vector fields than for scalar fields, as we now illustrate. Consider the correlation of the velocity at **r** with that at **r**′, and suppose the separation vector $\boldsymbol{\rho}$ is allowed to rotate about **r**. Construct a coordinate system that is rigidly attached to $\boldsymbol{\rho}$ and that, without loss of generality, has axes parallel and perpendicular to $\boldsymbol{\rho}$ (Fig. 10.1). A vector field is isotropic if the correlations of the various components (e.g., $\langle v_l(\mathbf{r}') v_l(\mathbf{r}) \rangle$) are independent of the orientation of $\boldsymbol{\rho}$. Although the component correlations in the coordinate system tied to $\boldsymbol{\rho}$ are independent of the direction of $\boldsymbol{\rho}$, it can be deduced that the correlation of components in the nonrotated frame does, in general, depend on direction of $\boldsymbol{\rho}$. Thus for an isotropic field $\langle u(\mathbf{r}) u(\mathbf{r}') \rangle$ may depend on the direction of $\boldsymbol{\rho}$, so u cannot be considered a scalar, for which correlation is a function only of the magnitude $|\boldsymbol{\rho}|$.

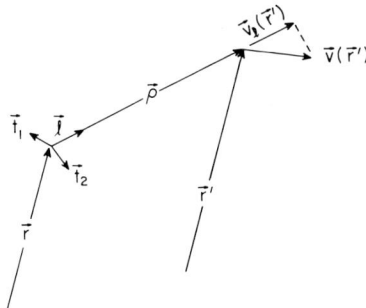

Fig. 10.1. Coordinate system of orthogonal unit vectors **l**, **t**$_1$, **t**$_2$ rigidly tied to the separation vector **ρ**. The projection $v_l(\mathbf{r}')$ of the velocity vector **v**(**r**') is onto the coordinate axis **l** that is parallel to **ρ**. Vectors are indicated by overarrows in figures and by bold letters in the text.

A homogeneous field need not be isotropic. For example, the field having the correlation function

$$R_{ij}(\mathbf{r} - \mathbf{r}') = R_{ij}[\alpha(x - x') + \beta(y - y') + \gamma(z - z')] \tag{10.3}$$

is homogeneous but not isotropic. Although a random vector field can be homogeneous but not isotropic, it can be shown that an isotropic field must be homogeneous.

For isotropic random vector fields, the correlation tensor is in general

$$R_{ij}(\mathbf{r}, \mathbf{r} + \boldsymbol{\rho}) = R_{ij}(\boldsymbol{\rho}) = P(\rho)\delta_{ij} + Q(\rho)\rho_i\rho_j/\rho^2 \tag{10.4}$$

(Tatarskii, 1971), which does not depend on the location of **r** but does depend on the direction and magnitude of **ρ**, where δ_{ij} is the Kronecker delta function and ρ_i is the projection of **ρ** on the ith axis. $P(\rho)$ and $Q(\rho)$ can be any functions of the magnitude ρ. If **ρ** lies along the kth coordinate of the fixed frame, it can be deduced from (10.4) that $R_{ii} = R_{jj} = P(\rho)$, which is then called the *transverse correlation function* $R_{tt}(\rho)$, and $R_{kk} = P + Q$ is R_{ll}, the *longitudinal correlation*. Thus (10.4) can be expressed

$$R_{ij}(\boldsymbol{\rho}) = (1/\rho^2)(R_{ll} - R_{tt})\rho_i\rho_j + R_{tt}\delta_{ij}. \tag{10.5}$$

Even though isotropic fields on scales commensurate with the radar resolution volume are an exception in the atmosphere, the mathematical handling of such fields is tractable, and general results are relatively simple to obtain.

The Fourier transform in vector wave space of the correlation function of a statistically homogeneous field is known as the *spectral density tensor*:

$$\Phi_{ij}(\mathbf{K}) \equiv \frac{1}{(2\pi)^3}\int R_{ij}(\boldsymbol{\rho})\exp(-j\mathbf{K}\cdot\boldsymbol{\rho})\,dV_\rho. \tag{10.6}$$

Its inverse retrieves the correlation function:

$$R_{ij}(\boldsymbol{\rho}) = \int \Phi_{ij}(\mathbf{K})\exp(j\mathbf{K}\cdot\boldsymbol{\rho})\,dV_K, \tag{10.7}$$

10.1 Statistical Theory of Turbulence

where V_ρ and V_K are volumes in lag distance ρ space and wavenumber **K** space, and the integrals are over these volumes. Introducing spherical coordinates and executing the integrations along the angular coordinates for an assumed isotropic scalar field, the transform pair, (10.6) and (10.7), reduces to the following:

For scalar fields,

$$\Phi(K) = \frac{1}{2\pi^2} \int_0^\infty \frac{\sin K\rho}{K\rho} R(\rho)\rho^2 \, d\rho, \tag{10.8}$$

$$R(\rho) = 4\pi \int_0^\infty \frac{\sin K\rho}{K\rho} \Phi(K) K^2 \, dK. \tag{10.9}$$

For an isotropic vector field, the spectral density tensor must, like the correlation tensor, be independent of rotation, so it can also be expressed in a form analogous to (10.4)

$$\Phi_{ij}(\mathbf{K}) = F(K)\delta_{ij} + G(K) K_i K_j / K^2, \tag{10.10a}$$

or to (10.5),

$$\Phi_{ij}(\mathbf{K}) = (1/K^2)(\Phi_{ll} - \Phi_{tt}) K_i K_j + \Phi_{tt}\delta_{ij}, \tag{10.10b}$$

where Φ_{ll} and Φ_{tt} are functions of the magnitude of the wavenumber and are called the longitudinal and transverse spectral densities. They are related to R_{ll} and R_{tt} through Eqs. (10.5)–(10.7) and (10.10a) and are not simply Fourier transform pairs of their corresponding correlation functions [e.g., $\Phi_{ll} \neq \mathcal{F}(R_{ll})$]. Panchev (1971, p. 102) gives the relations between Φ_{ll}, Φ_{tt} and R_{ll}, R_{tt} for isotropic vector fields. These relations show that, for example, Φ_{tt} is a function of both R_{ll} and R_{tt}. If the flow is incompressible, the continuity equation puts an additional constraint on the random vector field, resulting in

$$R_{tt} = \frac{1}{2\rho} \frac{d}{d\rho}(\rho^2 R_{ll}). \tag{10.11}$$

The condition of incompressibility results in $\Phi_{ll}(K) \equiv 0$ (Panchev, 1971, p. 108), so that (10.10) reduces to

$$\Phi_{ij}(\mathbf{K}) = \left(\delta_{ij} - \frac{K_i K_j}{K^2}\right) \frac{E(K)}{4\pi K^2}, \tag{10.12}$$

where

$$E(K) = 4\pi K^2 \Phi_{tt}(K) \tag{10.13}$$

is the single spectral density that characterizes isotropic incompressible turbulent flow. $E(K) \, dK$ is the contribution to the total kinetic energy per unit mass (or variance) from wave numbers of magnitude from K to $K + dK$.

Measurements of three-dimensional spectra in the past were quite impractical because the sensing instruments, such as towers or even aircraft, could at best estimate the one-dimensional counterpart $S_{ij}(K_1)$, where K_1 is the wavenumber along the direction of the $i = 1$ coordinate. In contrast, radars can in a very short time (i.e., a few minutes) obtain a volume of velocity data. Even though the three-dimensional spectrum can be obtained from radar data, it is considerably simpler to compute spectra along lines. Moreover, such one-dimensional spectra provide a convenient visual picture of the wave components along the line. By applying the formula $2\pi\delta(K) = \int \exp(jK\rho) d\rho$ to (10.7), we can derive the following relation:

$$S_{ij}(K_1) \equiv \frac{1}{2\pi} \int_{-\infty}^{\infty} R_{ij}(\rho_1, 0, 0) e^{-jK_1\rho_1} d\rho_1 = \int\int_{-\infty}^{\infty} \Phi_{ij}(\mathbf{K}) dK_2 dK_3, \quad (10.14)$$

where now the lag vector $\boldsymbol{\rho}$ is directed along the coordinate axis for which $i \equiv 1$. For isotropic scalar fields, the spectral density $\Phi(K)$ is a function of the magnitude of the wavenumber, so with a change to cylindrical coordinates and after integration, the double integral in (10.14) reduces to

$$S(K_1) = 2\pi \int_0^\infty \Phi[(K_1^2 + K'^2)^{1/2}] K' dK' = 2\pi \int_{K_1}^\infty \Phi(K) K dK, \quad (10.15)$$

where $K' = (K_2^2 + K_3^2)^{1/2}$, or

$$\Phi(K) = -\frac{1}{2\pi K} \frac{dS(K)}{dK}. \quad (10.16)$$

For incompressible flow, the one-dimensional longitudinal and transverse spectra of isotropic vector fields are related to $E(K)$ through (10.12)–(10.14), as

$$S_l(K_1) = \frac{1}{2} \int_{K_1}^\infty \left(1 - \frac{K_1^2}{K^2}\right) \frac{E(K)}{K} dK, \quad (10.17)$$

$$S_t(K_1) = \frac{1}{4} \int_{K_1}^\infty \left(1 + \frac{K_1^2}{K^2}\right) \frac{E(K)}{K} dK, \quad (10.18)$$

where $S_l(K_1) = S_{11}(K_1)$, whereas $S_t(K_1) = S_{22}(K_1) = S_{33}(K_1)$. Note that any wavenumber K_i can replace K_1 in (10.17) and (10.18), in conformity with the fact that isotropic turbulence is independent of direction.

Of considerable value in turbulence studies is the correlation function, given by

$$R(\rho) = \frac{R(0)}{2^{\nu-1}\Gamma(\nu)} \left(\frac{\rho}{\rho_0}\right)^\nu K_\nu\left(\frac{\rho}{\rho_0}\right) \quad \text{for} \quad \nu > -\tfrac{1}{2}, \quad (10.19)$$

which has the one-dimensional spectrum

$$S(K) = R(0)\Gamma(v + \tfrac{1}{2})\rho_0/\sqrt{\pi}\,\Gamma(v)[1 + (K\rho_0)^2]^{v+1/2}, \tag{10.20a}$$

where $K_v(\rho/\rho_0)$ is the Bessel function of the second kind of order v, and ρ_0 (the correlation length) is the distance beyond which variables can be considered to be uncorrelated. The Bessel correlation function is particularly attractive because it can, depending on the values selected for v, ρ_0, fit many experimental data for the correlations of the longitudinal and transverse components quite well. In fact, the value $v = \tfrac{1}{3}$ gives a spectrum that coincides with one predicted by turbulence theory (Batchelor, 1953, p. 114) for the range of K corresponding to $K\rho_0 \gg 1$ (i.e., the inertial subrange). $S(K)$ is the variance, or intensity of turbulence, per unit wavenumber. The three-dimensional spectrum corresponding to the correlation (10.19) of a scalar field is given by

$$\Phi(K) = \frac{\Gamma(v + \tfrac{3}{2})}{\pi\sqrt{\pi}\,\Gamma(v)} \frac{R(0)\rho_0^3}{[1 + (K\rho_0)^2]^{v+3/2}}. \tag{10.20b}$$

10.1.2 Structure Functions (Locally Homogeneous Fields)

The previous sections have dealt with homogeneous random fields. However, quite often the velocity, temperature, pressure, and humidity change gradually in space, and therefore it becomes necessary to decide whether these changes should be regarded as slow changes in the statistical moments (e.g., the mean and correlation) or as large-scale irregularities. It is evident that the mean square value or intensity of the fluctuations and the form of the autocorrelation function and its corresponding power spectrum depend on this decision.

For these situations it has been found convenient (Kolmogorov, 1941) to define increments that are the differences in the random $v_i(\mathbf{r}_1)$ taken at two points spaced $\mathbf{p}$ units apart:

$$v_i(\mathbf{r} + \mathbf{p}) - v_i(\mathbf{r}), \tag{10.21}$$

which for $\mathbf{p}$ not too large (i.e., small compared to the length of gradual changes) is a random variable that can be considered homogeneous. For example, if a random variable $v_i(\mathbf{r})$ has a spatial distribution like that sketched in Fig. 10.2 and has a mean value that changes with r as indicated by the dashed line, it is not a homogeneous process. However, if the mean value is linear over the range of r under consideration, it can be shown that the difference (10.21) is a homogeneous random function. Under these conditions

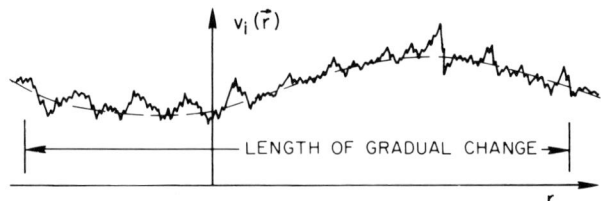

Fig. 10.2. Example of a random variable with a gradual change or trend.

$v_i(\mathbf{r})$ is referrred to as a *random variable with stationary first increments*. For example, with

$$\langle v_i(\mathbf{r}) \rangle = \mathbf{a} \cdot \mathbf{r} + b \tag{10.22}$$

it can be shown that

$$\langle v_i(\mathbf{r} + \mathbf{\rho}) - v_i(\mathbf{r}) \rangle = \mathbf{a} \cdot \mathbf{\rho}, \tag{10.23}$$

which is independent of $\mathbf{r}$. If the autocorrelation of (10.21) is now a function only of the difference $\mathbf{\rho} = \mathbf{r}' - \mathbf{r}$, then (10.21) is statistically (in the wide sense; see Papoulis, 1965, p. 302) homogeneous process.

We can represent the correlation function of the increment as

$$\langle [v_i(\mathbf{r} + \mathbf{\rho}) - v_i(\mathbf{r})][v_i(\mathbf{r}' + \mathbf{\rho}) - v_i(\mathbf{r}')] \rangle$$
$$= \tfrac{1}{2}\langle v_i(\mathbf{r} + \mathbf{\rho}) - v_i(\mathbf{r}')]^2 \rangle + \tfrac{1}{2}\langle [v_i(\mathbf{r}) - v_i(\mathbf{r}' + \mathbf{\rho})]^2 \rangle$$
$$- \tfrac{1}{2}\langle [v_i(\mathbf{r} + \mathbf{\rho}) - v_i(\mathbf{r}' + \mathbf{\rho})]^2 \rangle - \tfrac{1}{2}\langle [v_i(\mathbf{r}) - v_i(\mathbf{r}')]^2 \rangle. \tag{10.24}$$

The function

$$D_{ii}(\mathbf{r}_m, \mathbf{r}_n) \equiv \langle [v_i(\mathbf{r}_m) - v_i(\mathbf{r}_n)]^2 \rangle \tag{10.25}$$

that forms each term of (10.24) is called the *structure function* of the random process. If (10.24) depends only on the difference $\mathbf{r}_m - \mathbf{r}_n$, the random variable $v_i(\mathbf{r})$ is then said to form a *locally homogeneous process*. The structure function

$$D_{ii}(\mathbf{\rho}) = \langle [v_i(\mathbf{r} + \mathbf{\rho}) - v_i(\mathbf{r})]^2 \rangle \tag{10.26}$$

is the *basic characteristic* of a random process having stationary first increments. From (10.26) we see that $D_{ii}(\mathbf{\rho})$ is proportional to the intensity of the fluctuations with spatial length $\mathbf{\rho}$, small compared to the length of a gradual change (Fig. 10.2). If the random process under consideration has structure function (10.26) independent of $\mathbf{r}$ for the range $\mathbf{\rho}$ under consideration, then $v_i(\mathbf{r})$ is said to be locally homogeneous.

For a completely homogeneous random process (i.e., one without large-scale changes), there is a simple direct relation between the structure function

10.1 Statistical Theory of Turbulence

$D_{ii}(\boldsymbol{\rho})$ and the correlation function $R_{ii}(\boldsymbol{\rho})$. This is made evident by expanding (10.26):

$$D_{ii}(\boldsymbol{\rho}) = 2[R_{ii}(0) - R_{ii}(\boldsymbol{\rho})]. \tag{10.27}$$

Now, because v_i is assumed to have zero mean so that $R_{ii}(\infty) = 0$ [see Eq. (5.15)], it follows from (10.27) that we can express the autocorrelation function in terms of the structure function:

$$R_{ii}(\boldsymbol{\rho}) = \tfrac{1}{2}[D_{ii}(\infty) - D_{ii}(\boldsymbol{\rho})]. \tag{10.28}$$

The structure function is widely used in turbulence studies because large-scale irregularities are neither homogeneous nor isotropic. Thus, in beginning a study of a random process that we are not sure is completely homogeneous, it is prudent to construct the structure function and to test its statistics rather than to use directly the correlation function.

10.1.3 Structure and Correlation Functions from Similarity Assumptions

In this section the general relationships between the structure function and the spectra of the correlation function for locally isotropic scalar fields and those between the longitudinal and transverse components of vector fields are discussed. Formal relationships among the structure function, the correlation function, and their spectra developed in previous sections are used to derive the spectral densities. Often in applications of the statistical theory of turbulence that are relevant to the atmosphere, the structure function corresponding to the correlation function given by (10.19) is useful. Applying (10.27), we obtain

$$D_{ii}(\rho) = 2R_{ii}(0)\left[1 - \frac{1}{2^{\nu-1}\Gamma(\nu)}\left(\frac{\rho}{\rho_0}\right)^\nu K_\nu\left(\frac{\rho}{\rho_0}\right)\right], \tag{10.29}$$

where now the indices ii must denote either the transverse or the longitudinal component with respect to the lag vector $\boldsymbol{\rho}$. For small values of $\rho \ll \rho_0$ and $|\nu| < 1$, (10.29) can be shown to reduce to

$$D_{ii}(\rho) \approx 2R_{ii}(0)[\Gamma(1-\nu)/\Gamma(1+\nu)](\rho/2\rho_0)^{2\nu}. \tag{10.30}$$

With $\nu = \tfrac{1}{3}$ this becomes

$$D_{ii}(\rho) = C_{ii}^2 \rho^{2/3}, \tag{10.31}$$

where

$$C_{ii}^2 = [3 \times 2^{1/3}\Gamma(\tfrac{2}{3})/\Gamma(\tfrac{1}{3})]R_{ii}(0)\rho_0^{-2/3}. \tag{10.32}$$

A relationship exactly analogous to (10.11) ties D_{tt} to D_{ll}. From this relationship it follows that, for structure functions having any power law dependence, $C_{tt}^2 = \frac{4}{3} C_{ll}^2$. That is, for isotropic turbulence the constant C_{tt}^2 of the transverse velocity component is larger than the structure constant C_{ll}^2 for the longitudinal component. Because $R_{ll}(0) = R_{tt}(0)$, the ρ_0 for longitudinal structure constant differs from the one for the transverse.

Equation (10.31) was predicted by Kolmogorov, who used dimensional analysis and similarity arguments. Briefly, he hypothesized that there is a range of eddy sizes in which there is no creation or dissipation of energy. Only a cascade (flow) of energy from larger eddies to smaller ones occurs as they fragment. In equilibrium this continuous flux of energy numerically equals the energy dissipation rate ε due to viscosity at very small sizes. Dimensional analysis then reveals that in order for the longitudinal structure function $D_{ll}(\rho)$ to be independent of the kinematic viscosity ν, it must have the form (Tatarskii, 1971, p. 54).

$$D_{ll}(\rho) = C^2 \varepsilon^{2/3} \rho^{2/3}, \tag{10.33}$$

where C^2 is a dimensionless constant of the order of unity. The sizes ρ for which (10.33) holds satisfy

$$\rho_i < \rho < \rho_o, \tag{10.34}$$

where ρ_o is the outer scale and ρ_i is the inner scale of turbulence. It can be shown (Tatarskii, 1971) from turbulence theory that for $\rho < \rho_i$

$$D_{ll}(\rho) = \tfrac{1}{15}(\varepsilon/\nu)\rho^2. \tag{10.35}$$

The range given by Eq. (10.34) for which the power of $\frac{2}{3}$ law of Eq. (10.33) is applicable is known as the *inertial subrange*. Note that because of (10.27) the correlation function follows a power of $\frac{2}{3}$ law if turbulence is completely homogeneous. The region for which $\rho < \rho_i$ is known as the *dissipation range*.

The power spectrum for the inertial subrange, where $K\rho_0 \gg 1$, is obtained by substituting (10.32) in (10.20a):

$$S_i(K) = C_{ii}^2 \Gamma(\tfrac{5}{6}) K^{-5/3} / 3 \times 2^{1/3} \Gamma(\tfrac{2}{3}) \sqrt{\pi}. \tag{10.36}$$

Thus, the one-dimensional spectra of the velocity components parallel to and perpendicular to the separation vector $\boldsymbol{\rho}$ have a $K^{-5/3}$ dependence. C_{ii}^2 is a measure of the intensity of the fluctuations having scale sizes within the inertial subrange.

The general form of the correlation in the inertial subrange can be written

$$R_{ii}(\rho, \tau_1 = 0) = R_{ii}(0)[1 - (\rho/\rho_0)^{2/3}], \tag{10.37}$$

where $\rho \ll \rho_0$ and the time lag τ_1 is shown explicitly in (10.37) to enunciate a possible time dependence, which will be treated shortly. This well-known relationship assumes steady state conditions and was deduced from similarity considerations. Equation (10.37) is valid for isotropic turbulence in the inertial subrange, where viscous forces are negligible and energy cascades from large-scale eddies toward dissipation by viscosity at small scales (Tennekes and Lumley, 1972). In the lower atmosphere the inertial subrange consists of eddies ranging in size from centimeters to several tens of meters. However, the correlation function of the longitudinal and transverse components in planes parallel to the earth's surface sometimes exhibits the two-thirds behavior to much larger scales (Vinnichenko and Dutton, 1969; Doviak and Berger, 1980).

For isotropic scalar fields having a correlation of the form of (10.19), the spectrum of variance can be found by substituting (10.32) into (10.20b):

$$\Phi_n(K) = \frac{\Gamma(\frac{11}{6})C_n^2 K^{-11/3}}{\pi^{3/2} 3 \times 2^{1/3} \Gamma(\frac{2}{3})} = 0.033 C_n^2 K^{-11/3}, \qquad (10.38)$$

where the subscript ii has been replaced by n to represent the scalar field of, for example, the refractive index.

10.1.4 Chandrasekhar's Theory

Section 10.1.3 gave the spatial behavior of the correlation function and the spectra for homogeneous isotropic turbulence. The verified theory based on similarity predicts a power of $\frac{2}{3}$ law for spatial lags corresponding to scales in the inertial subrange but makes no prediction about the temporal dependence. A completely satisfactory and accepted theory of the joint temporal and spatial laws that correlation function should follow does not exist.

Chandrasekhar (1955, 1956) has derived the partial differential equation

$$\frac{\partial^2 R_{ll}}{\partial^2 \tau_n \, \partial \rho_n} = R_{ll} \frac{\partial}{\partial \rho} (D_5 R_{ll}), \qquad (10.39)$$

which the longitudinal correlation function must satisfy (for zero viscosity). It describes the spacetime velocity correlation function $R_{ll}(\rho_n, \tau_1)$ for isotropic turbulence. D_5 is the Lagrangian operator $\partial^2/\partial^2 \rho_n + (4/\rho_n)(\partial/\partial \rho_n)$, and the parameters $\rho_n = \rho/\rho_0$ and $\tau_n = \tau_1 R_{ll}^{1/2}(0)/\rho_0$ are normalized space and time coordinates, respectively. The parameter ρ_0 represents a characteristic length associated with the largest eddies present in the medium. Similarly, it is possible to define a parameter τ_0 to represent the characteristic lifetime of the eddies. In addition to assuming an isotropic, nonviscous, and incompressible medium in deriving (10.39), Chandrasekhar (1955, 1956) assumed that various fourth-order moments can be expressed as sums of products of second-order moments; he then produced the following solution to (10.39):

$$R_{11}(\rho_n, \tau_n) = [1 - \rho_n \psi(\tau_n/\rho_n)] R_{11}(0, 0) \qquad (10.40)$$

applicable when changes in $R_{11}(\rho_n, \tau_n)/R_{11}(0, 0)$ are small compared to unity and $\tau_n \ll 1$. In (10.40),

$$\psi(x) = \frac{1}{2(v + 2)} [|x - 1|^{v+2}(v + 2 + x) + |x + 1|^{v+2}(v + 2 - x)]. \qquad (10.41)$$

In both (10.40) and (10.41) v is a constant to be assigned a particular value and in (10.41) $x = \tau_n/\rho_n$.

There are two limiting cases of (10.41)—one in which $\tau_n \to 0$ when $\rho_n \neq 0$, the other such that $\tau_n \neq 0$ when $\rho_n \to 0$. These cases are

$$R_{11}(\rho_n, 0) = (1 - \rho_n) R_{11}(0, 0), \qquad \tau_n \to 0, \qquad (10.42)$$

and

$$R_{11}(0, \tau_n) = [1 - \tfrac{1}{3}(v + 1)(v + 3)\tau_n] R_{11}(0, 0), \qquad \rho_n \to 0, \qquad (10.43)$$

respectively.

With the presence of uncorrelated measurement errors, the correlation functions will contain both a signal and a noise portion at $\rho_n = 0$, $\tau_n = 0$. Then $R_{11}(0, 0)$ must be computed from

$$\sigma^2 = R_{11}(0, 0) + \sigma_e^2, \qquad (10.44)$$

where σ^2 is the measured variance and σ_e^2 is the variance due to measurement error.

Smythe and Zrnić (1983) report on the spatial and temporal behavior of the autocorrelation function. They give, as shown in Fig. 10.3, the spatial dependence of the correlation coefficients for velocity and reflectivity fields. The data used to generate Fig. 10.3 were obtained in clear air, and each measured point is an average obtained from over 20 estimates. A single autocorrelation estimate is calculated by lagging an array (9 km × 12°) of radial velocities (or reflectivities) on itself. The azimuthal spacing was 0.5° and the range gate separation was 150 m, so the array contained 60 × 25 velocity or reflectivity estimates. The lagging was along radials in regions where radials were oriented parallel to the mean wind. The orientation of observed horizontal rolls was parallel to the mean wind on that day, so contamination by azimuthally directed wavelike components associated with the rolls was significantly reduced. The least squares fitted curve [(10.42) for $v = \tfrac{2}{3}$] to the velocity autocorrelation matches the data well.

Most striking in comparing the two correlation functions is the considerably higher value of the autocorrelation for the velocity field. This implies that characteristic scale sizes are larger in the velocity field than in the reflectivity field. A possible explanation for the difference in scale sizes may be

10.1 Statistical Theory of Turbulence

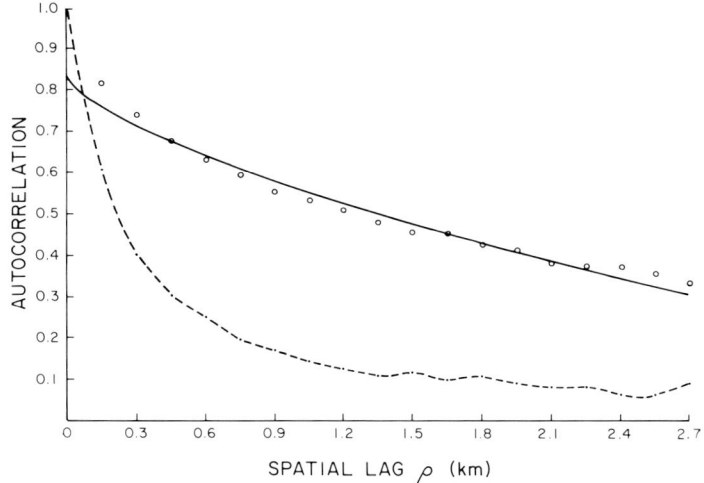

Fig. 10.3. Autocorrelation functions of the velocity and reflectivity obtained from single-radar data in the planetary boundary layer. Velocity data are least squares fitted with a two-thirds power law. (From Smythe and Zrnić, 1983.) 27 April 1977 from 14:27:07 to 14:27:31. ○, velocity; ——— two-thirds fit; - - -, reflectivity.

found in the mechanism generating these eddies. It seems that reflectivity fluctuations are not exclusively coupled to velocity fluctuations, which may, in part, be turbulence associated with the rolls. The reflectivity fluctuations may be primarily due to plumes and convective bubbles of small scale emanating from the solar-heated earth's surface.

The temporal behavior of the correlation functions (Fig. 10.4) is obtained from the maxima of cross correlation between arrays (6 km × 8°) of velocities at two different times. Data points on Fig. 10.4 are averages of such correlation maxima. Note again that the velocity field has higher correlation values. A least squares fit of velocity correlation vs (10.43) for $v = \frac{2}{3}$ is made to obtain the signal variance and correlation time $\tau_0 = (27/55)^{3/2} \rho_0 / R_{11}^{1/2}(0)$. We have no firm theoretical basis to choose a two-thirds power law time dependence, but the differential equation (10.39), which describes the time dependence of the velocity correlation, does admit a solution of the selected form [Eq. (10.43)] at small time lags.

Reflectivity was observed to decorrelate, in space, much more rapidly that radial velocity. As a consequence, the average decorrelation lengths have a ratio of 0.176. If scale sizes are defined by the length at which the correlation coefficient first drops to 0.5, then the velocity scales are about 2 km, vs 300 m for reflectivity. Corresponding lifetimes are about 500 s for velocities and less than 200 s for reflectivities.

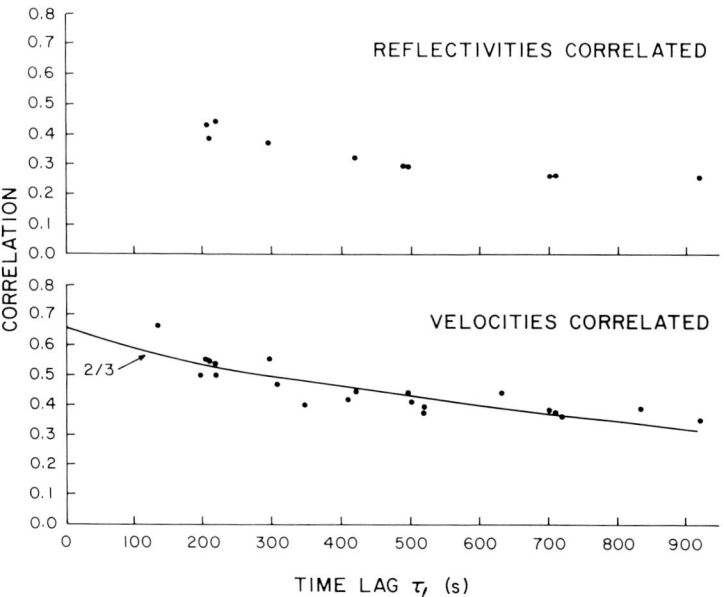

Fig. 10.4. Comparison of velocity and reflectivity temporal autocorrelations obtained by correlating 6 km × 8° arrays.

10.2 SPATIAL SPECTRA OF POINT AND AVERAGE VELOCITIES

10.2.1 Filtering by the Resolution Volume

In radar measurement the spatial spectra of turbulence are filtered by the resolution volume weighting function (5.40), and this produces attenuation in the observed turbulence intensity at scales smaller than V_6. To find quantitative relationships, we follow Srivastava and Atlas (1974), who use Eq. (5.48) as a starting point and assume that the normalized weighting function $I_n(\mathbf{r}, \mathbf{r}_1)$ [(5.46b)], depends only on $\mathbf{r} - \mathbf{r}_1$. With this, (5.48) becomes

$$\bar{v}(\mathbf{r}) = \int v(\mathbf{r}_1) I_n(\mathbf{r} - \mathbf{r}_1) \, dV_1. \tag{10.45}$$

It is assumed that the reflectivity $\eta(\mathbf{r}_1)$ is constant. Because (10.45) is a convolution product, the spatial spectrum $\Phi_{\bar{v}}(\mathbf{K})$ of averaged radial velocities equals

$$\Phi_{\bar{v}}(\mathbf{K}) = (2\pi)^6 \Phi_v(\mathbf{K}) |F(\mathbf{K})|^2, \tag{10.46}$$

10.2 Spatial Spectra of Point and Average Velocities

where $F(\mathbf{K})$ is the Fourier transform of the resolution volume weighting function I_n. A three-dimensional Gaussian shape for $|F(\mathbf{K})|^2$ is a good approximation to most resolution volume weighting functions. Thus we shall use

$$|F(\mathbf{K})|^2 = (2\pi)^{-6} \exp[-(K_2^2 + K_3^2)r^2\sigma_\theta^2 - K_1^2\sigma_r^2]. \qquad (10.47)$$

Equation (10.47) assumes that the wavenumber K_1 is along the beam axis, σ_θ^2 is the second central moment of the two-way antenna pattern (5.67), and σ_r^2 is the second central moment of the range resolution function $W^2(r)$. For a rectangular transmitted pulse and Gaussian receiver frequency response under matched conditions, the relationship between pulse duration and σ_r is given by (5.68).

The spectrum $\Phi_v(\mathbf{K})$ of point radial velocities can be related to the spectral density tensor Φ_{ii}. Assume that the size of the resolution volume is small compared to the range, so that the divergence of the radial velocities (within the resolution volume) may be ignored. Then let the beam direction coincide with x_1 to obtain

$$\Phi_v(\mathbf{K}) \equiv \Phi_{11}(\mathbf{K}). \qquad (10.48)$$

The combination of (10.12), (10.17) or (10.18), and (10.46) produces the measured (filtered by the resolution volume) spectra S_l^f and S_t^f of the longitudinal and transverse velocities:

$$S_l^f(K_1) = \frac{(2\pi)^6}{2} \int_{K_1}^\infty \left(1 - \frac{K_1^2}{K^2}\right) \frac{E(K)}{K} |F(K_1, K')|^2 \, dK, \qquad (10.49)$$

$$S_t^f(K_1) = \frac{(2\pi)^6}{4} \int_{K_1}^\infty \left(1 + \frac{K_1^2}{K^2}\right) \frac{E(K)}{K} |F(K_1, K')|^2 \, dK, \qquad (10.50)$$

where we have explicitly separated K_1 from $K' = (K^2 - K_1^2)^{1/2}$ in $F(K_1, K')$. This dependence on two wavenumbers results from circularly symmetrical antenna beams.

The spatial spectra (10.49) and (10.50) may be obtained from single or dual Doppler measurements. To help visualize the analysis for a single radar, Fig. 10.5 is presented. Mean radial velocities are obtained along and perpendicular to the mean wind, so that subsequent spectral analysis of data along these two directions yield $S_l^f(K_1)$ and $S_t^f(K_2)$. A third spectral component $S_t^f(K_1)$ can be computed from any one of the range locations along radials perpendicular to the mean wind. It suffices to record the temporal change of the velocity due to advection, transform time to space, and perform spectral analysis. Using (10.17) or (10.18), the energy spectrum $E(K)$ of isotropic turbulence can be obtained from either of the two one-dimensional measured spectra for scale sizes whose observed variance is not strongly attenuated by the filter.

Filtering effects on one-dimensional spectra of isotropic scalar fields smoothed by spherically symmetric three-dimensional filters can be assessed by expressing the filtered $S^f(K_1)$ and unfiltered $S(K_1)$ one-dimensional spectra in terms of their three-dimensional counterparts, as in (10.16). Substitution of (10.46) and inversion then produces

$$S(K_1) = -\frac{1}{(2\pi)^6} \int_{K_1}^{\infty} \frac{1}{|F(K)|^2} \frac{dS^f(K)}{dK} dK. \qquad (10.51)$$

Whereas the relationship between filtered and point three-dimensional spectra involves a simple multiplication with the transfer function [Eq. (10.46)], the corresponding one-dimensional spectra have a more complicated functional dependence, (10.49)–(10.51), which must be evaluated for each case if one wants to determine the effects of the filter.

A comprehensive illustration of the effects of resolution volume filtering on the longitudinal transverse one-dimensional spectra of turbulence is given by Srivastava and Atlas (1974). They consider a Kolmogorov–Obukhov spectrum with cutoff at wavenumber K_0:

$$E(K) = \begin{cases} \pi K^{-5/3} & \text{for } K > K_0, \\ 0 & \text{for } K < K_0. \end{cases} \qquad (10.52)$$

Then, from (10.17), they obtain the one-dimensional spectra of point velocities:

$$S_l(K_1) = \begin{cases} \frac{9}{55}\pi K_1^{-5/3} & \text{for } K_1 > K_0, \\ 0.3\pi K_0^{-5/3}[1 - \frac{5}{11}(K_1/K_0)^2] & \text{for } K_1 < K_0. \end{cases} \qquad (10.53)$$

Even though the three-dimensional spectrum $E(K)$ has no energy for $K < K_0$,

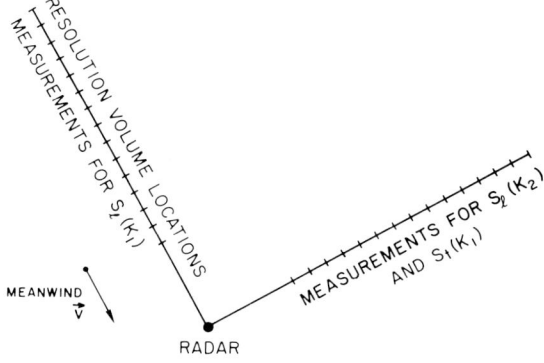

Fig. 10.5. Geometry of data collection with a single Doppler radar in order to compute $S_l(K_1)$, $S_t(K_2)$, and $S_t(K_1)$.

10.2 Spatial Spectra of Point and Average Velocities

the one-dimensional spectrum does have values for $K_1 < K_0$. Filtered spectra are obtained from (10.49) after (10.47) is substituted:

$$S_1^f(K_1) = \tfrac{1}{2} \exp[-K_1^2(\sigma_r^2 - r^2\sigma_\theta^2)] \int_a^\infty \left(1 - \frac{K_1^2}{K^2}\right) \frac{E(K)}{K} \exp(-K^2 r^2 \sigma_\theta^2) \, dK, \tag{10.54}$$

where $a = K_1$ for $K_1 > K_0$ and $a = K_0$ for $K_1 < K_0$.

Observe that the filtered one-dimensional spectrum is attenuated in a direction along which there may have been no filtering; i.e., even when $\sigma_r = 0$ in (10.54) there is degradation in the K_1 direction. This is due to filtering in the orthogonal (K_2, K_3) directions and the superposition (integration of the resulting attenuated three-dimensional spectral density.

A plot of $S_1^f(K_1)$, from Srivastava and Atlas (Fig. 10.6), illustrates the filtering effect. Equation (10.54) is graphed with $\sigma_r = 0$ for various $r\sigma_\theta$ and for $K_0 = \pi$ and 2π. Results for other σ_r are obtained by multiplying the curves by the Gaussian function $\exp(-K_1^2\sigma_r^2)$. Unfiltered spectra, from (10.53), are curves with $r\theta_1 = 0$ obeying the power of $\tfrac{5}{3}$ law when $K_1 > K_0$.

The energy at wavenumbers smaller than the cutoff ($K_1 < K_0$) is contributed by scales (K_1, K_2, K_3) such that $K^2 > K_0^2$; that is, by wavenumbers larger than K_0 whose projection along the K_1 axis is smaller than K_0. This is why dashed curves with the larger cutoff wavenumber have less power than the solid curves at long wavelengths.

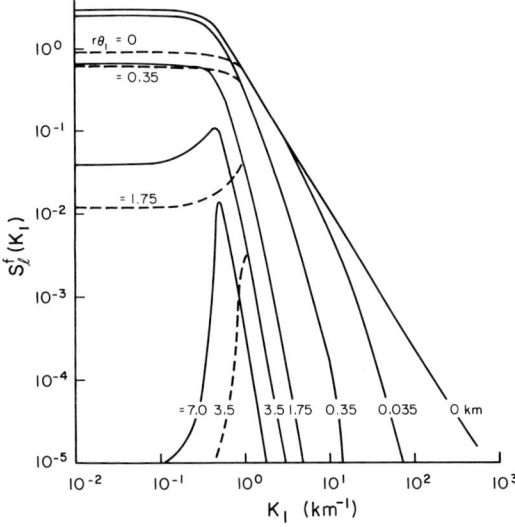

Fig. 10.6. Filtered longitudinal one-dimensional spectrum function. Equation (5.67) relates θ_1 to σ_θ. ——, $K_0 = \pi$ km^{-1}; ---, $K_0 = 2\pi$ km^{-1}; $\sigma_r = 0$. (From Srivastava and Atlas, 1974.)

An illustration of the one-dimensional spatial spectrum calculated from dual Doppler synthesized wind fields in the planetary boundary layer (Doviak and Berger, 1980) is shown in Fig. 10.7, together with the spectrum calculated from tower measurements. The only filter—other, of course, than the radar resolution volume—acting on the velocity field used for spectral analysis is the Cressman interpolation filter (Section 9.2.1).

The S_1 for the radar data shown in Fig. 10.7 is an average from 32 individual spectra obtained by a discrete Fourier transform of velocities parallel to the mean wind direction. Smoothed tower spectra were computed by weighting the autocovariances with an 85-lag (850 s) Tukey window (Jenkins and Watts, 1968), and space-to-time conversion (Taylor hypothesis) was used to compare the scales observed by radar and tower.

Plots of the tower wind spectra have shapes similar to the spectra of wind synthesized from radar data. Also, spectral densities and the total variance of the horizontal wind from both tower and radar data are comparable: $\sqrt{R(0)}(\text{tower}) = 2.1$ m s^{-1}. and $\sqrt{R(0)}(\text{radar}) = 1.7$ m s^{-1}. The lower variance observed with the radar must, at least in part, result from the reduction in variance by filtering due to the interpolation volume. The unfiltered spectra

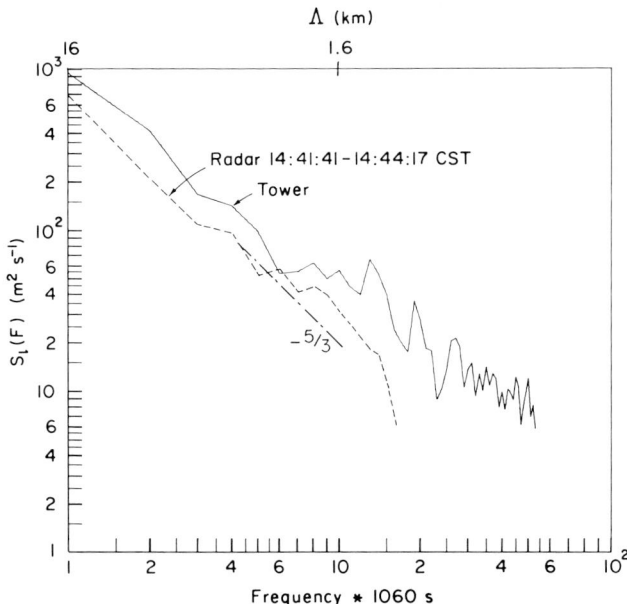

Fig. 10.7. Longitudinal spectra from KTVY TV tower and radar data. An advection speed of 15 m s^{-1} gives the wavelength scale on the upper part of the plot. Level 7 (444 m); 13:33:00–15:55:40 C.S.T.

10.2 Spatial Spectra of Point and Average Velocities

for the radar observations have more variance density at all wavelengths; however, the increase will be largest at shorter wavelengths, where the attenuation is strongest, and hence there would be even better agreement between radar and tower spectra. An example of how to account for the filtering effects of the radar's resolution volume and interpolation filters is presented by Doviak and Berger (1980). We note in these data that the $-\frac{5}{3}$ slope extends to long wavelengths well above the inertial subrange.

First measurements of turbulent velocity spectra were reported by Lhermitte (1968), who estimated the up- and downwind spectra from mean Doppler velocities along a radial aligned with the wind. His results in a snowstorm showed good agreement with the five-thirds power law. The up- and downwind spectra were quite similar, suggesting that turbulence was homogeneous.

Chernikov *et al.* (1969) have compared spatial spectra obtained by a Doppler radar to spectra measured by an in situ aircraft. The experiment consisted of a Doppler radar, instrumented aircraft, and tethered balloon for chaff release. The radar antenna beam was directed along the mean wind, and time variations in radial velocity in a single resolution cell were recorded. Longitudinal spectra in the frequency domain were converted to the spatial domain to compare the results with those obtained by an aircraft that flew up- and downwind collecting spatial spectra. The authors report good agreement in general, with resemblance even between small details (Fig. 10.8). At scales larger than 1 km, the radar spectra systematically contained

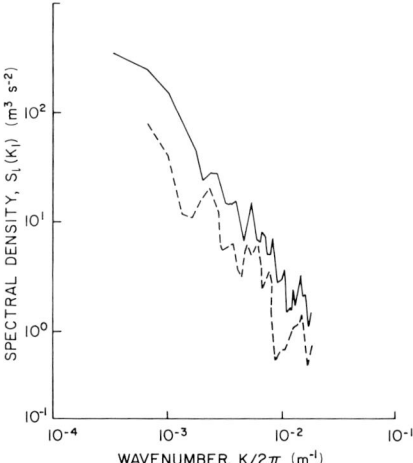

Fig. 10.8. An example of the comparison of radar (---) and aircraft-obtained (——) spectra that shows the agreement between small details of spectral curves. (The radar spectrum is plotted 4 dB below its real position.) (Adapted from Chernikov *et al.*, 1969, copyrighted by the American Geophysical Union.)

less energy than did the spectra obtained by aircraft. The authors attribute this to large-scale convective formations that moved slower than the mean wind speed because of their association with the mountainous features of the terrain. Some of this difference might also be attributed to the filtering of energy by the radar's resolution volume.

10.2.2 Variance of Point and Average Velocities

In this section we define the relationships between various variances of measured velocities (Rogers and Tripp, 1964). Let the variance of the velocity v at a point be σ_p^2. It is obtained from the ensemble average:

$$\sigma_p^2 = \langle v^2 \rangle - \langle v \rangle^2. \tag{10.55}$$

For uniform reflectivity the Doppler spectrum width σ_v is given by (5.51);

$$\sigma_v^2 = \overline{v^2} - (\bar{v})^2, \tag{10.56}$$

where the overbar denotes an average of velocities spatially weighted by the beam and range resolution functions. An additional equation is needed for the variance (ensemble average) of the mean Doppler velocity $\bar{v}$. By definition,

$$\sigma_{\bar{v}}^2 \equiv \langle (\bar{v})^2 \rangle - \langle \bar{v} \rangle^2. \tag{10.57}$$

Assuming turbulence to be locally homogeneous, resolution volume weighting functions symmetrical, and recognizing that ensemble and spatial averages commute, we can rewrite (10.57) as

$$\sigma_{\bar{v}}^2 \equiv \langle (\bar{v})^2 \rangle - \langle v \rangle^2. \tag{10.58}$$

Finally, the ensemble average of (10.56) added to (10.58), after commuting ensemble and spatial averages, produces

$$\sigma_p^2 = \langle \sigma_v^2 \rangle + \sigma_{\bar{v}}^2, \tag{10.59}$$

which indicates that the variance of the velocity at a point is equal to the sum of the ensemble average of the square of the Doppler spectrum width and the variance of the velocities spatially weighted by the resolution volume function.

This very general result is independent of the resolution volume weighting function but requires turbulence to be locally homogeneous, although not isotropic. In principle, ensembles can be obtained by repeating the measurement under the same circumstances, but this is practical only if the phenomena are stationary during the duration of the experiments. Another approach to obtaining the ensemble average is to assume statistically homogeneous

10.2 Spatial Spectra of Point and Average Velocities

media and to take measurements at points spaced far enough apart that samples are uncorrelated.

In addition to being proportional to the turbulent kinetic energy, the two variances $\langle \sigma_v^2 \rangle$ and $\sigma_{\bar{v}}^2$ have relative magnitudes that describe how the kinetic energy is partitioned between subresolution volume scales and scales larger than the resolution volume.

10.2.3 Turbulence Parameters from a Single Radar

A variation of a VAD technique (Section 9.3.3) can be applied to turbulent velocities to deduce some characteristics if turbulence is horizontally homogeneous. We shall examine two regimes of turbulent eddy scales: (1) scales large compared to the resolution volume V_6 and (2) scales small compared to V_6.

10.2.3.1 Large Scale Eddies

Equation (9.20) also applies to the resolution-volume-weighted radial velocities corrected for target terminal velocity. The variance of the resolution-volume-weighted velocities is obtained directly from (9.20):

$$\mathrm{var}(\bar{v}_r) = \sigma_{\bar{u}}^2 \cos^2 \theta_e \sin^2 \phi + \sigma_{\bar{v}}^2 \cos^2 \theta_e \cos^2 \phi + \sigma_{\bar{w}}^2 \sin^2 \theta_e$$
$$+ \mathrm{cov}(\bar{u}\bar{v}) \cos^2 \theta_e \sin 2\phi + \mathrm{cov}(\bar{v}\bar{w}) \sin 2\theta_e \cos \phi$$
$$+ \mathrm{cov}(\bar{u}\bar{w}) \sin 2\theta_e \sin \phi, \qquad (10.60)$$

where

$$\sigma_{\bar{u}}^2 \equiv \langle (\bar{u})^2 \rangle - \langle \bar{u} \rangle^2,$$
$$\mathrm{cov}(\bar{u}\bar{v}) \equiv \langle \bar{u}\bar{v} \rangle - \langle \bar{u} \rangle \langle \bar{v} \rangle$$

are typical of the variances and covariances. Now, if the turbulence is homogeneous such that the variances and covariances are independent of ϕ (i.e., horizontally homogeneous), then the various terms of (10.60) can be estimated using the techniques outlined in Section 9.3. If the turbulence is superimposed on a linear wind field, the radial velocity of the least-squares-fitted linear wind needs to be subtracted from the estimated $\hat{v}_r$ at each point. The residuals about the fitted curve then have a variance with angular dependence given by (10.60). Before the residuals can be fitted by an assumed homogeneous turbulent field, the variance in the estimates due to statistical uncertainty (Chapter 6) must be removed, as must velocity perturbations due to waves (Section 9.4.2).

Lhermitte (1969) first suggested that the difference in $\mathrm{var}(\bar{v}_r)$ at $\phi = 0$ and π could be used to estimate directly $\mathrm{cov}(\bar{v}\bar{w})$, which is related to shearing stress.

10.2.3.2 EDDIES SMALLER THAN THE RESOLUTION VOLUME

Taking the resolution volume average of the square of the difference between the point velocity [Eq. (9.20)] and the resolution-volume-averaged velocity one obtains an expression for the spectrum width σ_v exactly analogous to (10.60):

$$\sigma_v^2 = \sigma_{u'}^2 \cos^2\theta_e \sin^2\phi + \sigma_{v'}^2 \cos^2\theta_e \cos^2\phi + \sigma_{w'}^2 \sin^2\theta_e$$
$$+ \text{cov}(u'v') \cos^2\theta_e \sin 2\phi + \text{cov}(v'w') \sin 2\theta_e \cos\phi$$
$$+ \text{cov}(u'w') \sin 2\theta_e \sin\phi, \qquad (10.61)$$

which again is in terms of variances and covariances on scales small compared to V_6 and where, for example, $\sigma_{u'}^2 = \overline{(u - \bar{u})^2}$ and $\text{cov}(u'v') = \overline{(u - \bar{u})(v - \bar{v})}$. If turbulence is homogeneous, the various terms in (10.61), like in (10.60), can be determined by applying the techniques outlined in Section 9.3. A more instructive form of (10.61) is

$$\sigma_v^2 = \tfrac{1}{2}(\sigma_{u'}^2 + \sigma_{v'}^2)\cos^2\theta_e + \sigma_{w'}^2 \sin^2\theta_e + \text{cov}(u'v')\cos^2\theta_e \sin 2\phi$$
$$+ \text{cov}(v'w')\sin 2\theta_e \cos\phi + \text{cov}(u'w')\sin 2\theta_e \sin\phi$$
$$+ \tfrac{1}{2}(\sigma_{v'}^2 - \sigma_{u'}^2)\cos^2\theta_e \cos 2\phi. \qquad (10.62)$$

Therefore, harmonic analysis of $\sigma_v^2(\phi)$ (similar to that for v_r in Section 9.3) can be used to yield turbulent velocity parameters on subresolution volume scales. At low elevation angles the first term in (10.62) is an estimate of the horizontal turbulent kinetic energy. The last term is a measure of the horizontal isotropy of the turbulent field. Harris (1975) has applied these concepts to stratiform precipitation and has obtained realistic values for the variances and momentum fluxes.

10.2.4 Turbulence Parameters from Two Doppler Radars

Rabin *et al.* (1982), using data collected with two Doppler radars, observed the clear-air planetary boundary layer (PBL) in order to estimate $\bar{u}, \bar{v}, \bar{w}$, from which they obtained variances and covariances as functions of height (Fig. 10.9). Because dual Doppler observations require interpolation of radial velocities to grids common to the two radars, further filtering results. The spectral response of the interpolation filter is given by Doviak and Berger (1980). Thus in Fig. 10.9 the superscript r defines radar-"resolved" parameters (e.g., the covariance), which are analogous to the resolution volume parameters in (10.60). The angle brackets indicate that the parameter estimates at each grid point spaced 500 m apart were averaged over a domain $25 \times 25 \text{ km}^2$. The covariance for scales small compared to the interpolation

10.3 Doppler Spectrum Width and Eddy Dissipation Rate

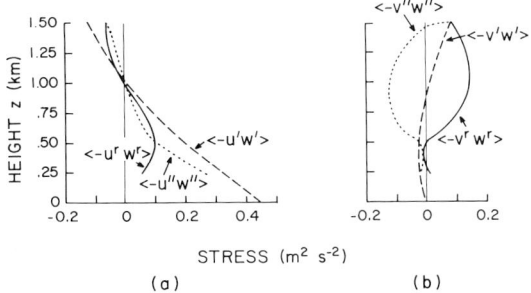

Fig. 10.9. Stress profiles in the surface-layer coordinate system; the longitudinal u and transverse v components are parallel and normal, respectively, to the surface wind (210°). (a) The longitudinal component of the stress. (b) The transverse component of the stress, 14:00 C.S.T., 27 April 1977. ——, radar; - - -, velocity defect equations; · · ·, inferred subgrid scales.

and resolution volumes was obtained from the difference between the total flux, computed from the velocity defect equations, and the radar-resolved fluxes. No attempt was made to use the spectrum width to estimate subgrid covariances.

The longitudinal flux $\langle(-u^r w^r)\rangle$ measured by radar crossed zero at the same height as the total flux $\langle(-u'w')\rangle$. Above 0.5 km, about half the subgrid flux $\langle(-u''w'')\rangle$ appears to be contained in the subgrid scales. However, below 0.5 km the subgrid flux appears to be much larger, which is to be expected from mixing-length theory.

The transverse flux $\langle(-v'w')\rangle$ inferred from the velocity defect equations has much less magnitude than that measured by radar, $\langle(-v^r w^r)\rangle$. The subgrid flux is of the same magnitude but opposite in sign to the radar-measured flux.

The sign of the subgrid flux, unlike the radar-measured transverse flux, is such that eddy diffusivity K_y is positive if the diffusion equations are restricted to apply to subgrid scales. Rabin *et al.* (1982) demonstrate that the profile of eddy diffusivity computed from the subgrid-scale flux agrees with numerical models.

10.3 DOPPLER SPECTRUM WIDTH AND EDDY DISSIPATION RATE

Equation (10.46) states that the spatial spectrum of the point Doppler velocity is filtered by the resolution functions. This, together with (10.59), will be used to obtain a relationship between the Doppler spectrum width and the spatial spectrum of turbulence. First, we express the variances of point and averaged velocities in terms of their spectra:

$$\sigma_p^2 = \int \Phi_v(\mathbf{K}) \, dV_K, \tag{10.63}$$

$$\sigma_{\bar{v}}^2 = \int \Phi_{\bar{v}}(\mathbf{K}) \, dV_K. \tag{10.64}$$

Now, Eq. (10.59) shows that the Doppler spectrum width $\langle \sigma_v^2 \rangle$ is the difference between (10.63) and (10.64). After this difference is taken and $\Phi_{\bar{v}}(\mathbf{K})$ substituted from (10.46), we get the following formula, which connects the Doppler spectrum width to $\Phi_v(\mathbf{K})$ (Frisch and Clifford, 1974):

$$\langle \sigma_v^2 \rangle = \int [1 - (2\pi)^6 |F(\mathbf{K})|^2] \Phi_v(\mathbf{K}) \, dV_K. \tag{10.65}$$

If some expression is found to relate $\Phi_v(\mathbf{K})$ to the turbulence parameters of the medium and Eq. (10.65) is inverted, those parameters can be estimated. This implies knowledge of $I_n(\mathbf{r} - \mathbf{r}_1)$. Frisch and Clifford (1974) assume that the reflectivity within the resolution volume is uniform and that $I_n(\mathbf{r} - \mathbf{r}_1)$ is a three-dimensional Gaussian function with one width parameter transverse to and the other along the beam, so that the squared magnitude of its Fourier transform is given by (10.47).

In order to treat Eq. (10.65) further, we consider the turbulence of an incompressible fluid so that Eqs. (10.12) and (10.13) are valid. Furthermore, we invoke Kolmogorov's hypothesis, which implies that kinetic energy is provided to the medium by the space Fourier components corresponding to large scales (for thunderstorm studies these are about a few hundred meters and larger) and is dissipated by viscosity at very small scales (of the order of a few centimeters to millimeters). Between these ranges there exists an *inertial subrange* in which energy cascades from the large to the smaller scales, with eventual dissipation by viscous forces. Under these conditions the energy spectrum function $E(K)$ becomes

$$E(K) = A\varepsilon^{2/3} K^{-5/3}. \tag{10.66}$$

A is a universal dimensionless constant between 1.53 and 1.68 (Gossard and Strauch, 1981, p. 264), and ε is the turbulent energy dissipation rate, normalized to unit mass.

To tie Eq. (10.66) to the Doppler spectrum width, it must be assumed that all σ_v contributions come from velocity scales within the inertial subrange. This is approximately true provided contributions from scales larger than V_6, which are not part of the inertial subrange, are removed. Such contributions can be lumped together in a shear term (corresponding to the small K values), which can be computed simply by assuming a linear variation in the mean radial velocity $\bar{v}$ throughout the resolution volume. If the outer scale of inertial subrange turbulence is known, it is possible to obtain a more accurate relationship between the dissipation rate and turbulence. Such a calculation has been made by Bohne (1982), who has also considered the influence of falling precipitation through isotropic turbulence. He has concluded that the imperfect response of precipitation to turbulent motion must be accounted for at short range (<20 km) and when the turbulent outer scale is ≤ 0.5 km. At larger range an estimate of the outer scale is necessary if one is concerned with classification of the severity of turbulence.

To obtain the variance associated with inertial subrange motion, the linear shear contribution σ_s^2 (Chapter 5) is subtracted from the square of the Doppler spectrum width σ_v^2. Equations (10.12) and (10.66) are substituted in Eq. (10.65) and the integration performed to obtain an expression giving ε as a function of σ_v and the parameters describing $I_n(\mathbf{r} - \mathbf{r}_1)$ (see Labitt, 1981; Gossard and Strauch, 1981). For a range resolution smaller than the beamwidth ($\sigma_r \leq r\sigma_\theta$), this expression is

$$\sigma_v^2 = A\Gamma(\tfrac{2}{3})(\varepsilon r\sigma_\theta)^{2/3} F(-\tfrac{1}{3}, \tfrac{1}{2}; \tfrac{5}{2}; 1 - \sigma_r^2/r^2\sigma_\theta^2), \tag{10.67}$$

where F is the hypergeometric function bounded between 0.918 and 1. Consequently, for $\sigma_r \ll r\sigma_\theta$, $F = 0.918$, and the dissipation rate is very well approximated by

$$\varepsilon \approx 0.72 \sigma_v^3 / r\sigma_\theta A^{3/2}. \tag{10.68}$$

When the beamwidth is smaller than pulse depth ($r\sigma_\theta \leq \sigma_r$), the hypergeometric function in (10.67) has different arguments and is bounded by

$$\tfrac{27}{55} < F(-\tfrac{1}{3}, 2; \tfrac{5}{2}; 1 - r^2\sigma_\theta^2/\sigma_r^2) < 1 \tag{10.69}$$

(Labitt, 1981). Using the series expansion for F, the dissipation rate can be approximated to first order in $r^2\sigma_\theta^2/\sigma_r^2$ by

$$\varepsilon \approx [\sigma_v^3/\sigma_r(1.35A)^{3/2}](\tfrac{11}{15} + \tfrac{4}{15}r^2\sigma_\theta^2/\sigma_r^2)^{-3/2}. \tag{10.70}$$

10.4 DOPPLER SPECTRUM WIDTH IN SEVERE THUNDERSTORMS

The two major broadening mechanisms of the Doppler spectrum width due to meteorological factors are shear and turbulence. It was shown in Chapter 5 that, because these spectral broadening mechanisms are independent of one another, the total spectrum width squared can be considered as a sum of the σ^2 contributed by each.

The radial velocity shear across the radar resolution volume can be determined directly from the spatial dependence of the mean radial velocity

$\bar{v}$. Then, to arrive at the component due to turbulence, one must extract the shear part from total spectrum width.

Thunderstorms contain a continuum of turbulence scales. Turbulence with scales larger than the radar resolution volume is most likely to consist of anisotropic eddies, which would appear in the data as radial velocity shear. An example of a radial velocity field for a thunderstorm exhibits areas of large shear (Fig. 10.10a). This is the shear that contributes to the measured spectrum width. A large shear region is about the mesocyclone centered at 97 km north and 36 km east of the Norman radar. Another shear region starting at 94 km north and 40 km east and extending to the bottom of the field, was identified from dual Doppler data to be the low-level boundary (gust front) between storm inflow to the east of the shear line and outflow to the west (also see Fig. 9.2a). Increased spectrum widths coincident with these larger shear regions are evident in Fig. 10.10b. Large widths farther

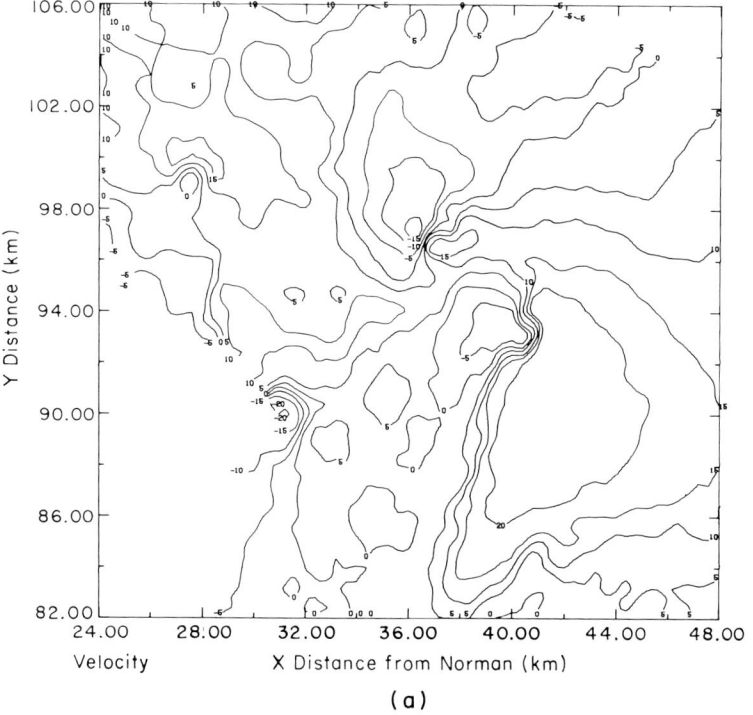

Fig. 10a. Doppler velocity field at a height of 1.5 km for the Stillwater tornadic storm at 17:42 on 13 June 1975. The grid spacing is 400 m. Velocities are in meters per second and contours (isodops) are in 5-m s^{-1} steps.

10.4 Doppler Spectrum Width in Severe Thunderstorms

north are where the tornado mesocyclone formed; the other region of large width is embedded in the gust front.

Measurements of spectrum widths σ_v observed in three severe tornadic storms (Fig. 10.11), one of whose field is shown in Fig. 10.10b (13 June 1975), exhibit a median width value of about 4 m s^{-1}, and about 20% of widths are larger than 6 m s^{-1}. These probabilities are derived from about 15,000 sample points for each storm; the sample points extend from near the ground to over 15 km. Because these data do not have the range dependence predicted for isotropic turbulence (10.68), the widths are probably due to turbulence that is neither homogeneous nor isotropic for scale sizes small compared to the radar's resolution volume. For these experiments $\theta_1 = 0.8°$ and $\tau = 1$ μs, so weather radars, not having better resolution, should obtain similar width distributions in severe storms.

An estimate of the angular shear across the resolution volume can be obtained by least squares fitting a surface that relates the mean radial velocity

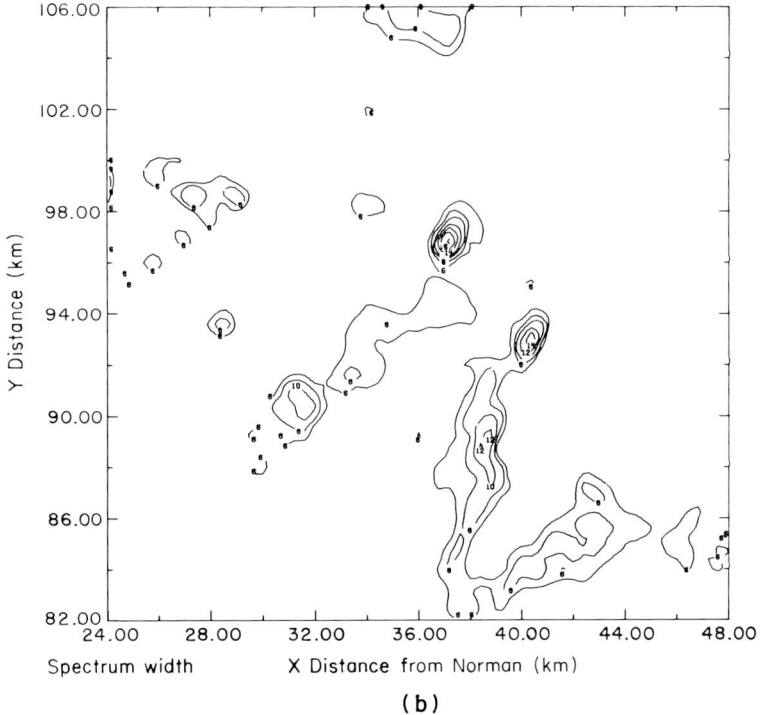

Fig. 10.10b. Contours of the spectral width at 1.5 km above ground (13 June 1975 at 17:42). Values greater than or equal to 6 m s^{-1}, in steps of 2 m s^{-1}, are displayed for visual clarity.

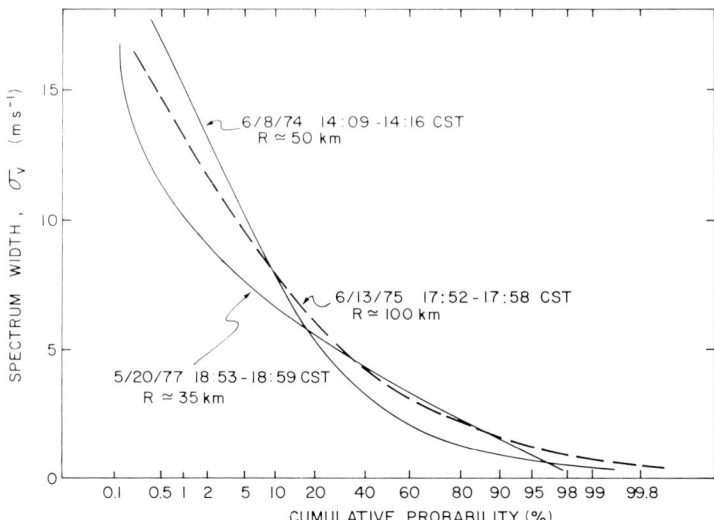

Fig. 10.11. Cumulative probability of spectrum widths for echoes from three tornadic storms. Spectrum widths are derived from spectra computed using discrete Fourier transforms of echo samples having SNR $\geq$ 15 dB. Antenna beam width $\theta_1 = 0.8°$.

to the azimuth and elevation position surrounding the resolution volume. The limitation of this assumption is that not all shear is removed. That is, if data were fitted to a quadratic or higher-order polynomial surface, a broader range of turbulence scales larger than the resolution volume would be represented by the surface.

The equation of the linear surface with superposed errors e_i reads

$$v_i = v_0 + k_\phi (\phi_i - \phi_0) r \cos \theta_i + k_\theta (\theta_i - \theta_0) r + e_i, \qquad (10.71)$$

where (θ_0, ϕ_0) is the origin of the surface.

The parameters of the surface, i.e., v_0, k_ϕ, k_θ, are determined through a matrix operation described by Neter and Wasserman (1974):

$$\begin{bmatrix} v_0 \\ k_\phi \\ k_\theta \end{bmatrix} = \begin{bmatrix} M & \sum l_{\phi i} & \sum l_{\theta i} \\ \sum l_{\phi i} & \sum l_{\phi i}^2 & \sum l_{\phi i} l_{\theta i} \\ \sum l_{\theta i} & \sum l_{\phi i} l_{\theta i} & \sum l_{\theta i}^2 \end{bmatrix} \begin{bmatrix} \sum v_i \\ \sum v_i l_{\phi i} \\ \sum v_i l_{\theta i} \end{bmatrix}, \qquad (10.72)$$

where

$$l_\phi = r(\phi_i - \phi_0) \cos \theta_i, \qquad l_\theta = r(\theta_i - \theta_0)$$

are the arc lengths from the origin of the fitted surface to a data point location

10.4 Doppler Spectrum Width in Severe Thunderstorms

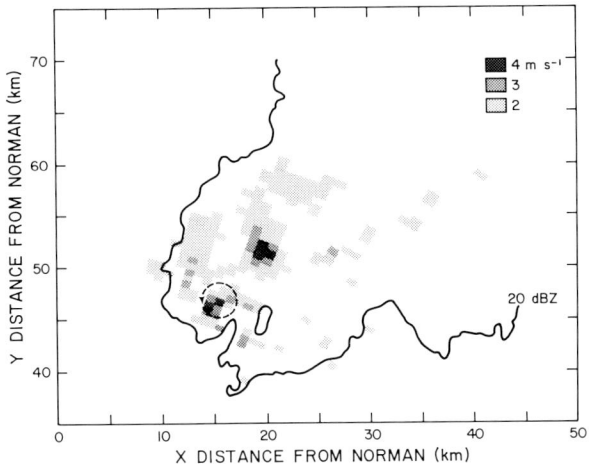

Fig. 10.12. Spectrum width due to radial velocity shear. (8 June 1974 at 14:20 C.S.T. Elevation angle: 5.1°.)

(θ_i, ϕ_i). M is the number of data points used to determine the fitted surface. Once the shears k_ϕ, k_θ are obtained, the spectrum width due to these linear shears can be computed using (5.66).

Istok (1981) has applied the procedure described here to data from a severe storm. One example from his analysis is shown in Fig. 10.12. Spectrum widths were calculated using stored time series data, on which the pulse pair algorithm (6.27) was applied and the contribution of the shear was then isolated. The circle in Fig. 10.12 indicates a mesocyclone in which the contribution of the shear is significant; there is one more region, 20 km east and 52 km north of Norman, where the width due to shear exceeds 3 m s^{-1}. Examination of dual Doppler wind fields revealed that this region is at the transition between updraft and downdraft. Otherwise, the contribution of shear is minimal, 0–1 m s^{-1}.

Figure 10.13 shows the width due to turbulence for the same data. Outside of the 20-dBZ contour, the signal-to-noise ratios are small and spectrum widths are large. The mesocyclone region has only moderate widths due to turbulence whose contribution is comparable to that from shear. The reflectivity field had a strong core (from 55 to over 60 dBZ) about 2 km to the east of the mesocyclone. We note large widths in this region as well as outside a 20-dBZ contour at the same range. These are from sidelobes that have illuminated the 60-dBZ core (see Section 7.7). Doppler spectra for the locations in question contained several peaks of comparable powers distributed around zero velocity. We believe that most of the other broad spectra from

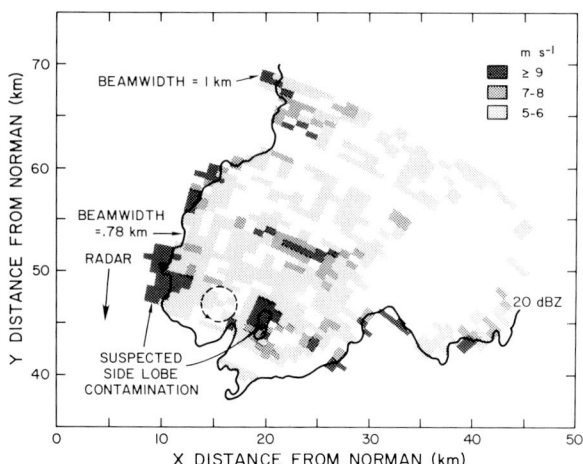

Fig. 10.13. Spectrum width due to turbulence. (8 June 1974 at 14:20 C.S.T. Elevation angle: 5.1°.)

Fig. 10.13 are due to eddies that were shed off the strong updraft by environmental winds (from the southwest).

According to Lee (1977), there is a strong connection between the spectrum width and aircraft penetration measurements of turbulence. His data show that when aircraft-derived gust velocities exceeded 6 m s^{-1}, corresponding to moderate or severe turbulence, the spectrum width exceeded 5 m s^{-1} in every case for aircraft within 1 km of the radar resolution volume. The cumulative probability of the total spectrum width and the width due to shear and turbulence, from Istok (1981), is presented in Fig. 10.14. Spectrum widths in excess of 5 m s^{-1} exist in 52% of the storm volume, suggesting that over one-half the storm volume contains moderate or severe turbulence. Most areas of large widths are found within the upper regions of the storm.

Large shear by itself is not necessarily very dangerous to aircraft except at takeoff or landing, because at higher altitudes there is ample space and time for the aircraft to recover. However, large shear may produce extreme turbulence, which is dangerous, and hence pilots should avoid it.

The cumulative probability of shears in azimuth k_ϕ and elevation k_θ from the least squares fit is plotted in Fig. 10.15. In general, a greater proportion of the storm contains larger shear in the elevation direction than in the azimuth direction. Only 19% of all azimuthal shears in the storm are in excess of 3×10^{-3} s^{-1}, while 35% of all elevation shears are larger than 3×10^{-3} s^{-1}. However, the largest values of shear are in the azimuthal direction. These shears (16–20×10^{-3} s^{-1}) are associated with the mesocyclone.

10.4 Doppler Spectrum Width in Severe Thunderstorms

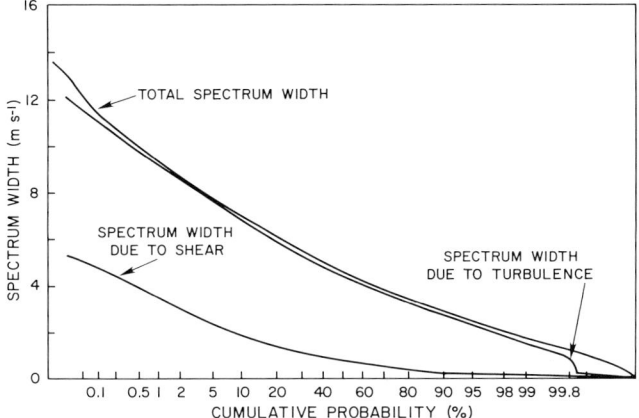

Fig. 10.14. Cumulative probability of the total spectrum width and the width due to linear radial velocity shear and turbulence (8 June 1974 at 14:20 C.S.T.).

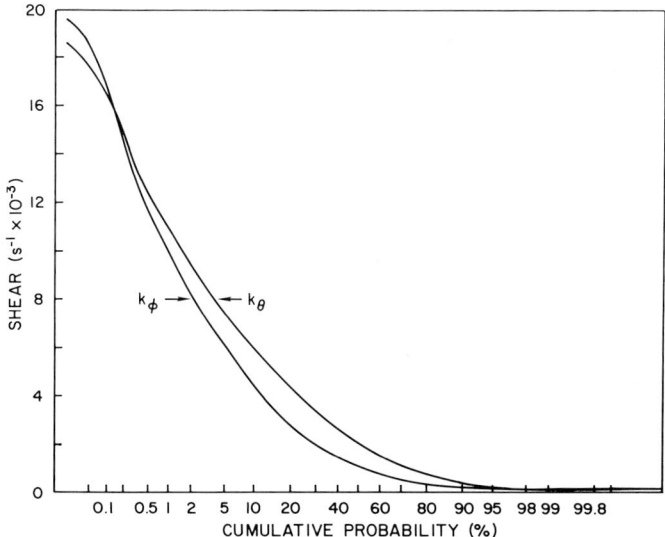

Fig. 10.15. Cumulative probability of the linear radial velocity shear in the elevation (k_θ) and azimuthal (k_ϕ) directions (8 June 1974 at 14:20 C.S.T.).

There is an inherent problem with using the Doppler spectrum width as a measure of turbulence: The spectrum width is a function of the radar resolution volume size in addition to the intensity of turbulence. The turbulent kinetic energy dissipation rate (10.68 or 10.70) that represents the energy flow across any given wavenumber is a more useful quantity for characterizing the intensity of turbulence within the inertial subrange. Within this subrange, turbulence is locally isotropic. Sinclair (1974) has observed the upper limit of the inertial subrange in a severe storm to vary from 150 to about 2000 m. It appeared to Sinclair that this variability is related to the storm intensity and the measurement altitude. That is, the upper limit of the inertial subrange is largest in the upper half of the storm, where vertical velocities are usually largest and turbulence is most intense. This variability of the upper limit is not uncommon. Other investigators (Rhyne and Steiner, 1964; MacCready, 1962, 1964; Reiter and Burns, 1966; Reiter, 1970) have shown or suggested that the upper limit of the inertial subrange may vary from 300 to 800 m, depending on the phenomena and the location of measurement.

The dissipation rate of the turbulent kinetic energy can be estimated from the Doppler spectrum width due to turbulence if and only if the largest scale of the inertial subrange is larger than the largest dimension of the radar resolution volume. If the largest dimension of the radar resolution volume is greater than the upper limit of eddy sizes within the inertial subrange, then the spectrum width after removal of the shear contribution would still be due to turbulence within the energy input and inertial subranges. If eddies within the energy input range are contained within the radar resolution volume, the spectrum width cannot be related to ε. This is because, for the input energy containing range, eddies might not be isotropic; they may have no known spectral form, and hence relating σ_v to ε can result in large errors. Furthermore, total spectrum width may become aspect sensitive, so that radars viewing the volume from different directions will detect different magnitudes of σ_v^2. However, the total spectrum width correctly characterizes the spread of reflectivity-weighted radial velocities within the radar resolution volume, although one cannot relate these to eddy dissipation rate.

Dissipation rates for the 8 June 1974 storm were computed from Eq. (10.68), which is valid at ranges larger than 30 km. Presented in Fig. 10.16 is a plot of the cumulative probability of ε. Dissipation rates less than 1 m^2 s^{-3} exist in 99% of the storm volume, and in 50% of the storm they are less than 0.1 m^2 s^{-3}.

Frisch and Strauch (1976) observed maximum dissipation rates of 0.06 m^2 s^{-3} in a Colorado convective storm. (This value is 5.2 times smaller from the one they actually reported, because we use a more recent estimate of the universal constant $A = 1.6$.) Such large rates were often noted between

10.4 Doppler Spectrum Width in Severe Thunderstorms

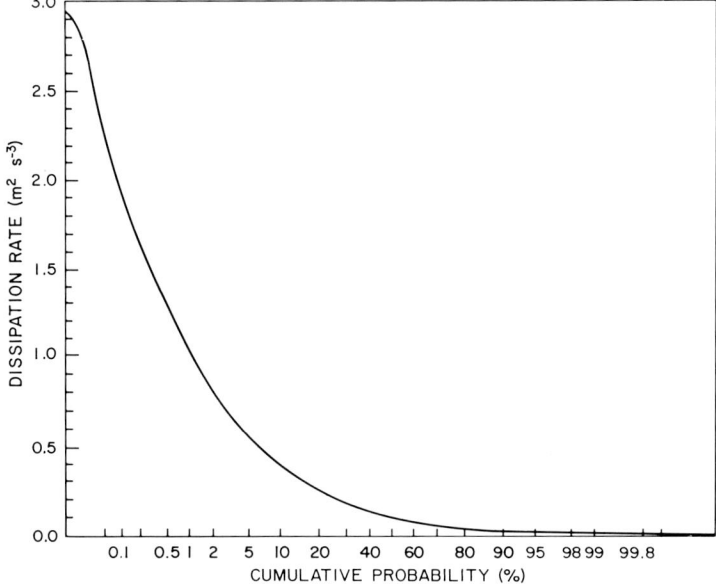

Fig. 10.16. Cumulative probability of the eddy dissipation rate ϵ (8 June 1974 at 14:20 C.S.T.). (From Bohne, 1981.)

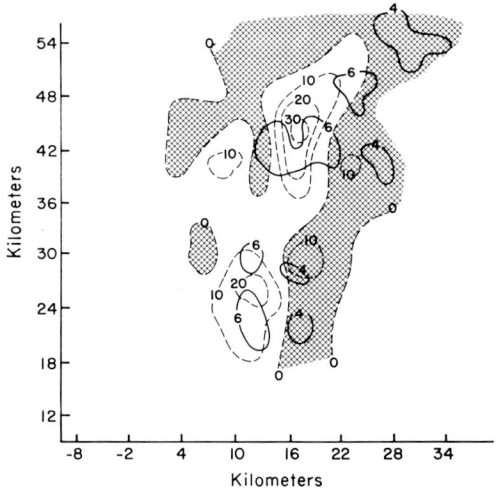

Fig. 10.17. Storm of 8 June 1974; the spectrum width (m s^{-1}) contours are solid lines; vertical motion (m s^{-1}) contours are dashed lines. The cross-hatched area is the downdraft. The origin of the coordinate system is at the radar site. The height is 5 km. (From Lee, 1977.)

updrafts and downdrafts. It seems reasonable to expect larger dissipation rates in the Oklahoma storm, which had an updraft of 40 m s^{-1}, than in the Colorado storm with an updraft of 20 m s^{-1}. It is also likely, however, that some of the largest dissipation rates are due to anisotropic eddies that are not part of the inertial subrange. In such cases the derived dissipation rate could be larger than the true dissipation rate of turbulent kinetic energy.

Insufficient Doppler data are available for final conclusions concerning the locations of broad Doppler spectra in a storm. Figure 10.17 is presented to show the correlation between turbulence and the up- (down-) drafts in a tornadic storm (Lee, 1977). The maximum reflectivity (not shown) is north of the updrafts. It is apparent in this case that areas of large spectrum width are on the edges of the updraft, with a preference for higher values when a downdraft is in close proximity. Thus, in this example, turbulence is probably produced by horizontal shear of the vertical wind.

Figure 10.18 compares the squares of the Doppler spectrum widths measured by radar and those deduced from aircraft penetrations. Radar-measured widths were computed from the Fourier transform of time series data, and an adaptive threshold proposed by Hildebrand and Sekhon (1974) was used to reduce noise effects. The contribution of the linear shear to the width has been taken into account. To obtain Fig. 10.18 the aircraft path was shifted until a maximum correlation between the two spectrum widths was achieved. The aircraft widths were obtained by weighting, with the radar resolution function, the point velocities the aircraft had recorded. To avoid contamination by scales larger than the outer scale of turbulence,

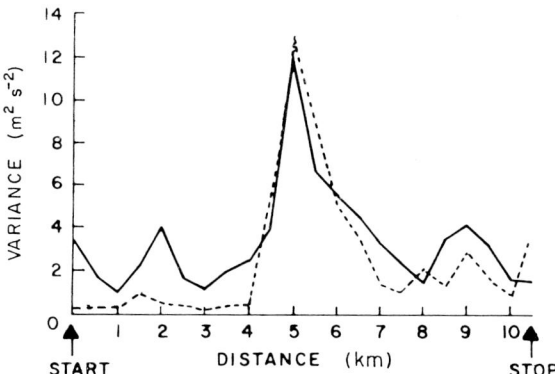

Fig. 10.18. Aircraft (dashed) and radar (solid) estimates of the square of the Doppler spectrum width (variances) at grid point locations, along with the best correlated aircraft tracks for one penetration. (From Bohne, 1981.)

Bohne (1981) subjectively detrended segments of aircraft data. He then obtained a correlation of 0.89 between the two curves. In this and other instances, agreement is best at large widths, whereas at smaller widths the radar indicates larger values. If this proves true in general, some false alarms could be expected, but warnings should be correct.

REFERENCES

Batchelor, G. K. (1953). "The Theory of Homogeneous Turbulence." Cambridge Univ. Press, London and New York.
Bohne, A. R. (1981). Estimation of turbulence severity in precipitation environments by radar. *Prepr., Radar Meteorol. Conf., 20th, 1981* pp. 446–453.
Bohne, A. R. (1982). Radar detection of turbulence in precipitation environments. *J. Atmos. Sci.* **39,** 1819–1837.
Chandrasekhar, S. (1955). A theory of turbulence. *Proc. R. Soc. London, Ser. A* **229,** 1–19.
Chandrasekhar, S. (1956). Theory of turbulence. *Phys. Rev.* **102,** 941–953.
Chernikov, A. A., Mel'nichuk, Yu. V., Pinus, N. K., Shmeter, S. M., and Vinnichenko, N. K. (1969). Investigations of the turbulence in convective atmosphere using radar and aircraft. *Radio Sci.* **4,** 1257–1260.
Doviak, R. J., and Berger, M. I. (1980). Turbulence and waves in the optically clear planetary boundary layer resolved by dual-Doppler radars. *Radio Sci.* **15,** 297–317.
Frisch, A. S., and Clifford, S. F. (1974). A study of convection capped by a stable layer using Doppler radar and acoustic echo sounders. *J. Atmos. Sci.* **31,** 1622–1628.
Frisch, A. S., and Strauch, R. G. (1976). Doppler radar measurements of turbulent kinetic energy dissipation rates in a northeastern Colorado convective storm. *J. Appl. Meteorol.* **15,** 1012–1017.
Gossard, E. E., and Strauch, R. G. (1981). "Radar Applications in Cloud and Clear Air Studies," NOAA Tech. Rep. ERL WPL-87. Natl. Tech. Inf. Serv., Operations Div., Springfield, Virginia.
Harris, F. I. (1975). Motion field characteristics of the evaporative base region of a stratiform precipitation layer as determined by ANDASCE. *Prepr., Radar Meteorol. Conf., 16th, 1975* pp. 225–230.
Hildebrand, P. H., and Sekhon, R. S. (1974). Objective determination of noise level in Doppler spectra. *J. Atmos. Sci.* **13,** 808–811.
Istok, M. (1981). Analysis of Doppler spectrum broadening mechanisms in thunderstorms. *Prepr., Radar Meteorol. Conf., 20th, 1981* pp. 454–458.
Jenkins, G. M., and Watts, D. G. (1968). "Spectral Analysis and Its Application." Holden Day, San Francisco, California.
Kolmogoroff, A. N. (1941). Dissipation of energy in the locally isotropic turbulence. *Dokl. Akad. Nauk SSSR* **32,** 16–18.
Labitt, M. (1981). "Coordinated Radar and Aircraft Observations of Turbulence," Proj. Rep. ATC 108. Massachusetts Institute of Technology, Lincoln Lab., Cambridge.
Lee, J. T. (1977). "Application of Doppler Radar to Turbulence Measurements Which Affect Aircraft," Final Rep. No. FAA-RD-77-145. FAA Syst. Res. Dev. Serv., Washington, D.C.
Lhermitte, R. M. (1968). Turbulent air motion as observed by Doppler radar. *Prepr., Radar Meteorol. Conf., 13th, 1968* pp. 14–17.
Lhermitte, R. M. (1969). Note on the observation of small-scale atmospheric turbulence by Doppler radar technique. *Radio Sci.* **4,** 1241–1246.

MacCready, P. B., Jr. (1962). Turbulence measurements by sailplane. *J. Geophys. Res.* **67**, 1041–1050.

MacCready, P. B., Jr. (1964). Standardization of gustiness values from aircraft. *J. Appl. Meteorol.* **3**, 439–449.

Neter, J., and Wasserman, W. (1974). "Applied Linear Statistical Models." Richard D. Irwin, Inc., Homewood, Illinois.

Panchev, S. (1971). "Random Functions and Turbulence." Pergamon, Oxford.

Papoulis, A. (1965). "Probability, Random Variables, and Stochastic Processes." McGraw-Hill, New York.

Rabin, R. M., Doviak, R. J., and Sundara-Rajan, A. (1982). Doppler radar observations of momentum flux in a cloudless convective layer with rolls. *J. Atmos. Sci.* **39**, 851–863.

Reiter, E. R. (1970). Recent advances in the study of clear-air turbulence (CAT). *Riv. Meteorol. Aeronaut.* **30**, 10–13.

Reiter, E. R., and Burns, A. (1960). The structure of clear-air turbulence derived from "TOPCAT" aircraft measurements. *J. Atmos. Sci.* **23**, 206–212.

Rhyne, R. H., and Steiner, R. (1964). Power spectral measurements of atmospheric turbulence in severe storms and cumulus clouds. *NASA Tech. Note* **NASA TN D-2469**, 1–48.

Rogers, R. R., and Tripp, B. R. (1964). Some radar measurements of turbulence in snow. *J. Appl. Meteorol.* **3**, 603–610.

Sinclair, P. C. (1974) Severe storm turbulent energy structure. *Conf. Aerosp. Aeronaut. Meteorol., 6th, 1974, El Paso, Texas* (unpublished manuscript).

Smythe, G. R., and Zrnić, D. S. (1983). Correlation analysis of Doppler radar data and retrieval of the horizontal wind. *J. Climate Appl. Meteorol.* **22**, 297–311.

Srivastava, R. C., and Atlas, D. (1974). Effect of finite radar pulse volume on turbulence measurements. *J. Appl. Meteorol.* **13**, 472–480.

Tatarskii, V. I. (1971). "The Effects of the Turbulent Atmosphere on Wave Propagation" (translated by Israel Program for Scientific Translations Ltd.). (Available from U.S. Dept. of Commerce, Natl. Tech. Inf. Serv., Springfield, Virginia.)

Tennekes, H., and Lumley, J. L. (1972). "A First Course in Turbulence." M.I.T. Press, Cambridge, Massachusetts.

Vinnichenko, N. K., and Dutton, J. A. (1969). Empirical studies of atmospheric structure and spectra in the free atmosphere. *Radio Sci.* **4**, 1115–1126.

11

Echoes from the Precipitation-Free Turbulent Troposphere

In previous chapters we have discussed the properties of signals scattered from a distribution of point targets acting as tracers of laminar or turbulent air motion. However, it is known that radar also senses returns from air in which there is no precipitation or other point targets. It is the purpose of this chapter to develop, from basic theory, a relationship between the turbulent characteristics of the atmosphere and the Doppler-shifted signals sensed by the weather radar. Wind profiling applications are also discussed.

11.1 REFLECTION, REFRACTION, AND SCATTER: COHERENCE

The permittivity $\varepsilon(\mathbf{r}, t)$ of the troposphere has, in general, both spatial and temporal variation. Even though the averaged permittivity $\langle \varepsilon(\mathbf{r}, t) \rangle$ may not contain small-scale spatial irregularities, it always exhibits large-scale spatial inhomogeneities. The spectrum of scales that remain is a function of integration time, so the longer the latter, the smoother is the average permittivity. However, for the troposphere, because of the omnipresence of gravity and solar heat, there is a limit beyond which, no matter how long we integrate, or at least for any practical time interval, $\varepsilon(\mathbf{r})$ does not become smoother and spatial variations, particularly along the vertical, remain. As discussed in Chapter 2, these spatial inhomogeneities of the average ε cause ray paths to be refracted from their normal straight-line course.

Studies of the reflection and refraction of waves in inhomogeneous nonturbulent media have been carried out by many authors (Budden, 1961; Wait, 1962; Brekhovskikh, 1960), neglecting the turbulent nature of the media. Other authors (Tatarskii, 1961; Wheelon, 1959; Booker and Gordon, 1950) have considered the scattering properties of turbulent media, but most ignore the spatial inhomogeneity of the average permittivity. Du Castel *et al.* (1962) have studied the reflection process when both turbulent fluctuations and inhomogeneity in average values are present. These give rise to waves comprising (1) a coherent component caused by reflection or refraction and (2) an incoherent component caused by scatter.

Waves that are reflected from sharp quasipermanent changes in ε, as well as those refracted from a transmitter to a remote receiver by gradual spatial changes in ε, form the coherent component of the signal, whereas waves scattered from turbulent media give incoherent signals. Signals are completely time coherent when the scattering medium does not modulate the amplitude or phase of the wave. However, it should be understood that spatial variations of the amplitude and phase can exist, and furthermore, these spatial variations can also be described in terms of spatial coherence. In reality, the atmosphere is always nonstationary, and signals change, slowly, in amplitude and phase.

In practice we can distinguish between reflection or refraction and scatter when the signal spectrum exhibits two distinct distributions—one peaked and narrow, the other broad, as sketched in Fig. 11.1a. The narrow distribution is associated with coherent signals and usually with refracted waves, and the broad distribution is associated with incoherent turbulent scatter. For example, over-the-horizon communication links contain both coherent

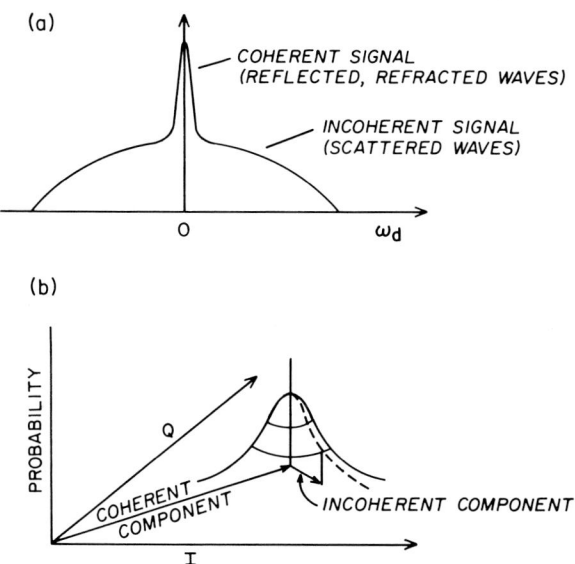

Fig. 11.1. (a) The spectrum of signals from waves simultaneously both reflected or refracted and scattered. (b) The distribution of samples of I and Q. Samples are taken over a time interval small compared to the coherence time of the coherent signal.

and incoherent signals. Vertical gradients of the mean refractive index can refract the transmitted signal to the receiver, as shown in Chapter 2. This refracted signal will exhibit slow changes in amplitude and phase. Simultaneously, turbulent irregularities in the refractive index, which change much more quickly, will scatter signals to the receiver, and these signals will produce rapid variations in amplitude and phase superimposed on the slow variations caused by the gradual changes in refractive index. In such cases the distribution of I and Q is as shown in Fig. 11.1b. If many independent scattering elements are effective (see Section 11.5.1), there is a Gaussian distribution centered about a mean value (when the mean is taken over a time short compared to the coherence time of the coherent signal component). Signals backscattered by raindrops are incoherent and have an I, Q distribution centered about zero, but when the receiver is separate from the transmitter, both coherent (from direct line-of-sight propagation or from refracted waves) and incoherent signals may be present.

Even though scatter produces signals that are considered incoherent, the signal remains coherent for durations small compared to the time it takes for the scattering medium to alter significantly. Only when the coherence time is sufficiently long, so that samples of the signal are highly correlated, can a meaningful Doppler spectrum measurement be made.

The following development, which describes the propagation of waves through inhomogeneous turbulent media, will be framed so as to embrace both refraction or reflection and scatter. The amplitude and phase of refracted waves are usually found through exact solution of Maxwell's equations, when certain vertical profiles (e.g., linear, parabolic, or exponential) of ε are considered or by geometric optics ray-tracing calculations used when the spatial variations cannot be described by simple functions but are sufficiently gradual, as demonstrated in Chapter 2. The random or scatter contribution to the fields are usually solved by invoking the Born approximation. In this approximation it is assumed that the scatter contribution to the electro-magnetic field in the region of scattering is small compared to the unperturbed refracted field (i.e., the field that would exist if turbulence were nonexistent).

11.2 FORMULATION OF THE WAVE EQUATION FOR INHOMOGENEOUS AND TURBULENT MEDIA

Starting with Maxwell's equations,

$$\mathbf{V} \times \mathbf{E} = -\mu_0 \, \partial \mathbf{H}/\partial t, \tag{11.1}$$

$$\mathbf{V} \times \mathbf{H} = \partial \varepsilon \mathbf{E}/\partial t, \tag{11.2}$$

for media in which the permeability μ is a constant μ_0, and taking the curl of (11.1) and substituting into (11.2), we obtain

$$\nabla(\nabla \cdot \mathbf{E}) - \nabla^2 \mathbf{E} = -\mu_0 \varepsilon \frac{\partial^2 \mathbf{E}}{\partial t^2} \tag{11.3}$$

when the time scales from changes in the permittivity $\varepsilon(\mathbf{r}, t)$ are long compared to changes in $\mathbf{E}$ at the transmitted frequency. Taking the divergence of (11.2),

$$\nabla \cdot \nabla \times \mathbf{H} \equiv 0 = \frac{\partial}{\partial t} \nabla \cdot (\varepsilon \mathbf{E}) = \frac{\partial}{\partial t} (\varepsilon \nabla \cdot \mathbf{E} + \mathbf{E} \cdot \nabla \varepsilon), \tag{11.4}$$

we find that

$$\varepsilon \nabla \cdot \mathbf{E} + \mathbf{E} \cdot \nabla \varepsilon = \text{const},$$

independent of time. Because we seek solutions that are wavelike and therefore time varying, we can set the constant equal to zero and solve the resulting equation to obtain

$$\nabla \cdot \mathbf{E} = -(\mathbf{E} \cdot \nabla \ln \varepsilon/\varepsilon_0). \tag{11.5}$$

Substitution of (11.5) into (11.3) produces the wave equation:

$$\nabla^2 \mathbf{E} - \mu_0 \varepsilon \frac{\partial^2 \mathbf{E}}{\partial t^2} = -\nabla(\mathbf{E} \cdot \nabla \ln \varepsilon/\varepsilon_0), \tag{11.6}$$

in which the right side is considered as a source distribution determined by the resultant electric field and permittivity. Equation (11.6), a formulation of the wave equation for an inhomogeneous random medium, is a set of three coupled linear differential equations with nonconstant coefficients and, as such, has no known general analytic solution. In order to find an acceptable approximation, let us express the relative permittivity as

$$\varepsilon/\varepsilon_0 = \varepsilon_a/\varepsilon_0 + \Delta\varepsilon/\varepsilon_0, \tag{11.7}$$

where ε_a is an ensemble average of the permittivity taken at any point. In general ε_a may be a function of the coordinates of the point, but its spatial changes are deterministic so that nonstatistical algebraic or numerical solutions can be obtained. If spatial changes of ε_a are sufficiently gradual, then geometric optics ray tracing solutions outlined in Chapter 2 can be used to solve for the refracted fields. $\Delta\varepsilon$ represents a zero-mean random process and is the fluctuation produced by atmospheric turbulence. The refractive index will, for convenience, be defined as follows:

$$n^2 = n_a^2 + 2n_a \Delta n = \varepsilon_a/\varepsilon_0 + \Delta\varepsilon/\varepsilon_0, \tag{11.8}$$

where

$$n_a \equiv (\varepsilon_a/\varepsilon_0)^{1/2}, \qquad \Delta\varepsilon/\varepsilon_0 = 2n_a \Delta n, \tag{11.9}$$

11.2 Formulation of the Wave Equation for Inhomogeneous and Turbulent Media

and Δn is, to a good approximation if $\Delta n \ll n_a$, the deviation of the refractive index from its average value n_a. Substituting (11.7) and (11.8) into (11.6), we obtain

$$\nabla^2 \mathbf{E} - \frac{n_a^2}{c^2} \frac{\partial^2 \mathbf{E}}{\partial t^2} = \frac{2n_a \Delta n}{c^2} \frac{\partial^2 \mathbf{E}}{\partial t^2}$$

$$- 2\nabla[\mathbf{E} \cdot \nabla \ln(n_a)] - \nabla\left[\mathbf{E} \cdot \nabla \ln\left(1 + \frac{2\Delta n}{n_a}\right)\right]. \quad (11.10)$$

We seek a perturbation solution of the form

$$\mathbf{E} = \mathbf{E}_0 + \mathbf{E}_1, \quad (11.11a)$$

where $\mathbf{E}_0$ is the solution in the absence of turbulence (i.e., for $\Delta n = 0$).

Substituting (11.11) into (11.10) after expanding

$$\ln\left(1 + \frac{2\Delta n}{n_a}\right) = \frac{2\Delta n}{n_a} - \frac{1}{2}\left(\frac{2\Delta n}{n_a}\right)^2 + \frac{1}{3}\left(\frac{2\Delta n}{n_a}\right)^3 - \cdots$$

in a power series, we obtain

$$\nabla^2 \mathbf{E}_0 + \nabla^2 \mathbf{E}_1 - \frac{n_a^2}{c^2} \frac{\partial^2}{\partial t^2}(\mathbf{E}_0 + \mathbf{E}_1)$$

$$= \frac{2n_a \Delta n}{c^2} \frac{\partial^2}{\partial t^2}(\mathbf{E}_0 + \mathbf{E}_1) - 2\nabla\{(\mathbf{E}_0 + \mathbf{E}_1) \cdot [\nabla \ln(n_a)]\}$$

$$- \nabla\left\{(\mathbf{E}_0 + \mathbf{E}_1) \cdot \nabla\left[\frac{2\Delta n}{n_a} - \frac{1}{2}\left(\frac{2\Delta n}{n_a}\right)^2 + \cdots\right]\right\}. \quad (11.11b)$$

Because we assume $\Delta n \ll n_a$, we shall ignore higher-order terms in $\Delta n / n_a$. Furthermore, we shall also ignore terms that contain products of $\mathbf{E}_1$ and Δn or its derivatives. In this way we are in effect neglecting multiple scattering, that is, the interaction of the scattered field $\mathbf{E}_1$ with refractive index perturbations producing secondary scattered fields. Usually these are small compared to the first-order scattered fields for $\Delta n \ll n_a$. Under these assumptions, the previous equation reduces to

$$\nabla^2 \mathbf{E}_0 + \nabla^2 \mathbf{E}_1 - \frac{n_a^2}{c^2} \frac{\partial^2}{\partial t^2}(\mathbf{E}_0 + \mathbf{E}_1)$$

$$= \frac{2n_a \Delta n}{c^2} \frac{\partial^2 \mathbf{E}_0}{\partial t^2} - 2\nabla[\mathbf{E}_0 \cdot \nabla \ln(n_a) + \mathbf{E}_1 \cdot \nabla \ln(n_a)]$$

$$- \nabla\left[\mathbf{E}_0 \cdot \nabla\left(\frac{2\Delta n}{n_a}\right)\right]. \quad (11.12)$$

Note that $\mathbf{E}_1$ is not necessarily small compared to $\mathbf{E}_0$. There can be regions of space where the scattered field $\mathbf{E}_1$ is larger than $\mathbf{E}_0$, but this occurs only outside the main lobe of the transmitter beam. Zero-order terms are obtained when $\Delta n \to 0$ (i.e., $\mathbf{E}_1 \to 0$), and from (11.12) we have $\mathbf{E}_0$ satisfying

$$\nabla^2 \mathbf{E}_0 - \frac{n_a^2}{c^2} \frac{\partial^2}{\partial t^2} \mathbf{E}_0 = -2\, \nabla[\mathbf{E}_0 \cdot \nabla \ln(n_a)]. \qquad (11.13\text{a})$$

If the transmitter produces harmonic waves with radian frequency ω_0, the refracted wave $\mathbf{E}_0(\mathbf{r}, t)$ will also be harmonic, and (11.13a) can be written

$$\nabla^2 \mathbf{E}_0(\mathbf{r}) + k_0^2 n_a^2 \mathbf{E}_0(\mathbf{r}) = -2\, \nabla[\mathbf{E}_0(\mathbf{r}) \cdot \nabla \ln(n_a)], \qquad (11.13\text{b})$$

where $\mathbf{E}_0(\mathbf{r}, t) = \mathbf{E}_0(\mathbf{r}) e^{j\omega_0 t}$ and $k_0^2 \equiv \omega_0^2/c^2$.

Because $\Delta n(\mathbf{r}, t)$ is time varying, it produces a scattered field $\mathbf{E}_1(\mathbf{r}, t)$ that is not a pure harmonic signal. Thus, we cannot in general express the second-order time derivative of $\mathbf{E}_1$ as a product of it and ω_0^2.

Outwardly (11.13) is similar to (11.6), but, by definition, $\mathbf{E}_0$ of (11.13) is the coherent part of the total solution. Subtracting (11.13a) from (11.12), we obtain the wave equation whose solution gives the incoherent field that results from scatter of the incident refracted field $\mathbf{E}_0$:

$$\nabla^2 \mathbf{E}_1 - \frac{n_a^2}{c^2} \frac{\partial^2 \mathbf{E}_1}{\partial t^2}$$
$$= -2 k_0^2 n_a\, \Delta n\, \mathbf{E}_0 - 2\, \nabla \left[\mathbf{E}_1 \cdot \nabla \ln(n_a) + \mathbf{E}_0 \cdot \nabla \left(\frac{\Delta n}{n_a} \right) \right]. \qquad (11.14)$$

Thus, from (11.13) and (11.14) we see that $\mathbf{E}_0$ is the field due to the refractive effects of n_a, and $\mathbf{E}_1$ is the contribution from the turbulent irregularities in the structure of n.

Solutions of (11.13) for plane-stratified media are discussed extensively in the literature by many authors (Budden, 1961; Brekhovskikh, 1960; Wait, 1962). In their texts exact analytic expressions are obtained for specialized cases of the wave polarization and spatial variations of n_a. However, when the change of n_a within a wavelength is small (Appendix D), that is, when

$$|\nabla n_a|/2\pi \ll 1/\lambda \qquad \text{everywhere,} \qquad (11.15)$$

then Eq. (11.13b) can be approximated by

$$\nabla^2 \mathbf{E}_0 + k_0^2 n_a^2 \mathbf{E}_0 \approx 0. \qquad (11.16)$$

It should also be noted from (11.13) that for media in which the stratification is in one direction, as may occur whenever a field of mechanical force along a dominant direction is exerted on the medium, (11.16) becomes exact for polarization of $\mathbf{E}$ perpendicular to the direction of stratification. Thus, we would expect (11.16) to be exact for horizontally polarized waves propagating in the atmosphere. At short radio wavelengths, such as those in the decimetric-to-centimetric bands, it is unlikely that large gradients in n would be sustained to violate (11.15). Thus, (11.16), should be valid for UHF and microwave weather radars.

The solution of Eq. (11.14) is known as the Born approximation and is valid whenever multiple scattering [i.e., higher-order terms in Eq. (11.11b)] can be neglected. In a medium in which the average refractive index profile varies smoothly [i.e., where (11.15) holds], the term

$$2\,\mathbf{V}[\mathbf{E}_1 \cdot \mathbf{V}\ln(n_a)] \qquad (11.17)$$

is negligible with respect to $(n_a^2/c^2)\,\partial^2 \mathbf{E}_1/\partial t^2$, as can be deduced from the arguments put forth in Appendix D. We need only make the added assumption that the time rate of change of the amplitude and phase of $\mathbf{E}_1$ is much slower than the harmonically related changes (i.e., changes at the rate ω_0). Thus, the differential equation that gives the first-order solution to the fields scattered by irregularities in an otherwise smoothly changing medium is

$$\nabla^2 \mathbf{E}_1 - \frac{n_a^2}{c^2}\frac{\partial^2 \mathbf{E}_1}{\partial t^2} = -2k_0^2 n_a^2 \left(\frac{\Delta n}{n_a}\right) \mathbf{E}_0 - 2\,\mathbf{V}\left[\mathbf{E}_0 \cdot \mathbf{V}\left(\frac{\Delta n}{n_a}\right)\right], \qquad (11.18)$$

where $\mathbf{E}_0$ is the solution of (11.16). It will be shown in Section 11.3 that the second term on the right-hand side, $2\,\mathbf{V}[\mathbf{E}_0 \cdot \mathbf{V}(\Delta n/n_a)]$, does not contribute to the first-order transverse fields but does cancel a first-order longitudinal field arising from the first term. Even though we are usually interested only in the transverse field components, because they are associated with power transfer, we cannot ignore the second term on the right side of (11.18) unless we ignore the first-order longitudinal fields.

11.3 SOLUTION FOR FIELDS SCATTERED BY IRREGULARITIES

Although the focus of this book is on weather radars, in this section we consider the transmitter to be separate from the receiver. This will allow us (1) to demonstrate the relation between the scales of refractive index irregularities effective in the scattering process and transmitter–receiver separation, (2) to determine the transmission loss of a forward-scatter link given the spectrum of turbulence in the illuminated volume, and (3) to examine the Doppler shift due to propagating irregularities such as those generated by acoustic waves (Fetter, 1961; Smith, 1961). In a combined *radioacoustic sounding system* (RASS), acoustic waves cause traveling sinusoidal perturbations in the refraction index that are tracked by a bistatic radar to measure temperature profiles in the first 1–2 km of the troposphere (Frankel and Peterson, 1976).

Because narrow-beam antennas are typically used in remote sensing and also in forward-scatter communication links, the dimensions of the volume contributing to the scattered field are sufficiently small that variations in r_t and r_r over this volume of integration are much smaller than the respective distances r_{t0} and r_{r0} (Fig. 11.2). This condition allows simplifying assumptions to be applicable to the development of a solution to (11.18). We also assume the origin of our coordinate system to be at the intersection of the main beam axes, which lie along $\mathbf{r}_{t0}, \mathbf{r}_{r0}$.

In the troposphere $n_a \simeq 1$, so Eq. (11.18) can be approximated by

$$\nabla^2 \mathbf{E}_1(\mathbf{r}, t) - \frac{1}{c^2} \frac{\partial^2 \mathbf{E}_1(\mathbf{r}, t)}{\partial t^2} = -\mathbf{f}(\mathbf{r}, t), \tag{11.19}$$

where

$$\mathbf{f}(\mathbf{r}, t) \equiv (2k_0^2 \, \Delta n(\mathbf{r}, t)\mathbf{E}_0(\mathbf{r})$$
$$+ 2\, \nabla\{\mathbf{E}_0(\mathbf{r}) \cdot \nabla[\Delta n(\mathbf{r}, t)]\})e^{j\omega_0 t} \tag{11.20}$$

and $\mathbf{r}$ is the coordinate of any observation point, which in this case would be $\mathbf{r}_{r0}$ (Fig. 11.2).

Pulse-modulated transmitters give

$$\mathbf{E}_0(\mathbf{r}, t) = \mathbf{E}_0(\mathbf{r}) e^{j\omega_0(t - r/c)} U(t - r/c),$$
$$U(t^*) = 1, \quad 0 \leq t^* \leq \tau \tag{11.21}$$
$$= 0 \quad \text{otherwise,}$$

where τ is the pulse width, $U(t^*)$ is a unit pulse function, and $t^* \equiv t - r/c$ is the retardation time. In order to account for pulse modulation we shall now make an assumption that represents a compromise between the ease of following a reasonable physical argument and the pursuit of a less-tractable mathematical method. In Section 4.3.1 it was shown that sampled backscattered signals came, for all practical purposes, from a thin spheroidal shell of scatters whose location and contribution to the signal are determined by the sample time delay, the transmitted pulse shape, and the receiver frequency transfer function. It can be shown analogously that for a spaced transmitter–receiver the scatter volume that contributes to the sampled received signal is confined to a prolate spheroidal shell whose foci are at the transmitter and receiver locations. The range r_r from the receiver to an element of the scatter volume V is

$$r_r = \frac{1}{2} \frac{(ct_s)^2 - (2d)^2}{ct_s - 2d \cos \theta_r}, \quad t_s \geq 2d/c, \tag{11.22}$$

11.3 Solution for Fields Scattered by Irregularities

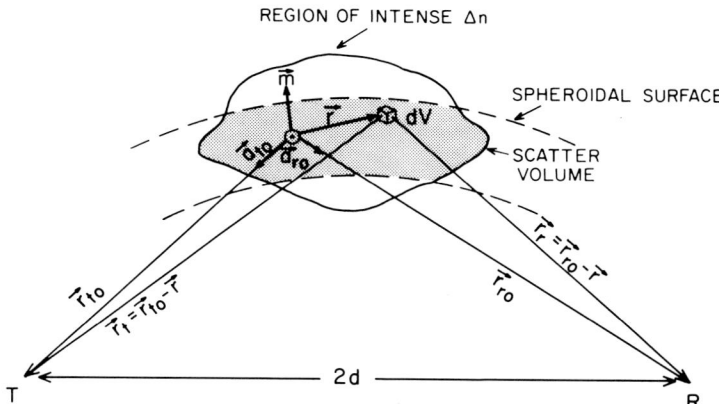

Fig. 11.2. Illustration of the scatter geometry and the spheroidal shell within which scattering irregularities contribute to the received signal sampled at some time delay after the emission of a transmitter pulse. Line TR is the axis of rotation for the spheroid. $\mathbf{a}_{t0}$ and $\mathbf{a}_{r0}$ are unit vectors at the origin. Vectors are indicated by overarrows in figures and by bold letters in text.

where θ_r is the receiver elevation angle and $2d$ is the distance between transmitter and receiver. The sample time t_s is the time after the transmitter pulse is radiated at which $\mathbf{E}_1$ is sampled. Obviously we cannot receive echoes for $t_s < 2d/c$. As in the backscatter case, it appears reasonable to assume that the weight across the spheroidal shell is uniform when the pulse is rectangular. The separation Δs between the inner and outer boundaries of the spheroidal shell is determined by the pulse width τ and t_s:

$$\Delta s \simeq (c\tau/2)[1 + 4d^2 \sin^2 \theta_r/(ct_s - 2d \cos \theta_r)^2]^{1/2}. \quad (11.23)$$

When $d \to 0$ (i.e., for backscatter), $\Delta s \to c\tau/2$, as it should. The shell has maximum thickness (largest scatter volume) when the beams intersect at a point equidistant from the transmitter and receiver. Furthermore, the shell thickness for forward scatter is always larger than for backscatter if the pulse widths are the same.

We can therefore dispense with the unit pulse functions in (11.21) if, when integrating contributions from illuminated scatter volumes, we uniformly add all and only the contributions from scatter volume elements within the spheroidal shell of thickness Δs. This procedure gives the field $\mathbf{E}_1$ incident on the receiver antenna. However, to determine the signal at the output of the receiver, we need also to consider the receiver transfer function. Therefore, to compute the output signal for a matched filter receiver, we must weight the source function $\mathbf{f}(\mathbf{r}, t)$ with a normalized triangular function whose base equals $2\Delta s$.

It can be shown that when the period of the harmonic vibrations is much shorter than the time scales for changes in Δn, the scattered field $\mathbf{E}_1(\mathbf{r}, t)$ is a narrowband signal spread about ω_0, and (11.19) can be then approximated by

$$\nabla^2 \mathbf{E}_1(\mathbf{r}, t) + k_0^2 \mathbf{E}_1(\mathbf{r}, t) = -\mathbf{f}(\mathbf{r}, t). \tag{11.24}$$

Using Green's function, developed in Appendix E, for the general time-varying function, we obtain the solution of the scattered field at the receiver location:

$$\mathbf{E}_1(\mathbf{r}_{r0}, t) = \frac{1}{4\pi} \int_{-\infty}^{+\infty} \int_V \frac{\mathbf{f}(\mathbf{r}, t')\delta(t - t' - r_r/c)}{r_r} dV \, dt', \tag{11.25}$$

where $r_r \equiv |\mathbf{r}_{r0} - \mathbf{r}|$.

The integration with respect to t' can be carried out to obtain

$$\int_{-\infty}^{+\infty} \mathbf{f}(\mathbf{r}, t')\delta(t - t' - r_r/c) \, dt'$$
$$= (2k_0^2 \mathbf{E}_0(\mathbf{r}) \Delta n(\mathbf{r}, t^*) + 2 \nabla\{\mathbf{E}_0(\mathbf{r}) \cdot \nabla[\Delta n(\mathbf{r}, t^*)]\}) e^{j\omega_0 t - jk_0 r_r}$$
$$= \mathbf{f}(\mathbf{r}, t^*), \tag{11.26}$$

where $t^* \equiv t - r_r/c$ is the retardation time. For typical scatter volume dimensions, the time shifts relative to r_{r0}/c are less than milliseconds and can be assumed to be short compared to the time scales for changes in Δn. Then

$$\Delta n(\mathbf{r}, t^*) \simeq \Delta n(\mathbf{r}, t - r_{r0}/c) \equiv \Delta n(\mathbf{r}, t_0^*),$$

so the solution to the scattered field is

$$\mathbf{E}_1(\mathbf{r}_{r0}, t) = \frac{1}{4\pi} \int_V \frac{\mathbf{f}(\mathbf{r}, t_0^*) e^{-jk_0 r_r}}{r_r} dV. \tag{11.27}$$

With the proviso that receivers are in the far field of the scatter volume (the conditions for this are given in Section 11.4), the signals at two receivers spaced along the direction $\mathbf{a}_{r0}$ will have the same time dependence, except that one will be delayed with respect to the other by a time difference equal to the difference in retardation times. However, receivers located at the same range r_{r0} but along a different direction $\mathbf{a}_{r0}$ will not necessarily have the same time dependence, because the exponential term in (11.27) will have different phases for the same $\mathbf{r}$, and hence the integral (11.27) will have different values. Because we shall be considering scatter from random media, the detailed relation between time change in Δn and the corresponding time change in $\mathbf{E}_1$ is of no concern; rather, it is the statistical properties of the echoes that count. Therefore, the constant retardation time delay r_{r0}/c in (11.27) can be ignored without affecting the statistics of the received signal.

11.3 Solution for Fields Scattered by Irregularities

All Doppler shifts are contained in the complex function $\mathbf{E}_1(\mathbf{r}_{r0}, t)$. We shall show in Section 11.4 how irregularities of Δn that drift with uniform velocity produce a mean Doppler shift. If there is no mean motion, then the time variations of $\mathbf{E}_1$ produce a spectrum whose broadness about ω_0 depends on the rapidity of changes in $\mathbf{E}_1$ due to changing $\Delta n(t)$, which is the only parameter in $\mathbf{f}(\mathbf{r}, t)$ that has other than harmonic time variation.

For narrow-beam antennas and the typical ranges used in remote sensing, we can use the approximation

$$r_r \approx r_{r0} \tag{11.28}$$

in the denominator of (11.27). For $\mathbf{E}_0$ in (11.26) we shall use

$$\mathbf{E}_0 = \mathbf{A}_0(\mathbf{r}) e^{-jk_0 r_t} / r_t, \tag{11.29}$$

where $\mathbf{A}_0$ contains the polarization and angular dependence (i.e., the radiation pattern) of the transmitted field. For reasons already cited, we can also approximate the denominator of (11.29) by

$$r_t = |\mathbf{r}_{t0} - \mathbf{r}| \approx r_{t0}. \tag{11.30}$$

Substitution of (11.29) and (11.30) into (11.20) and the result into (11.27) gives the scatter field intensity:

$$\mathbf{E}_1(\mathbf{r}_{r0}, t) = \frac{1}{2\pi r_{t0} r_{r0}} (\mathbf{I}_1 + \mathbf{I}_2), \tag{11.31a}$$

where

$$\mathbf{I}_1 = k_0^2 \int_V \Delta n \, \mathbf{A}_0 e^{-jk_0(r_t + r_r)} \, dV, \tag{11.31b}$$

$$\mathbf{I}_2 = \int_V \nabla [e^{-jk_0 r_t} \mathbf{A}_0 \cdot \nabla(\Delta n)] e^{-jk_0 r_r} \, dV. \tag{11.31c}$$

We now apply the identity (Johnson, 1965, p. 444)

$$\int_V u \, \nabla \phi \, dV \equiv \int_S u\phi \, d\mathbf{S} - \int_V \phi \, \nabla u \, dV \tag{11.32}$$

to (11.31c), where we define

$$u \equiv e^{-jk_0 r_r}, \tag{11.33a}$$

$$\phi \equiv \mathbf{A}_0 \cdot \nabla(\Delta n) e^{-jk_0 r_t}. \tag{11.33b}$$

The surface integral can be set equal to zero by letting the surface of integration move beyond the limits of the scatter volume, i.e., the region of space that contributes significantly to the scattered field. Equation (11.31c) is therefore reduced to

$$\mathbf{I}_2 = -\int_V \{\mathbf{A}_0 \cdot \mathbf{V}_r(\Delta n)\} e^{-jk_0 r_t} \mathbf{V}_r e^{-jk_0 r_r} \, dV, \tag{11.34}$$

where, in order to emphasize that $\mathbf{V}$ operates only on the $\mathbf{r}$ and not the $\mathbf{r}_{r0}$ dependence of the functions, we have added the subscript r. Now,

$$\mathbf{V}_r e^{-jk_0 r_r} = -jk_0 e^{-jk_0 r_r} \mathbf{V}_r(r_r), \tag{11.35}$$

and it can be shown that, without approximation,

$$\mathbf{V}_r(r_r) = \frac{r_{r0}}{2} \left(\frac{2r\mathbf{a}_r}{r_{r0}^2} - \frac{2\mathbf{a}_{r0}}{r_{r0}} \right) \left(1 - \frac{2\mathbf{a}_{r0} \cdot \mathbf{r}}{r_{r0}} + \frac{r^2}{r_{r0}^2} \right)^{-1/2}, \tag{11.36}$$

where $\mathbf{a}_r$ is the unit vector in the direction of $\mathbf{r}$. Because we have assumed scatter volume dimensions small compared to r_{r0} (e.g., $r/r_{r0} \ll 1$),

$$\mathbf{V}_r(r_r) \simeq -\mathbf{a}_{r0}. \tag{11.37}$$

Thus $\mathbf{I}_2$ can be approximated by

$$\mathbf{I}_2 = -jk_0 \mathbf{a}_{r0} \int_V [\mathbf{A}_0 \cdot \mathbf{V}_r(\Delta n)] e^{-jk_0(r_t + r_r)} \, dV. \tag{11.38}$$

Using the identity (11.32) and the argument that follows it, the equation

$$\int_V \mathbf{u} \cdot \mathbf{V} \phi \, dV = -\int_V \phi \, \mathbf{V} \cdot \mathbf{u} \, dV \tag{11.39}$$

can be derived. Assigning

$$\mathbf{u} = \mathbf{A}_0 e^{-jk_0(r_t + r_r)}, \tag{11.40a}$$

$$\phi = \Delta n, \tag{11.40b}$$

and applying (11.39) to (11.38), we obtain

$$\mathbf{I}_2 = j\mathbf{a}_{r0} k_0 \int_V \Delta n \, \mathbf{V}_r \cdot [\mathbf{A}_0 e^{-jk_0(r_t + r_r)}] \, dV. \tag{11.41}$$

Because the pattern weighting function varies slowly over distances comparable to a wavelength, (11.41) can be reduced to

$$\mathbf{I}_2 = -\mathbf{a}_{r0} k_0^2 (\mathbf{a}_{r0} + \mathbf{a}_{t0}) \cdot \int_V \mathbf{A}_0 \, \Delta n \, e^{-jk_0(r_t + r_r)} \, dV \tag{11.42}$$

by applying (11.35) and (11.37).

11.4 Fraunhofer Scatter

For the condition of a narrow beam of transmitted radiation, $\mathbf{A}_0$ is nearly perpendicular to $\mathbf{a}_{t0}$ over the region where $\mathbf{A}_0$ is significantly large. Therefore, the second term of (11.42) is negligibly small. Furthermore, observe that the first portion of (11.42) exactly cancels the longitudinal component (i.e., the component in the direction of $\mathbf{a}_{r0}$) of (11.31b). Therefore only the transverse component (i.e., transverse to $\mathbf{a}_{r0}$) of (11.31b) contributes to the scattered field, whereas all the contribution of $\mathbf{I}_2$ is canceled. This transverse component is obtained from (11.31b) by forming the vector product $\mathbf{a}_{r0} \times \mathbf{A}_0$ with $\mathbf{a}_{r0}$. Therefore the solution (11.27) for the scattered field reduces to

$$\mathbf{E}_1(\mathbf{r}_{r0}, t) = -\frac{k_0^2 \mathbf{a}_{r0}}{2\pi r_{r0} r_{t0}} \times \int (\mathbf{a}_{r0} \times \mathbf{A}_0) \Delta n \, e^{-jk_0(r_t + r_r)} \, dV. \quad (11.43)$$

Although (11.43) is our basic solution to the scattered field the integration is not easily performed, because Δn is usually a random function whose statistical properties are only known. We therefore need to cast the solution in terms of the (supposedly known) statistical properties of Δn and relate these to the statistics of $\mathbf{E}_1$. In Section 11.4 we shall make simplifying assumptions to show clearly how the electromagnetic field samples certain scale sizes of the spectrum of scales contained in the field of irregularities Δn. Furthermore, these assumptions will simplify the development of the solution for the statistical properties of the random variable $\mathbf{E}_1$.

11.4 FRAUNHOFER SCATTER

We now introduce Fraunhofer conditions, which will specify that the phase fronts of the refracted incident wave are essentially planar over the scatter volume and that the receiver antenna is in the far field of this volume. These conditions are derived by expanding the exponential terms of (11.43) in a Taylor series:

$$r_r \equiv |\mathbf{r}_{r0} - \mathbf{r}| = r_{r0} - \mathbf{a}_{r0} \cdot \mathbf{r} + (1/2r_{r0})[r^2 - (\mathbf{a}_{r0} \cdot \mathbf{r})^2] + \cdots \quad (11.44)$$

and retaining only the first-order terms. If we are to neglect the second-order terms, it is necessary that the third term of (11.44) multiplied by k_0 satisfy

$$(k_0/2r_{r0})[r^2 - (\mathbf{a}_{r0} \cdot \mathbf{r})^2] < 1. \quad (11.45)$$

Condition (11.45) stipulates that the maximum dimension of the scatter volume transverse to $\mathbf{a}_{r0}$ must satisfy

$$r_{\max,r} < \sqrt{\lambda r_{r0}/\pi}. \quad (11.46a)$$

Equation (11.46a) specifies that the receiver is in the far field of the scatter volume. In a like manner we obtain, on expansion of r_t, the condition

$$r_{\max,t} < \sqrt{\lambda r_{t0}/\pi}, \tag{11.46b}$$

which limits the incident phase fronts to be essentially planar over the scatter volume.

When $r_{\max,r}$ is determined by the field of view $r\theta_1$ of the receiving antenna (Fig. 11.3), as is usual for troposcatter communication links, and not by the spatial extent of Δn, then

$$r_{\max,r} \approx \theta_1 r_{r0}/2 \approx 0.6\lambda r_{r0}/D, \tag{11.47}$$

where D is the diameter of the receiving antenna aperture. However, when (11.47) is substituted into (11.46a), we obtain

$$r_{r0} < D^2/\lambda, \tag{11.48}$$

which stipulates the common volume V_c of the beams to be in the near zone of the receiver antenna. However, if V_c were in the near field of the antenna, the dimension $r_{\max}$ would be equal approximately to the diameter D of the antenna, and hence condition (11.46a) would be violated. Thus, it can be concluded that conditions (11.46a) and (11.48) are contradictory, and therefore the scatter volume for which the first-order condition (11.45) holds cannot be V_c.

It seems that we have derived conditions for which the ensuing solutions have limited applicability. However, it can be shown (Section 11.5.1) that if V_c has dimensions much larger than the correlation length of Δn, then V_c may be divided into smaller subvolumes each of which is to be many correlation

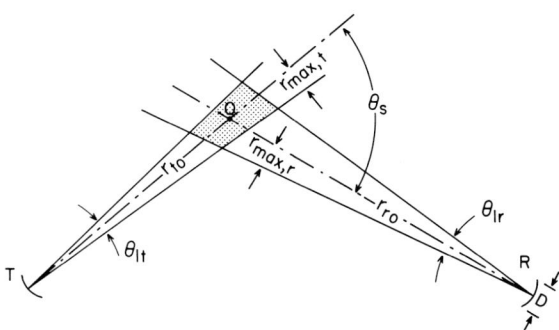

Fig. 11.3. Scatter volume size determined by the beamwidths θ_{1t}, θ_{1r}. The common beam volume V_c is shaded, and θ_1 is the one-way 3-dB beamwidth of an assumed circularly symmetric antenna aperture.

11.4 Fraunhofer Scatter

distances on a side. Each subvolume size, however, must not exceed the limits stipulated by condition (11.46), where now r_{t0}, r_{r0} are the ranges to the origin of each subvolume ΔV_{si}. The contributions of these subvolumes to the received signal are then incoherent, and hence the total average received power is an integral of the power contributed by each ΔV_{si}. For backscatter the volume ΔV_{si} degenerates into a section (of length $c\tau/2$) of a cone having an apex angle $2r_{max}/r_0$. Furthermore, the perpendicular distance to $\mathbf{r}_0$, given by the right side of (11.46), nearly equals the first Fresnel zone radius, $\sqrt{\lambda r_0/2}$. Thus, Fraunhofer scatter requires correlation lengths to be smaller than the Fresnel radius. The solution for scattered fields is more difficult when correlation lengths are comparable to or larger than the Fresnel radius. Approximate solutions for this backscatter case are developed in Section 11.5.2.

Assuming that condition (11.46) is satisfied,

$$e^{-jk_0 r_r} \approx e^{-jk_0 r_{r0} + jk_0 \mathbf{a}_{r0} \cdot \mathbf{r}}. \tag{11.49}$$

Similar arguments can be developed for r_t, and so

$$e^{-jk_0 r_t} \approx e^{-jk_0 r_{t0} + jk_0 \mathbf{a}_{t0} \cdot \mathbf{r}}, \tag{11.50}$$

which implies that the phase fronts of the incident field are planar (to within a fraction of a wavelength) across the scatter volume V_s. In these approximations, V_s is small compared to V_c, so $\mathbf{A}_0(\mathbf{r})$ is nearly constant over the scatter volume. This, and substitution of (11.49) and (11.50) into (11.43), result in the field intensity at the receiver location:

$$\mathbf{E}_1(\mathbf{r}_{r0}, t) = -\frac{k_0^2 \mathbf{a}_{r0} \times (\mathbf{a}_{r0} \times \mathbf{A}_0)}{2\pi r_{r0} r_{t0}} e^{-jk_0(r_{t0} + r_{r0})} C_1, \tag{11.51}$$

where

$$C_1 = \int_V \Delta n(\mathbf{r}, t) e^{-jk_0 \mathbf{m} \cdot \mathbf{r}} \, dV, \tag{11.52}$$

$$\mathbf{m} = -\mathbf{a}_{t0} - \mathbf{a}_{r0}, \quad |\mathbf{m}| = 2\sin(\theta_s/2), \tag{11.53}$$

θ_s, the scatter angle, is the angle between $-\mathbf{a}_{t0}$ and $\mathbf{a}_{r0}$, and $k_0 \mathbf{m}$ is termed a mirror wavenumber. In developing this solution, we noted that the term $2\mathbf{V}[\mathbf{E}_0 \cdot \mathbf{V} \Delta n(\mathbf{r}, t)]$ on the right-hand side of (11.20) contributes only a component of the field intensity parallel to $\mathbf{a}_{r0}$, canceling a like component resulting from the first term. Because the term $2\mathbf{V}[\mathbf{E}_0 \cdot \mathbf{V} \Delta n(\mathbf{r}, t)]$ does not contribute to the first-order transverse field components at $\mathbf{r}_{r0}$, we may write, for a first-order scatter theory, the approximate differential equation

$$\nabla^2 \mathbf{E}_1 + k_0^2 \mathbf{E}_1 = -2k_0^2 \, \Delta n \, \mathbf{E}_0, \tag{11.54}$$

provided one ignores the longitudinal component of the resulting solution.

11.4.1 Discussion and Examples

Equation (11.52) is the basic integral for Fraunhofer scatter. For the purpose of better understanding this integral, we shall discuss some of its properties by considering simplified examples. The vector **m** is dimensionless, and its magnitude is given by (11.53). The direction of **m** defines a mirror direction. Planes of discontinuity or of rapid change of Δn perpendicular to this direction are able to reflect the incident ray of direction $-\mathbf{a}_{t0}$ into its mirror direction $\mathbf{a}_{r0}$. The integral (11.52) gives the Fourier composition of Δn along the mirror direction, and hence the intensity of the scattered signal depends directly on the Fourier content of the irregularities in this direction.

In a rectangular coordinate system chosen such that the direction of the z axis lies along the mirror direction, (11.52) reduces to

$$C_1 = \int_V \Delta n(\mathbf{r}, t) e^{-j2k_0 z \sin(\theta_s/2)} \, dx \, dy \, dz. \tag{11.55}$$

Thus in order that C_1 be relatively large in value, the irregularities should have significant Fourier components with a spatial wavelength (structure wavelength) in the mirror direction. From (11.55) it can be deduced that the structure wavelength Λ that produces constructive interference at the receiver is

$$\Lambda = \lambda/2 \sin(\theta_s/2) \tag{11.56}$$

measured along the vector direction **m**. To give this result a physical interpretation, consider a point scatterer located at a point 1 and another located at a point 2, a distance δ directly above 1. In order for the fields scattered from 2 to interfere constructively *at the receiver* with the fields scattered from 1, the path length difference $2L$ must be equal to λ or to some integer multiple of λ. From the geometry depicted in Fig. 11.4 we find that

$$\delta = L/\sin(\theta_s/2) = \lambda/2 \sin(\theta_s/2), \tag{11.57}$$

which corresponds to Eq. (11.56). This equation is Bragg's Law, which gives the condition for waves reflected from adjacent scattering planes to be in phase.

Example 1: Frozen irregularities. Let us now consider a simple example in which $\Delta n(\mathbf{r}, t)$ is given by

$$\Delta n(\mathbf{r}) = \Delta n \, \cos(2\pi x/\Lambda_x) \cos(2\pi z/\Lambda_z), \tag{11.58}$$

where Λ_x, Λ_z are the structure wavelengths of $\Delta n(\mathbf{r})$ along the coordinate axes x, z, respectively, and Δn is the amplitude of variation. Thus the irregularities are sinusoidal and not random. Assume a symmetrical forward-scatter

11.4 Fraunhofer Scatter

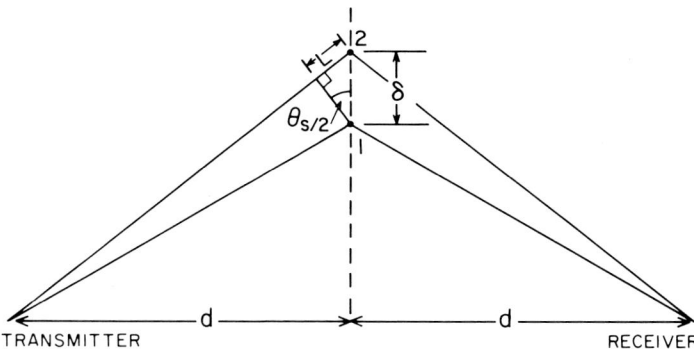

Fig. 11.4. Geometry showing the distance between scattering centers that produces constructive interference.

link for which $\mathbf{m} = 2\mathbf{a}_z \sin\theta_s/2$, where $\mathbf{a}_z$ is a unit vector in the z direction. Consider that the variation of $\Delta n(\mathbf{r})$ given by (11.58) is contained in a box of dimensions as shown in Fig. 11.5 and that outside the box $\Delta n(\mathbf{r}) = 0$. For these stipulated conditions, Eq. (11.52) becomes

$$C_1 = (4 \Delta n \, L_y \Lambda_x/\pi) \sin(2\pi L_x/\Lambda_x) \int_0^{L_z} \cos(K_z z) \cos(k'_0 z) \, dz, \quad (11.59)$$

where $K_z \equiv 2\pi/\Lambda_z$ and $k'_0 \equiv 2k_0 \sin(\theta_s/2)$. We note immediately from (11.59) that C_1 has a periodic variation of intensity as L_x changes. Let us assume some fixed value for L_x that gives a nonzero value to C_1 and define

$$A = (4 \Delta n \, L_y \Lambda_x/\pi) \sin(2\pi L_x/\Lambda_x). \quad (11.60)$$

Performing the integration in (11.59) yields

$$C_1 = \frac{AL_z}{2} \left[\frac{\sin(K_z + k'_0)L_z}{(K_z + k'_0)L_z} + \frac{\sin(K_z - k'_0)L_z}{(K_z - k'_0)L_z} \right]. \quad (11.61)$$

The dependence of the two terms in the brackets is sketched in Fig. 11.6 using k'_0 as the independent variable. It is then immediately obvious that the maximum contribution to the scattered signal occurs when

$$k'_0 = K_z \quad \text{or} \quad \lambda = 2\Lambda_z \sin(\theta_s/2), \quad (11.62)$$

in agreement with (11.56). We note that if (11.62) is satisfied, then the scatter intensity is proportional to L_z, and not periodic in L_z as it is for the case in which we vary L_x.

Example 2: Advecting frozen irregularities—the RASS. In this example we show that mean Doppler shifts are contained in the derived formulas when frozen irregularities advect with the mean wind. An apparent advection

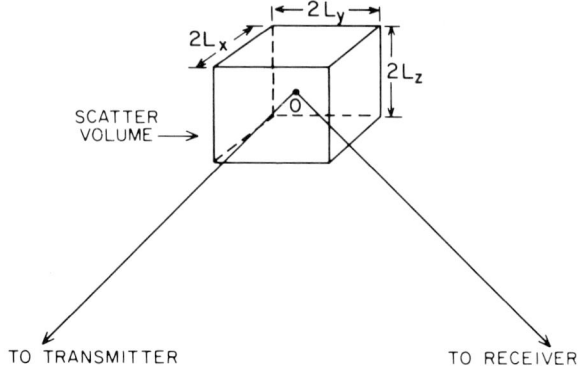

Fig. 11.5. Scatter from a rectangular box with sinusoidal variations of the refractive index.

of the medium also results with RASS (radio acoustic sounding system). Briefly, RASS consists of an acoustic source that beams toward the zenith a short burst of intense sound waves. The upward speed v_a of the acoustical vibrations is a function of air temperature. The intense pressure fluctuations of the acoustic wave modulates the radio refractive index, producing a traveling wave of sinusoidal perturbations in n:

$$\Delta n(\mathbf{r}, t) = \Delta n \cos K_z(z - v_a t). \tag{11.63}$$

A bistatic radar receives echoes from this traveling disturbance, and, as we shall show, a measurement of the Doppler shift will measure v_a. The weak echoes scattered by the perturbations are maximized when the Bragg resonance condition (11.56) is met.

We substitute (11.63) into (11.55) and execute the integration to obtain

$$C_1 = \tfrac{1}{2} V \Delta n \left[\frac{\sin(K_z + k'_0)L_z}{(K_z + k'_0)L_z} + \frac{\sin(K_z - k'_0)L_z}{(K_z - k'_0)L_z} \right] e^{-jk'_0 v_a t}, \tag{11.64}$$

where V is the volume containing the disturbance and $k'_0 v_a$ is the Doppler frequency. As shown in Fig. 11.6, the scattered field at the receiver is maximum when $k'_0 = K_z$, in which case

$$C_1 \approx \tfrac{1}{2} V \Delta n \, e^{-j\omega_s t}, \tag{11.65}$$

where ω_s is the sonic frequency. However, in general, the scattered field obtained from (11.51) is, to a good approximation,

$$\mathbf{E}_1(\mathbf{r}_{r0}, t) \simeq - \frac{k_0^2 \mathbf{a}_{r0} \times (\mathbf{a}_{r0} \times \mathbf{A}_0) V}{4\pi r_{r0}^2} \Delta n \left[\frac{\sin(K_z - k'_0)L_z}{(K_z - k'_0)L_z} \right]$$
$$\times e^{j(\omega_0 - k'_0 v_a)t - jk_0(2r_{r0})}, \tag{11.66}$$

11.4 Fraunhofer Scatter

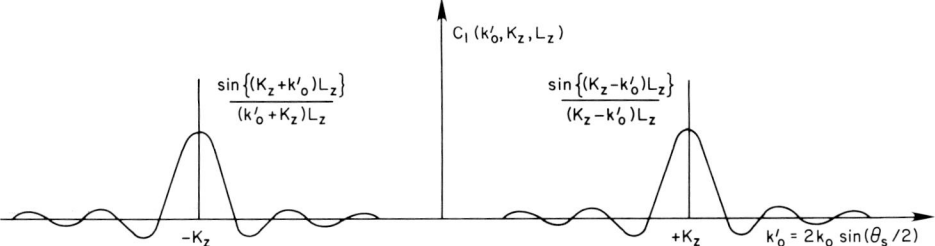

Fig. 11.6. Scatter intensity $C_1(k'_0, K_z, L_z)$ dependence on the electromagnetic wavenumber k_0 for sinusoidal refractive index fluctuations.

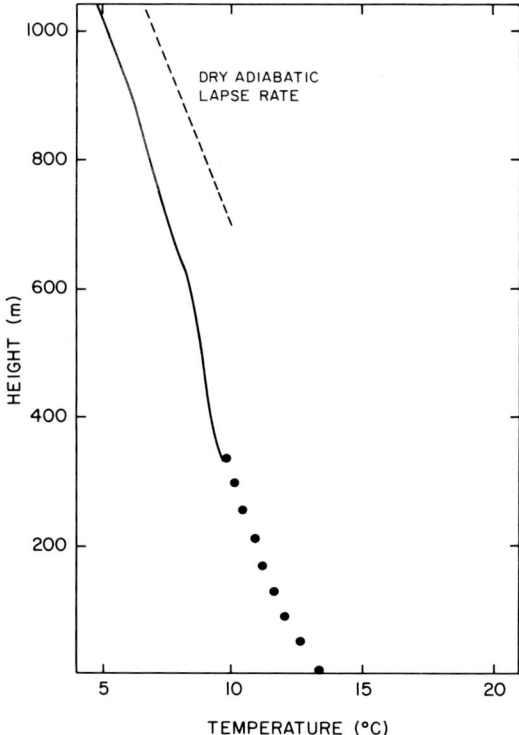

Fig. 11.7. Distribution of temperature deduced from soundings on 6 March 1979 at 16:50 L.S.T. ●, radiosonde; ——, RASS. (From Bonino et al., 1979.)

for k'_0 near K_z. Thus a negative Doppler shift is expected for perturbations moving away from the radar. To measure v_a at some height (and thus air temperature), the radar wavelength is set to produce a maximum return (i.e., k'_0 near K_z). The Doppler shift

$$f_d = -(2v_a/\lambda)\sin(\theta_s/2), \qquad (11.67)$$

then gives v_a from which temperature is deduced. A comparison of temperature profiles obtained with RASS and radiosonde is shown in Fig. 11.7.

11.4.2 Expected Scattered Power Density

Refractive index irregularities usually do not have such well-behaved deterministic variations, as illustrated in the example of Section 11.4.1. It is more likely that the statistical properties of the irregularities will only be known. Therefore in this section we relate the expected (i.e., time or ensemble average) power density incident at the receiver location $\mathbf{r}_{r0}$ to these statistical properties. The power density (Poynting vector), averaged over a cycle of the radio wave, is

$$\mathbf{S} = \tfrac{1}{2}\operatorname{Re}(\mathbf{E}_1 \times \mathbf{H}_1^*), \qquad (11.68a)$$

where, from (11.1),

$$\mathbf{H}_1 = -\nabla \times \mathbf{E}_1/j\omega\mu_0. \qquad (11.68b)$$

In computing $\mathbf{H}_1$ we assume that, at the receiver location, the primary field $\mathbf{E}_0$ is negligible compared to $\mathbf{E}_1$. Since the $\mathbf{E}_1$ and $\mathbf{H}_1$ fields are transverse to $\mathbf{a}_{r0}$ and to each other, as can be seen from (11.51) and (11.68b), it can be shown that

$$|\mathbf{H}_1| \approx |\mathbf{E}_1|\bigg/\sqrt{\frac{\mu_0}{\varepsilon_0}}.$$

Furthermore,

$$|-\mathbf{a}_{r0} \times (\mathbf{a}_{r0} \times \mathbf{A}_0)| = A_0 \sin\chi, \qquad (11.69)$$

where χ is the angle between the vectors $\mathbf{A}_0$ and $\mathbf{a}_{r0}$. Thus the Poynting vector $\mathbf{S}$ is directed along $\mathbf{a}_{r0}$ and has intensity

$$S = \tfrac{1}{2}|\mathbf{E}_1|^2/\eta_0, \qquad \eta_0 \equiv \sqrt{\mu_0/\varepsilon_0}.$$

Substituting (11.51) for $\mathbf{E}_1$ results in

$$S = (A_0^2 k_0^4 \sin^2\chi/8\eta_0\pi^2 r_{r0}^2 r_{t0}^2) C_1 C_1^*, \qquad (11.70)$$

which, upon substitution for C_1, becomes

$$S = \frac{A_0^2 k_0^4 \sin^2\chi}{8\eta_0\pi^2 r_{t0}^2 r_{r0}^2}\left|\int_V \Delta n(\mathbf{r},t) e^{-j k_0 \mathbf{m}\cdot\mathbf{r}}\,dV\right|^2. \qquad (11.71)$$

11.4 Fraunhofer Scatter

Since $\Delta n(\mathbf{r}, t)$ is a random variable, so is S. We shall determine its expected or mean value $\langle S \rangle$, which is related to the statistical characteristics of Δn. Before we proceed, however, let us write

$$\mathscr{I}(t)\mathscr{I}^*(t') = \int_V \Delta n(\mathbf{r}, t) e^{-jk_0 \mathbf{m} \cdot \mathbf{r}} \, dV \int_V \Delta n(\mathbf{r}', t') e^{jk_0 \mathbf{m} \cdot \mathbf{r}'} \, dV',$$

where $\mathbf{r}$ and $\mathbf{r}'$ are variables of integration and $\mathscr{I}(t)$, $\mathscr{I}(t')$ are samples of the complex (i.e., I and Q) signal at times t, t', respectively. Taking an ensemble average, we obtain

$$\langle \mathscr{I}(t)\mathscr{I}^*(t') \rangle = \int_V \int_{V'} \langle \Delta n(\mathbf{r}, t) \Delta n(\mathbf{r}', t') \rangle e^{-jk_0 \mathbf{m} \cdot (\mathbf{r} - \mathbf{r}')} \, dV \, dV'. \quad (11.72)$$

The term

$$\langle \Delta n(\mathbf{r}, t) \Delta n(\mathbf{r}', t') \rangle \equiv R(\mathbf{r}, \mathbf{r}', t, t') \quad (11.73a)$$

is the autocorrelation of the random variable $\Delta n(\mathbf{r}, t)$. In the case of temporally and spatially homogeneous turbulence,

$$R(\mathbf{r}, \mathbf{r}', t, t') = R(\mathbf{r} - \mathbf{r}', t - t') = R(\boldsymbol{\rho}, \tau_1). \quad (11.73b)$$

We briefly digress to discuss the relation of $\mathscr{I}(t)$ to the Doppler spectrum and correlation functions developed in earlier chapters. The received complex signals $V(t)$ and $V(t')$ discussed in Chapter 5 are directly proportional to $\mathscr{I}(t)$ and $\mathscr{I}(t')$. Hence $\langle \mathscr{I}(t)\mathscr{I}^*(t') \rangle$ is proportional to the received signal correlation $\langle V(t)V^*(t') \rangle$, and its Fourier transform is the Doppler spectrum. Thus, given the time dependence of the correlation function for refractive index irregularities, we can directly derive the Doppler spectrum. However, relating this spectrum to the turbulent velocity field is not so straightforward. In Chapter 10 we discussed elements of turbulence theory to derive the statistical characteristics of velocity irregularities under the assumption of isotropic turbulence. We illustrated that the space-time correlation function for the velocity field requires solving a rather difficult differential equation, Eq. (10.39). Even though we may find a solution to this equation, we are still confronted with the problem of relating the statistical properties of the velocity field to those for the refractive index irregularities generated by the turbulence.

Although we offer no rigorous solution that relates the statistical properties of the velocity field to the space–time correlation of Δn, we can, through heuristic arguments and assumptions, relate our earlier results for point targets in turbulent flow to this case. Later in this section we demonstrate that, given a broad spectrum of irregularities, only those having scales near $\Lambda_0 = \lambda/2 \sin(\theta_s/2)$ contribute to the scattered field. We assume that these irregularities act as tracers of larger-scale motion. Hence, we can apply our earlier results (e.g., Sections 5.2 and 5.3) that relate Doppler spectra to the distribution of point target velocities within the resolution volume.

Even though the refractive index irregularities have finite lifetimes and thus do not possess the long coherence time inherent in point targets such as hydrometeors, the finiteness of the lifetimes can be ignored if they are longer than the period required to decorrelate signals by the reshuffling of Λ_0 scales. In order to estimate the lifetimes of irregularities on the scale of Λ_0, we shall assume that scales larger than Λ_0 transport (reshuffle) smaller scales and that scales $\Lambda \leq \Lambda_0$ deform the shape and destroy irregularities of size Λ_0. However, new irregularities of scale Λ_0 are created as old ones are destroyed if steady state turbulence is assumed. The lifetime τ_0 of irregularites on the scale of Λ_0 is roughly equal to Λ_0/v, where v is the rms velocity of scales $\leq \Lambda_0$. In order to calculate v we integrate the energy spectral density $E(K)$ from $K_0 = 2\pi/\Lambda_0$ to infinity, which gives the velocity variance of scales smaller than Λ_0. If $K_0 \ll K_i$ where K_i is the wavenumber of the inner scale (i.e., where dissipation of kinetic energy due to viscosity begins), we can approximately estimate v by taking the square root of the integral of $E(K)$ given by (10.66) between K_0 and infinity assuming $A = 1.6$. Thus $\tau_0 \simeq 1.19 \Lambda_0^{2/3} \varepsilon^{-1/3}$, in agreement with the estimate made by Kristensen (1979). Thus the stronger the turbulence (i.e., the larger ε), the shorter is the lifetime of irregularities. For example, if $\Lambda_0 = 0.05$ m and turbulence is moderate ($\varepsilon \approx 10^{-2}$ m^2 s^{-3}), then $\tau_0 \simeq 0.8$ s, which is much longer than the time usually required to reshuffle irregularities a distance $\Lambda_0/2 \simeq 0.02$ m. Thus, the formulas previously derived that relate the velocity field to the Doppler spectrum (e.g., spectrum widths and eddy dissipation rate) also apply to the case in which refractive index irregularities are the targets. Henceforth we assume $\tau_1 = 0$ and proceed to work with $R(\mathbf{\rho}, 0)$ alone to determine the average power scattered by Δn.

To cast the integral (11.72) in a form that depends only on the difference variable, $\mathbf{r} - \mathbf{r}'$, consider for the sake of illustration a parallelopiped of dimensions $2L_x$, $2L_y$, $2L_z$ along the x, y, z axes, respectively. The integral (11.72) can be then written

$$\int_{-L_x}^{L_x} \int_{-L_y}^{L_y} \int_{-L_z}^{L_z} R(\mathbf{\rho}) \exp[-jk_0(\mathbf{m} \cdot \mathbf{\rho})] \, dx \, dx' \, dy \, dy' \, dz \, dz'. \quad (11.74)$$

Now,

$$\mathbf{\rho} = \mathbf{a}_x(x - x') + \mathbf{a}_y(y - y') + \mathbf{a}_z(z - z') = \mathbf{a}_x \alpha + \mathbf{a}_y \beta + \mathbf{a}_z \gamma, \quad (11.75)$$

where $\mathbf{a}_x$, $\mathbf{a}_y$, $\mathbf{a}_z$ are unit vectors along x, y, z, respectively. On substituting

11.4 Fraunhofer Scatter

(11.75) into (11.74), we obtain

$$\langle|\mathscr{I}|^2\rangle = \int_{-L_x}^{L_x}\int_{-L_y}^{L_y}\int_{-L_z}^{L_z}\int R(\alpha,\beta,\gamma)e^{-jk_0(m_x\alpha+m_y\beta+m_z\gamma)}\,dV\,dV', \quad (11.76)$$

where $k_0 m_x$, etc., are the component directions of the mirror wavenumber. Consider first the integration over x and x'. Using the identity (Papoulis, 1965)

$$\int_{-L_x}^{L_x}\int R(x-x')\{\exp[-js(x-x')]\}\,dx\,dx'$$

$$= \int_{-2L_x}^{2L_x} R(\alpha)e^{-js\alpha}(2L_x - |\alpha|)\,d\alpha, \quad (11.77)$$

we can reduce the double integral over $2L_x$ to a single integral over α, which results in (11.76) taking the form,

$$\langle|\mathscr{I}|^2\rangle = \int_{-L_y}^{L_y}\int_{-L_z}^{L_z}\int\left[\int_{-2L_x}^{2L_x} R(\alpha,\beta,\gamma)e^{-jk_0 m_x\alpha}(2L_x - |\alpha|)\,d\alpha\right]$$

$$\times e^{-jk_0 m_y\beta - jk_0 m_z\gamma}\,dy\,dy'\,dz\,dz'. \quad (11.78)$$

Interchanging the order of integration and applying (11.77) again, to the y and z integrals, (11.78) reduces to

$$\langle|\mathscr{I}|^2\rangle = \int\int\int_{V_2} \exp(-jk_0\mathbf{m}\cdot\boldsymbol{\rho})R(\boldsymbol{\rho})(2L_x - |\alpha|)$$

$$\times (2L_y - |\beta|)(2L_z - |\gamma|)\,dV_\rho, \quad (11.79)$$

where $dV_\rho \equiv d\alpha\,d\beta\,d\gamma$. If the dimensions in any direction of the volume are much larger than the correlation lengths α_c, β_c, and γ_c, the product terms $(2L_x - \alpha)(2L_y - \beta)(2L_z - \gamma)$ can be approximated by the volume $8L_x L_y L_z = V$ and the limits reduced from V_2 to V, and the expected power density is then

$$\langle S\rangle = \frac{A_0^2 k_0^4 \sin^2\chi}{8\eta_0\pi^2 r_{t0}^2 r_{r0}^2} V \int_V \exp(-jk_0\mathbf{m}\cdot\boldsymbol{\rho})R(\boldsymbol{\rho})\,dV_\rho. \quad (11.80)$$

The correlation length is taken to be that value of α (or β or γ) for which $R(\alpha,\beta,\gamma)$ is insignificantly small.

It is instructive to compare (11.80) to (11.71), the latter giving the power density of any sample of the ensemble or, in other words, the power density averaged over a cycle of ω_0 at any instant of time. Thus from (11.71) we note that samples of S depend on the square of the total volume V of integration, and $\langle S \rangle$ is linearly dependent on the product of V and some fraction of V over which the autocorrelation $R(\rho)$ is significant.

Let us consider the case in which dimensions of V are finite and satisfy the Fraunhofer conditions (11.46), although we shall assume that the volume size is large enough that (11.80) is a good approximation to (11.79). We now use the spectral representation of the correlation function, as defined in Chapter 10:

$$R(\rho) = \int\int\int_{-\infty}^{+\infty} \exp(j\mathbf{K} \cdot \rho)\Phi_n(\mathbf{K}) \, dV_K.$$

Substituting this into the following integral, obtained from (11.80),

$$\langle |\mathscr{I}_1|^2 \rangle \equiv \frac{1}{8\pi^3} \int_V R(\rho) \exp(-jk_0 \mathbf{m} \cdot \rho) \, dV_\rho,$$

we have

$$\langle |\mathscr{I}_1|^2 \rangle = \int\int\int_{-\infty}^{+\infty} \Phi_n(\mathbf{K}) \left[\frac{1}{8\pi^3} \int_V \exp[j(\mathbf{K} - k_0\mathbf{m}) \cdot \rho] \, dV_\rho \right] dV_K. \quad (11.81)$$

Let us now examine the inner integral and define a sampling function

$$F(\mathbf{q}) \equiv \frac{1}{8\pi^3} \int_V \exp(j\mathbf{q} \cdot \rho) \, dV_\rho, \qquad \mathbf{q} \equiv \mathbf{K} - k_0\mathbf{m}. \quad (11.82)$$

When the limits of integration can be extended to infinity, the integral in (11.82) becomes a delta function,

$$F(\mathbf{q}) = \delta(\mathbf{q}), \quad (11.83)$$

and so (11.81) reduces to

$$\langle |\mathscr{I}_1|^2 \rangle = \Phi_n(k_0\mathbf{m}). \quad (11.84)$$

Thus the expected scatter intensity is proportional to the power spectrum of refractive turbulence at the structure wavenumber $\mathbf{K}$ equal to the mirror wavenumber $k_0\mathbf{m}$.

In the usual case of finite scatter volume V, the function $F(\mathbf{q})$ has its maximum values in the region near the point $\mathbf{q} = \mathbf{0}$, and outside this region its amplitude oscillates about zero and diminishes in intensity (Fig. 11.8). Again as an example, consider the finite volume to be a parallelopiped with sides $2L_x, 2L_y, 2L_z$. In this case

$$F(\mathbf{q}) = L_x L_y L_z \frac{\sin q_x L_x}{\pi q_x L_x} \frac{\sin q_y L_y}{\pi q_y L_y} \frac{\sin q_z L_z}{\pi q_z L_z}, \quad (11.85\text{a})$$

11.4 Fraunhofer Scatter

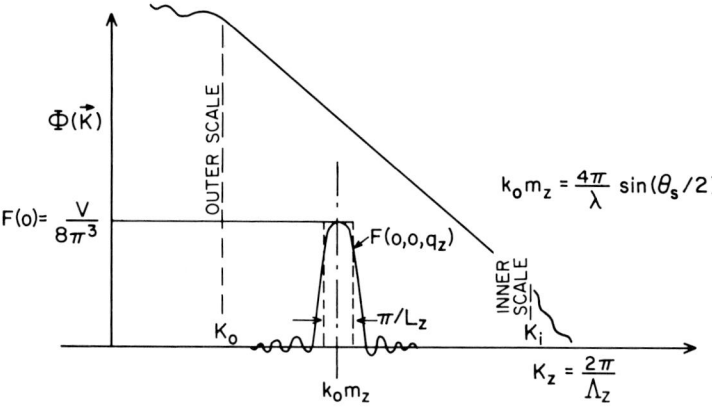

Fig. 11.8. Spectrum of the refractive index variations with imposed sampling function $F(\mathbf{q})$.

whose first zeros are at

$$q_{x0} = \pi/L_x, \qquad q_{y0} = \pi/L_y, \qquad q_{z0} = \pi/L_z. \tag{11.85b}$$

The region of wave vector space $\mathbf{q}$ over which $F(\mathbf{q})$ is appreciable is of the order of

$$q_{x0}q_{y0}q_{z0} \approx 8\pi^3/V \equiv \Upsilon. \tag{11.86}$$

Of course, the behavior of $F(\mathbf{q})$ in each direction of wave vector space depends on the shape of the real space volume V. From the foregoing discussion we recognize that the integral $\langle|\mathscr{I}_1|^2\rangle$ can then be approximated by

$$\langle|\mathscr{I}_1|^2\rangle \approx \int\int\int_\Upsilon \Phi_n(\mathbf{K})F(0)\, dV_K = \int_\Upsilon \Phi_n(\mathbf{K})(V/8\pi^3)\, dV_K$$

$$= \frac{1}{\Upsilon}\int_\Upsilon \Phi_n(\mathbf{K})\, dV_K \equiv \overline{\Phi_n(k_0\mathbf{m})}, \tag{11.87}$$

where $\overline{\Phi_n(k_0\mathbf{m})}$ is the mean value of the function $\Phi_n(\mathbf{K})$ obtained by averaging over a region of wavenumber space with a volume $8\pi^3/V$ surrounding the point $k_0\mathbf{m}$. In most cases $\Phi_n(\mathbf{K})$ can be considered constant over the small volume $8\pi^3/V$ of wavenumber space. In arriving at (11.87) we have essentially approximated $F(\mathbf{q})$ by a pulse-type sampling function of width q_{i0} ($i = x, y,$ or z) for each direction of wave vector space and integrated $\Phi_n(\mathbf{K})$ over this range of $\mathbf{K}$ centered about the point $k_0\mathbf{m}$. We now substitute (11.87) into (11.80) to obtain for the expected average power density (i.e., the expected value of the average over a cycle of ω_0)

$$\langle S\rangle = (\pi A_0^2 k_0^4 \sin^2\chi\, V/\eta_0 r_{t0}^2 r_{r0}^2)\overline{\Phi_n(k_0\mathbf{m})}. \tag{11.88}$$

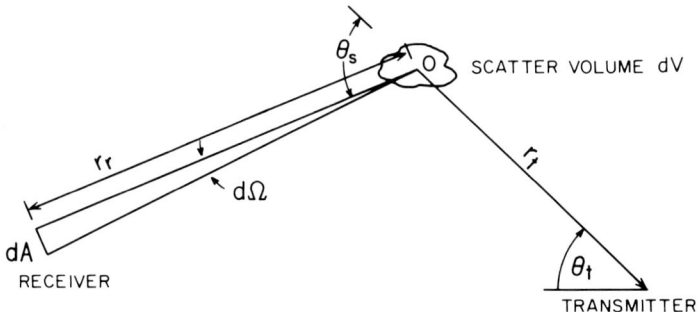

Fig. 11.9. Geometry used to define the scatter cross section per unit volume.

It follows from (11.88) that the power scattered in a particular direction $\mathbf{a}_{r0}$, where $\mathbf{a}_{r0}$ makes an angle θ_s with respect to the incident direction $-\mathbf{a}_{t0}$, is determined only by a narrow portion of the spectrum of irregularities near the wavenumber $\mathbf{K} = k_0 \mathbf{m}$. Thus only a small group of irregularity scales participate in scattering energy in a given direction (see Section 11.5.3 for more discussion on spatial sampling functions).

11.4.2.1 SCATTER CROSS SECTION PER UNIT VOLUME (REFLECTIVITY)

The reflectivity of precipitation defined in Chapter 4 is simply the sum of scatter cross section of each particle within an elemental volume. Although there are no point targets in clear air, we can nevertheless still define a cross section per unit volume. The power density arriving at the receiver site a distance r_r from the scatter volume dV is the Poynting vector dS, which is defined as the power dP incident per unit area subtended by the solid angle $d\Omega$ (see Fig. 11.9). That is,

$$dS = dP/dA. \tag{11.89}$$

The differential scatter cross section $d\sigma_s$ of dV is an equivalent area that intercepts a power $S_i \, d\sigma_s$ (where S_i is the magnitude of the incident Poynting vector), which, if radiated isotropically, produces in the direction of the receiver a power density

$$dS = S_i \, d\sigma_s / 4\pi r_r^2 \tag{11.90}$$

equal to that actually observed. Thus

$$d\sigma_s = (4\pi)^2 r_r^2 r_t^2 \, dS / P_t g_t f_t^2(\theta_t, \phi_t), \tag{11.91}$$

where $f_t^2(\theta, \phi)$ is the normalized angular pattern of the transmitted power density. After substituting the peak signal amplitude squared,

$$A_0^2 = P_t g_t f_t^2 \eta_0 / 2\pi, \tag{11.92}$$

11.4 Fraunhofer Scatter

into (11.88) and using (11.88) to compute dS (where V is replaced by dV), which in turn is substituted into (11.91), the expected scatter cross section per unit volume (reflectivity) is obtained:

$$\langle \eta \rangle \equiv d\sigma_s/dV = 8\pi^2 k_0^4 \sin^2 \chi \overline{\Phi_n(k_0 \mathbf{m})}. \tag{11.93}$$

11.4.2.2 Discussion and an Example

To clarify how the sampling function (11.82) determines $\Phi(\mathbf{K})$, consider as an example the case of statistically homogeneous and isotropic refractive index irregularities. We choose rectangular coordinates such that z lies parallel to $\mathbf{m}$. Thus,

$$m_z = |\mathbf{m}| = 2\sin(\theta_s/2), \qquad \Phi_n(\mathbf{K}) = \Phi_n(K_x, K_y, K_z). \tag{11.94}$$

Because $m_x = m_y = 0$, $K_x = K_y = 0$ and $K_z = k_0|\mathbf{m}|$ locate the position where $F(\mathbf{q})$ is maximum, as can be deduced from (11.82) and (11.83). Assuming that the width of $\Phi_n(\mathbf{K})$ is large compared to the width of $F(\mathbf{q})$ transverse to the K_z axes, we can replace K with

$$K = |\mathbf{K}| = K_z, \tag{11.95}$$

and $\overline{\Phi_n(k_0 m_z)}$ in the expression for reflectivity (11.93) is then

$$\overline{\Phi_n(k_0 m_z)} = \frac{1}{\Upsilon} \int_\Upsilon \Phi_n(K_z)\, dK_x\, dK_y\, dK_z. \tag{11.96}$$

Integrating over K_x, K_y (see Fig. 11.8), we obtain

$$\overline{\Phi_n(k_0 m_z)} \approx \frac{\pi^2}{\Upsilon L_x L_y} \int_{a^-}^{a^+} \Phi_n(K_z)\, dK_z, \tag{11.97}$$

where $a^\pm \equiv k_0 m_z \pm \pi/2L_z$. Substituting for Υ from (11.86), this reduces to

$$\overline{\Phi_n(k_0 m_z)} = \frac{L_z}{\pi} \int_{a^-}^{a^+} \Phi_n(k_z)\, dK_z. \tag{11.98}$$

Equation (11.98) shows that, in addition to the prime contribution from the spectral component $k_0 m_z$, spectral components with structure wavelengths in the interval

$$\frac{1}{k_0 m_z + \pi/2L_z} \leq \frac{\Lambda}{2\pi} \leq \frac{1}{k_0 m_z - \pi/2L_z} \tag{11.99}$$

also contribute significantly to the power scattered in the direction θ_s. Thus, the function $F(\mathbf{q})$ samples a band of spectral components of Δn centered about $k_0 m_z$. The width of the band is π/L_z, and in the usual case of scatter at microwave frequencies and of large scatter volume, the width is much smaller

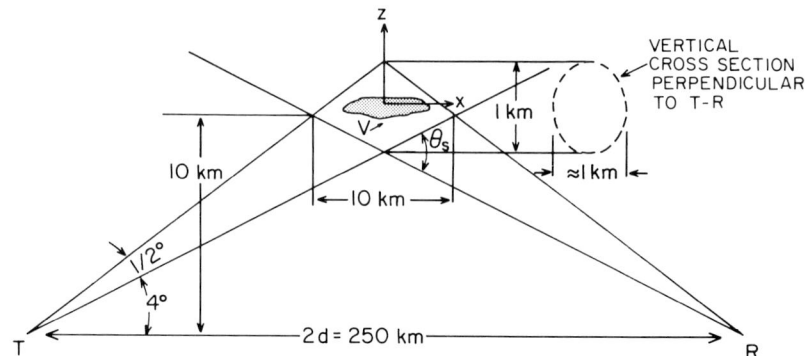

Fig. 11.10. A forward-scatter link ($2d = 250$ km) for a scatter volume V.

than the central value $k_0 m_z$. This is particularly true if $k_0 \mathbf{m}$ is at structure wavenumbers larger than the reciprocal of the correlation length and if the scatter volume is large compared to this length. This is consistent with our basic assumption that the dimensions of V are much larger than the correlation length. However, the dimension of V must not exceed r_{max} given by (11.46). Only for these cases can we approximate $\overline{\Phi_n(k_0 \mathbf{m})}$ by $\Phi_n(k_0 \mathbf{m})$.

For the purpose of estimating the order of magnitude of some of the parameters we have defined, consider a forward-scatter link of 250 km having at each terminal an 18-m-diam antenna pointed at an angle of 4° above the straight line connecting the terminals (Fig. 11.10). Assuming a transmitted wavelength of 10 cm, the beamwidth would be about 0.5° and $g_r = g_t \simeq 1.6 \times 10^5$.

The common beam volume (i.e., where illumination is contained within the field of view) has the size indicated in Fig. 11.10. For the sake of simplicity and in order to satisfy the conditions of our formula, we shall assume the size of the scatter volume V to be equal to limits imposed by the Fraunhofer condition (11.46). Thus, the lengths are about 200 m along y, 3000 m along x, and 200 m along z. Furthermore, a pulse width τ is assumed to be sufficiently large that the distance Δs [see Eq. (11.23)] within the spheroidal shells of constant time delay is larger than the vertical thickness of V. Assume that this volume contains refractive index irregularities having a power spectrum as predicted by turbulence theories [Eq. (10.38)]. In order to obtain some idea of the magnitude of C_n^2, consider the following categories, defined by Davis (1966) to be associated with different levels of turbulence:

$$C_n^2 \approx 6 \times 10^{-17} \text{ m}^{-2/3} \quad \text{(weak)}, \qquad (11.100a)$$

$$C_n^2 \approx 2 \times 10^{-15} \text{ m}^{-2/3} \quad \text{(intermediate)}, \qquad (11.100b)$$

$$C_n^2 \approx 3 \times 10^{-13} \text{ m}^{-2/3} \quad \text{(strong)}. \qquad (11.100c)$$

11.4 Fraunhofer Scatter

However, as we shall see in Section 11.6.2, the value of C_n^2 depends not only on the level of turbulence but also on the temperature and humidity gradients where turbulence occurs. Thus we consider (11.100) to define categories of refractive index irregularities that have been measured in the atmosphere. Substituting the value (11.100c) for C_n^2 into (10.38) and the result into (11.98) gives

$$\overline{\Phi_n(k_0 m_z)} \simeq 3 \times 10^{-13} \frac{0.033 L_z}{\pi} \int_{a^-}^{a^+} K_z^{-11/3} \, dK_z, \quad (11.101)$$

where $k_0 m_z = 8.8 \text{ m}^{-1}$. As mentioned previously, we note that the range of k_z over which the integration in (11.101) is performed is much smaller than the central value $k_0 m_z$. Thus (11.101) is, to a good approximation, equal to

$$\overline{\Phi_n(k_0 m_z)} \simeq 9.9 \times 10^{-15} (k_0 m_z)^{-11/3}. \quad (11.102)$$

Substituting (11.102) into (11.93) and assuming $\chi = \pi/2$, we obtain the reflectivity of this scatter volume:

$$\eta = 7.8 \times 10^{-13} k_0^4 [2k_0 \sin(\theta_s/2)]^{-11/3}. \quad (11.103a)$$

For the values used in the example,

$$\eta = 4.2 \times 10^{-9} \text{ m}^{-1}, \quad (11.103b)$$

Eq. (11.103a) shows that, over the range of wavelengths for which (10.38) is valid, the wavelength dependence of η is $\lambda^{-1/3}$. This wavelength dependence has been first verified by experiments with high-resolution multiwavelength radars located at Wallops Island, Virginia (Atlas et al., 1966). In that experiment simultaneous backscatter measurements were made from a volume at an altitude of 12 km common to three radars operating on three different wavelengths.

From (11.56) we see that refractive index irregularities having a structure wavelength $\Lambda = 71$ cm contribute most to η in the forward-scatter link on Fig. 11.10. However, for the case of backscatter at $\lambda = 10$ cm, the irregularities having $\Lambda = 5$ cm contribute more to the scattered signal. Applying the equations leading to (11.103a) to the backscatter case, it can be shown that

$$\eta = 0.38 \lambda^{-1/3} C_n^2. \quad (11.104)$$

The wavelength dependence in (11.104) was verified by Kropfli et al. (1968) in a carefully controlled experiment in which an airborne microwave refractometer measured nearly simultaneously the C_n^2 along the flight path through the volume in which η was measured by a high-resolution radar.

Although the cross section per unit volume, (11.103b), appears small, the total scattered power depends upon the size of the scatter volume. It therefore is instructive to continue the computations to determine the average received power $\langle P_r \rangle$ given that P_t watts are transmitted. The received power is given by

$$\langle P_r \rangle = A_e \langle S \rangle, \tag{11.105}$$

where A_e is the effective aperture area [Eq. (3.21)] of the receiving antenna:

$$A_e = \lambda^2 g_r / 4\pi \approx 1.3 \times 10^2 \quad \text{m}^2.$$

Substituting our cited values into the appropriate formula, we obtain

$$\langle P_r \rangle / P_t \simeq 5.3 \times 10^{-17}. \tag{11.106a}$$

Thus for $P_t = 10^6$ W

$$\langle P_r \rangle = 5.3 \times 10^{-11} \quad \text{W}, \tag{11.106b}$$

which is a signal power level that should be easily detected. Of course, the (11.106b) was estimated under the assumptions that the entire scatter volume was filled with irregularities Δn and that C_n^2 was large. However, these are not unreasonable assumptions, and hence $\langle P_r \rangle$ is correct to within an order of magnitude. The presence of persistent turbulence is the reason that forward-scatter communication links have been successfully employed.

11.5 FRESNEL SCATTERING

So far our derivation of reflectivity has followed that of Tatarskii (1961, Section 4.2) where it is assumed that the scatter volume V_s has dimensions that satisfy the Fraunhofer condition (11.46). These conditions could imply that the dimensions of V_s cannot be determined by the volume V_c common to the intersecting beams of the transmitting and receiving antennas and the spheroidal shell whose location and thickness are given by (11.22) and (11.23). These are highly restrictive conditions, and therefore we must extend the formulations. When scattering occurs over the entire common volume V_c, we can still apply, as we suggested in Section 11.4, the derived formulas to elements ΔV_s large compared to the correlation lengths that satisfy the Fraunhofer conditions.

11.5.1 Correlation Length Shorter Than the Fresnel Length

In a later publication Tatarskii (1971) derived a fairly general solution for the scattered field with less restrictive assumptions concerning the geometry of the incident wave, the size of the scattering volume, and its distance from the observation point. Equation (11.88), can also be used to derive an ex-

11.5 Fresnel Scattering

pression for the total scatter cross section for the case in which the scatter volume is determined by the common volume V_c. The assumption needed to extend the formulation to the case of a scatter volume having dimensions V_c larger than the subvolumes ΔV_s is that the dimensions of each ΔV_s must satisfy

$$\rho_{c,\parallel} < L_\parallel < \frac{1}{\sin(\theta_s/2)} \left[\frac{\lambda r_t r_r}{\pi(r_t + r_r)}\right]^{1/2} \equiv \frac{f_\parallel}{\sqrt{\pi}}, \quad (11.107a)$$

$$\rho_{c,m} < L_m < \frac{1}{\cos(\theta_s/2)} \left[\frac{\lambda r_t r_r}{\pi(r_t + r_r)}\right]^{1/2} \equiv \frac{f_\parallel}{\sqrt{\pi}} \tan(\theta_s/2), \quad (11.107b)$$

$$\rho_{c,\perp} < L_\perp < \left[\frac{\lambda r_t r_r}{\pi(r_t + r_r)}\right]^{1/2} \equiv \frac{f_\perp}{\sqrt{\pi}}, \quad (11.107c)$$

where r_t, r_r are ranges to the origin of each subvolume; ρ_c is the correlation length; $L_\parallel$ is the dimension of ΔV_s perpendicular to the mirror direction $\mathbf{m}$ but in the plane of $\mathbf{a}_t$, $\mathbf{a}_r$; L_m is the dimension along the mirror direction; and $L_\perp$ is the dimension perpendicular to the plane $\mathbf{a}_t$, $\mathbf{a}_r$ (Fig. 11.11). The lengths $f_\parallel$, $f_\perp$ are the Fresnel lengths perpendicular to $\mathbf{m}$ along the respective directions. These lengths limit the dimensions of ΔV_s to those satisfying

$$\frac{k_0}{2r_r}[r^2 - (\mathbf{a}_r \cdot \mathbf{r})^2] + \frac{k_0}{2r_t}[r^2 - (\mathbf{a}_t \cdot \mathbf{r})^2] < 1,$$

which makes the phase term in (11.43) linear in $\mathbf{r}$. With the condition implied by (11.107), the field scattered from each elemental scatter volume ΔV_s is incoherent, and hence the total received power is given by the sum of the contributions from each volume or by an integral over V_c with ΔV_s representing an elemental volume dV_c. If there are many subvolumes within V_c,

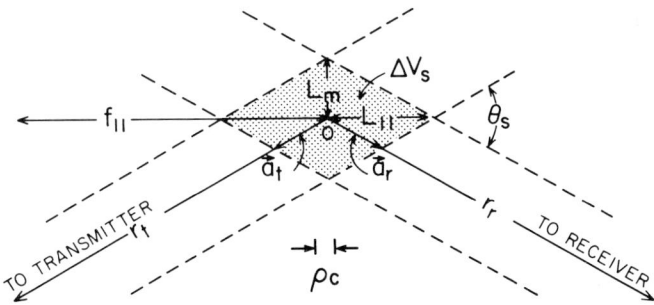

Fig. 11.11. Construction of an elemental scatter volume of largest dimensions subject to (11.46a) and (11.46b).

then the amplitude distribution of I and Q will be Gaussian. The case of either $f_\parallel$ or $f_\perp \leq \rho_c$ is solved in Section 11.5.2.

With reference to Fig. 11.12, $\mathbf{r}_{t0}$ and $\mathbf{r}_{r0}$ are the respective distances from the transmitter and receiver to the intersection O of the axes of the main beams of each respective antenna, and $\mathbf{r}$ is the vector distance from O to the elemental scatter volume dV_c. The elemental power dP received at $\mathbf{r}_{r0}$ because of scatter from dV_c located at $\mathbf{r}$ is

$$dP = A_e(\mathbf{r})\,dS(\mathbf{r}), \tag{11.108}$$

where $A_e(\mathbf{r})$ is the effective reception area of the receiver antenna for energy incident from the direction of dV_c, and $dS(\mathbf{r})$ is the Poynting vector magnitude due to scattering from volume element dV_c. Using the equivalent differential scatter cross section $d\sigma_s$, assumed to be located at the origin 0 we obtain

$$[S_i(0)\,d\sigma_s/4\pi r_{r0}^2]A_e(0) = dP = A_e(\mathbf{r})\,dS(\mathbf{r}), \tag{11.109}$$

where $S_i(0)$ is the magnitude of the Poynting vector along the transmitter main beam axis. Substituting (3.4) and (3.21) into (11.109) and solving for the mean value $\langle d\sigma_s \rangle$, we obtain

$$\langle d\sigma_s \rangle = \frac{(4\pi)^2 r_{t0}^2 r_{r0}^2 f_r^2(\theta, \phi)}{P_t g_t} \langle dS(\mathbf{r}) \rangle. \tag{11.110}$$

The power density scattered from dV_c and incident on the receiver is given by (11.88) with V replaced by dV_c, r_{t0} and r_{r0} replaced by $r_t(\mathbf{r})$ and $r_r(\mathbf{r})$, respectively, and (11.92) substituted for A_0^2. When all this is substituted into

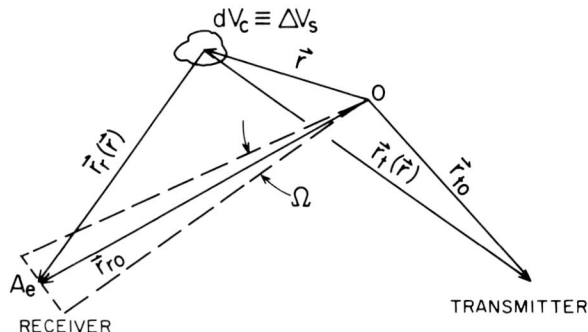

Fig. 11.12. Scatter geometry for a transmitter and receiver in the Fresnel zone of a scatter volume.

11.5 Fresnel Scattering

(11.110) and the result is integrated over the common volume V_c, we obtain the total scatter cross section of the common volume V_c:

$$\langle \sigma_s \rangle = 8\pi^2 r_{t0}^2 r_{r0}^2 k_0^4 \int_{V_c} \frac{f_r^2(\mathbf{r}) f_t^2(\mathbf{r}) \sin^2 \chi(\mathbf{r})}{r_t^2(\mathbf{r}) r_r^2(\mathbf{r})} \Phi_n(k_0 \mathbf{m}(\mathbf{r}), \mathbf{r}) \, dV, \quad (11.111)$$

where V_c is the entire region common to both antennas and the spheroidal shell whose location and thickness is given by (11.22) and (11.23). However, most contribution to σ_s come from the region about the main beam axis of each antenna. Sidelobes can contribute significantly only if the intensity of Φ_n at the sidelobe location is strong relative to the intensity of Φ_n at the intersection of the main beams, as observed by Doviak *et al.* (1971).

11.5.2 Correlation Length Comparable to or Larger Than the Fresnel Length

Although (11.111) is a useful formula, it does contain the implicit assumption that the correlation length of the irregularities must be small compared to the dimensions given by (11.107). There is evidence that scattering irregularities can be highly anisotropic with correlation lengths along the horizontal that can be comparable to or larger than is stipulated by the Fraunhofer conditions (Gage *et al.*, 1981). This is especially true for radars used in wind profiling, where the range is often short (several kilometers), so that the Fresnel lengths are comparable to horizontal correlation lengths. On the basis of angle-of-arrival observations of satellite emissions at 7.3 GHz, Crane (1976) deduced that the power spectral density along K_z was several orders of magnitude larger than along K_x or K_y at the same scale size ($\Lambda \approx 70$ m). This deduction and propagation path geometry suggested the presence of highly anisotropic irregularities having long correlation lengths along the horizontal and short ones along the vertical in the stably stratified inversion layer capping the planetary boundary layer. We shall now focus our study on backscatter because (1) it simplifies the mathematics and (2) it covers the case of most importance in remote sensing.

Ignoring the vector direction of $\mathbf{E}$, the backscattered field intensity, obtained from (11.43), is

$$E_1(\mathbf{r}_0, t) = \frac{k_0^2}{r_0^2} \left[\frac{P_t \eta_0 g}{(2\pi)^3} \right]^{1/2} \int_{V_s} f_\theta(\mathbf{r}) \Delta n(\mathbf{r}, t) e^{-j2k_0 r_t} \, dV, \quad (11.112)$$

where (11.92) has been substituted for A_0, and $f_\theta(\mathbf{r})$ is the one-way electric field angular pattern, which will be assumed to be circularly symmetric about the beam axis. Using (3.21), which expresses the dependence of the effective area of the receiving antenna on the position of the elemental scatter volume dV, the increment of current $|dI|$ produced by dE in a matched filter receiver having an internal resistance R is

$$|dI| = |dE_1|\lambda W(\mathbf{r})[gf_\theta^2(\mathbf{r})/4\pi\eta_0 R]^{1/2}, \quad (11.113)$$

where dE_1 is the increment of field intensity scattered from dV. $W(\mathbf{r})$ is the range weighting function (see Chapter 4), which we now introduce explicitly into the equation for the scattered field. The integration now extends over all $\mathbf{r}$ for which $W f_\theta \Delta n$ is significant. For a receiver filter matched to a rectangular transmitted pulse, the range weighting function is

$$W(\mathbf{r}) = \begin{cases} 1 - 2|\mathbf{r}\cdot\mathbf{a}_0|/c\tau, & |\mathbf{r}\cdot\mathbf{a}_0| \le c\tau/2, \\ 0 & \text{otherwise}, \end{cases} \quad (11.114)$$

where $\mathbf{a}_0$ is the unit vector from the origin in the center of the resolution volume to the radar.

The time-averaged received power (i.e., averaged over a cycle of ω_0) is

$$P_r = \tfrac{1}{2} II^* R, \quad (11.115a)$$

where from the integral of (11.113)

$$I = \frac{\lambda k_0^2 g}{(2\pi)^2 r_0^2}\left(\frac{P_t}{2R}\right)^{1/2}\int W(\mathbf{r})f_\theta^2(\mathbf{r})\,\Delta n(\mathbf{r},t)e^{-j2k_0 r_t}\,dV. \quad (11.115b)$$

Substituting (11.115b) into (11.115a), taking the ensemble average, and using (11.73a), we obtain for the expected received power

$$\langle P_r\rangle = \frac{P_t g^2}{4\lambda^2 r_0^4}\int\int R(\mathbf{r},\mathbf{r}')W(\mathbf{r})W(\mathbf{r}')f_\theta^2(\mathbf{r})f_\theta^2(\mathbf{r}')e^{-j2k_0(r_t-r_t')}\,dV\,dV'. \quad (11.116)$$

Let us assume that the irregularities have homogeneous statistical properties so that $R(\mathbf{r},\mathbf{r}') \approx R(\mathbf{r}-\mathbf{r}')$ and, for a circularly symmetric antenna, the two-way pattern function $f_\theta^2(\mathbf{r})$ is

$$f_\theta^2(\mathbf{r}) = e^{-\theta^2/4\sigma_\theta^2}, \quad (11.117a)$$

where σ_θ^2 is the second central moment of the two-way power pattern and θ is the angular displacement, measured at the radar site, of $\mathbf{r}$ from the origin of V_6. In terms of the 3-dB beamwidth θ_1 of the one-way pattern $f_\theta^2(\mathbf{r})$, $\theta_1 = 3.33\sigma_\theta$. On the assumption of narrow beams, (11.117a) can be approximated by

$$f_\theta^2 \approx e^{-(x^2+y^2)/4r_0^2\sigma_\theta^2}, \quad (11.117b)$$

where $(x^2+y^2)^{1/2}$ is the projection of $\mathbf{r}$ onto the plane perpendicular to the beam axis, which is directed along z at a zenith angle ψ. The Gaussian matched filter provides the best resolution of all receivers having the same

11.5 Fresnel Scattering

bandwidth (Zrnić and Doviak, 1978). Because of this, and because practical matched filters used in Doppler weather radars are Gaussian, we assume that $W(\mathbf{r})$ is well approximated by

$$W(\mathbf{r}) \approx e^{-(\mathbf{a}_0 \cdot \mathbf{r})^2/4\sigma_r^2}, \tag{11.118a}$$

where σ_r^2 is the second central moment of the weighting function $W^2(\mathbf{r})$ and [see Eq. (5.68)]

$$\sigma_r = 0.35c\tau/2 = 0.30r_6 \tag{11.118b}$$

for a Gaussian filter matched to a rectangular pulse of width τ.

We now use the Taylor expansion (11.44), where now $r_t \equiv r_r$:

$$r_t \equiv |\mathbf{r}_0 - \mathbf{r}| \approx r_0 - \mathbf{a}_0 \cdot \mathbf{r} + (1/2r_0)[r^2 - (\mathbf{a}_0 \cdot \mathbf{r})^2] \tag{11.119a}$$

for terms up to second order in r^2. We note that the second and third terms of this expansion are the projections of $\mathbf{r}$ in the $\mathbf{a}_0$ direction and onto the x, y plane, respectively. Thus, in terms of the Cartesian coordinates centered in the resolution volume

$$r_t \approx r_0 + z + (1/2r_0)(x^2 + y^2). \tag{11.119b}$$

This quadratic expansion is valid (i.e., third-order terms in r are negligible) provided that the scatter volume size is limited by

$$l_\perp^2 < 2r_0[(z^2 + \lambda r_0/2\pi)^{1/2} - |z|], \tag{11.119c}$$

where $l_\perp$ is the length perpendicular to the beam axis and $z = -(\mathbf{a}_0 \cdot \mathbf{r})$ is the length measured from the resolution volume center. Condition (11.119c) assumes $z < r_0$. As can be seen, the farther the integration variable $\mathbf{r}$ is displaced from the plane $z = 0$, the smaller must be the scatter volume size perpendicular to the beam axis. To simplify the condition, z in (11.119c) can be set equal to the smaller of $r_6/2$ or the projection $l_z/2 = (l/\cos\psi + r_0\theta_1 \tan\psi)/2$ along the beam of a horizontal layer of thickness l containing refractive irregularities. For a 6-m radar pointed vertically at a 6-km-high scattering layer, $l_\perp$ must be less than 600 m if a layer has thickness (or range resolution if the reflectivity is uniform with height) of 120 m and less than 300 m if the thickness is 600 m. Usually, the antenna pattern weighting function limits $l_\perp$ and for these two cases the beamwidths must be less than 12° and 6°, respectively. Substituting (11.119b), the integral in (11.116) becomes

$$\mathscr{I} \equiv \int\int R(\mathbf{r} - \mathbf{r}')W(z)W(z') \exp\left[-(x^2 + x'^2)/2\sigma_\perp^2 - (y^2 + y'^2)/2\sigma_\perp^2 \right.$$
$$\left. - j2k_0\left(z - z' + \frac{x^2 - x'^2 + y^2 - y'^2}{2r_0}\right)\right] dV \, dV', \tag{11.120}$$

where $\sigma_\perp \equiv \sigma_\theta r_0/\sqrt{2}$. We shall now find it convenient to define new coordinate axes:

$$x - x' \equiv \delta_x, \qquad y - y' \equiv \delta_y, \qquad z - z' \equiv \delta_z,$$
$$\tfrac{1}{2}(x + x') \equiv \sigma_x, \qquad \tfrac{1}{2}(y + y') \equiv \sigma_y, \qquad \tfrac{1}{2}(z + z') \equiv \sigma_z,$$

so that the x, x' component of (11.120) can be written

$$\mathscr{I}_x = \iint R(\mathbf{r} - \mathbf{r}') \exp[-(\sigma_x^2 + \delta_x^2/4)/\sigma_\perp^2 - j2k_0\sigma_x\delta_x/r_0] \, d\sigma_x \, d\delta_x. \quad (11.121\text{a})$$

The transformation from (11.120) to (11.121a) is valid provided that, as is assumed here, the limits of integration cover the entire volume where the integrand has significant value. Thus, executing the integration over σ_x,

$$\mathscr{I}_x = \sigma_\perp\sqrt{\pi} \int R(\mathbf{r} - \mathbf{r}') \exp[-(k_0\delta_x\sigma_\perp/r_0)^2 - \delta_x^2/4\sigma_\perp^2] \, d\delta_x. \quad (11.121\text{b})$$

Applying similar procedures to the y and z coordinate integrations yields

$$\mathscr{I} = \sqrt{2\sigma_r\sigma_\perp^2\pi^{3/2}} \int\!\!\int\!\!\int_{-\infty}^{+\infty} R(\boldsymbol{\rho})$$

$$\times \exp\left[-\underbrace{\left(\frac{\delta_x^2 + \delta_y^2}{4\sigma_\perp^2} + \frac{\delta_z^2}{8\sigma_r^2}\right)}_{\text{resolution volume weight}} - \underbrace{\frac{\pi^2\sigma_\perp^2}{f^4}(\delta_x^2 + \delta_y^2) - j2k_0\delta_z}_{\text{Fresnel term}}\right] dV_\rho, \quad (11.122)$$

where $\boldsymbol{\rho} \equiv \mathbf{a}_{x0}\delta_x + \mathbf{a}_{y0}\delta_y + \mathbf{a}_{z0}\delta_z$ and $f = \sqrt{\lambda r_0/2}$ is the radius of the first Fresnel zone.

The solution (11.122) is acceptable provided the second-order expansion of r_t in (11.119b) is valid. Inequality (11.119c) ensures that the quadratic expansion is valid. However for the case in which the lateral dimension $l_\perp$ of the scatter volume is so large that (11.119c) is not satisfied, we can still use (11.122) if $R(\boldsymbol{\rho})$ is small when the lag vector $\boldsymbol{\rho}$ has magnitude comparable to or larger than the right side of (11.119c). For this condition to hold, the correlation length $\rho_{c\perp}$ perpendicular to the beam axis must be

$$\rho_{c\perp}^2 < 2r_0\{[(l_z/2)^2 + f^2/\pi]^{1/2} - |l_z/2|\}. \quad (11.123)$$

If $r_6 < l_z$ then l_z in (11.123) is replaced by r_6. When (11.119c) holds, then there is no restriction on $\rho_{c\perp}$. Comparing (11.123) with (11.107), we can see the second-order expansion relaxes the limits placed on the scatter volume size $l_\perp$ and correlation length $\rho_{c\perp}$. These limits are increased by the factor $(8\pi r_0/\lambda)^{1/4}$.

11.5 Fresnel Scattering

Now compare the integral (11.122) to (11.80). We have the same form, but the correlation function is multiplied by two additional exponential functions:

(1) The first weights $R(\rho)$ in proportion to the beamwidth and range resolution (the larger $\sigma_\perp$ and σ_r compared to the correlation lengths transverse and parallel, respectively, to the beam axes, the less is the importance of the beamwidth and range resolution to the integral).

(2) The second term gives a weight along the x, y direction that depends on the ratio $f/\sigma_\perp$ of the Fresnel radius to the beamwidth.

Only when the radius of the Fresnel zone is large compared to $\sqrt{\sigma_\perp \rho_{c\perp} \pi}$ can the Fresnel term in (11.122) be ignored. Therefore, both ratios of beamwidth and correlation length to Fresnel radius are important, but because $\sigma_\perp$ is a function of λ and r_0 [Eqs. (3.2b) and (5.67)],

$$\sigma_\perp = \frac{0.45}{\sqrt{\ln 2}} \frac{\lambda r_0}{D} = \frac{0.9}{\sqrt{\ln 2}} \frac{f^2}{D}, \qquad (11.124)$$

where D is the antenna diameter, the Fresnel term can be ignored only if the correlation length $\rho_{c\perp}$ satisfies

$$\rho_{c\perp} < (\sqrt{\ln 2}/0.9\pi)D. \qquad (11.125a)$$

On the other hand, because f is always smaller than $\sigma_\perp$, the Fresnel term in (11.122) will have more weight than the beam width part of the resolution volume term. Thus situations that allow us to neglect the Fresnel term will also permit us to ignore beamwidth influence. This can occur when the scatter volume V_s is smaller than the resolution volume V_6. However, if (11.125a) is satisfied, we can still use 11.122 (without the beamwidth and Fresnel terms) to obtain the scattered field, even though V_s is larger than V_6; then we need to sum incoherent echo power from elemental volumes large compared to ρ_c^3 but small compared to V_6. We call this case incoherent Fraunhofer scatter. But Hodara (1966) shows that within the lower troposphere, the correlation length has the following height dependence:

$$\rho_c \simeq 0.4h/(1 + 0.01h) \quad \text{m}, \qquad (11.125b)$$

where h is in meters. Furthermore VHF backscatter data fitted to the theory in Section 11.5.3 suggest that $\rho_{c\perp} \simeq 20$ m for irregularities in the lower stratosphere. Thus, unless the antenna diameter is of the order of 100 m or more, the Fresnel term will be important in determining the field scattered by refractive irregularities. If the scattering volume contains many subvolumes for which (11.122) applies, but (11.125a) is not satisfied, we have a situation of incoherent Fresnel scatter. When $l_\perp < \sqrt{2r_0 f}$ (from Eq. 11.119c), then we have coherent Fresnel scatter. If $l_\perp < f$, then the echo is coherent irrespective of the transverse reshuffling of refractive index irregularities.

Now we shall examine the effects of the weighting functions (due to the resolution volume, the Fresnel term, and geometry) on the received power using spatial spectra. From Eq. (11.8), (11.122) can be written

$$\mathscr{I} = \sqrt{2\sigma_r}\sigma_\perp^2 \pi^{3/2} \int\!\!\int\!\!\int_{-\infty}^{+\infty} \Phi_n(\mathbf{K}) \left[\int\!\!\int\!\!\int_{-\infty}^{+\infty} \exp[j(\mathbf{K} - 2k_0 \mathbf{a}_{z0}) \cdot \boldsymbol{\rho}] H(\boldsymbol{\rho}) \, dV_\rho \right] dV_K, \quad (11.126)$$

where

$$H(\boldsymbol{\rho}) \equiv \exp\left[-\left(\frac{\delta_x^2 + \delta_y^2}{4\sigma_\perp^2} + \frac{\delta_z^2}{8\sigma_r^2} \right) - \frac{\pi^2 \sigma_\perp^2}{f^4} (\delta_x^2 + \delta_y^2) \right]$$

is the weighting function in lag space. Now the bracketed term in (11.126) is a sampling function, similar to that of (11.81) but modified by the weighting function $H(\boldsymbol{\rho})$. We therefore define a normalized spectral sampling function:

$$F(\mathbf{K}) \equiv \frac{1}{8\pi^3} \int\!\!\int\!\!\int \exp[j(\mathbf{K} - 2k_0 \mathbf{a}_{z0}) \cdot \boldsymbol{\rho}] H(\boldsymbol{\rho}) \, dV_\rho, \quad (11.127)$$

which has a peak value at $\mathbf{K} = 2k_0 \mathbf{a}_{z0}$. As the width of $H(\boldsymbol{\rho})$ becomes broader, the spectral sampling function becomes narrower. The integrations along $\boldsymbol{\rho}$ can be performed to give

$$F(\mathbf{K}) = \frac{\pi^{-3/2} \sigma_\perp^2 \sqrt{2\sigma_r}}{1 + 4\pi^2 \sigma_\perp^4/f^4} \exp\left[-2\sigma_r^2 (K_z - 2k_0)^2 - \frac{\sigma_\perp^2 (K_x^2 + K_y^2)}{1 + 4\pi^2 \sigma_\perp^4/f^4} \right]. \quad (11.128)$$

In order for the Fresnel term to be negligible, $4\pi^2 \sigma_\perp^2/f^4$ would have to be small relative to 1. However, for r_0 in the far field of the antenna, it can be shown that

$$4\pi^2 \sigma_\perp^4/f^4 \gg 1. \quad (11.129)$$

Thus, for resolution volumes in the antenna far field, a common situation in remote sensing, the Fresnel term in $F(\mathbf{K})$ cannot be ignored. Assuming far-field conditions, (11.128) reduces to

$$F(\mathbf{K}) = \frac{0.44 D^2 \sigma_r \ln 2}{\pi^{7/2}} \exp\left[-2\sigma_r^2 (K_z - 2k_0)^2 - \frac{D^2 \ln 2 (K_x^2 + K_y^2)}{3.24\pi^2} \right], \quad (11.130)$$

in which we have substituted (11.124) for $\sigma_\perp$. Thus, the larger the antenna diameter D, the narrower is the sampling function. A surprising result is that the sampling function shape is independent of r_0. For a given antenna diameter, the spectrum $\Phi_n(\mathbf{K})$ of irregularities is weighted equally for all resolution volumes in space. However, condition (11.119c) must be satisfied.

11.5 Fresnel Scattering

11.5.3 Backscattering from Anisotropic Irregularities

As an example, let us consider the angular dependence of the echo power when the scattering medium has a strong anisotropic component. We shall assume that the correlation function can be decomposed into a part R_i due to isotropic turbulence and an anisotropic component R_a (i.e., $R = R_i + R_a$). We shall further assume that R_a is two dimensional and isotropic in a plane that is rotated by $\psi = \pi/2 - \theta$ about the δ_x axis (which is perpendicular to the beam). To obtain order-of-magnitude effects, we shall take R_a to be of the form

$$R_a = \langle \Delta n_a^2 \rangle \exp(-\delta_x'^2/2\rho_h^2 - \delta_y'^2/2\rho_h^2)\delta(\delta_z'), \quad (11.131)$$

where ρ_h is the horizontal correlation length and $\delta(\delta_z')$ is a Dirac delta function.

The resolution volume coordinates δ_x, δ_y, δ_z are related to the primed coordinates aligned with the axis of the correlation function by

$$\delta_x' = \delta_x, \quad \delta_y' = \delta_y \cos\psi - \delta_z \sin\psi, \quad \delta_z' = \delta_y \sin\psi + \delta_z \cos\psi. \quad (11.132)$$

After introducing (11.132) in (11.131) and the correlation (11.131) into (11.122), integration along δ_z is performed so that the three-dimensional integral reduces to a two-dimensional one. For anisotropic turbulence alone (i.e., $R_i = 0$) and for far-field conditions, this leads to the ratio of powers at an angle ψ and at zenith ($\psi = 0$):

$$\frac{\langle P_a(\psi) \rangle}{\langle P_a(0) \rangle} = \cos\psi \sqrt{\frac{Q(0)}{Q(\psi)}} \exp\left[-\frac{k_0^2 \tan^2\psi}{Q(\psi)}\right], \quad (11.133a)$$

where

$$Q(\psi) = \frac{\sec^2\psi}{2\rho_h^2} + \frac{\tan^2\psi}{8\sigma_r^2} + \frac{\pi^2\sigma_\perp^2}{f^4}.$$

The term $\cos\psi$ results when we account for the decrease in power at constant height h due to the range and for the $\sigma_\perp(\psi) = \sqrt{2h\sigma_\theta}/\cos\psi$ increase with tilt away from the vertical. To $\langle P_a(\psi) \rangle$ we must add the power due to isotropic turbulence. Assume $A = P_i(0)/P_a(0)$ is the ratio of the powers due to isotropic and anisotropic scatter at $\psi = 0$. The theoretical formula for normalized power is then

$$(\langle P_a(\psi) \rangle/\langle P_a(0) \rangle + A\cos^2\psi)/(1 + A), \quad (11.133b)$$

which was subjectively fitted to data from Röttger et al. (1981) (Fig. 11.13). Pertinent parameters for these data are: a wavelength of $\lambda = 6.4$ m, height of 17.5 km, beamwidth $\theta_1 = 1.7°$, and range resolution of 300 m. We find the value $\rho_h = 20$ m for the width of the correlation function, and $A = 0.04$

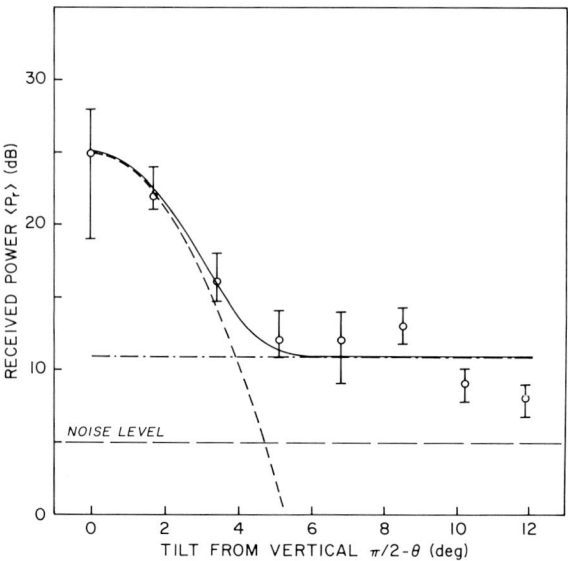

Fig. 11.13. Angular dependence of the observed mean backscatter power (open circles) from anisotropic irregularities as the radar beam axis is tilted away from the vertical. (From Röttger, 1981.) Fitted to the data is a model that consists of anisotropic turbulence with a two-dimensional (horizontal) correlation function in an isotropic background. -·-, isotopic; ---, anisotropic; —, total power. Height = 16.9–18.1 km. $\theta_1 = 1.7°$. $\rho_h = 20$ m. $A = 0.04$ (-14 dB). $\lambda = 6.4$ m. Reprinted with permission from *J. Atmos. Terr. Phys.* **43**, Röttger *et al.* Copyright 1981, Pergamon Press, Ltd.

fits these data well. Although the Fresnel term does not contribute significantly in this case, we see that scattering from turbulent anisotropic irregularities can account for the observed angular dependence of the returned power.

To provide physical insight, we now consider a spectral description for the angular dependence of scatter from anisotropic turbulence. From (11.126), (11.127), and (11.116), the backscattered power becomes

$$\langle P_r \rangle = \frac{2\sqrt{2}(0.45)^2 P_t g^2 \sigma_r \pi^{9/2}}{r_0^2 D^2 \ln 2} \iiint \Phi_n(\mathbf{K}) F(\mathbf{K}) \, dV_K. \qquad (11.134)$$

Equation (11.134) shows that the echo power is proportional to the integral of the product of the spectral intensity $\Phi_n(\mathbf{K})$ of the refractive index irregularities and the normalized sampling function $F(\mathbf{K})$. Now, if the horizontal correlation lengths are large compared to the vertical ones, $\Phi_n(\mathbf{K})$ will be sharply peaked in the K_x, K_y directions and less peaked along the K_z axis. If the irregularities can be roughly described as oblate spheroids, then the correlation $R(\mathbf{\rho})$ has a similar form, but $\Phi_n(\mathbf{K})$ is prolate spheroidal in shape, as shown in Fig. 11.14a.

11.5 Fresnel Scattering

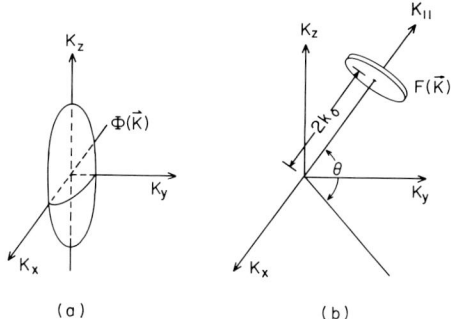

Fig. 11.14. (a) Contour surface of constant spectra intensity $\Phi(\mathbf{K})$ for irregularities having symmetric correlation lengths along x and y that are longer than the correlation length along z. (b) Contour surface of sampling weight $F(\mathbf{K})$ for a beam axis at elevation angle θ.

In deriving $F(\mathbf{K})$ we assumed the z axis to be along the beam axis but made no assumption as to the direction of z in space relative to the earth. Furthermore, the shape of the sampling function depends only on the range resolution and antenna diameter and is thus independent of the location of the resolution volume V_6 in real space. Therefore, in K space $F(\mathbf{K})$ is invariant under rotation of the K coordinate axes, and the sampling function (11.130) can be formulated more generally as

$$F(\mathbf{K}) = \frac{0.44 D^2 \sigma_r \ln 2}{\pi^{7/2}} \exp\left[-2\sigma_r^2 (K_\parallel - 2k_0)^2 - \frac{D^2 K_\perp^2 \ln 2}{3.24 \pi^2}\right], \quad (11.135)$$

where $K_\parallel$ is the wave number magnitude along $\mathbf{k}_0$ and $K_\perp$ is the perpendicular distance of $\mathbf{K}$ from $\mathbf{k}_0$. Thus, if z is along the vertical and the beam axis direction (i.e., $\mathbf{k}_0/k_0$) is rotated by $\pi/2 - \theta$ from it, then $F(\mathbf{K})$ must be rotated by $\pi/2 - \theta$ from the K_z axis. Equation (11.135) reveals that whenever the range resolution $r_6 = 3.33\sigma_r > 0.34D$, a common situation, $F(\mathbf{K})$ will be narrower along $K_\parallel$ than along $K_\perp$. Figure 11.14b shows the contours of $F(\mathbf{K})$.

Obviously, we obtain maximum $\langle P_r \rangle$ for $F(\mathbf{K})$ centered where $\Phi_n(\mathbf{K})$ is a maximum. However, we do not have the freedom to choose arbitrarily the location of $F(\mathbf{K})$, because the radar wavelength centers the $F(\mathbf{K})$ peak at a wavenumber magnitude $2k_0$. For the conditions shown in Fig. 11.14, maximum $\langle P_r \rangle$ occurs when $\theta \to \pi/2$. The angular dependence of $\langle P_r \rangle$ on θ is strongly influenced by the sharpness of $\Phi_n(\mathbf{K})$ and $F(\mathbf{K})$. If $\Phi_n(\mathbf{K})$ is highly anisotropic, as shown in Fig. 11.14a, then $\langle P_r \rangle$ will decrease from its peak value at $\theta = \pi/2$ following an angular dependence of the antenna radiation pattern. Furthermore, it is evident that, whenever $F(\mathbf{K})$ is much narrower than $\Phi_n(\mathbf{K})$ along every direction, $\langle P_r \rangle$ will be linearly dependent on the range resolution but proportional to the square of the antenna diameter D, because g^2 is proportional to D^4 and the integral in (11.134) is independent of D.

To measure $\Phi_n(\mathbf{K})$ accurately with radar requires $F(\mathbf{K})$ to be sharp compared to the shape of $\Phi_n(\mathbf{K})$. Thus, for anisotropic irregularities we need $D > \rho_h$ and $\sigma_r > \rho_v$, where ρ_v is the vertical correlation length. For spectral samples along the horizontal direction, we must have $D > \rho_v$ and $\sigma_r > \rho_h$. That the antenna diameter instead of the linear beamwidth $(r_0 \theta_1)$ enters into (11.135) illustrates the importance of the Fresnel term in (11.128). Only when this term is negligible does the sampling function narrow as the linear beamwidth widens. Furthermore, the Fresnel term results because the irregularities have their axis of symmetry along Cartesian coordinates, whereas the phase fronts of the radar are spherical. If the irregularities had shapes that were concave downward with radius of curvature equal to r_0, then it could be shown that there would be no Fresnel term, and the width of $F(\mathbf{K})$ would be inversely proportional to the linear beamwidth as well as to the range resolution.

11.6 STRUCTURE CONSTANT OF THE REFRACTIVE INDEX

In Section 11.5 a relation between the spatial spectrum of refractive index irregularities and the scattered field has been derived. When, radars sample $\Phi_n(\mathbf{K})$ in the inertial subrange of turbulence the spectral density of variance in Δn has the well-known eleven-thirds power law dependence on K and is proportional to the refractive index structure constant C_n^2. (The span of wavelengths for which this assumption of an inertial range is valid will be discussed in Section 11.6.2.) Because the backscatter is proportional to the intensity of irregularities at a scale size $\lambda/2$, specification of only C_n^2, the term commonly used to describe the refractively turbulent condition of the atmosphere, is usually sufficient to describe the reflecting power of clear air. The reflectivity is then related to C_n^2 by Eq. (11.104). Thus it will be instructive to know how C_n^2 depends on the intensity of turbulence and the state variables of the atmosphere. In the following discussion we assume that condensation does not occur, so that water remains in the gaseous state during displacements of air parcels.

Tatarskii (1971, p. 73) shows that the structure constant C_p^2 of any conservative passive additive is

$$C_p^2 = a^2 \varepsilon^{-1/3} K_p \left(\frac{d}{dz} \langle p \rangle \right)^2, \tag{11.136}$$

where $\langle p \rangle$ is the ensemble average of the additive, which is assumed to have

11.6 Structure Constant of the Refractive Index

only a height dependence; K_p is its coefficient of turbulent diffusion; ε is the eddy rate at which turbulent energy per unit mass is being dissipated and a^2 is a dimensionless constant, which needs to be found from experiments. It appears at present that a^2 is in the range from 3.2 to 4.0 (Gossard et al., 1982).

Equation (11.136) is valid for *any* conservative passive additive p that is carried along by the turbulent atmosphere. The potential temperature (i.e., $p \equiv \theta$), specific humidity q ($p \equiv q$), and possibly even density of particulate matter such as chaff may be considered as conservative passive additives. An additive is considered conservative when its characteristic (e.g., chaff density) does not change when the volume containing the additive is advected. It is passive if it does not affect the dynamics of turbulence. Equation (11.136) has been derived under the assumption that the turbulence is in a state of equilibrium. That is, the energy required to produce large-scale inhomogeneities is equal to the rate at which viscous forces remove the energy associated with small-scale inhomogeneities, eventually converting all the input energy into heat. Thus, although turbulent mixing tends to reduce the gradient $\nabla\langle p \rangle$, it is assumed that external sources maintain this gradient. The eddy coefficient of diffusion K_p for the additive is analogous to the molecular diffusion coefficient D. The latter is a measure of the rate at which gradients of additives are smoothed out by the thermal motion of the molecules, whereas K_p is a measure of the smoothing of gradients by turbulence. K_p usually exceeds the molecular diffusion coefficient D by several orders of magnitude.

The refractive index in N units [see Eq. (2.19)] is

$$N = (77.6/T)(P + 4810 P_w/T), \qquad (11.137)$$

where P is the total atmospheric pressure in millibars, P_w is the water vapor pressure in millibars, and T is the absolute temperature. As discussed here, (11.136) is valid for conservative passive additives. However, it is well known that as small parcels of air are displaced vertically by turbulent motion, their pressure is continuously brought into equalization with that of the surrounding air, resulting in a change of parcel temperature and water vapor pressure (a rising parcel of air cools 1°K for every 100 m of altitude change). Therefore, T and P_w, and thus N, are not conserved properties. Although the refractive index of an air parcel changes while the parcel moves vertically, irregularities Δn are produced only when there is change relative to the surrounding environment, not necessarily when there is change in the parcel's N. For example, if the environment had an adiabatic lapse rate and P_w, decreased with height at the same rate as P then N of a displaced parcel would always be the same as the surrounding N, and hence no irregularities in refractive index would be created even if there were a gradient of $\langle N \rangle$

and the parcel's N changed. Thus radio scientists were led to define the potential refractive index ϕ (Bean and Dutton, 1966, p. 17):

$$\phi \equiv (77.6/\theta)(P_0 + 4810 P_{w0}/\theta), \quad (11.138)$$

where

$$\theta = T(P_0/P)^\alpha, \qquad \alpha = (C_p - C_v)/C_p = 0.286, \quad (11.139a)$$

$$P_{w0} = P_w(P_0/P) \quad (11.139b)$$

are the potential temperature and potential water vapor pressure, respectively, when the reference pressure $P_0 = 1000$ mbar; and C_p, C_v are the specific heat capacities at constant pressure and volume, respectively. The mixture of water vapor with dry air does not seriously alter the heat capacity of the mixture from the value for dry air (Hess, 1959, p. 42), so α can be considered a constant. Because θ and P_{w0} are conserved properties of the air parcel, ϕ is also a conserved property. Thus we can use (11.136) to determine the structure constant from gradients of ϕ:

$$C_\phi^2 = a^2 \varepsilon^{-1/3} K_\phi \left(\frac{d}{dz} \langle \phi \rangle \right)^2. \quad (11.140)$$

However, the scattered field intensity is a function of C_n^2, so it is necessary to relate C_ϕ^2 to C_N^2 where $C_N^2 = 10^{12} C_n^2$. This relation can be obtained by dividing (11.137) by (11.138) and substituting (11.139) and $P_w = 1.608 Pq$ (Hess, 1959, p. 44) for T and P_w so that

$$N = \phi \left\{ \left[1 + \frac{7733q}{\theta} \left(\frac{P}{P_0} \right)^{-\alpha} \right] \middle/ \left(1 + \frac{7733q}{\theta} \right) \right\} \left(\frac{P}{P_0} \right)^{1-\alpha}, \quad (11.141)$$

where q is the specific humidity, which is the ratio of the mass of water to the mass of moist air (water vapor plus dry air) per unit volume. The quantity q is conserved if there is no condensation of water vapor as the parcel is displaced. If one is to relate C_ϕ^2 to C_N^2, the variance of N must be expressed in terms of the variance of ϕ and covariances of ϕ and q, of ϕ and θ, and of q and θ using (11.141). However, we can simplify this relation by assuming a reference height at the level of the scattering layer; then $P/P_0 \approx 1$, and hence

$$C_N^2 \approx C_\phi^2. \quad (11.142)$$

In the remainder of this chapter we assume the reference level to be at the height of observation, so the variables θ, q, and ϕ [the generalized potential temperature, specific humidity, and potential refractive index (Ottersten,

11.6 Structure Constant of the Refractive Index

1969b), respectively] will be taken to be at this height as well. With this understanding,

$$C_n^2 = a^2 \varepsilon^{-1/3} K_\phi \times 10^{-12} \left(\frac{d}{dz}\langle\phi\rangle\right)^2. \qquad (11.143\text{a})$$

Sometimes the structure constants C_θ^2, C_q^2 of θ, q and the constant associated with the structure function of the product θq can be measured separately or estimated in the convective boundary layer using parametrizations based on observation (Gossard, 1960; Lenschow and Wyngaard, 1980). In this case (Gossard, 1977; Gossard and Strauch, 1983)

$$C_n^2 = a^2 C_\theta^2 + b^2 C_q^2 - 2ab C_{\theta q}^2, \qquad (11.143\text{b})$$

where the constants depend on the properties of the air mass. For tropical maritime air, for example,

$$a^2 = 2.24 \times 10^{-12}, \qquad b^2 = 17.8 \times 10^{-12}, \qquad 2ab = 12.6 \times 10^{-12},$$

whereas for warm continental air

$$a^2 = 1.25 \times 10^{-12}, \qquad b^2 = 16.2 \times 10^{-12}, \qquad 2ab = 9.01 \times 10^{-12}.$$

Under the assumption of steady state turbulence, horizontally homogeneous perturbation statistics, and a mean velocity horizontally directed and dependent only on height, the turbulent energy budget equation (Tennekes and Lumley, 1972, p. 97) can be reduced to

$$\varepsilon = \langle -u'w'\rangle \frac{\partial \langle u \rangle}{\partial z} + \langle -v'w'\rangle \frac{\partial \langle v \rangle}{\partial z} + \frac{g}{\theta_0}\langle w'\theta'\rangle \qquad (11.144)$$

by neglecting pressure perturbations (parcels are assumed to be continuously brought into equilibrium with the environment so that pressure fluctuations are not generated, or are at least negligibly small) and terms of order higher than the second in the perturbation quantities. The equations

$$\langle -u'w'\rangle \equiv K_m \partial\langle u\rangle/\partial z, \qquad \langle -v'w'\rangle \equiv K_m \partial\langle v\rangle/\partial z, \qquad (11.145)$$

$$\langle -\theta'w'\rangle \equiv K_H \partial\langle\theta\rangle/\partial z \qquad (11.146)$$

define the diffusion coefficients of momentum K_m and heat K_H given the mean gradients and the transfer rates of turbulent momentum $\langle -\rho u'w'\rangle$ and heat $\langle \rho C_p \theta' w'\rangle$. We assume that the diffusion coefficients for u' and v' are equal, although this may not always be the case (Rabin et al., 1982, and Fig. 10.9). Because the reference level is at the height of observation, the potential temperature θ_0 is simply the mean temperature T_1 of the turbulent layer. Combining (11.144)–(11.146), we obtain

$$\varepsilon = K_m |d\mathbf{v}_h/dz|^2 (1 - R_f), \qquad (11.147)$$

where R_f is the flux Richardson number

$$R_f = \left(g\frac{\partial\langle\theta\rangle}{\partial z}\bigg/T_1\left|\frac{d\langle v_h\rangle}{dz}\right|^2\right)\left(\frac{K_H}{K_m}\right) = R_g\left(\frac{K_H}{K_m}\right), \quad (11.148)$$

and R_g is the gradient Richardson number. The eddy diffusion coefficient K_ϕ of the refractive index can be defined analogously to Eq. (11.146) for heat. However, because K_ϕ is expected to be nearly equal to K_H, we write (11.143a) in an alternative form using (11.147) and (11.148):

$$C_n^2 = \frac{a^2\varepsilon^{2/3}T_1 R_f}{(1-R_f)g\,d\langle\theta\rangle/dz}\left(\frac{K_\phi}{K_H}\right) \times 10^{-12}\left(\frac{d\langle\phi\rangle}{dz}\right)^2. \quad (11.149)$$

Although the ratio K_ϕ/K_H might be well approximated by unity, the ratio K_H/K_m is much larger than unity when buoyancy-generated turbulence dominates shear production, as can be expected in the boundary layer for sunny afternoon weather (Tennekes and Lumley, 1972, p. 102).

Carefully controlled laboratory experiments with statically stable shear flow in aqueous solutions (Thorpe, 1969) and wind tunnels (Scotti and Corcos, 1969) have established convincingly that disturbances become dynamically unstable when R_g is reduced to $\frac{1}{4}$, at which value waves appear at a wavelength having the maximum growth rate. Most interestingly, Thorpe's data suggest how fine-scale three-dimensional turbulence might be generated from large-scale two-dimensional waves. After all, it is the fine-scale irregularities that are responsible for scatter at microwave frequencies. Turbulence is generated as the waves roll up, producing a spiraling layer of fluid across which $R_g \approx \frac{1}{4}$. The transition to turbulence occurs after the rolling up has begun, probably resulting from gravitational instability as the denser air becomes superimposed over the lighter air in the spiral structure of the breaking waves. The spiraling structure generates patches of turbulence which then are elongated in the shear flow and coalesce, eventually forming a quasiuniform layer. This layer contains the turbulent motion and irregularities similar to those often observed in the atmosphere with radar. The experiments of Thorpe (1973) suggest not only that $R_g = \frac{1}{4}$ is required for the generation of unstable waves but also that $R_f \simeq \frac{1}{4}$ during the transition from initial to final flow after the turbulence has smoothed the initial gradients of $\langle\theta\rangle$ and $\langle v_h\rangle$. Furthermore, comparison of typical values of quantities in the laboratory, ocean, and atmosphere suggest that the wavelength of the waves of maximal growth rate are about seven times the thickness of the turbulent layer.

Clearly (11.149) stresses that C_n^2 is not only a function of the turbulence intensity K_m [through ε in (11.147)] but also depends on the gradient of $\langle\phi\rangle$. Thus turbulence can be strong and yet C_n^2 be undetectably small. This is

11.6 Structure Constant of the Refractive Index

particularly true in the lower atmosphere, where gradients of specific humidity can dominate those of potential temperature. However, in the drier atmosphere above the first kilometer or two, where the contribution of water vapor to the refractive index irregularities can often be ignored, C_n^2 can be expressed more simply as

$$C_n^2 \approx \frac{a^2 R_f (77.6 P_1)^2 \varepsilon^{2/3} \times 10^{-12}}{(1 - R_f) g T_1^3} \left(\frac{K_\phi}{K_H} \right) \frac{d\langle\theta\rangle}{dz}. \qquad (11.150)$$

It is not necessary to take the absolute value of the gradient term, because the upper atmosphere is, on the average, stable, and thus $d\langle\theta\rangle/dz$ is positive. Under the assumptions used to derive (11.150), both C_n^2 and ε can be estimated from Doppler radar measurements. The relation between ε and spectrum width σ_v is given by Eqs. (10.68) or (10.70). The mean layer temperature T_1 can be estimated with reasonable accuracy from climatological tables. The values $R_f \approx \frac{1}{4}$, $3.2 \leq a^2 \leq 4.0$, and $K_\phi/K_H \approx 1$ specify all parameters except for $d\langle\theta\rangle/dz$ and the radar-measurable ones, C_n^2 and ε. Measurements of C_n^2 and ε can therefore be used in (11.150) to estimate $d\langle\theta\rangle/dz$, the stability of the atmosphere. The location of temperature inversions (i.e., layers of large $d\langle\theta\rangle/dz$) are useful in retrieving temperature profiles from radiometric measurements (Westwater and Grody, 1980). Gossard et al. (1982) have used radar measurements of ε and C_n^2 to estimate the mean gradients of the potential refractive index in a moist boundary layer. These results compare well with those calculated from rawinsonde observations. Turbulence usually occurs in layers of small vertical extent (e.g., 10–1000 m) and might be small compared to the resolution volume size. Thus, unless a high-resolution (large-bandwidth) radar is used pointing nearly in a vertical direction, the C_n^2 measured will be resolution volume averaged, $\overline{C_n^2}$, and hence $d\langle\theta\rangle/dz$ can be underestimated.

On the other hand, Weinstock (1981) points out that ε (and hence K_m if $d\langle v_h \rangle/dz$ is also measured) can be estimated solely from C_n^2 with relatively high accuracy (i.e., to within a factor of 2) if C_n^2 is accurately measured. This is a result of a combination of factors:

(1) The relation between shear and the $\langle\theta\rangle$ gradient required for shear production of turbulence causes a weak dependence of C_n^2 on the gradients of θ.

(2) The layer thickness, although smaller than the resolution volume, is approximately equal to the outer limit of the inertial subrange scales, which in turn is dependent on ε and $d\langle\theta\rangle/dz$.

(3) The range of variation in $d\langle\theta\rangle/dz$ in the stably stratified layers above the lower troposphere is limited.

Thus the layer of turbulence need not fill the resolution volume in order to make reasonably accurate estimates of ε within the layer.

11.6.1 Dependence of the Structure Constant on the Height

When there is an abundance of moisture near the earth's surface, the reflectivity of the lower atmosphere is significantly larger than for the drier air of the upper troposphere because of the contribution to C_n^2 from gradients of P_{w0}. There is usually a marked decrease of P_{w0} at the top of the convectively mixed boundary layer (CBL) accompanied by an increase in θ, (i.e., a stable layer). Because ϕ depends linearly on P_{w0} but inversely on $\langle \theta \rangle$, the opposite gradients of P_{w0} and $\langle \theta \rangle$ in this transition region result in large values of $\partial \langle \phi \rangle / \partial z$ and hence enhanced values of C_n^2. This enhancement is resolved quite well by high-resolution radars (Fig. 11.15).

Above the CBL, refractive index irregularities are found in layered zones of dry air where shear generates turbulence and potential temperature gradients are slightly positive. These zones usually have a small vertical extent (several tens of meters) separated by regions of little or no turbulent activity (see Fig. 11.16a). The data of Fig. 11.16a are obtained with a 10.7-cm radar, whereas for Fig. 11.16b a 70-cm radar was used. The differences between these data are discussed in Section 11.6.2.

Whenever vertical profiles of C_n^2 are shown, they usually represent spatial and temporal averages, whereas instantaneous values are highly variable, change by orders of magnitude over a few hundred meters altitude, and depend on the properties of the air mass and turbulence. Nevertheless, it is useful to estimate expected values of C_n^2 vs height in order to gauge the feasibility of using weather radars, if observations are made over sufficiently long periods of time, to detect backscatter with sufficient strength to enable measurements of wind and turbulence.

Gossard (1977) has determined the variation of C_n^2 with height for different air masses using the mean properties of the air for each category (e.g., maritime air). He assumed that $\varepsilon^{-1/3} K_m$ is neither systematically related to height nor correlated with $\nabla \langle \phi \rangle$ and estimated the mean value of the unknown factor $a^2 \varepsilon^{-1/3} K_m$ from the mean values of the measured optical structure constant, which is relatively insensitive to the amount of moisture in the air. Figure 11.17 shows the range of C_n^2 deduced by Gossard for Summer and Winter maritime air of both tropical and polar origin. Maritime air originates over oceans and is of particular interest to radar meteorologists because it is the air that usually participates in the development of severe thunderstorms. Continental air originating over land has slightly lower (by about a factor of 4) values of C_n^2 below 2 km, but there is a dramatic decrease

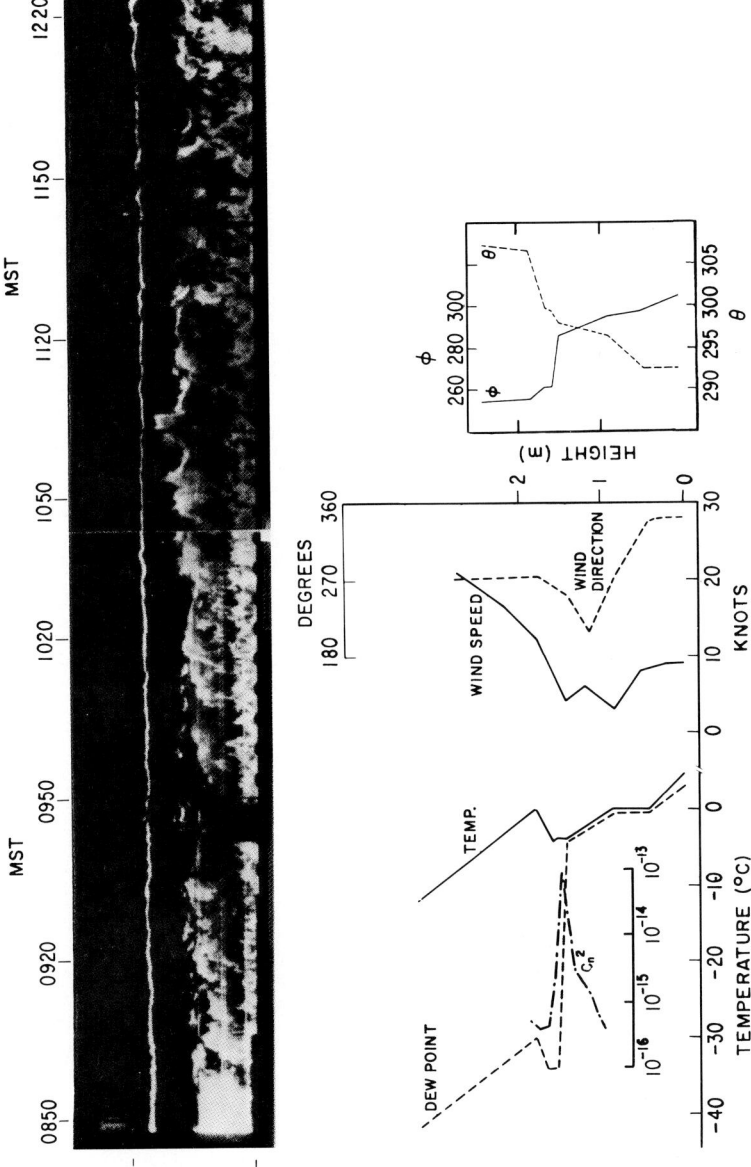

Fig. 11.15. Vertically pointing FM cw radar and balloon-sounding data near Denver, Colorado. (From Gossard, 1981.)

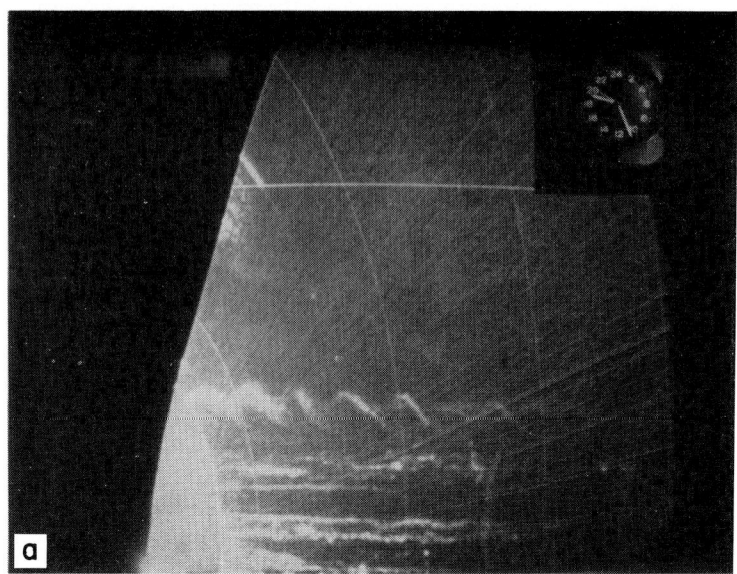

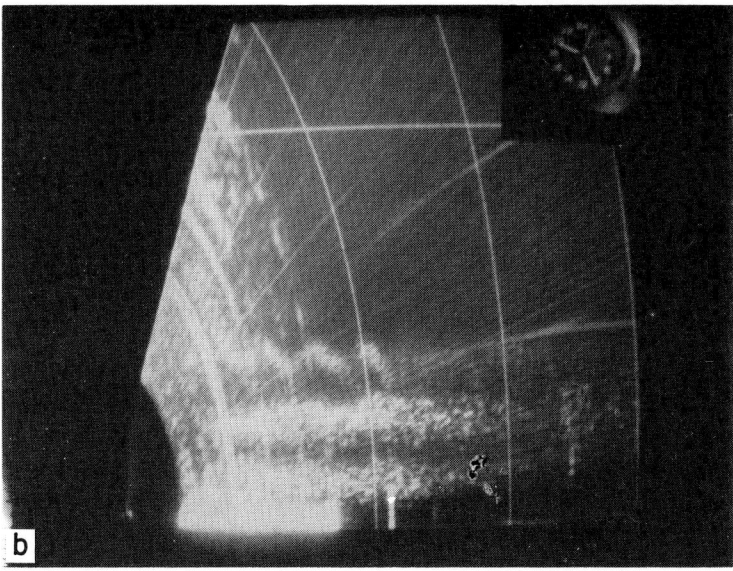

Fig. 11.16. (a) Range height indicator (RHI) photograph at 120° azimuth taken at 14:26 E.S.T. on 17 March 1969 with the 10.7-cm radar at Wallops Island, Virginia. (b) The same situation except that data are obtained with a 70-cm radar. The height mark is at 12.2 km and the range marks are at 9.3-km intervals. More than seven separate horizontally stratified clear-air layers are visible below 6 km. Note the strong wave in the top layer. (Courtesy of Jack Howard, NASA Wallops Island, Virginia.)

11.6 Structure Constant of the Refractive Index

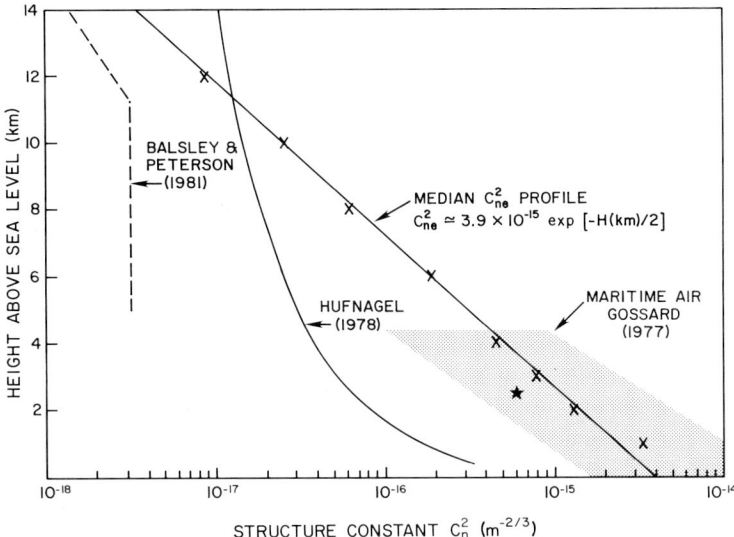

Fig. 11.17. Structure constant C_n^2 estimates versus height. [×, values computed by $L^{(101)} = L^t$ where $L^{(101)}$ is the median transmission loss measured from forward-scatter data and L^t is the loss computed assuming turbulence uniform within the scatter volume. Star denotes the point from Chadwick and Moran (1980).]

in C_n^2 above. During midday, when surface heating and convection is vigorous, Gossard suggests that in the CBL C_n^2 is increased by an order of magnitude. Burk (1978) used a second-moment turbulence closure model to show that the maritime boundary layer can have C_n^2 larger than 10^{-12} m$^{-2/3}$ for all heights up to 1 km. Aircraft and radar measurements in maritime air over Oklahoma compared well and gave a mean $C_n^2 = 5 \times 10^{-13}$ m and peak values as high as 3×10^{-12} m$^{-2/3}$ for midafternoon clear skies (Doviak and Berger, 1980). Chadwick and Moran (1980) made long-term measurements of C_n^2 in the boundary layer using an FM cw 10-cm wavelength radar. For the thunderstorm months between March and October, they found a median value of C_n^2 of about 6×10^{-16} m$^{-2/3}$ for data 800 m above the high plains of Colorado. This value is plotted as the star in Fig. 11.17.

The reflectivity is considerably less in the upper troposphere, on the average, than in the CBL, and weather radars rarely receive echoes from refractive index irregularities at these high altitudes. In the late 1960s, high-power large-aperture incoherent 10-cm radars established that refractive index irregularities in the upper troposphere are usually located in widely spaced stratified layers and are associated with shear induced turbulence (Hardy and Katz, 1969). Doppler radars have increased capability over incoherent ones and should sense more of these layers, so that height continuous wind profiles might be obtained with reasonable vertical resolution (e.g., 1 km).

VHF radar measurements in the upper troposphere suggest that C_n^2 may, on the average, be about 10^{-17} m$^{-2/3}$ or larger (Green et al., 1979). Gage et al. (1980) show C_n^2 profiles with values above 10^{-18} m$^{-2/3}$ throughout the upper troposphere and lower stratosphere (<15 km). However, caution should be exercised before extending these observations to the UHF wavelength of weather radars because the larger-scale refractive irregularities that echo the VHF waves are likely to be anisotropic, in which case (11.150) would not be applicable. It is also probable that the UHF sampling wavelength may be outside the inertial subrange, as discussed in Section 11.6.2.

The earliest reflectivity measurements in the UHF band were made during the 1940s and 1950s using troposcatter communication links. The National Bureau of Standards' Technical Note 101 (Rice et al., 1966) synthesizes data from hundreds of links and gives the median conditions of atmospheric reflectivity for various link geometries. By equating the measured median transmission loss with the transmission loss computed assuming the resolution volume to be uniformly filled with refractive index fluctuations (having a spectral distribution corresponding to the inertial subrange), the height dependence of an effective structure constant C_{ne}^2 is easily derived. In Fig. 11.17 C_{ne}^2 can be assumed to represent a diffusely scattering troposphere.

Values of C_n^2 obtained from an empirical formula deduced by Hufnagel (1978) for the structure constant of the optical refractive index are also plotted, for comparison, in Fig. 11.17. It is evident that the values derived from 10-cm median transmission loss data are larger in the lower troposphere, consistent with the fact that moisture irregularities, more intense in the first few kilometers, contribute significantly to the C_n^2 deduced from UHF observations while giving negligible contribution at optical frequencies.

C_{ne}^2 in Fig. 11.17 is derived for a wavelength of 10 cm and should correspond to scatter from nearly isotropic irregularities. However, forward-scatter links operating at this wavelength scatter from irregularities of larger scales than backscatter radar. Because forward-scatter angles are usually less than $10°$, refractive irregularities effective in scattering are at scales significantly larger than $\lambda/2$ and hence might lie within the inertial subrange, whereas the 5-cm scales effective in 10-cm backscatter could fall into the dissipation range where the intensity of refractive index variations are much smaller (see Section 11.6.2). If these C_{ne}^2 values are used in (11.104), the median radar reflectivity may be overestimated. Nevertheless, C_{ne}^2 values correspond quite closely to the mean values measured with a 10-cm radar by Chadwick and Moran (1980) and those computed by Gossard (1977) suggesting that, at least in the first kilometer or two of the troposphere, the inertial subrange extends to scales as small as a few centimeters.

11.6 Structure Constant of the Refractive Index

Radar measurements of the structure constant in the upper troposphere have been made at a wavelength of 23 cm by Crane (1977), who observed $C_n^2 \geq 10^{-17}$ m$^{-2/3}$ for heights below 15 km on 13 days of clear air between January and July in New England. Balsley and Peterson (1981), using the 23-cm radar in Chatanika, Alaska, found that $C_n^2 \geq 10^{-18}$ m$^{-2/3}$ more than 50% of the time over several days at heights below 16 km. A subjective fit of their median data is plotted in Fig. 11.17. The large dispersion of median C_n^2 values in Fig. 11.17 may be the result of (1) the different locations of the radar sites and thus different meteorological conditions, (2) seasonal variation, and (3) insufficiently large data samples to produce a statistically stable median value.

11.6.2 Inertial Subrange

Equation (11.104) is appropriate for UHF radars, and weather radars in particular, only if the refractive index scales of size Λ, equal to half the radar wavelength, lie within the inertial subrange of the spectrum of turbulence. In this range the rate at which turbulent energy is transferred to smaller scales as large eddies fragment depends only on the dissipation rate ε of turbulent energy. However, as the scales become smaller, the air's viscosity directly affects the intensity of turbulence, and the power density spectrum decreases more rapidly than for scales within the inertial subrange. The small-scale eddies affected by viscous forces are then within the dissipation range of the scales, and it is there, ultimately, that turbulent energy is dissipated as heat. Therefore, we need to examine the conditions for which UHF radiation, and in particular radiation at weather radar wavelengths, scatters from irregularity scales that fall into the dissipative spectral range where reflectivity would be significantly less than that predicted by application of (11.104). Hill (1978) has developed a model that predicts the behavior of the spectra for the transition region between the inertial and dissipation range of scales. For refractive index spectra dominated by temperature fluctuations, Hill's spectra show that when

$$\lambda/2 \geq \Lambda_t \equiv 5\pi(v^3/\varepsilon)^{0.25} \tag{11.151}$$

(where v is the kinematic viscosity), the spectral density at $K = 4\pi/\lambda$ is at least as large as that predicted for the inertial subrange scales. The use of (11.104) and C_n^2 data to estimate the radar performance is then justified. The kinematic viscosity v has a well-defined height dependence, but the eddy dissipation rate ε may vary by three or more orders of magnitude at any height. Based on model predictions of Hill that lead to (11.151), the maximum height $H_{\max}$ to which (11.104) is applicable is plotted against λ_t, the radar wavelength at the transition point, in Fig. 11.18, assuming a standard atmosphere (Shea, 1965), and treating ε as a parameter. The radar receives

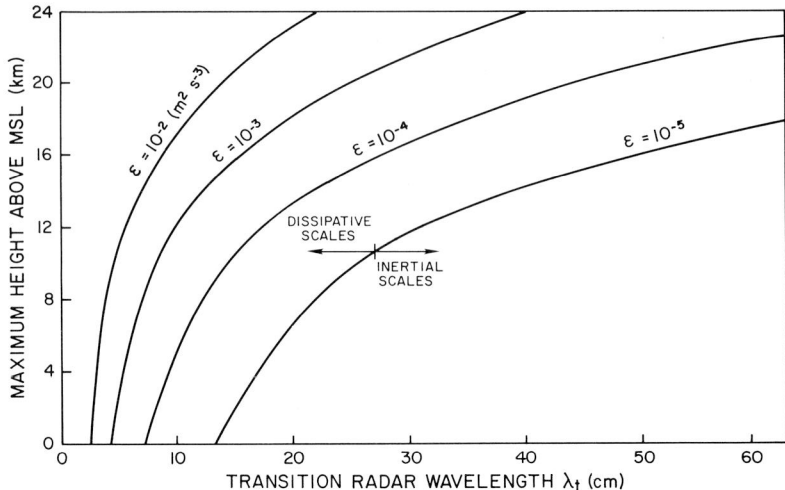

Fig. 11.18. Maximum height to which radar echoes are returned from irregularities of size within the inertial subrange for a given level of the eddy dissipation rate ε. The values of ε corresponding to various levels of turbulence are gauged by its effects on aircraft: severe turbulence: $\varepsilon = 6.75 \times 10^{-2}$; moderate turbulence: $\varepsilon = 8.5 \times 10^{-3}$; light turbulence: $\varepsilon = 3.0 = 10^{-3}$ (Trout and Panofsky, 1969). (Courtesy of R. G. Strauch, WPL/NOAA.)

echoes from refractive index irregularities at wavelengths $\Lambda = \lambda/2$ that lie within the dissipative range whenever H_{max} is exceeded. In this range the intensity of the irregularities, and hence the radar reflectivity η, decrease exponentially as $K = 2\pi/\Lambda$ increases (Tatarskii, 1971, p. 65). Thus, for radars operating at wavelengths to the right of the line of constant ε, η is proportional to $\lambda^{-1/3}$, whereas for those operating to the left η decreases exponentially for $\lambda < \lambda_t$.

For example the spectra of Hill (1978) indicate that echoes from regions of light turbulence ($\varepsilon = 3 \times 10^{-3}$ m^2 s^{-3}) at 20 km would be about 9 dB below the power that would have been scattered if the scattering scales ($\lambda/2 = 5$ cm) were within the inertial subrange. Above 14 km, turbulence needs to be moderate to severe in order that ε be sufficiently large (i.e., $\varepsilon \geq 8 \times 10^{-3}$) so that the 5-cm scattering scales lie within the inertial subrange.

Shear-generated turbulence occurs in layers having various levels of turbulence intensity ε, and when $\lambda > \lambda_t$, η is proportional to $\varepsilon^{2/3}$ [e.g., Eq. (11.150)]. However, the strong exponential decrease of η for $\lambda \leq \lambda_t$ might give the appearance that ε is more strongly stratified than it really is if (11.150) and (11.104) are used to estimate ε from η. This may explain the layered appearance of turbulence (Fig. 11.16) seen by radars operating at the upper end (i.e., 10 cm) of the UHF band. However, the often coarser resolution of the

11.6 Structure Constant of the Refractive Index

longer-wavelength radars would smear the appearance of layers. Furthermore, whereas radars operating at long wavelengths (e.g., VHF) typically show a continuum of η versus height, high-power weather radars (e.g., $\lambda \simeq 10$ cm) show layers of turbulence separated by regions of what appears to be no turbulence.

The appearance of a height continuum of echoes observed with VHF radars may be a result of their capability to detect layers of weaker ε for which $\lambda_{\text{VHF}} > \lambda_{\text{t}}$, so that irregularities of n are still within the inertial subrange. However, it is unlikely that all layers are turbulent all the time, a deduction supported by the study of Lilly *et al.* (1974), although their results apply to turbulence measurements above 14 km. Thus the apparent height continuum of echoes observed with long-wavelength radar may signify the presence of fossil turbulence in which irregularities in n remain although velocity perturbations have vanished (Woods *et al.*, 1969). On the other hand, if radar does not have adequate resolution, the intermittent turbulence occuring in various layers could appear continuous in height. High-resolution VHF observations by Röttger and Schmidt (1979) clearly show the layered structure of η.

Lilly *et al.* (1974) plotted the cumulative probability of turbulence intensity for lower stratospheric data (14 km $\leq H \leq$ 21 km) which suggests an exponentially decreasing cumulative probability with increasing turbulence. Extrapolation of their results indicates that zero turbulence occurs 90% of the time. Although extrapolation is risky, the data show that $\varepsilon \geq 10^{-4}$ m^2 s^{-3} less than 2% of the time and, furthermore, that zero turbulence occurs most of the time. However, turbulence is expected to be much more frequent in the troposphere, where the air is less stable, and R_g is more likely to fall to values less than the critical $\frac{1}{4}$.

Inferences of ε from the diffusion of smoke puffs, direct measurements of turbulent shearing stress and wind shear, and measurements of the spectra of turbulent velocities (e.g., Ball, 1961) suggest $\varepsilon < 10^{-5}$ m^2 s^{-3} for heights above 3 km. However, radar-inferred values appear to be larger than 10^{-5} or even 10^{-4} for most heights up to 12 km on a day of strong shear (Gage *et al.*, 1980). The eddy dissipation rate can change by several orders of magnitude (Chen, 1974), and differences from radar-measured values can result. Radar-inferred values of greater than 10^{-3} m^2 s^{-3} were found to be common for all heights up to 12 km over a period of 12 hr during which the atmosphere was disturbed by the passage of a polar jet stream. If prestorm atmospheres have such large values of ε, then Doppler weather radar has the potential to give profiles of winds preceding significant weather. Numerical models give evidence that the vertical shear of horizontal wind influences the intensity of storms that develop in a sheared environment (Weisman and Klemp, 1982; Schlesinger, 1982).

11.6.3 Criteria for Measurement of the Velocity of Refractive Index Irregularities

The SNR is a parameter of paramount importance to the measurement of the clear air velocity. For a Gaussian receiver–filter matched to the transmitted pulse, the per-pulse SNR referred to the receiver input is

$$\text{SNR} = \frac{0.21 P_t l_t \varepsilon_a A r_6^2}{4\lambda^{1/3} \pi c r_0^2 k T_{sy}} C_n^2, \tag{11.152}$$

where P_t is the peak power at the transmitter; l_t is the total loss factor, equal to the product of the transmitter and receiver transmission line losses (including radome loss) as well as the atmospheric loss factor; $\varepsilon_a = A_e/A$ is the antenna efficiency relating the aperture's physical area A to A_e [Eq. (3.21)] and accounting for nonuniform illumination (Section 3.1.1) as well as ohmic losses in the reflector; k is the Boltzmann constant; c is the speed of light; T_{sy} is the system noise temperature (Section 3.5.1); and r_6 is the range resolution. In deriving (11.152), the weather radar equation (4.25) is used, but line and radome losses are deleted from the gain g. Then g is related to θ_1 as $g = (16 \ln 2)/\theta_1^2$ and combining (4.25) with Eqs. (3.21), (3.39), and (11.104) leads to (11.152). The SNR is weakly dependent on λ. Thus, the only radar parameters in (11.152) that can produce a proportional increase in the SNR without changing the range resolution are P_t, A, and T_{sy}. In the UHF band, T_{sy} is usually most dependent on the receiver noise figure, so the use of low-noise preamplifiers can significantly reduce T_{sy}. At the high elevation angles suitable for obtaining wind profiles, a representative value of T_{sy} is 195 K. Longer-wavelength radars have higher T_{sy} because of intense radiation from space at those wavelengths. P_t can be effectively increased by transmitting longer coded pulses so that r_6 remains the same (see Chapter 7).

Criteria for the detection of scatter from turbulent air under the assumption that the resolution volume is completely filled with refractive irregularities and that these are statistically homogeneous, stationary, and isotropic have been given by Hennington et al. (1976). The criteria are exactly the same as those specified by Gage and Balsley (1978). We make similar assumptions but derive criteria that incorporate a parameter important to Doppler measurements—the precision of the Doppler velocity estimates.

In order to determine how well we can estimate the Doppler velocity given the radar system parameters, we use the minimum variance bound from (6.38b). The minimum SNR that can be tolerated when M samples, uniformly spaced by T_s, are processed to produce an estimate of Doppler velocity with a standard deviation $\text{SD}(\hat{v})$ is

$$\text{SNR(min)}_{\text{best}} = \frac{\pi^{1/4} 2^{3/2} \sigma_v^{3/2} T_s}{\sqrt{\lambda T_d}\, \text{SD}(\hat{v})}, \tag{11.153}$$

11.6 Structure Constant of the Refractive Index

where $T_d = MT_s$ is the dwell time and σ_v the Doppler velocity spectrum width, a parameter that is principally controlled by the intensity of shear and turbulence weighted by the resolution functions and reflectivity (see Chapter 5). Equation (11.153) assumes that T_d is sufficiently long that the spectral resolution $\lambda/2T_d$ is significantly finer than σ_v. Equating (11.153) and (11.152), we can determine the minimum structure constant $C_n^2(\min)$ required to produce velocity estimates with precision $SD(v)$ for a given set of radar parameters:

$$C_n^2(\min)_{\text{best}} = \frac{385\sigma_v^{3/2} r_0^2 k T_{sy}}{\lambda^{1/6} \, SD(\hat{v}) P_{av} A \varepsilon_a l_t r_6 \sqrt{T_d}}, \qquad (11.154)$$

where P_{av} is the average transmitted power. The average power antenna area product $P_{av}A$ is paramount in determining the radar sensitivity, although T_{sy} is also significant. Because $C_n^2(\min)$ depends weakly on the wavelength, there should be no significant difference in the performance of radars operating at different wavelengths if all other parameters are constant and if the scattering irregularities are within the inertial subrange for all λ. However, when receivers have low noise figures, the receiver temperature is controlled by the radiation emanating from outer space, and T_{sy} increases at a rate of about $\lambda^{2.4}$ when λ is larger than about 20 cm (Blake, 1970). T_{sy} has a broad minimum in the microwave ($\lambda = 3$–10 cm) region. Thus, short-wavelength UHF radars appear attractive because they may require the least power–area product $P_{av}A$. For example, a 10-cm UHF radar requires only $\frac{1}{40}$th the antenna area needed by a 6-m VHF radar with the same transmitted power. This is expected to be true when VHF radars are pointed away from the vertical (i.e., $\geq 15°$) in order to use the Doppler beam swinging method (Röttger, 1981; also see Section 9.3) to sound the wind.

Although Eq. (11.154) gives an important lower bound on detectable C_n^2, it does not specify an algorithm that can achieve this optimum condition (i.e., minimize the variance of $\hat{v}$). Thus we can consider (11.154) only as a goal. Because covariance methods (pulse pair processing) are commonly used in Doppler weather radar, and because Miller and Rochwarger (1972) have shown this algorithm to be a maximum likelihood one when pairs are independent, we now examine how well this popular algorithm works on correlated pairs and compare these results with the best given by (11.154).

When signals are weaker than noise and the spectrum width is narrow compared to the Nyquist interval, the minimum SNR that can be tolerated to produce velocity estimates with a precision of $SD(\hat{v})$ is obtained from (6.22a) by retaining only the term that is multiplied by N^2/S^2:

$$\min(SNR_{pp}) = \frac{\lambda \exp[8(\pi\sigma_v T_s/\lambda)^2]}{4\pi\sqrt{2T_d T_s} \, SD(\hat{v})}. \qquad (11.155)$$

A minimum of minima is obtained when the system period T_s is selected to be

$$T_s = \lambda/4\pi\sigma_v\sqrt{2}. \qquad (11.156)$$

In deriving (11.156) we assumed the dwell time to be fixed and P_{av} to be kept constant (at the highest permissible level). However, if r_6 is constrained to be a constant, and if the radar is peak power limited, pulse coding techniques (see Section 7.10) may be necessary to maintain P_{av} at its level when T_s is set by (11.156).

Doppler spectra of clear-air echoes observed with the NSSL's radar usually have σ_v ranging between 1 and 2 m s^{-1}. Crane (1980) observed Doppler spectra of clear-air turbulence over the Marshall Islands and estimated spectrum widths on the order of 1 m s^{-1}. Thus, for 10-cm-wavelength radars, T_s needs to be about 3 ms, with a corresponding Nyquist interval of 16 m s^{-1}. Although this may appear small, the radar beam in a profiling mode could be directed at large elevation angles (i.e., $\geq 70°$) so that high-speed horizontal winds do not cause Doppler shifts that alias more than once or twice. These could easily be dealiased.

Using (11.153), (11.155), and (11.156), we find that

$$C_n^2(\min)_{pp} = 1.43 C_n^2(\min)_{best}, \qquad (11.157)$$

which means that by selecting the proper T_s we can come very close to the minimum variance bound. We do not suggest that pulse pair processing is preferred. Spectral processing may do as well or better and, moreover, have other decisive advantages (e.g., easier identification of artifacts and anomalous signals). However, it is instructive to note that the minimum variance bound can be closely approached using known algorithms.

11.7 OBSERVATIONS OF CLEAR-AIR REFLECTIVITY

Although clear air does not contain such awesome phenomena as tornados or the baseball-sized hail found in severe storms, its structure is nevertheless rich in meteorological events that can lead to the development of these storms. The spectra of clear weather phenomena span scales from centimeters to thousands of kilometers and embrace such occurrences as convective cells that spring from heated surfaces and waves generated by shear instabilities. Many of these can be identified from observations of the reflectivity fields.

In the 1960s ultrasensitive incoherent radars at Wallops Island, Virginia, routinely mapped echoes from the convective boundary layer and shear zones at higher altitudes. Moderately sensitive Doppler weather radars, which have the advantage of coherent processing, may also have the capability of mapping echoes from the clear air. The structure of the reflectivity

11.7 Observations of Clear-Air Reflectivity

fields clearly shows the presence of convective cells in conditions of light wind (Fig. 11.19) and streets (i.e., alignment of echoes along parallel lines) of reflectivity associated with convective rolls (Fig. 11.20) when wind is strong. Cloudless skies and curved flow (gradients of shear) are responsible for the elongation of reflectivity in the direction of the mean wind (see Section 11.8.2). Convective cells are clearly delineated by 10-cm radar in the range height indicator (RHI) presentations shown in Fig. 11.21. Note the enhanced intensity of echoes at the cell top and sides, which can cause doughnut-shaped patterns when the radar beam scans azimuthally, as seen in Fig. 11.19. This increase in C_n^2 at the upper boundaries can be explained using (11.143b) and the following argument: As buoyant air is warmed by the heated ground and moistened by evaporation, it rises and cools adiabatically. At the lowest levels the cell is warm and moist relative to the environment, so it will continue to rise until the temperature of the cell air equals the ambient temperature, which defines the level of zero buoyancy. However, the momentum of the rising air is sufficient to carry it above the level of equal temperature, and if we assume the ambient air above to be stable (i.e., to have a positive gradient of θ) and to have decreasing q with height, air parcels at the cell top will be

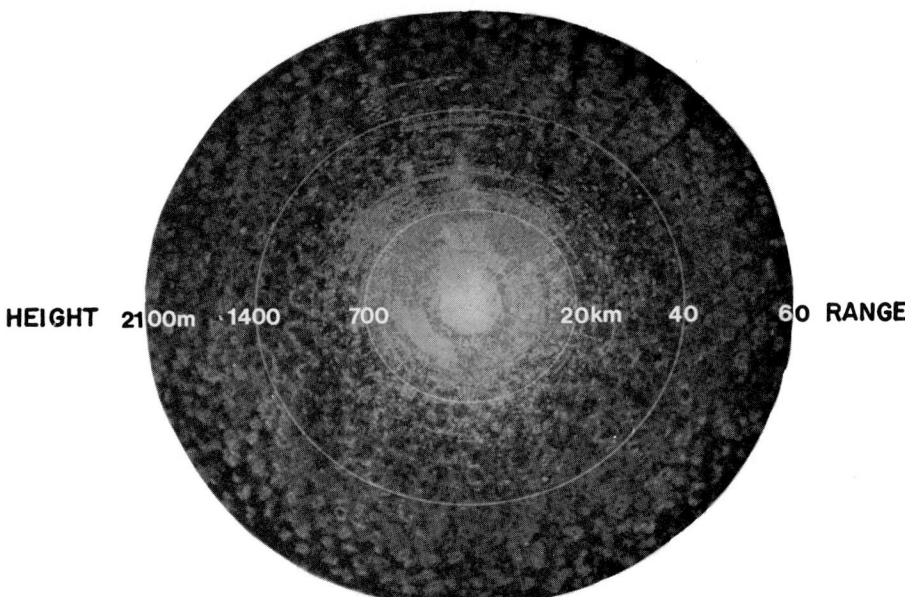

Fig. 11.19. Plan position display of a uniform pattern of convective cells observed with NSSL's 10-cm Norman Doppler weather radar on 17 August 1970 at 13:12 C.S.T. (Norman Doppler reflectivity data; cumulus humilis clouds). Elevation: 2.0°.

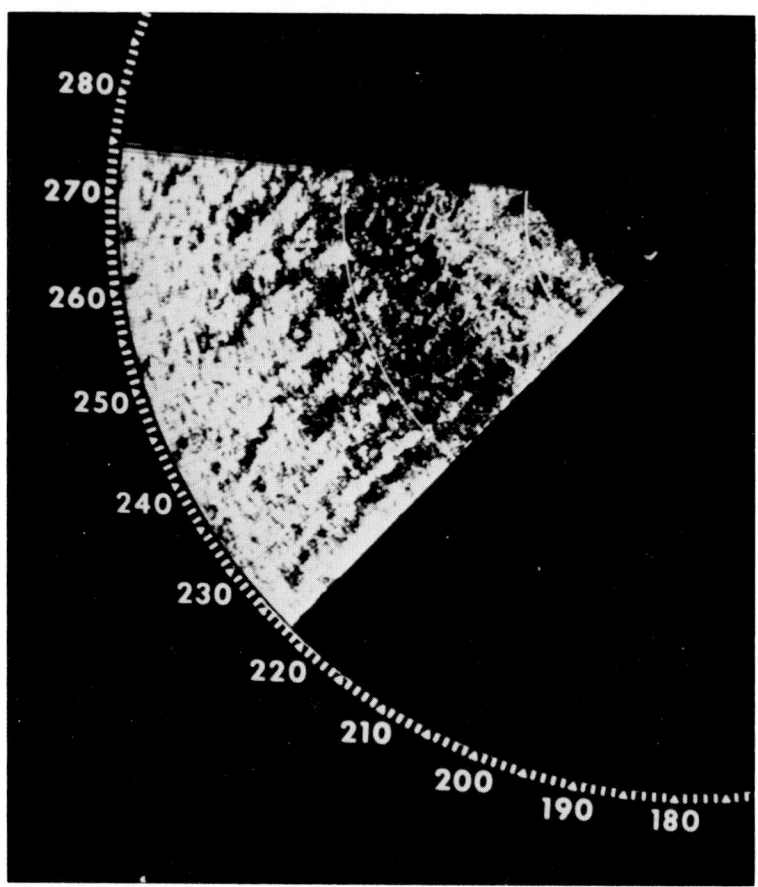

Fig. 11.20. Plan position indicator contour display of ηr^{-2} (where r is the range) from the NSSL's Norman Doppler radar. The bright areas of higher reflectivity delineate convective rolls aligned parallel to the mean wind, and bands are separated by about 4 km. Range marks are 20 km apart; the maximum display range is 65 km.

cool and moist relative to the environment. The decrease in temperature associated with increased moisture causes the temperature and specific humidity to be negatively correlated so that $C_{\theta q}^2$ in (11.143b) adds to the first two terms of (11.143b). Inside the cell, θ and q are positively correlated (i.e., an increase in the temperature is accompanied by an increased moisture), so $C_{\theta q}^2$ must be subtracted from the first two terms. Therefore, C_n^2 inside the cell is expected to be less than at its top surface, producing the results seen in Fig. 11.21.

The location of inversions can also be found, because enhanced reflectivity

11.7 Observations of Clear-Air Reflectivity

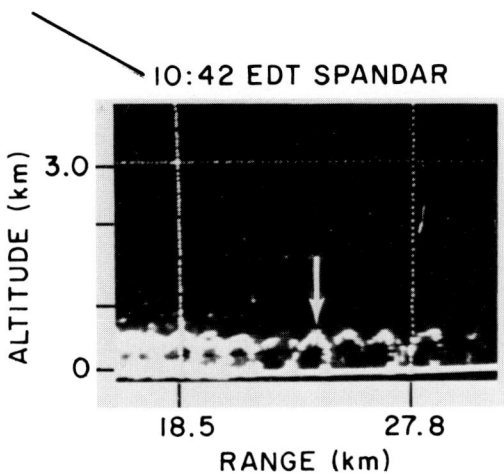

Fig. 11.21. Vertical cross section of the reflectivity structure produced by convective cells. (From Rowland and Arnold, 1974.)

usually is associated with large gradients of $\langle \theta \rangle$ and P_w. The thin continuous line on a time height indicator display (Fig. 11.15) is an example of such an inversion layer, one increasing in height. The structure constant C_n^2 is also plotted in Fig. 11.15, along with the rawinsonde data that show the high correlation between the gradient of the potential refractive index $\langle \phi \rangle$ and the peak of C_n^2. However, the gradient of $\langle \phi \rangle$ is controlled by both P_w and $\langle \theta \rangle$, so the location of the peak gradient of $\langle \phi \rangle$ is lower than the inversion layer, which is centered at the height of maximum gradient of $\langle \theta \rangle$. In this case $\mathbf{V}\langle \phi \rangle$ is most strongly correlated with $\mathbf{V}\langle q \rangle$.

The presence of frontal boundaries can also be resolved by observing the reflectivity structure of clear air. Figure 11.22a gives the reflectivity field with enhanced echoes along the boundary of an advancing cold front. The reflectivity enhancement was evident within the radar range (100 km) 2 hr before the clouds (Fig. 11.22b) developed. Notice the enhanced reflectivity to the north of the cold front, whereas reflectivity is significantly lower in the warm sector. The cold air advancing over the warm surface creates instabilities, which result in increased turbulence and enhanced C_n^2. The markedly enhanced reflectivity along the frontal boundary may be associated with more vigorous mixing of the air between the two contrasting air masses. However, the gusts associated with fronts have been observed to sweep up substantial amounts of debris that also contribute to the reflectivity. There is no way that single-wavelength single-polarization radar can be used to distinguish between returns from refractive index and debris that has a

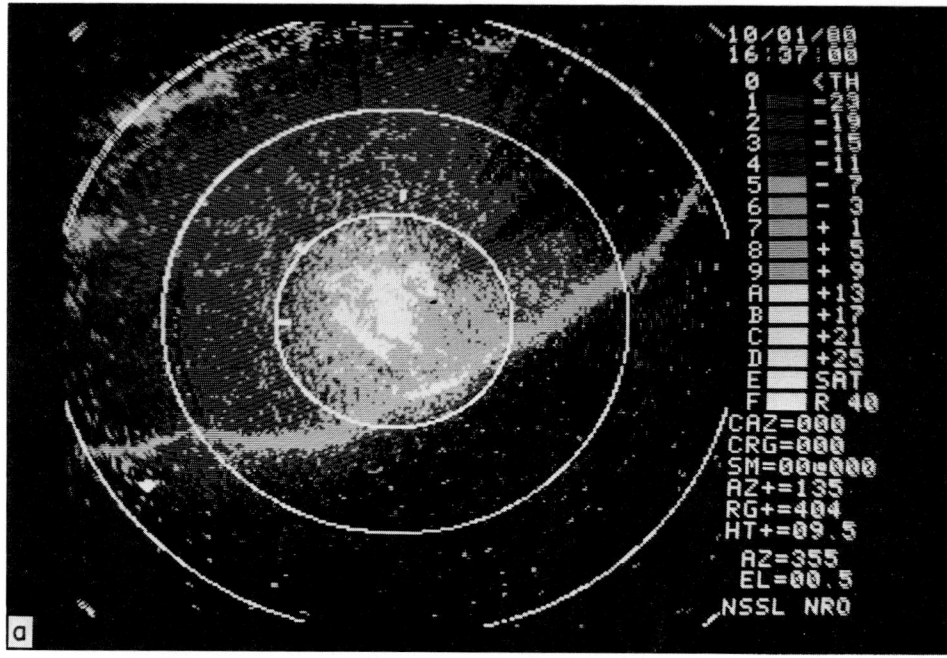

Fig. 11.22a. The reflectivity field with enhanced backscatter on the cold side (north) and along the frontal boundary.

nearly uniform distribution throughout the lower atmosphere. Only when debris is so sparsely scattered that there is at most one or two targets per resolution volume can debris be distinguished from refractive index irregularities (if scatter volumes are much larger than the correlation volume). The amplitudes of echos from debris then no longer have the Rayleigh statistical distribution.

The ultrasensitive radars at Wallops Island regularly detect horizontally stratified reflective layers in the troposphere (Fig. 11.16). The multiwavelength radars at Wallops Island showed the wavelength dependence of the reflectivity within the layers to be associated with scatter from irregularities having scales within the inertial subrange (Hardy and Glover, 1966). The scattering layers are often quite thin; the vertical depth can be as small as a few tens of meters and rarely exceeds a few hundred meters. Although water vapor fluctuations may dominate the contribution to backscatter, there is mounting evidence that the layers of enhanced reflectivity in the upper troposphere are associated with layers of enhanced stability so that temperature fluctuations may be the dominant contribution (Green and Gage, 1980).

11.7 Observations of Clear-Air Reflectivity

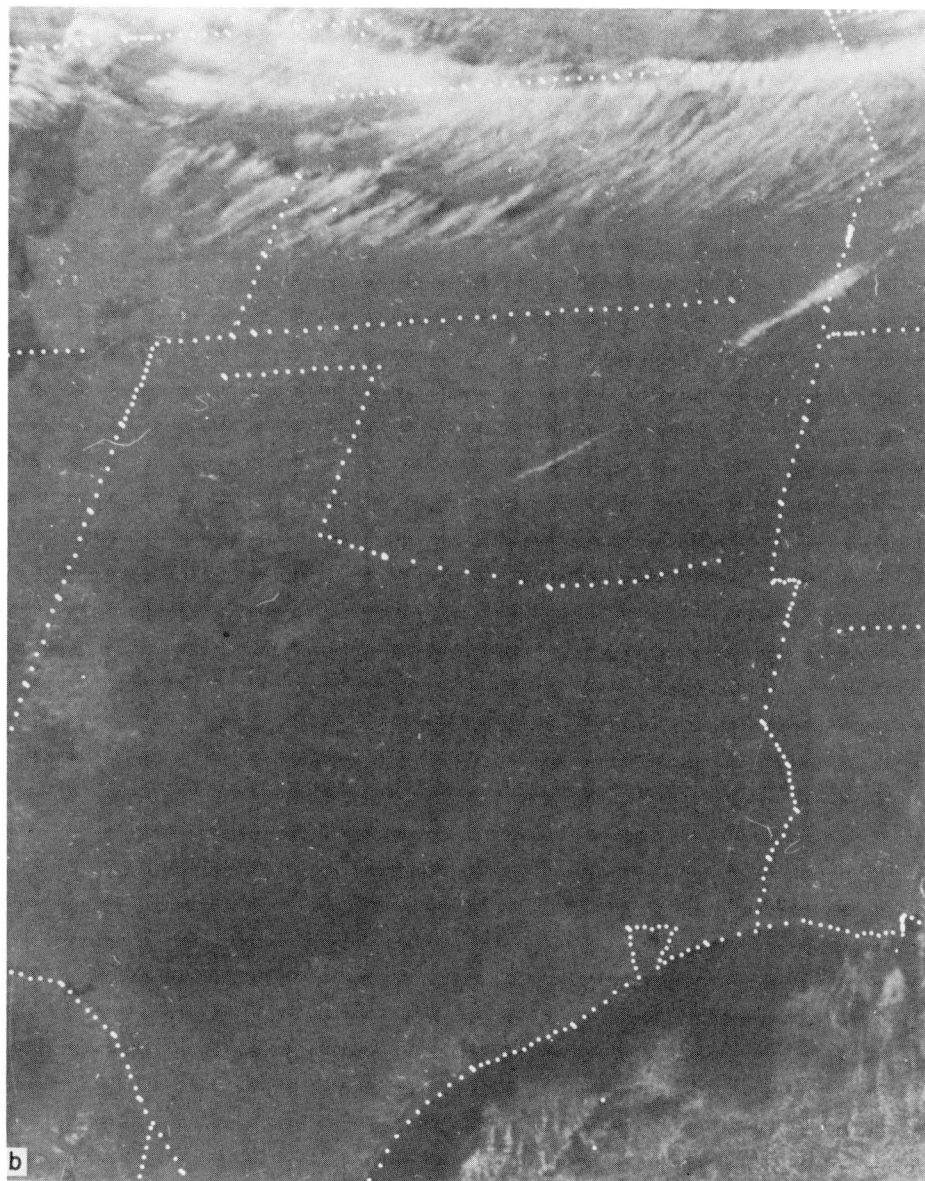

Fig. 11.22b. Satellite picture showing clouds that developed along the cold front as it passed through central Oklahoma.

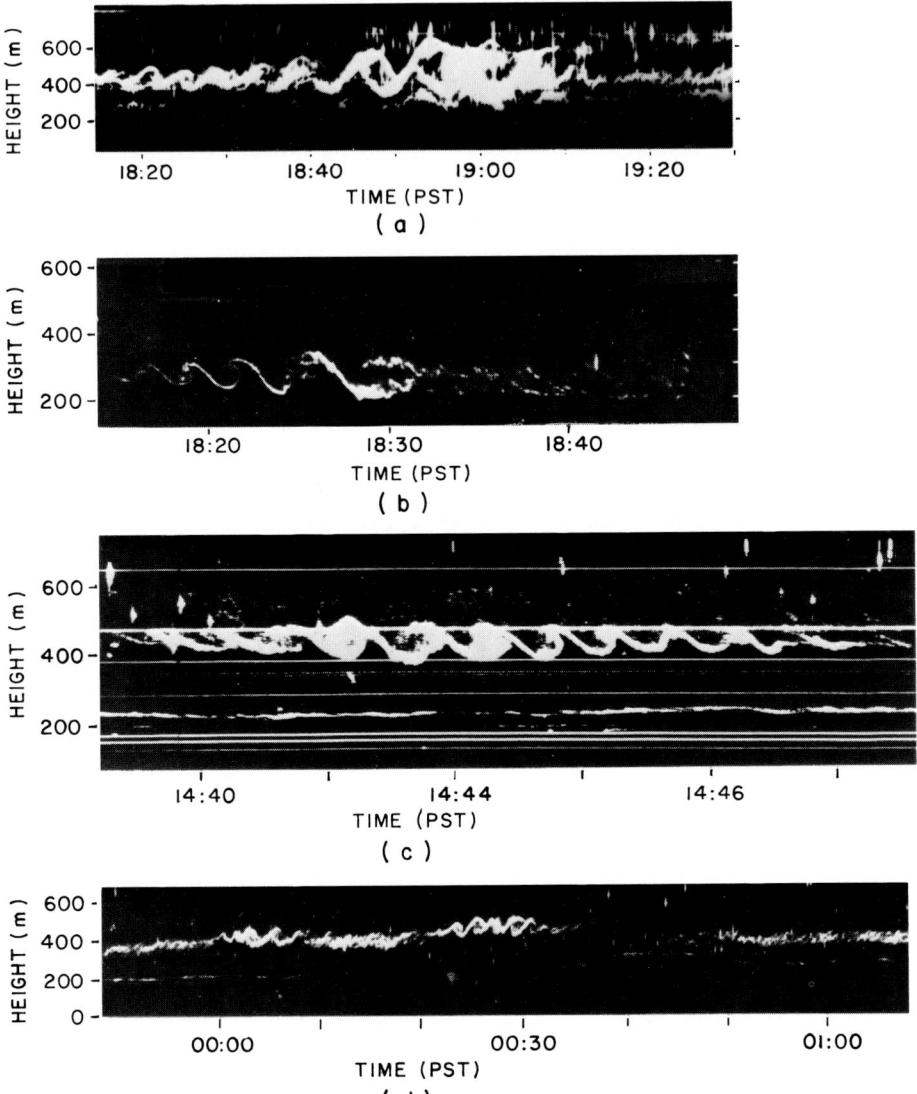

Fig. 11.23. Examples of wave perturbations of atmospheric layers observed by the FMcw radar at San Diego. The remarkable resolution of this radar is evident from the thin small-scale perturbations. (a) 28 September 1971. (b) 6 August 1969. (c) 24 August 1970. (d) 14 July 1969. (From Gossard and Hooke, 1975.)

The tropopause marks the location of the base of the stratosphere, which is a region of persistent static stability. Figure 11.16a shows a thin-layered reflectivity structure that exhibits a wavelike shape. Although the layer seen in Fig. 11.16a is not at the tropopause height, Ottersten (1969a) gives an example of one at the tropopause. The breaking waves in the lower atmosphere are well illustrated by the reflectivity field of high-resolution FM cw radars (Fig. 11.23).

11.8 OBSERVATIONS OF WIND, WAVES, AND TURBULENCE IN CLEAR AIR

Although the reflectivity field obtained by high-resolution radars facilitates the identification of many normally invisible meteorological phenomena, the Doppler radar provides two additional fields of information. The mean Doppler velocity and spectrum width can be used to infer the wind and turbulence and hence to measure important kinematic properties of the atmosphere.

There are two observational modes that can generate data on winds:

(1) With Doppler beam swinging at high-elevation angles (i.e., $>45°$), the upper regions of the troposphere, which have weak reflectivity, are close to the radar. By assuming a linear wind within the small circles in which the beam intersects the layers of constant height, profiles can be obtained of horizontal wind, divergence, and deformation (Section 9.3).

(2) By beam swinging at low elevation angles (i.e., $<10°$), where larger volumes of the more reflecting CBL produce radial (Doppler) wind fields, the spatial dependence of these fields on the radar's resolution volume position can be analyzed to produce maps of wind speed, direction, divergence, and deformation (e.g., Figs. 9.12 and 9.13).

11.8.1 Wind Profiling

Using NSSL's weather radar parameters (Table 11.1), the maximum range to which wind can be measured assuming $C_n^2 = 10^{-18}$ m$^{-2/3}$ is less than 2.0 km when $T_d \approx 100$ s. This dwell time implies that velocity estimates are made with SNR $= -25$ dB. Although T_d can, in principal, be increased indefinitely to improve detection sensitivity, it becomes difficult to obtain a radar system free of artifacts that could allow for the processing of signals 20 dB or more below the receiver noise. As shown in Fig. 11.24, present NSSL weather radars have little or no capability to sound wind much above the

TABLE 11.1 NSSL Doppler Weather Radar Parameters (1980)

average transmitter power[a]	$P_{av} = 1000$ W (N); 650 W (P)
pulse width[a]	$\tau = 1$ μs (N) 3 μs (P)
receiver bandwidth	$B_6 = 8.5 \times 10^5$ Hz
repetition period[a]	$T_s = 768$ μs (N); 2304 μs (P)
wavelength	$\lambda = 10.52$ cm
antenna area	$A = 65$ m^2
system noise temperature	$T_{sy} = 315$ K
aperture efficiency	$\varepsilon_a \simeq 0.60$
transmitter line loss factor[b]	$l_{lt} = 0.63$
receiver line loss factor[b]	$l_{lr} = 0.50$
dwell time[a]	$T_d \simeq 50$ ms (N); 150 ms (P)

[a] N: normal mode for storm data collection; P: prestorm clear air mode.
[b] Includes randome loss.

CBL. If C_n^2 values of 10^{-17} m$^{-2/3}$ are highly probable in the lower troposphere, then wind profiles to 5–6-km altitudes can be routinely acquired with the present system if prestorm parameters (labeled P in Table 11.1) are used.

However, if the average power–aperture area product were increased by a factor of 10, line losses minimized, and range resolution r_6 coarsened to

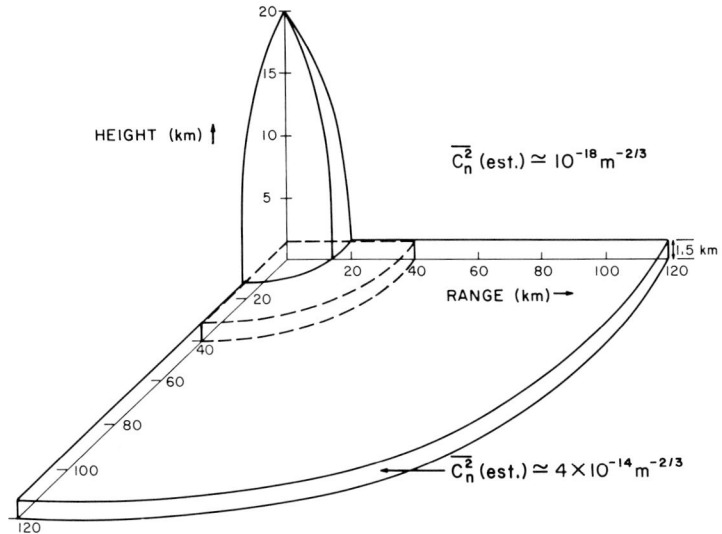

Fig. 11.24. Range of wind profiling and boundary layer clear-air wind mapping with an existing 10-cm weather radar system (see Table 11.1) (dashed lines) and with an upgraded radar (solid lines) for the two-step C_n^2 profile shown.

11.8 Observations of Wind, Waves, and Turbulence in Clear Air

about 700 m, the radar would then be able to profile winds to an altitude of 15–20 km if $C_n^2 \geq 10^{-18}\,m^{-2/3}$, assuming that $\lambda \geq \lambda_t$ and that the resolution volume is filled with this level of C_n^2. The profiling performance of such an upgraded weather radar is also shown in Fig. 11.24. Thus, Doppler weather radars might serve three functions: (1) storm observations, (2) obtaining clear-air wind profiles, and (3) mapping the kinematic structure of the CBL.

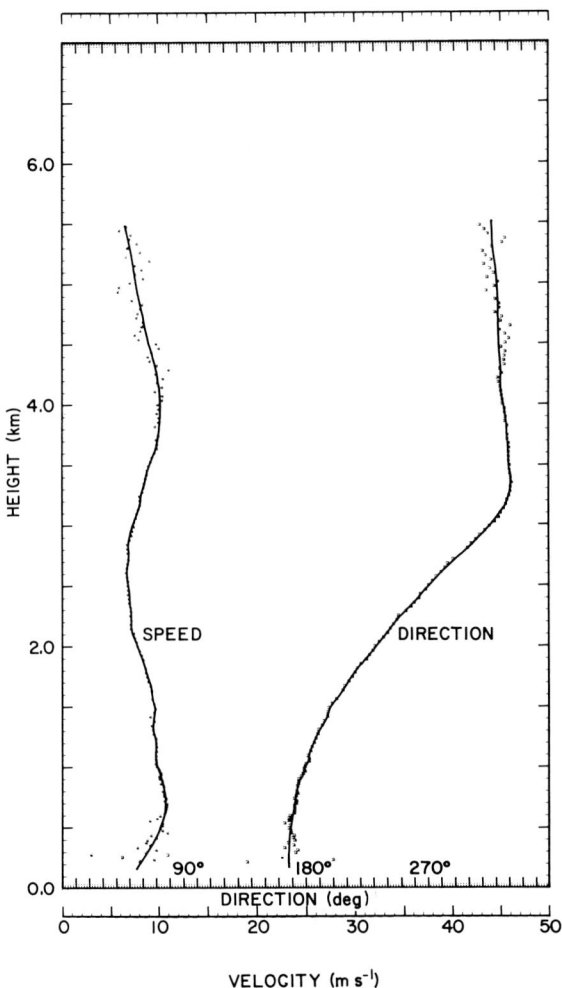

Fig. 11.25. An example of an NSSL Doppler-derived wind profile obtained in a prestorm nonprecipitating environment (21 June 1979; 15:34:45–15:35:45 C.S.T.).

We should recall that the computations used to generate Fig 11.24 assume that the scales of refractive index irregularities are within the inertial subrange. As discussed in Section 11.6.2, this condition is expected to hold for light turbulence up to heights of 14 km. If light turbulence is a common occurrence within the troposphere, then more powerful upgraded weather radars should be able routinely to obtain profiles of the wind throughout the troposphere.

In 1979 NSSL's Norman Doppler radar was modified to increase its performance to detect clear-air echoes. In summary, the pulse width and T_s were increased while the average power P_{av} was kept constant. Thus, range resolution was sacrificed (r_6 was increased) for greater sensitivity. Furthermore, operation at longer T_s allows processing at lower SNR to produce quality estimates of velocity, whereas longer τ increases the per pulse SNR [see Eq. (4.31)]. The first operation with the modified radar occurred in the prestorm clear skies of 21 June 1979. Detectable targets in precipitation-free air almost completely filled the volume from the surface up to 5 km and out to a range of 120 km. This suggests that the structure constant was as large as 10^{-13} m$^{-2/3}$ up to these heights. It is likely that a deep moist layer streaming in from the south at about 2 km (seen on synoptic charts) contributed to these unusually large C_n^2 values. However, clouds were also present and, although no precipitation was recorded on the ground, may have contributed to the observed reflectivity.

The vertical profile of horizontal wind obtained from the Doppler radar data on this day is given in Fig. 11.25. Below 3 km the wind turns strongly clockwise with height while above it turns counterclockwise. This suggests warm air advection overlaid by cold air advection, an interpretation confirmed by examination of the synoptic charts. This condition contributes to the destabilization of the atmosphere and may have been responsible for the severe storm that passed through central Oklahoma a few hours later.

11.8.2 Kinematic Structure of the Convective Boundary Layer

The second mode of clear-air radar surveillance pertains to obervations of the wind field in the CBL. Some applications were already discussed in Section 9.3.6. Measurements to ranges of about 120 km, nearly the practical limit set by earth's curvature, would certainly enhance the weather radar's capability to sense precursors to the development of thunderstorms. Assuming that $C_n^2 \approx 4 \times 10^{-14}$ m$^{-2/3}$ [from Fig. 11.17, adding 10 dB to the C_n^2 value, as suggested by Gossard (1977), to account for midday convection], $\sigma_v = 1.5$ m s^{-1}, $T_{sy} = 315$ K, $l_t = 0.22$, $\varepsilon_a = 0.6$ and $r_6 = 450$ m, Eq. (11.154) shows that the $P_{av}A$ product required for estimates with accuracy of 1 m s^{-1} at $r_0 = 120$ km is

$$P_{av}A = 7 \times 10^4 \quad Wm^2 \qquad (11.158)$$

if $T_d \simeq 0.15$ s (e.g., 7° s^{-1} rotation rate with a 1° angular resolution). With the present NSSL radar parameters in Table 11.1 (P, receiver not matched to the pulse) and pulse pair processing, the maximum range would be about 40 km, as indicated in Fig. 11.24, but during the 1980 storm period, a continuum of targets was observed to ranges in excess of 100 km on 10 of the 12 prestorm days during the observational period (15 April–19 June; Doviak et al., 1983). These observations usually began 2–3 hr prior to the forecast time for storm initiation, and so most observations took place under midafternoon sunny skies (if storms are forecast, an observation is prestorm, even if storms did not form). Therefore C_n^2 in the Oklahoma prestorm CBL appears to be somewhat larger than that depicted on Fig. 11.24.

Based on these preliminary observations it appears that weather radars could be capable of mapping the kinematic structure of the CBL during prestorm weather conditions.

11.8.2.1 Observations with Two Doppler Radars

Figure 11.26 shows the locations of the radars at Cimarron Airport and Norman. Synthesized dual Doppler radar winds were analyzed in detail in a 625-km^2 region (the square region with solid outline in Fig. 11.26), although winds were mapped over a 150-km square (Berger and Doviak, 1979). It would have been preferable to synthesize the wind in an area over the 500-m-tall, meteorologically instrumented tower, but ground clutter from Oklahoma City excessively contaminated our air velocity estimates in that region. Ridges (see Fig. 11.26) to the south of Cimarron and west of Norman shadowed the ground beneath the synthesis area and provided velocity data relatively free of ground-clutter contamination.

The winds were fairly uniform from the southwest on 27 April 1977, but there were small perturbations from the mean wind of about one order of magnitude less. As is evident in Fig. 11.26, the x direction and u component of the wind are along the mean wind, and the y direction and v component are normal to it. The mean CBL wind in Fig. 11.26 is the vector average of the horizontal wind over the height interval 0–1.25 km in the synthesis area during the period 1426–1450 CST. The wind is displayed versus height in Fig. 11.27. Although the change in wind direction is not large, considerable speed shear is noticeable in the lowest 400 m. The winds plotted from 0.5 to 1.5 km were determined from the Doppler radars. They represent the spatial average of u and v over the synthesis area (625 km^2). The two lower curves in Fig. 11.27 are each determined from averaging winds at the KTVY tower for 29 min. The dotted curve corresponds to the same time at which the radar

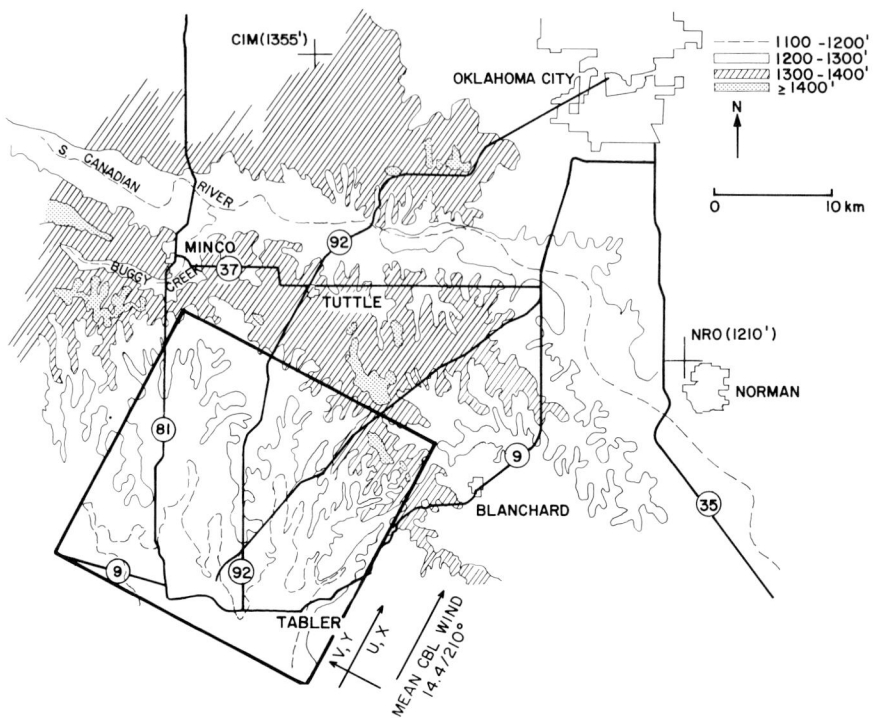

Fig. 11.26. Topographic map with locations of the two Doppler radars at Norman (NRO) and Cimarron Airport (CIM) in central Oklahoma and the wind synthesis area (outlined by the square). Contours are height (in feet) of ground above mean sea level (MSL). A 500-m meteorologically instrumented (KTVY) tower is at 356° and 37 km from NRO.

winds are presented, yet the winds appear discontinuous near 0.5 km. However, because the air over the synthesis area at radar observation time only reaches the tower 75 min later, comparison must be made with that time lag. The solid curve corresponds to this later time, and winds appear nearly continuous with height.

Caution must be exercised in making a comparison between a spatial average and a temporal average at a point because the existence of stationary helical circulations can make the wind comparison at a point unrepresentative for an area. Although the spatial variations due to turbulence average out in time, longitudinal waves have circulations with axes nearly parallel to the mean wind, and a temporal average might not erase their spatial variation. Brown (1970) used a numerical model to deduce significant spatial variability in the time-averaged wind following flow disturbances caused by convective rolls. The model shows that time-averaged wind speed at a point can be as

11.8 Observations of Wind, Waves, and Turbulence in Clear Air

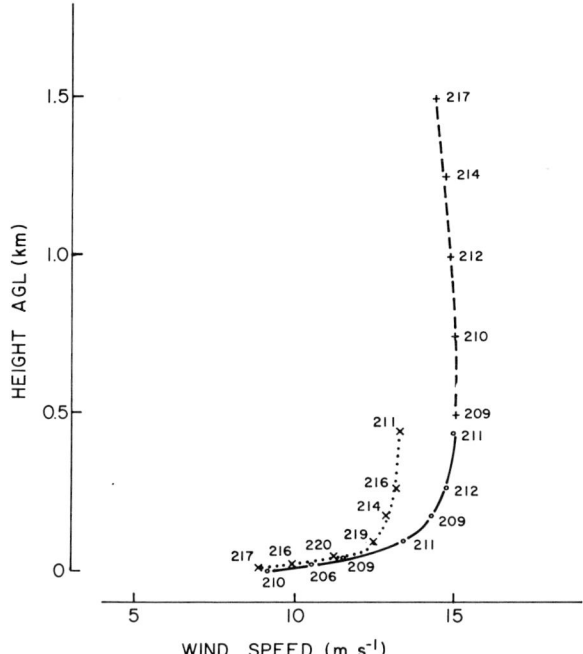

Fig. 11.27. Mean wind speed and direction above ground level (AGL) from tower and Doppler data. Wind obtained from two Doppler radar data (---) is a spatial average of u and v over the synthesis region analyzed at time 14:30–14:33 C.S.T. ···, 14:15–14:45 C.S.T. (tower); ———, 15:30–16:00 C.S.T. (tower).

much as 1.6 times the areal average, and the direction changes up to 10° when an observation point is moved 1 km perpendicular to the longitudinal roll circulations. Hence secondary circulations in the field can contribute to differences between spatially averaged radar winds and time-averaged winds measured at the tower.

The Norman and Cimarron Doppler velocities at common grid points are combined vectorially to synthesize the wind components u and v. The vertical velocity contributions are negligible because beam elevation angles are less than 4°. Variations in the mean flow are small, and Fig. 11.28 shows the perturbation winds at two times separated by 3 min. Arrow lengths measure the wind speed relative to the scale at the upper left corner. Distances x and y are measured from the Norman radar site. The CBL mean wind has a magnitude twice that indicated by the arrow at the upper right corner. The perturbation fields appear noiselike and random, but if one overlays the two fields while shifting the earlier one downstream by an amount that the eddies would have drifted owing to advection, one can immediately see that the fields are well correlated. These two fields were numerically correlated by lagging one data field with respect to the other in increments of 500 m. The correlation as a function of lag in the x and y directions is plotted in Fig. 11.29, where it is seen that the correlation peaks at a lag equal to the distance that perturbations would be advected by the mean wind in the time difference between the two data fields.

In order to evaluate the dominant scales of motion and to resolve waves in the turbulent wind, spectral power densities of the synthesized u and v fluctuations along both the x and y directions were examined for scales of 1–16-km wavelength. One-dimensional spatial spectra S as a function of wavenumber K are displayed in Fig. 11.30 for synthesized data from one tilt

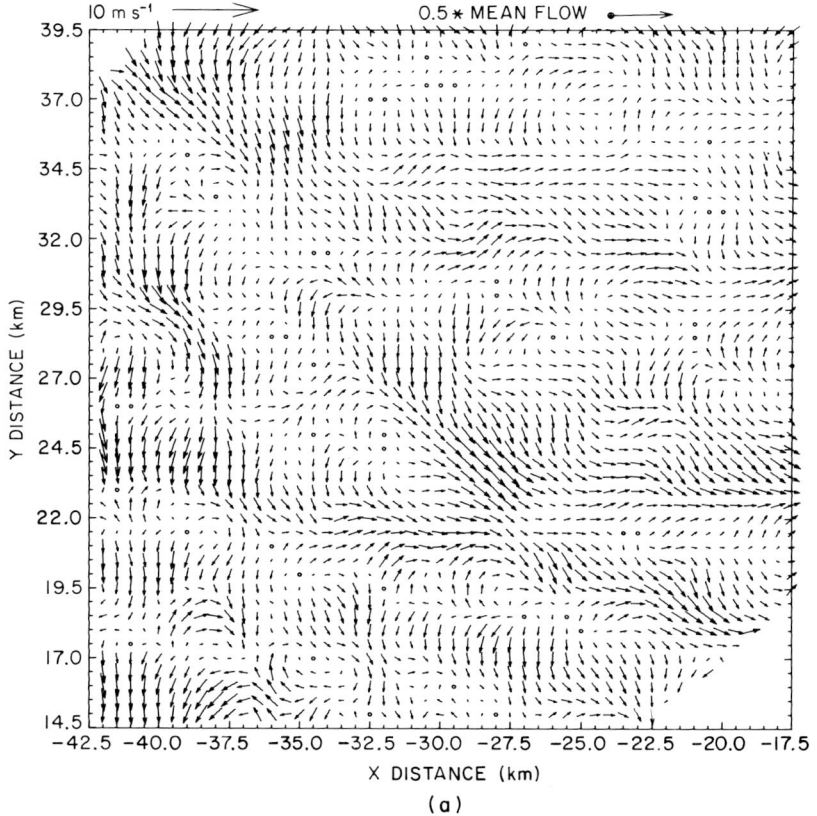

Fig. 11.28. Dual Doppler radar winds at (a) 14:38 C.S.T. and (b) 14:41 C.S.T. on 27 July 1977 with mean CBL wind removed. Solid circles indicate wind speed of less than 0.1 m s^{-1}. Height: 1.0 km.

11.8 Observations of Wind, Waves, and Turbulence in Clear Air

sequence at the indicated heights. The only filter acting on the data set used for spectral analyses is the Cressman interpolation filter [Eq. (9.3)]. However, because interpolation alters the spectral shape, it is necessary to estimate its influence before comparisons can be made with in situ point sensors. This was done by Doviak and Berger (1980), who showed good agreement between radar- and tower-derived spectra (see Fig. 10.7).

The values of each $S(K)$ in Fig. 11.30 are averages from 32 individual spectra along the K_x and K_y directions obtained by a discrete Fourier transform. Here $S(K)$ is defined as the power (or mean square velocity) at a wavenumber $K = n \Delta K$, divided by the wavenumber interval ΔK, where n is an integer and $\Delta K = (\pi/8) \times 10^{-3}$ m^{-1}. Although we have placed a power of $-\frac{5}{3}$ line in Fig. 11.30, we are not implying that these spectra should necessarily follow this law. On the other hand, Panofsky (1969) has observed

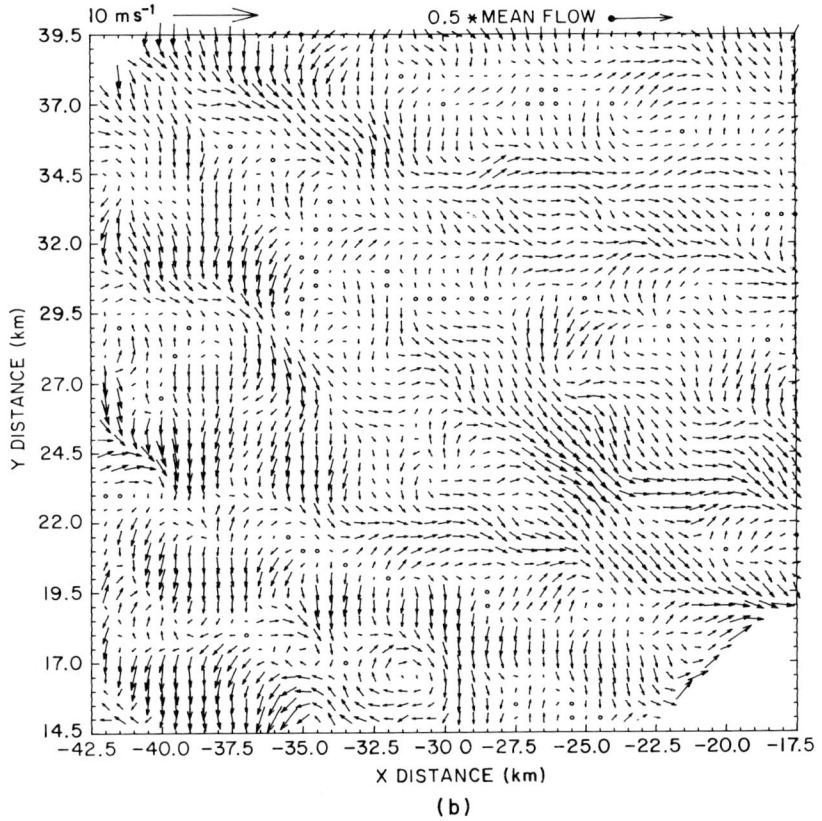

Fig. 11.28. (*continued*)

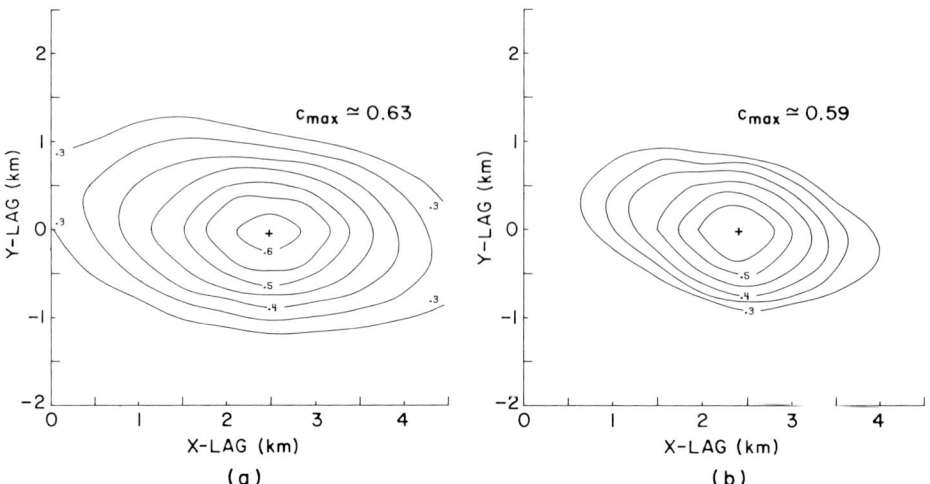

Fig. 11.29. Cross correlation function c versus the horizontal spatial lag of (a) the u component and (b) the v component for two wind fields 3.5 min apart at 1.0 km in height. The mean wind speed at $H = 1.0$ km is 14.8 m s^{-1}, and the median times for the two fields are 14:48:10 and 14:51:40 C.S.T. (27 July 1977).

that the law is often followed to wavelengths as long as five times the measurement height. As might be expected, our data follow the law better at the highest wavenumbers.

There is a hint of a spectral peak at a wavelength of 4 km for v in the K_y direction at a height of 1 km. Supportive evidence of this wave is seen in the PPI photo of echo power (Fig. 11.20). It is apparent that reflectivity streets are aligned roughly parallel to the mean wind in the CBL and have a separation of about 4 km. In order to get a better indication of which scale sizes in the x and y direction are significant, $KS(K)$ was plotted versus log K for six tilt sequences covering a time interval of about $\frac{1}{2}$ hr. These data are presented by Berger and Doviak (1979). The most consistent and prominent feature found was the large power of v in the y direction centered around a wavelength $\Lambda = 4$ km at a height of about 1.0 km, a result that was confirmed by in situ aircraft observations (Reinking et al., 1981). As can be seen from Fig. 11.30, this feature does not appear at $H = 1.5$ or 0.5 km.

The 4-km wave is probably an indication of horizontal roll vortices having axes roughly parallel to the mean wind and spaced 4 km apart. In fact, the absence of the 4-km wave in the crosswind v spectra at heights of 0.5 and 1.5 km suggests that the roll is centered at about 0.5 km and has peak v wind at about 1 km, with another peak near the surface where radar data are unavailable. Other peaks appear in these spectral plots but with less consistency in time or height and are thus difficult to interpret.

References

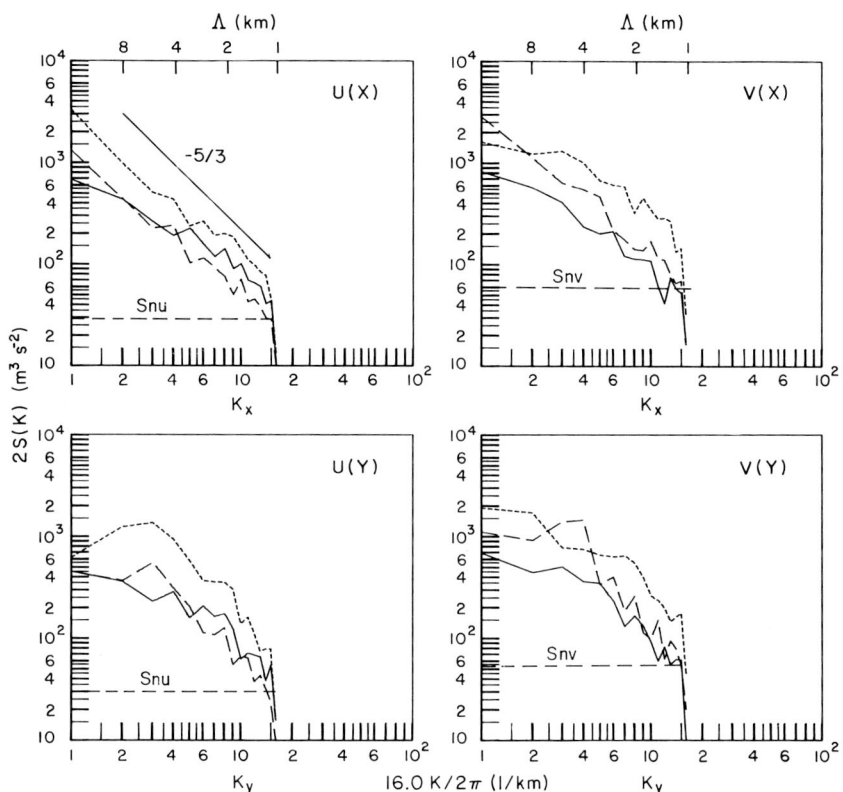

Fig. 11.30. Averaged power density $S(K)$ along the K_x and K_y directions for u and v components synthesized from Doppler velocities. Spectra are for winds synthesized at heights of 0.5 (---), 1.0 (-- --), and 1.5 (——) km. The horizontal lines through the base of the spectra are estimated noise levels at 1.0 km. Λ is the wavelength of the eddies. (14:41:41–14:44:17 C.S.T., 27 April 1977.)

REFERENCES

Atlas, D., Hardy, K. R., Glover, K. M., Katz, I., and Konrad, T. G. (1966). Tropopause detected by radar. *Science* **153**, 1110–1112.

Ball, F. K. (1961). Viscous dissipation in the atmosphere. *J. Meteorol.* **18**, 553–557.

Balsley, B., and Peterson, V. L. (1981). Doppler-radar measurements of clear air atmospheric turbulence at 1290 MHz. *J. Appl. Meteorol.* **20**, 266–274.

Bean, B. R., and Dutton, E. J. (1966). "Radio Meteorology," NBS Monogr. 92, U.S. Govt. Printing Office, Washington, D.C.

Berger, M., and Doviak, R. J. (1979). "An Analysis of the Clear Air Planetary Boundary Layer Wind Synthesized from NSSL's Dual Doppler-radar Data," NOAA Tech. Memo ERL NSSL-87. Natl. Severe Storms Lab., Norman, Oklahoma.

Blake, L. V. (1970). Prediction of radar range. *In* "Radar Handbook" (M. I. Skolnik, ed.), pp. 2-1–2-73. McGraw-Hill, New York.

Bonino, G., Lombardini, P. P., and Trivero, P. (1979). A metric wave radio-acoustic tropospheric sounder. *IEEE Trans. Geosci. Electron.* **GE-17,** 179–181.
Booker, H. G., and Gordon, W. E. (1950). A theory of radio scattering in the troposphere. *Proc. IRE* **38,** 401.
Brekhovskikh, L. M. (1960). "Waves in Layered Media." Academic Press, New York.
Brown, R. A. (1970). A secondary flow model for the planetary boundary layer. *J. Atmos. Sci.* **27,** 742–757.
Budden, K. G. (1961). "Radio Waves in the Ionosphere." Cambridge Univ. Press, London and New York.
Burk, S. D. (1978). "Use of a Second-moment Turbulence Closure Model for Computation of Refractive Index Structure Coefficients," Tech. Rep. TR-78-04. Naval Environmental Prediction Research Facility, Monterey, California.
Chadwick, R. B., and Moran, K. P. (1980). Long-term measurements of C_n^2 in the boundary layer. *Radio Sci.* **15,** 355–362.
Chen, W. Y. (1974). Energy dissipation rates of free atmospheric turbulence. *J. Atmos. Soc.* **31,** 2222–2225.
Crane, R. K. (1976). "Low Elevation Angle Measurement Limitations Imposed by the Troposphere: An Analysis of Scintillation Observations Made at Haystack and Millstone," Tech. Rep. 518. Lincoln Lab., Massachusetts Institute of Technology, Lexington.
Crane, R. K. (1977). "Stratospheric Turbulence," Rep. No. AFGL TR 77-0207. Air Force Geophys. Lab., Hanscom AFB, Massachusetts.
Crane, R. K. (1980). Radar measurements of wind at Kwajalein. *Radio Sci.* **15,** 383–394.
Davis, J. I. (1966). Consideration of atmospheric turbulence in laser systems design. *Appl. Opt.* **5,** 139.
Doviak, R. J., and Berger, M. J. (1980). Turbulence and waves in the optically clear planetary boundary layer resolved by dual-Doppler radars. *Radio Sci.* **15,** 297–317.
Doviak, R. J., Goldhirsh, J., and Miller, A. (1971). Simultaneous bistatic and monostatic detection of tropospheric layers. *IEEE Trans. Antennas Propag.* **AP-19,** 714–716.
Doviak, R. J., Rabin, R. M., and Koscielny, A. J. (1983). Doppler weather radar for profiling and mapping winds in the prestorm environment. *IEEE Trans. Geosci. Remote Sens.* **21,** 25–33.
DuCastel, F., Misme, P., Spizzichino, A., and Voge, J. (1962). On the role of the process of reflection in radio wave propagation. *J. Res. Natl. Bur. Stand., Sect. D* **66,** 273–284.
Fetter, R. W. (1961). Remote measurement of wind velocity by the electromagnetic-acoustic probe II: Experimental system. *Proc. Natl. Conv. Mil. Electron., 5th, 1961* pp. 54–59.
Frankel, M. S., and Peterson, A. M. (1976). Remote temperature profiling in the lower troposphere. *Radio Sci.* **11,** 157–166.
Gage, K. S., and Balsley, B. (1978). Doppler radar probing of the clear atmosphere. *Bull. Am. Meteorol. Soc.* **59,** 1074–1094.
Gage, K. S., Green, J. L., and Van Zandt, T. E. (1980). Use of Doppler radar for the measurement of atmospheric turbulence parameters from the intensity of clear-air echoes. *Radio Sci.* **15,** 407–416.
Gage, K. S., Balsley, B. B., and Green, J. L. (1981). Fresnel scattering model for the specular echoes observed by VHF radar. *Radio Sci.* **16,** 1447–1453.
Gossard, E. E. (1960). Power spectra of temperature, humidity and refractive index from aircraft and tethered balloon measurements. *IRE Trans. Antennas Propag.* **AP-8,** No. 2, 186–201.
Gossard, E. E. (1977). Refractive index variance and its height distribution in different air masses. *Radio Sci.* **12,** 89–105.

References

Gossard, E. E. (1981). Clear weather meteorological effects on propagation at frequencies above 1 GHz. *Radio Sci.* **16**, 589–608.

Gossard, E. E., and Hooke, W. H. (1975). "Waves in the Atmosphere; Atmospheric Infrasound and Gravity: Their Generation and Propagation." Elsevier, Amsterdam.

Gossard, E. E., and Strauch, R. G. (1982). "Radar Observation of Clear Air and Clouds." Elsevier, Amsterdam.

Gossard, E. E., Chadwick, R. B., Neff, W. D., and Moran, K. P. (1982). The use of ground-based Doppler radars to measure gradients, fluxes and structure parameters in elevated layers. *J. Appl. Meteorol.* **21**, 211–226.

Green, J. L., and Gage, K. S. (1980). Ovservations of stable layers in the troposphere and stratosphere using VHF radar. *Radio Sci.* **15**, 395–406.

Green, J. L., Gage, K. S., and Van Zandt, T. E. (1979). Atmospheric measurements by VHF pulsed Doppler radar. *IEEE Trans. Geosci. Electron., Spec. Issue Radio Meteorol.* **GE-17**, 262–280.

Hardy, K. R., and Glover, K. M. (1966). Twenty-four-hour history of radar angel activity at three wavelengths. *Proc. Radar Meteorol. Conf., 12th, 1966* pp. 269–274.

Hardy, K. R., and Katz, I. (1969). Probing the clear atmosphere with high power, high resolution radars. *Proc. IEEE* **57**, 469–480.

Hennington, L., Doviak, R. J., Sirmans, D., Zrnić, D., and Strauch, R. G. (1976). Measurement of winds in the optically clear air with microwave pulse-Doppler radar. *Prepr., Radar Meteorol. Conf., 17th, 1976* pp. 342–348.

Hess, S. (1959). "Introduction to Theoretical Meteorology." Holt, New York.

Hill, R. J. (1978). Spectra of fluctuations in refractivity, temperature, humidity, and the temperature-humidity cospectrum in the inertial and dissipation ranges. *Radio Sci.* **13**, 953–961.

Hodara, H. (1966). Laser wave propagation through the atmosphere. *Proc. IEEE* **54**, 308.

Hufnagel, R. E. (1978). "The Infrared Handbook" (W. L. Wolfe and G. J. Zissis, eds.), Libr. Congr. Cat. No. 77-90786, Chapter VI, U.S. Govt. Printing Office, Washington, D.C.

Johnson, C. C. (1965). "Field and Wave Electrodynamics." McGraw-Hill, New York.

Kristensen, L. (1979). On longitudinal spectral coherence. *Boundary-Layer Meteorol.* **16**, 145–153.

Kropfli, R. A., Katz, I., Konrad, T. G., and Dobson, E. B. (1968). Simultaneous radar reflectivity measurements and refractive index spectra in the clear atmosphere. *Prepr., Radar Meteorol. Conf., 13th, 1968* pp. 270–271.

Lenschow, D. H., and Wyngaard, J. C. (1980). Mean-field and second-moment budgets in a baroclinic, convective boundary. *J. Atmos. Sci.* **37**, 1313–1326.

Lilly, D. K., Waco, D. E., and Adelfang, S. I. (1974). Stratospheric mixing estimated from high-altitude turbulence measurements. *J. Appl. Meteorol.* **13**, 488–493.

Miller, K. S., and Rochwarger, M. C. (1972). A covariance approach to spectral moment estimation. *IEEE Trans. Inf. Theory* **IT-18**, 558–596.

Ottersten, H. (1969a). Atmospheric structure and radar backscattering in clear air. *Radio Sci.* **4**, 1179–1193.

Ottersten, H. (1969b). Mean gradient of potential refractive index in turbulent mixing and radar detection of CAT. *Radio Sci.* **12**, 1247–1249.

Panofsky, H. A. (1969). Spectra of atmospheric variables in the boundary layer. *Radio Sci.* **4**(12), 1101–1109.

Papoulis, A. (1965). "Probability, Random Variables, and Stochastic Processes." McGraw-Hill, New York.

Rabin, R. M., Doviak, R. J., and Sundara-Rajan, A. (1982). Doppler radar observations of momentum flux in a cloudless convective layer with rolls. *J. Atmos. Sci.* **39**, 851–863.

Reinking, R. F., Doviak, R. J., and Gilmer, R. O. (1981). Clear-air roll vortices and turbulent motions as detected with an airborne gust probe and dual-Doppler radar. *J. Appl. Meteorol.* **20**, 678–685.

Rice, P. L., Langley, A. G., Norton, K. A., and Barses, A. P. (1966). "Transmission Loss Predictions for Tropospheric Communication Circuits," *Tech. Note 101*. Natl. Bur. Stand., Boulder, Colorado.

Röttger, J. (1981). Investigations of lower and middle atmosphere dynamics with spaced antenna drift radars. *J. Atmos. Terr. Phys.* **43**, 277–292.

Röttger, J., and Schmidt, G. (1979). High-resolution VHF radar sounding of the troposphere and stratosphere. *IEEE Trans. Geosci. Electron.* **GE-17**, 182–189.

Röttger, T., Czechowsky, P., and Schmidt, G. (1981). First low-power VHF radar observations of tropospheric, stratospheric and mesospheric winds and turbulence at the Arecibo observatory. *J. Atmos. Terr. Phys.* **43**, 789–800.

Rowland, J. R., and Arnold, A. (1975). Vertical velocity structure and the geometry of clear air convective elements. *Prepr. Radar Meteorol. Conf., 16th, 1975* p. 292.

Schlesinger, R. E. (1982). Effects of mesoscale lifting, precipitation and boundary-layer shear on severe storm dynamics in a three dimensional numerical modeling study. *Prepr., Severe Local Storms Conf., 12th, 1982* pp. 536–541.

Scotti, R. S., and Corcos, G. M. (1969). Measurements on the growth of small disturbances in a stratified shear layer. *Radio Sci.* **4**, 1309–1313.

Shea, L. V. (1965). "Handbook of Geophysics and Space Environments," pp. 2–19. Air Force Cambridge Res. Lab., Office Aerosp. Res. USAF, Washington, D.C.

Smith, P. L. (1961). Remote measurement of wind velocity by the electromagnetic-acoustic probe I: System analysis. *Proc. Natl. Conv. Mil. Electron., 5th, 1961* pp. 48–53.

Tatarski, V. I. (1961). "Wave Propagation in a Turbulent Medium." McGraw-Hill, New York.

Tatarskii, V. I. (1971). "The Effects of the Turbulent Atmosphere on Wave Propagation" (translation from Russian edition by the Israel Program for Scientific Translations Ltd. 1PST Cat. No. 5319). (Available from the U.S. Dept. of Commerce, UDC 551.510, ISBN 07065 0680 4, Natl. Tech. Inf. Serv., Springfield, Virginia.)

Tennekes, H., and Lumley, J. L. (1972). "A First Course in Turbulence." MIT Press, Cambridge, Massachusetts.

Thorpe, S. A. (1969). Experiments on the stability of stratified shear flows. *Radio Sci.* **12**, 1327–1331.

Thorpe, S. A. (1973). Turbulence in stably stratified fluids: A review of laboratory experiments. *Boundary Layer Meteorol.* **5**, 95–119.

Trout, D., and Panofsky, H. A. (1969). Energy dissipation near the tropopause. *Tellus* **21**, 355–358.

Wait, J. R. (1926). "Electromagnetic Waves in Stratified Media." Macmillan, New York.

Weinstock, J. (1981). Using radar to estimate dissipation rates in thin layers of turbulence. *Radio Sci.* **16**, 1401–1406.

Weisman, M. L., and Klemp, J. B. (1982). The effects of directional turning of the low level wind shear vector on modeled multicell and supercell storms. *Prepr. Severe Local Storms Conf., 12th, 1982* pp. 528–531.

Westwater, E. R., and Grody, N. C. (1980). Combined surface- and satellite-based microwave temperature profile retrieval. *J. Appl. Meteorol.* **19**, 1438–1444.

Wheelon, A. D. (1959). Radio wave scattering by tropospheric irregularities. *J. Res. Natl. Bur. Stand., Sect. D* **63**, 205–233.

Woods, J. D., Hogström, V., Misme, P., Ottersten, H., and Phillips, O. M. (1969). Fossil turbulence. *Radio Sci.* **4**, 1365–1367.

Zrnić, D. S., and Doviak, R. J. (1978). Matched filter criteria and range weighting for weather radar. *IEEE Trans. Aerosp. Electron. Syst.* **AES-14**, 925–930.

APPENDIX A

Geometric Relations for Rays in the Troposphere

A.1 INTEGRAL SOLUTION FOR RAY PATH IN A SPHERICALLY STRATIFIED MEDIUM

In this appendix we derive an integral equation for the path of a ray in a spherically stratified medium. The determination of electromagnetic energy paths can be obtained from ray theory under the condition (Stratton, 1941, p. 343) that

$$\lambda \, dn/dR \ll 1, \qquad (A.1)$$

where λ is the wavelength of the electromagnetic radiation and dn/dR is the vertical gradient of refractive index n. One can easily show for most meteorological conditions existing in the troposphere (A.1) is well satisfied at weather radar wavelengths. Consider the ray path geometry shown in Fig. A.1. Snell's law for spherically stratified media (Bean and Dutton, 1966, p. 87) is

$$Rn(R) \cos \theta = \text{const} \equiv C, \qquad (A.2)$$

which applies to any point along the ray path $R(\psi)$. Now,

$$dl = [(R \, d\psi)^2 + (dR)^2]^{1/2}$$

and

$$\cos^2 \theta = \frac{(R \, d\psi)^2}{(R \, d\psi)^2 + (dR)^2} = \frac{R^2}{R^2 + (dR/d\psi)^2}. \qquad (A.3)$$

Solving for $d\psi/dR$, we obtain

$$d\psi/dR = (1/R) \cot \theta.$$

Therefore

$$\psi(R) = \int_{R_1}^{R} [(1/R) \cot \theta] \, dR + \psi(R_1), \qquad (A.4)$$

A Geometric Relations for Rays in the Troposphere

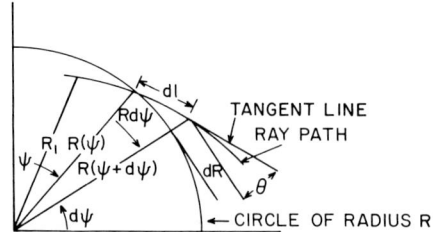

Fig. A.1. Ray path in a spherically stratified medium.

where $\psi(R_1)$ is any known ray path location (e.g., the transmitter location). Using (A.2) we can express the integrand of (A.4) entirely in terms of functions dependent on R. For a transmitter on the surface of the earth, whose true radius we label a, the integral solution of (A.4) becomes

$$s(h) = \int_0^h \frac{aC\,dh}{(a+h)[(a+h)^2 n^2(h) - C^2]^{1/2}}, \qquad \text{(A.5a)}$$

where $s(h)$ is the great circle distance from the transmitter and

$$C = an(0)\cos\theta_e \qquad \text{(A.5b)}$$

is specified by the ray's initial elevation angle θ_e and refractive index $n(0)$ at the transmitter. It can be shown that (A.5a) satisfies

$$\frac{d^2 h}{ds^2} - \left(\frac{2}{R} + \frac{1}{n}\frac{dn}{dh}\right)\left(\frac{dh}{ds}\right)^2 - \left(\frac{R}{a}\right)^2\left(\frac{1}{R} + \frac{1}{n}\frac{dn}{dh}\right) = 0, \qquad \text{(A.6)}$$

given by Hartree *et al.* (1946) as the exact differential equation that specifies the path of a ray in a spherically stratified medium.

A.2 RELATING A TARGET'S APPARENT RANGE AND ELEVATION ANGLE TO ITS TRUE HEIGHT AND GREAT CIRCLE DISTANCE

Although (A.5) is useful for plotting ray paths in the troposphere, it does not explicity relate target height to radar measurements of apparent range r_a and elevation angle θ_e. The great circle distance s is not known, but we can seek a solution to h in terms of the measured parameters r_a and θ_e.

In a time interval t an electromagnetic packet (pulse) of energy travels a distance

$$l = vt = ct/n.$$

The radar measures time intervals, and we assign an apparent range r_a assuming that the electromagnetic packets travel with the velocity of light. Thus

$$r_a = ct$$

is the apparent range that energy travels in time t. Therefore

$$dr_a = n\, dl, \quad \text{or} \quad r_a = \int_0^l n\, dl \tag{A.7}$$

is the *relation between the true path length l of the ray and the radar-measured length or range r_a*.

Now, referring to Fig. A. 1, we deduce that $dh = dl \sin \theta$, so that

$$r_a = \int_0^h \frac{n(h)\, dh}{\sin \theta}, \tag{A.8}$$

and as before we apply Snell's law [(A.2)] to (A.8) to obtain

$$r_a = \int_0^h \frac{(a+h)n^2(h)\, dh}{[(a+h)^2 n^2(h) - a^2 n^2(0) \cos^2 \theta_e]^{1/2}}. \tag{A.9}$$

Solving (A.9) gives a formula that relates the target height h to the two measurable parameters, the (apparent) range r_a and elevation angle θ_e. Of course, one needs to know $n(h)$. Once we find h for a given r_a and θ_e, we can substitute it into (A.5) to obtain the great circle distance to the target.

REFERENCES

Bean, B. R., and Dutton, E. J. (1966). "Radio Meteorology," NBS Monogr. 92. U.S. Govt. Printing Office, Washington, D.C.
Hartree, D. R., Michel, J. G. L., and Nicolson, P. (1946). Practical methods for the solution of the equations of tropospheric refraction. In "Meteorological Factors in Radio Wave Propagation," pp. 127–168. Physical Society, London.
Stratton, J. A. (1941). Electromagnetic Theory." McGraw-Hill, New York.

APPENDIX B

Correlation between Echo Samples as a Function of Sample Time

Consider the in-phase and quadrature components, from Eq. (4.1):

$$V(\tau_s, 0) = I_0 + jQ_0 = \frac{1}{\sqrt{2}} \sum_i A_i W_i e^{-j\phi_i} \tag{B.1}$$

at a sample time 0, where $\phi_i = 4\pi r_i/\lambda$, and subsequent echoes at the same range time τ_s but mT_s seconds later.

$$V(\tau_s, mT_s) = I_m + jQ_m = \frac{1}{\sqrt{2}} \sum_i A_i W_i e^{-j\zeta_i}, \tag{B.2}$$

where $\zeta_i = \phi_i + 4\pi v_i mT_s/\lambda$. It is important to bear in mind that the ϕ_i have a uniform distribution over the interval of width 2π, while the v_i are usually concentrated around the mean velocity. Furthermore, we assume the complex voltages A_i are independent of the phases ϕ_i and velocities v_i (i.e., the scatterer's size and phase shift due to scattering should not be dependent on its position or velocity). The complex amplitudes are random variables because drops may oscillate or change their canting angle (i.e., the angle between the vertical and axis of symmetry). Thus the autocorrelation function is

$$R(mT_s) = E[V^*(\tau_s, 0)V(\tau_s, mT_s)]$$
$$= \tfrac{1}{2} \sum_i \sum_k E(A_i^* A_k W_i^* W_k) E\{\exp[j(\phi_i - \phi_k - 4\pi v_k mT_s/\lambda)]\}$$
$$= \tfrac{1}{2} \sum_k E(|A_k|^2 |W_k|^2) E(e^{-j4\pi v_k mT_s/\lambda}). \tag{B.3}$$

The double summation reduces to a single one because the expectations with respect to the exponential argument are zero except at $i = k$. As we can see, the autocorrelation is totally independent of the initial phases ϕ_i. Had we not used the conjugate of $V(\tau_s)$, we would have obtained

$$E[V(\tau_s, 0)V(\tau_s, mT_s)] = 0 \tag{B.4}$$

Correlation between Echo Samples as a Function of Sample Time

because $E(e^{-j2\phi_i}) = 0$. Expressing (B.3) in terms of the in-phase and quadrature components, we obtain

$$E(I_0 I_m + Q_0 Q_m) = \text{Re}[R(mT_s)], \quad \text{(B.5a)}$$

$$E(I_0 Q_m - I_m Q_0) = \text{Im}[R(mT_s)]. \quad \text{(B.5b)}$$

We shall prove that

$$E(I_0 I_m) = E(Q_0 Q_m), \quad \text{(B.6a)}$$

$$E(I_0 Q_m) = -E(I_m Q_0). \quad \text{(B.6b)}$$

Start with

$$E(I_0 I_m) = \tfrac{1}{2} \sum_i \sum_k |W_i W_k| E(|A_i A_k|) E(\cos \gamma_{i0} \cos \gamma_{km}), \quad \text{(B.7a)}$$

$$E(Q_0 Q_m) = \tfrac{1}{2} \sum_i \sum_k |W_i W_k| E(|A_i A_k|) E(\sin \gamma_{i0} \sin \gamma_{km}), \quad \text{(B.7b)}$$

where γ_{im} is the phase shift for lag mT_s [in (4.3b)], and subtract (B.7b) from (B.7a) to get

$$E(I_0 I_m) - E(Q_0 Q_m) = \tfrac{1}{2} \sum_i \sum_k |W_i W_k| E(|A_i A_k|)$$
$$\times E[\cos(\phi_i + \phi_k + \psi_i + \psi_k + 4\pi v_k mT_s/\lambda)]. \quad \text{(B.8)}$$

Since the ϕ_i are uniformly distributed, it follows that the argument of the cosine in (B.8) is likewise uniform, making the expected value zero, and (B.6a) results. A similar derivation establishes (B.6b).

Next we shall prove that

$$E(I_m Q_m) = 0 \quad \text{(B.9)}$$

by considering

$$E(I_m Q_m) = \tfrac{1}{2} \sum_i \sum_k |W_i W_k| E[(|A_i A_k|) \cos \gamma_{im} \sin \gamma_{km}]. \quad \text{(B.10)}$$

Again, because we can assume that the A_i are independent of the γ_{im},

$$E[I_m Q_m] = \tfrac{1}{4} \sum_i \sum_k |W_i W_k| E(|A_i A_k|) E[\sin(\gamma_{im} + \gamma_{km}) + \sin(\gamma_{km} - \gamma_{im})], \quad \text{(B.11)}$$

and since γ_{im} and γ_{km} are uniformly distributed across 2π, the expectation of the trigonometric functions is zero.

Finally, we demonstrate that the correlation between two samples I_0, I_m is zero only if the distribution of $4\pi v_k mT_s$ is uniform over the interval of 2π.

We start with (B.7) and expand the product of cosine terms:

$$E(I_0 I_m) = \tfrac{1}{4} \sum_i \sum_k |W_i W_k| E(|A_i A_k|)$$
$$\times \{ E[\cos(\phi_i + \phi_k + \psi_i + \psi_k + 4\pi v_k m T_s/\lambda)]$$
$$+ E[\cos(\phi_k - \phi_i + \psi_k - \psi_i + 4\pi v_k m T_s/\lambda)] \}. \quad (B.12)$$

For $i \neq k$ both expected values of the cosine term are zero, but when $i = k$, (B.12) becomes

$$E(I_0 I_m) = \tfrac{1}{4} \sum_k |W_k|^2 E(|A_k|^2) E[\cos(4\pi v_k m T_s/\lambda)], \quad (B.13)$$

which in general is nonzero except when the cosine argument is uniform across the interval of 2π. The reader can verify that a result identical to (B.13) is obtained for $E(Q_0 Q_m)$.

APPENDIX C

Correlation of Echoes from Spaced Resolution Volumes

In this appendix we derive the correlation of the echo voltage at the receiver input for samples from resolution volumes both (1) spaced in range by $c\,\delta\tau_s/2$, where $\delta\tau_s$ is the sample spacing, and (2) spaced in azimuth (or elevation) by $\Delta\phi$. The latter spacing is caused when the radar scans its beam and $\Delta\phi$ is the azimuthal displacement between successive echo samples taken at constant range delay τ_s but at sample times differing by T_s.

Assume the scatterers are randomly distributed in space but have uniform statistical properties (e. g., η, the cross section per unit volume, is constant). The echo voltage increments dV from elemental volumes large compared to the wavelength but small compared to the resolution volume are then independent, but the expected elemental powers $E(dP)$ are equal. We assume the transmitted pulse to be rectangular and the angular pattern of radiation to be Gaussian. Scatterers in either case (1) or (2) are assumed to be fixed in place although they have random placement. The assumption of frozen scatterers is not strictly true for weather targets; it is a good approximation when a correlation in range is sought because the time separation between echoes from overlapping range intervals is very small (microseconds). However, there may be appreciable reshuffling of scatterers during the time T_s between two successive azimuth samples. In order to arrive at the decorrelation (and spectrum broadening) due solely to antenna motion, and to keep the derivation simple, we chose to treat frozen scatterers.

C.1 ECHO SAMPLE CORRELATION VERSUS RANGE DIFFERENCE $c\delta\tau_s/2$

For convenience, assume that echo voltages from elemental volumes have been summed over θ, ϕ ($r = $ const), producing a linear array of spherical shells of thickness dr along the range coordinate. One still obtains from each element of this radial array a voltage increment dV_m independent of all others. Thus the echo voltage sampled at τ_s is

$$V_x(\tau_s) = \sum_{m=1}^{M} dV_m, \qquad (C.1)$$

where we have assumed M elemental shells contained in the range interval $c\tau/2$. Because the echo input voltages are not filtered by the receiver, contributions to echo samples are confined to a range interval $c\tau/2$ (we use the subscript x to differentiate the input voltage from receiver output voltage). An echo voltage sampled at $\tau_s + \delta\tau_s$ has a value

$$V_x(\tau_s + \delta\tau_s) = \sum_{m=l+1}^{M+l} dV_m, \qquad (C.2a)$$

where

$$l = \delta\tau_s M/\tau, \qquad (C.2b)$$

as can be seen in the following schematic for $M = 8$ and $l = 6$:

```
                     V_x(τ_s)
              ⌢⎯⎯⎯⎯⎯⎯⎯⎯⎯⎯⎯⌢
dV_m:  |  |  |  |  |  |  |  |  |  |  |  |  |  |
       1  2  3  4  5  6  7  8  9 10 11 12 13 14
       →|dr|←         ⌣⎯⎯⎯⎯⎯⎯⎯⎯⎯⎯⎯⎯⎯⎯⎯⎯⎯⎯⎯⌣       r →
                          V_x(τ_s + δτ_s)
```

which depicts the elemental contributions dV_m to voltages sampled at τ_s and $\tau_s + \delta\tau_s$; dV_m is an elemental voltage contributed by scatterers located in a shell of thickness dr.

It is quickly realized that if $\delta\tau_s > \tau$, then the echo voltage samples are independent because no shell has contributed a common dV_m to both samples. We now form the product

$$V_x^*(\tau_s) V_x(\tau_s + \delta\tau_s) = \sum_{m=1}^{M} dV_m^* \sum_{m=l+1}^{M+l} dV_m. \qquad (C.2c)$$

Taking the expectation of (C.2), we have

$$E[V_x^*(\tau_s) V_x(\tau_s + \delta\tau_s)] = \sum_{m=1}^{M} \sum_{n=l+1}^{M+l} E(dV_m^* \, dV_n). \qquad (C.3)$$

Equation (C.3) gives a sum of elements in a matrix of products $E(dV_m^* dV_n)$. There are M^2 terms, but because the dV_m are independent and have zero mean, only the $M - l$ terms that have common elements differ from zero.

Therefore,

$$E(dV_m^* dV_n) = \begin{cases} 0, & m \neq n, \\ 2\sigma_r^2 \, dr, & m = n, \end{cases} \quad (C.4)$$

where $2\sigma_r^2$ is the power per unit length (see Section 4.2 for the factor of 2), which is multiplied by dr to give the power contributed by the elemental shell. Thus (C.3) reduces to

$$E[V_x^*(\tau_s) V_x(\tau_s + \delta\tau_s)] = 2(M-l)\sigma_r^2 \, dr. \quad (C.5)$$

Recognizing that

$$\bar{P}(r) = 2\sigma_r^2 M \, dr \quad (C.6a)$$

is the expected (mean) power of the samples at τ_s (if $r_6 \ll r$), where

$$M \equiv c\tau/2 \, dr, \quad (C.6b)$$

we can rewrite (C.5) as

$$E[V_x^*(\tau_s) V_x(\tau_s + \delta\tau_s)] \equiv R_{xx}(\delta\tau_s) = (1 - |l|/M)\bar{P}(r) \quad \text{for} \quad |l| \leq M. \quad (C.7)$$

The absolute value $|l|$ is used because $R_{xx}(\delta\tau_s)$ is symmetric in l. Substituting (C.6b) and (C.2b) in (C.7), we obtain the correlation of the input signal:

$$R_{xx}(\delta\tau_s) = (1 - |\delta\tau_s|/\tau)\bar{P}(r) \quad \text{for} \quad \delta\tau_s \leq \tau. \quad (C.8)$$

C.2 CORRELATION OF ECHOES FROM AZIMUTHALLY SPACED RESOLUTION VOLUMES

The correlation of echoes sampled when the antenna beam is pointed at different azimuths can be deduced by applying the principles derived for range correlation, but in a slightly different way. The input receiver voltage sampled at τ_s is

$$V_x(\tau_s, \theta_e, \phi_0) = \iint V(\theta, \phi) f^2(\theta - \theta_e, \phi - \phi_0) \sin\theta \, d\theta \, d\phi \quad (C.9)$$

when the antenna beam axis is pointed in the direction ϕ_0, θ_e, and where $V(\theta, \phi) \sin\theta \, d\theta \, d\phi$ is the elemental voltage contributed by the volume $(c\tau/2) r^2 \sin\theta \, d\theta \, d\phi$ of scatterers. The voltage for a beam position at $\phi_0 - \Delta\phi, \theta_e$ is

$$V_x(\tau_s, \theta_e, \phi_0 - \Delta\phi) = \iint V(\theta, \phi) f^2(\theta - \theta_e, \phi - \phi_0 + \Delta\phi) \sin\theta \, d\theta \, d\phi. \quad (C.10)$$

Recognizing that we can set $\phi_0 = 0$ without loss of generality, we form the product

$$V_x^*(\theta_e)V_x(\theta_e, \Delta\phi) = \iiiint V^*(\theta, \phi)V(\theta', \phi')f^2(\theta - \theta_e, \phi)$$
$$\times f^2(\theta' - \theta_e, \phi' + \Delta\phi)\sin\theta \sin\theta' \, d\theta' \, d\theta \, d\phi \, d\phi', \quad \text{(C.11)}$$

where θ', ϕ' are dummy variables of integration. The expectation of the product is

$$E[V_x^*(\theta_e)V_x(\theta_e, \Delta\phi)] = 2\sigma_\Omega^2 \iint f^2(\theta - \theta_e, \phi)$$
$$\times f^2(\theta - \theta_e, \phi + \Delta\phi)\sin\theta \, d\theta \, d\phi, \quad \text{(C.12a)}$$

where

$$E[V^*(\theta, \phi)V(\theta', \phi')] = 2\sigma_\Omega^2 \delta(\theta - \theta', \phi - \phi') \quad \text{(C.12b)}$$

and $2\sigma_\Omega^2$ is the power per unit solid angle.

Equation (C.12b) signifies that signals received from different elemental volumes are independent. The Dirac delta function $\delta(\theta - \theta', \phi - \phi')$ has the property

$$g(\theta, \phi) \equiv \iint g(\theta', \phi')\delta(\theta - \theta', \phi - \phi')\sin\theta' \, d\theta' \, d\phi', \quad \text{(C.12c)}$$

which, when used with (C.11) and (C.12b), produces (C.12a). From (C.12a) one obtains the total power when the antenna is stationary ($\Delta\phi = 0$):

$$\bar{P} = E[V_x^*(\theta_e)V_x(\theta_e)] = 2\sigma_\Omega^2 \iint f^4(\theta - \theta_e, \phi)\sin\theta \, d\theta \, d\phi. \quad \text{(C.13)}$$

Thus, solving for $2\sigma_\Omega^2$ and substituting in (C.12a), we have the correlation

$$R_{xx}[\theta_e, \Delta\phi] \equiv E[V_x^*(\theta_e)V_x(\theta_e, \Delta\phi)]$$
$$= \bar{P} \iint f^2(\theta - \theta_e, \phi)f^2(\theta - \theta_e, \phi + \Delta\phi)\sin\theta \, d\theta \, d\phi$$
$$\times \left[\iint f^4(\theta - \theta_e, \phi)\sin\theta \, d\theta \, d\phi\right]^{-1}. \quad \text{(C.14)}$$

On the assumption that the antenna pattern is product separable, (C.14) reduces to

$$R_{xx}[\theta_e, \Delta\phi] = \bar{P} \int f^2(\phi)f^2(\phi + \Delta\phi) \, d\phi \Big/ \int f^4(\phi) \, d\phi. \quad \text{(C.15)}$$

C.2 Correlation of Echoes from Azimuthally Spaced Resolution Volumes

To proceed further we need to define an antenna pattern. Let us assume that the two-way *electric field* pattern can be approximated by the Gaussian function

$$f^2(\phi) = \exp(-\phi^2 \cos^2 \theta_e / 4\sigma_\theta^2), \tag{C.16}$$

where σ_θ^2 is the second central moment of the two-way power pattern and the polar coordinate axis is perpendicular to the tangent plane at the radar location. The $\cos \theta_e$ term is required because the antenna pattern must be invariant to the direction of the beam axis. In terms of the one-way half-power width θ_1,

$$\sigma_\theta^2 = \theta_1^2 / 16 \ln 2, \tag{C.17}$$

so

$$f^2(\phi) = \exp[-4(\ln 2)\phi^2 \cos^2 \theta_e / \theta_1^2]. \tag{C.18}$$

Substitution of (C.18) into (C.15) and integration leads to

$$R_{xx}(\theta_e, \Delta\phi) = \bar{P} \exp[-2(\ln 2)(\Delta\phi)^2 \cos^2 \theta_e / \theta_1^2], \tag{C.19}$$

where we have assumed

$$4\sigma_\theta^2 / \cos^2 \theta_e \ll 2\pi \tag{C.20}$$

in order to simplify the integration.

The change $\Delta\phi$ that occurs between samples spaced mT_s apart when the antenna rotates in azimuth at a rate α is $mT_s\alpha$, so

$$R_{xx}(\theta_e, mT_s) = \bar{P} \exp[-2(\ln 2)(mT_s)^2 \alpha^2 \cos^2 \theta_e / \theta_1^2]. \tag{C.21}$$

Thus, even though scatterers are fixed in space, the successive echo samples are decorrelated if the antenna beam is moving. The width of the Gaussian correlation function (C.21) is

$$\sigma_\tau = \theta_1 / 2\alpha \cos \theta_e \sqrt{\ln 2}. \tag{C.22}$$

Because successive samples are decorrelated, we have spectral broadening. (If the antenna beam were stationary, the spectrum would contain a single line at zero Doppler velocity.) The spectral width caused by this decorrelation is

$$\sigma_\alpha = \lambda/4\pi\sigma_\tau = \alpha\lambda \cos \theta_e \sqrt{\ln 2}/2\pi\theta_1, \tag{C.23}$$

in velocity units.

APPENDIX D

Geometric Optics Approximation to the Wave Equation

From Eq. (11.13b) we have, in absence of turbulence,

$$\nabla^2 \mathbf{E}_0 + k_0^2 n_a^2 \mathbf{E}_0 = -2\nabla[\mathbf{E}_0 \cdot \nabla \ln(n_a)] \tag{D.1}$$

as the equation satisfied by the electric field in a region of space in which n_a, the average refractive index, is a function only of the spatial coordinate $\mathbf{r}$. We shall determine the condition under which

$$\nabla(\mathbf{E}_0 \cdot \nabla \ln n_a) \ll k_0^2 n_a^2 \mathbf{E}_0 \tag{D.2}$$

everywhere, so that the term on the right of (D.1) can be neglected. Assume that

$$n_a = 1 + n_1(\mathbf{r}), \tag{D.3}$$

where $n_1(\mathbf{r}) \ll 1$, typical of the troposphere. Thus expanding $\ln[1 + n_1(\mathbf{r})]$ in a power series and retaining only first-order terms in $n_1(\mathbf{r})$, we obtain

$$\ln n_a \simeq n_1(\mathbf{r}),$$

and the inequality (D.2) reduces to

$$\nabla(\mathbf{E}_0 \cdot \nabla n_1) \ll k_0^2 n_a^2 \mathbf{E}_0. \tag{D.4}$$

The term on the left cannot be larger than $k_0|\mathbf{E}_0|\,|\nabla n_1| + |\mathbf{E}_0|\,\nabla^2 n_1$, where

$$\mathbf{E}_0 = \mathbf{A}_0(\mathbf{r}) \exp(-j\mathbf{k}_0 \cdot \mathbf{r}) \tag{D.5}$$

has been assumed as a solution to (D.1) in which $\mathbf{A}_0(\mathbf{r})$ has variations changing spatially much more slowly than $\exp(-j\mathbf{k}_0 \cdot \mathbf{r})$. We now assume that the spatial Fourier spectrum of n_a contains structure wavelength sizes no smaller than l_0 and that $\lambda \ll l_0$. This value l_0 defines an effective cutoff of the spatial Fourier spectrum of time-invariant changes in the refractive index. Thus

$$\nabla^2 n_1 \leq (2\pi)^2 (n_1/l_0^2) \ll k_0 |\nabla n_1|,$$

Geometric Optics Approximation to the Wave Equation

and if the inequality

$$k_0 |\mathbf{E}_0\| \nabla n_1| < k_0^2 n_a^2 |\mathbf{E}_0| \tag{D.6}$$

is satisfied, then (D.4) and therefore (D.2) must also be satisfied.

Condition (D.6) can be further reduced to the form

$$\lambda |\nabla n_a|/2\pi n_a^2 \ll 1. \tag{D.7}$$

Condition (D.7) is imposed everywhere upon the spatial variation of the refractive index n_a in order that the equation of geometric optics,

$$\nabla^2 \mathbf{E}_0 + k_0^2 n_a^2 \mathbf{E}_0 = 0,$$

is a good approximation to the wave equation, (D.1).

APPENDIX

E

Derivation of Green's Function

In this appendix we derive Green's function for a general time-varying source $f(r, t)$, as given in Eq. (11.20). The scattering field $\mathbf{E}_1(\mathbf{r}, t)$ and $f(\mathbf{r}, t)$ have the following Fourier transform pairs:

$$\mathbf{E}_1(\mathbf{r}, \omega) = \int_{-\infty}^{\infty} \mathbf{E}_1(\mathbf{r}, t) e^{-j\omega t} \, dt, \tag{E.1a}$$

$$\mathbf{F}(\mathbf{r}, \omega) = \int_{-\infty}^{\infty} f(\mathbf{r}, t) e^{-j\omega t} \, dt, \tag{E.1b}$$

$$\mathbf{E}_1(\mathbf{r}, t) = \frac{1}{2\pi} \int_{-\infty}^{\infty} \mathbf{E}_1(\mathbf{r}, \omega) e^{j\omega t} \, d\omega, \tag{E.1c}$$

$$\mathbf{f}(\mathbf{r}, t) = \frac{1}{2\pi} \int_{-\infty}^{\infty} \mathbf{F}(\mathbf{r}, \omega) e^{j\omega t} \, d\omega. \tag{E.1d}$$

If (E.1c) and (E.1d) are substituted into (11.19), it is seen that *each* frequency component obeys the differential equation

$$\nabla^2 \mathbf{E}_1(\mathbf{r}, \omega) + k^2 \mathbf{E}_1(\mathbf{r}, \omega) = -\mathbf{F}(\mathbf{r}, \omega), \tag{E.2}$$

where $k^2 \equiv \omega^2/c^2$. The solution for $\mathbf{E}_1(\mathbf{r}, \omega)$ is given in many textbooks on electromagnetism (e.g., Morse and Feshbach, 1953, Vol. II, pp. 1769–1778). Its solution for the assumption that all waves are outgoing from the scatter volume V is

$$\mathbf{E}_1(\mathbf{r}_{r0}, \omega) = \frac{1}{4\pi} \int_V \frac{\mathbf{F}(\mathbf{r}, \omega) \varepsilon e^{-jk|\mathbf{r}_{r0} - \mathbf{r}|}}{|\mathbf{r}_{r0} - \mathbf{r}|} \, dV, \tag{E.3}$$

where $\mathbf{r}_{r0}$ is the observation point (receiver location) and r is the distance from the origin to the elemental scatter volume dV (Fig. 11.2). The parameter ε is the unit dyadic or idemfactor. The solution (E.3) is an integral of the vector Green's function

$$(\mathbf{a}_x + \mathbf{a}_y + \mathbf{a}_z) e^{-jk(|\mathbf{r}_{r0} - \mathbf{r}|)} / |\mathbf{r}_{r0} - \mathbf{r}|, \tag{E.4}$$

weighted by the source function $\mathbf{F}(\mathbf{r}, \omega)$ where $\mathbf{a}_x$, $\mathbf{a}_y$, and $\mathbf{a}_z$ are unit vectors. The vector Green's function is a solution of (E.2) for a unit vector source $(\mathbf{a}_x + \mathbf{a}_y + \mathbf{a}_z)\delta(\mathbf{r}_{r0} - \mathbf{r})$. The operator $\mathbf{V}$ in $\mathbf{F}(\mathbf{r}, \omega)$ of (E.3) operates on the source location $\mathbf{r}$ dependence.

The scattered field in the time domain is found by applying (E.1c) to (E.3):

$$\mathbf{E}_1(\mathbf{r}_{r0}, t) = \frac{1}{8\pi^2} \int_{-\infty}^{+\infty} \int_V \mathbf{F}(\mathbf{r}, \omega) \frac{\exp[j(\omega t - k|\mathbf{r}_{r0} - \mathbf{r}|)]}{|\mathbf{r}_{r0} - \mathbf{r}|} dV\, d\omega, \quad (E.5)$$

in which we have made use of the identity $\mathbf{F}(\mathbf{r}, \omega)\varepsilon \equiv \mathbf{F}(\mathbf{r}, \omega)$. Now, $\mathbf{F}(\mathbf{r}, \omega)$ can be expressed in terms of the source function $\mathbf{f}(\mathbf{r}, t)$ through use of (E.1b), so

$$\mathbf{E}_1(\mathbf{r}_{r0}, t) = \frac{1}{8\pi^2} \int_{-\infty}^{+\infty} \int_{-\infty}^{+\infty} \int_V \mathbf{f}(\mathbf{r}, t')$$
$$\times \frac{\exp[j\omega(t - t') - j(\omega/c)|\mathbf{r}_{r0} - \mathbf{r}|]}{|\mathbf{r}_{r0} - \mathbf{r}|} dV\, d\omega\, dt', \quad (E.6)$$

where t' gives the time variation of the source $\mathbf{f}(\mathbf{r}, t')$. The integration over ω can be executed easily by using the identity

$$\int_{-\infty}^{+\infty} \exp\{j\omega[t - t' - |\mathbf{r}_{r0} - \mathbf{r}|/c]\}\, d\omega \equiv 2\pi\delta(t - t' - |\mathbf{r}_{r0} - \mathbf{r}|/c),$$

so that

$$\mathbf{E}_1(\mathbf{r}_{r0}, t) = \frac{1}{4\pi} \int_{-\infty}^{+\infty} \int_V \frac{\mathbf{f}(\mathbf{r}, t')\delta(t - t' - |\mathbf{r}_{r0} - \mathbf{r}|/c)}{|\mathbf{r}_{r0} - \mathbf{r}|} dV\, dt'. \quad (E.7)$$

Therefore the Green's function for the general time-varying source is

$$G(\mathbf{r}_{r0} - \mathbf{r}, t - t') = \frac{\delta(t - t' - |\mathbf{r}_{r0} - \mathbf{r}|/c)}{4\pi|\mathbf{r}_{r0} - \mathbf{r}|}. \quad (E.8)$$

REFERENCE

Morse, P. M., and Feshbach, H. (1953). "Methods of Theoretical Physics," Vols. I and II. McGraw-Hill, New York.

Index

A

Air, refractive index of, 7
Aircraft
    gust-front effects on, 304, 310–311
    radar detection of, history, 24
Ambiguities
    range, 121
    velocity, 129
Antenna
    effective aperture area, 33
    effective pattern for scanning radar, 146–150
    side-lobe echoes, 150–153
Antenna gain, 26
Antenna–reflector, 24–26
Anvil cloud, in thunderstorms, 242, 254, 255
Atmosphere
    refractivity of, 9
    spherically stratified, 11–18
Attenuation, 29–33
    by cloud droplets, 30–32
    cross section, 195
    gaseous, 32–33
    by rain, 30
    by snow, 32
Autovariance processing of meteorological radar signal processing, 103–107

B

Backscatter cross section, 27
Bandwidth, 42
Beamwidth
    apparent, 149
    3-dB width, 24
Bessel correlation function, 329
Born approximation, 365

C

Chandrasekhar's theory, 333–335
Clear air
    convective boundary layer in, 426–433
    reflectivity of, 416–423
    turbulence in, 423–433
    waves in, 423–433
    wind in, 423–426
Cloud drop distribution, 181–184
Cloud droplets, radar attenuation by, 30–32
Clutter, definition of, 47
Coherency of echos, 130, 360
Complementary codes in Doppler radar, 167–168
COPLAN technique, 250
Correlation function, turbulence spectra and, 325–329
Cramer–Rao bounds, treatment of, 113–114
Cressman function, 251
Cyclones
    Doppler radar tracking of, 124
    tornado-type, Doppler-radar observations of, 281
    in thunderstorms, 253

D

Dirac delta function, 446
Discrete Fourier transform (DFT) of weather signal, 62–67, 88
Dissipation of energy in turbulence, 332
Distributed targets, definition of, 47
Doppler radar and Doppler radar spectra, 1, 2
    ambiguities in, 44–46
    antenna gain in, 26
    antenna side-lobe echoes in, 150–153
    bandwidth of, 42

Doppler radar and Doppler radar spectra (*Cont.*)
  block diagram of, 22
  complementary codes in, 167–168
  Doppler shift in, 28
  echo coherency in, 130–132
  effective antenna pattern of, 146–150
  effects due to amplitude and phase imbalances, 161
  electromagnetic beam from, 24–26
  filtered waveform in, 42–44
  FM cw-type, 168–177
  ground clutter in and its suppression, 153–156
  methods to decrease acquisition time, 142–146
    frequency diversity, 143
    random signal transmission, 143–146
  observations
    of tornadic storms, 253–260
    of tornadoes, 78, 294–299
    of weakly scattering weather targets, 163–164
    of weather echoes, 62–90
  power-weighted distribution of velocities in, 81–87
  prestorm observations by, 274–278
  pulse compression in, 164–166
  quantization and saturation noise in, 158–162
  of rain, 179–239
  range ambiguities in, 121–129
  received waveform in, 35–38
  receiver, block diagram of, 92
  receiving aspects of, 33–38
  signal-to-noise ratio in, 44
  spectral artifacts in, 156
  system noise temperature in, 38–42
  techniques to reduce information loss, 132–142
    correction of aliased velocities, 140–142
    interlaced sampling, 139–140
    phase diversity, 132–133
    spaced pairs with polarization coding, 133–135
    staggering the PRT, 135–139
  transmitting theory of, 21–26
  velocity ambiguities in, 129–130

  velocity spectrum width from shear and turbulence, 87–90
  vortex observation with, 281–285
Downdrafts in thunderstorms, 246–249
Drizzle, Z characteristics of, 223
Drop size distributions, 179–187
Dual Doppler radars, 249–260
Dual polarization technique in radar measurements of rain, 212–231
Dual wavelength radar, 208
Ducts, ground-based, 15–18

E

Earth, effective radius of, 12–15
Echo(es)
  correlation of, from spaced resolution volumes, 443–447
  from Doppler radar targets, time delay and strengths of, 23
  samples, correlation between as function of time, 440–442
Echo coherency in Doppler radar, 130–132
Electromagnetic beam from Doppler radar, 24–26
Electromagnetic spectrum, 5
Electromagnetic waves, 4–20
  definition of, 4
  theory and propagation of, 3–40
  time-averaged power density of, 5
Ergodic ensemble members, 70
Eye of hurricane, description of, 318

F

Fast fourier transforms (FFT), use in spectral analysis of weather signals, 67
FM cw Doppler radar, 168–177
Fog, Z characteristics of, 223
Fourier spectral representations, 3
Fraunhofer scatter
  basic integral for, 373, 374
  turbulent-troposphere echoes and, 371–373
Frequency diversity, use in Doppler radar to decrease acquisition time, 143
Fresnel scattering, turbulent-troposphere echoes and, 388–400

Index

G

Gain, antenna, 26
Gans theory of backscatter cross section of oblate spheroids, 213
Gas, permittivity of, 8
Graupel, Z characteristics of, 223
Green's function, 368
    derivation of, 450–451
Ground clutter in Doppler radar, suppression of, 153–156
Gust fronts, 304–312
    clear-air type, 306
    from hurricanes, 318

H

Hail
    effect on radar measurements, 58, 208, 232, 233
    from hurricanes, 318
    Z characteristics of, 223
Hamming window, 79
Hurricanes
    Doppler radar observations of, 317–319
    eye of, 318
Hydrometeors
    absorption by, 1, 2
    distribution of, from Doppler spectra, 231–233
    Z characteristics and type of, 223

I

Ice
    crystals, in thunderstorms, 219
    spheres, Doppler radar studies on, 233
Independence of signal samples, 97
Inertial subrange
    in turbulence theory, 332–346
    for troposphere, 411–413
In-phase signal component, 35
Interlaced sampling in Doppler radar, 139–140
Isodop pattern, idealized, 282
Isotropic field, definition of, 325

K

Klystron amplifiers, development of, 24
Kolmogorov–Obukhov spectrum, 338
Kolmogorov's hypothesis, 346
Kronecker delta function, 326

L

Lightning
    Doppler radar spectra of, 312–317
        characteristics, 313–315
        rain gush after, 315–316
        storm structure and, 315–317
Listening period of Doppler radar, 21
Longitudinal function, 326
Lorentz–Lorenz formula, 8
Low-noise amplifier (LNA) of radar receiver, 39, 40

M

Magnetron, development of, 24
Marshall–Palmer relations, 195, 201
Matched filter, 44
Maxwell's equations, 361
Mean frequency estimators for meteorological radar signal processing, 103–108
Median volume diameter, 189
Mesocyclones
    Doppler-radar observations of, 85, 253, 281, 289, 351
        multimoment display, 288
        tornado embedded in, 256
Microwaves
    definition of, 4
    for weather radars, 6

N

Noise
    bandwidth, 42
    sky, 41
    system temperature, 38
Nyquist frequency, 64–67, 98, 103, 107, 111, 135, 156, 157, 299

O

Obscuration, probability of, 121–129
  single cell, 124
  squall line, 126

P

Phase jitter, in Doppler radar, 163
Point targets, definition of, 47
Polarization, dual, 212
Polar molecules, 7
Precipitation echoes, 1
Prestorm observations of winds by Doppler radar, 274–278
Propagation paths, theory of, 7
Pulse compression in Doppler radar, 164–166
Pulse repetition time (PRT) of Doppler radar, 21
  staggering of, 135–139

Q

Quadrature signal component, 35

R

Radar, *see also* Doppler radar
  principles of, 21–46
  propagation paths of, 7
  uses of, 1
  weather mapping by, 1
Radar environment, definition of, 1
Radar equation, derivation of, 34–35
Radar meteorology, 24
Radioacoustic sounding system (RASS), 365–366, 375–376
Radome, losses from, 41–42
Rain, radar measurements of, 58, 179–239
  attenuation rate, 30, 195–199
  drop size distributions, 179–187
  dual polarization in, 212–231
  hydrometeor distribution from spectra of, 231–233
  liquid-water content effects, 189–190
  rainfall rate, 188–189, 193–195
    attenuation method, 202–205
    dual parameter measurement, 205–231
    single-parameter measurement, 199–205
  rain gauge combination with, 226–231
  reflectivity factor relations, 201
  terminal velocity of drops of, 187–188
  Z characteristics of, 223
Raindrops, scattering of electromagnetic energy by, 7
Rainfall, depth measurements of, 224
Rain gauge, radar combined with, 226–231
Rain gush, after lightning, 315–316
Rain parameter diagram, 206
Random field, definition of, 325
Random signal transmission, use in Doppler radar to decrease acquisition time, 143–146
Random variable with stationary first increments, 330
Range ambiguities, in Doppler radar, 121–129
Range weighting function, derivation of, 52–56
Rankine vortex model for tornadoes, 295
Ray path, in spherically stratified atmosphere, 11
Rays, in the troposphere, geometric relations for, 437–439
Receiver loss factor due to finite bandwidth, 55
Reflectivity, differential, 215
Reflectivity factor, derivation of, 58
Refractive index, structure constant of, 400–416
Refractivity of atmosphere, 9
Resolution volume, derivation of, 57
Richardson number, 404

S

Satellites, radar combined with, 233
Scalar field, definition of, 325
Scattering cross section, 26
Severe storms, Doppler-radar observations of, 285–294
Severe weather warnings, Doppler radar use for, 281
Shear
  Doppler radar spectra of, 347–348
  effect on aircraft, 352
Signal processing, 91–120
  autocovariance processing in, 103, 108–110
  echo sample intensity estimation, 94–103
  mean frequency estimators in, 103–108

Index

method for, 93–94
minimum variance bounds in, 113–115
performance on data, 115–119
range time averaging in, 102–103
spectral-moment estimation in, 91–92
spectral processing in, 107–108, 111–113
spectrum-width estimation, 108–110
Signal-to-noise ratio
  for distributed targets, 59, 414
  in Doppler radar, 44
Sky noise
  sources of, 40
  temperature, of idealized antenna, 41
Snell's law for spherically stratified media, 437
Snow
  radar attenuation by, 32
  wet or dry, Z characteristics of, 223
Spectral analysis of weather echoes, 62–81
  bias, variance, and window effect in, 73–75
    window weights and their spectra, 77
  convolution and correlation in, 67–70
  discrete Fourier transform in, 62–67
  expression of spectral estimates in terms of true spectrum, 75–80
  periodogram variance in, 80–81
  power spectrum of random sequences in, 71–73
Spectral artifacts, 156
Spectral density tensor, derivation of, 326
Spectral moments, estimation of, 91–92
Spectral window effect, 77
Spectrum width, 84, 342
  due to antenna rotation, 89
  due to drop size distribution, 88
  due to shear, 88–89
  due to turbulence, 344–347
Squall line, Doppler radar observation of, 126–129
Stabilized local oscillator (STALO) of Doppler radar, 35–36, 45
Staggered PRT, 135
Storms
  Doppler radar observations of, 240–323
    severe storms, 285–294
  lightning and structure of, 315–317
  radar detection of, history, 24
  structure of, 2
Structure constant
  of passive additive, 400
  of potential refractive index, 402

Structure functions
  in turbulence theory, 329–331
  from similarity assumptions, 331–333
Superheterodyne receivers, for Doppler radar, 36
Surface weather, plan view of, 247

T

Terminal velocity, 187–188
Thunderstorm(s)
  Doppler radar observations of, 115–118, 241–249
  energy and water released by, 3
  ice crystals in, 219
  severe, turbulence in, 347–357
Tornado(s)
  cyclones, Doppler-radar observations of, 281–285
  Doppler spectra of, 78, 79, 85, 121–124, 126, 253–260, 294–301
  model for, 295
  from hurricanes, 318
  warnings, comparison of types of, 300–301
Tornado spectral signature (TSS), definition of, 295
Transmit–receive (TR) switch of Doppler radar, 21
Transmitting aspects of Doppler radar, 21–26
Transverse correlation function, 326
Troposphere
  geometric relations for rays in, 437–439
  turbulence in, echoes from, 359–436
    backscattering in, 397–400
    expected scattered power density, 378–388
    of fields scattered by irregularities, 365–371
    Fraunhofer scatter in, 371–373
    Fresnel scattering in, 388–400
    structure constant of refractive index and, 400–416
Turbulence
  Chandrasekhar's theory and, 333–335
  in clear air, 423–433
  eddies, 343–344
    dissipation rate and spectrum width, 345–347
  measurement of, 324–358
  in severe thunderstorms, 347–357

Turbulence (*Cont.*)
    shear from, and aircraft, 352
    spatial spectra of point and average velocities of, 332–345
        filtering by resolution volume, 336–342
        from single radar, 343–344
        from two radars, 344–345
    spectra, and correlation function, 325–329
    structure functions for, 329–331
        from similarity assumptions, 331–333
    theory of, 3
        statistical, 324–336
    in troposphere, 359–436

U

Updrafts in thunderstorms, 246

V

Vector field, definition of, 325
Velocity ambiguities in Doppler radar, 129–130
Velocity azimuth display (VAD) of Doppler radar data, 260, 268
    for turbulence, 343
Volume velocity processing (VVP) of Doppler radar data, 261, 273–274
    prestorm observations, 274–278
von Hann weight, 295
von Hann window in spectral analysis of weather signals, 78, 79, 115
Vortices, Doppler-radar observations of, 281–285

W

Wave equation
    geometric optics approximation to, 448–449
    for inhomogeneous and turbulent media, 361–365
Waves of wind, Doppler-radar observations of, 280–281
Weather, radar applications to, 1
Weather echo(es)
    correlation along the range-time axis, 60–61
    Doppler spectra of, 62–90
    range weighting function of, 52–56
    reflectivity factors in, 58
    resolution volume of, 57
    sample, 48–49
    sample intensity estimation of, 94–103
    signal processing of, 91
    signals, 47–61
        statistics for, 49–51
    signal-to-noise ratio for distributed targets of, 59
Weather radar equation, derivation of, 51–52
Wind(s)
    analyses of
        on circular arc, 265–266
        on complete circle, 266–272
        on sections of a conical surface, 273
    Doppler radar observations of, 240–323
    linear measurements with single Doppler radar, 260–278
    nonlinear wind, 278–281
        vertical wind, 278–280
        waves, 280–281
    prestorm, 274–278
Wind fields
    reconstruction of, 250–253, 256
        errors in, 260
        least squares fitting of, 264–265
Window effect
    in spectral analysis of weather signals, 73–75
    rules governing use of, 80